Editorial Board Members' Collection Series in "Bioinorganic Chemistry of Copper"

Editorial Board Members' Collection Series in "Bioinorganic Chemistry of Copper"

Guest Editors

Christelle Hureau
Ana Maria Da Costa Ferreira
Gianella Facchin

Basel • Beijing • Wuhan • Barcelona • Belgrade • Novi Sad • Cluj • Manchester

Guest Editors

Christelle Hureau
CNRS Laboratoire de Chimie
de Coordination (LCC)
Toulouse
France

Ana Maria Da Costa Ferreira
Instituto de Química
Universidade de São Paulo
São Paulo
Brazil

Gianella Facchin
Departamento Estrella Campos
Universidad de la República
Montevideo
Uruguay

Editorial Office
MDPI AG
Grosspeteranlage 5
4052 Basel, Switzerland

This is a reprint of articles from the Special Issue published online in the open access journal *Inorganics* (ISSN 2304-6740) (available at: www.mdpi.com/journal/inorganics/special_issues/D2U9CXRHNY).

For citation purposes, cite each article independently as indicated on the article page online and using the guide below:

Lastname, A.A.; Lastname, B.B. Article Title. *Journal Name* **Year**, *Volume Number*, Page Range.

ISBN 978-3-7258-1960-7 (Hbk)
ISBN 978-3-7258-1959-1 (PDF)
https://doi.org/10.3390/books978-3-7258-1959-1

© 2024 by the authors. Articles in this book are Open Access and distributed under the Creative Commons Attribution (CC BY) license. The book as a whole is distributed by MDPI under the terms and conditions of the Creative Commons Attribution-NonCommercial-NoDerivs (CC BY-NC-ND) license (https://creativecommons.org/licenses/by-nc-nd/4.0/).

Contents

About the Editors vii

Ana Maria Da Costa Ferreira, Christelle Hureau and Gianella Facchin
Bioinorganic Chemistry of Copper: From Biochemistry to Pharmacology
Reprinted from: *Inorganics* 2024, 12, 97, doi:10.3390/inorganics12040097 1

Zenayda Aguilar-Jiménez, Adrián Espinoza-Guillén, Karen Resendiz-Acevedo, Inés Fuentes-Noriega, Carmen Mejía and Lena Ruiz-Azuara
The Importance of Being Casiopeina as Polypharmacologycal Profile (Mixed Chelate–Copper (II) Complexes and Their In Vitro and In Vivo Activities)
Reprinted from: *Inorganics* 2023, 11, 394, doi:10.3390/inorganics11100394 6

Celisnolia M. Leite, João H. Araujo-Neto, Adriana P. M. Guedes, Analu R. Costa, Felipe C. Demidoff, Chaquip D. Netto, et al.
Copper(I)/Triphenylphosphine Complexes Containing Naphthoquinone Ligands as Potential Anticancer Agents
Reprinted from: *Inorganics* 2023, 11, 367, doi:10.3390/inorganics11090367 29

Carlos Y. Fernández, Analu Rocha, Mohammad Azam, Natalia Alvarez, Kim Min, Alzir A. Batista, et al.
Synthesis, Characterization, DNA Binding and Cytotoxicity of Copper(II) Phenylcarboxylate Complexes
Reprinted from: *Inorganics* 2023, 11, 398, doi:10.3390/inorganics11100398 47

Doti Serre, Sule Erbek, Nathalie Berthet, Christian Philouze, Xavier Ronot, Véronique Martel-Frachet and Fabrice Thomas
Anti-Proliferation and DNA Cleavage Activities of Copper(II) Complexes of N_3O Tripodal Polyamine Ligands
Reprinted from: *Inorganics* 2023, 11, 396, doi:10.3390/inorganics11100396 59

Vasilii Graur, Irina Usataia, Ianina Graur, Olga Garbuz, Paulina Bourosh, Victor Kravtsov, et al.
Novel Copper(II) Complexes with N^4,S-Diallylisothiosemicarbazones as Potential Antibacterial/Anticancer Drugs
Reprinted from: *Inorganics* 2023, 11, 195, doi:10.3390/inorganics11050195 80

Amanda A. Silva, Silmara C. L. Frajácomo, Állefe B. Cruz, Kaio Eduardo Buglio, Daniele Daiane Affonso, Marcelo Cecconi Portes, et al.
Copper(II) and Platinum(II) Naproxenates: Insights on Synthesis, Characterization and Evaluation of Their Antiproliferative Activities
Reprinted from: *Inorganics* 2023, 11, 331, doi:10.3390/inorganics11080331 97

Evariste Umba-Tsumbu, Ahmed N. Hammouda and Graham Ellis Jackson
Evaluation of Membrane Permeability of Copper-Based Drugs
Reprinted from: *Inorganics* 2023, 11, 179, doi:10.3390/inorganics11050179 118

Alberto Rovetta, Laura Carosella, Federica Arrigoni, Jacopo Vertemara, Luca De Gioia, Giuseppe Zampella and Luca Bertini
Oxidation of Phospholipids by OH Radical Coordinated to Copper Amyloid-β Peptide—A Density Functional Theory Modeling [†]
Reprinted from: *Inorganics* 2023, 11, 227, doi:10.3390/inorganics11060227 125

Arian Kola, Ginevra Vigni and Daniela Valensin
Exploration of Lycorine and Copper(II)'s Association with the N-Terminal Domain of Amyloid β
Reprinted from: *Inorganics* **2023**, *11*, 443, doi:10.3390/inorganics11110443 . . . . **140**

Marcelo Cecconi Portes, Grazielle Alves Ribeiro, Gustavo Levendoski Sabino, Ricardo Alexandre Alves De Couto, Leda Quércia Vieira, Maria Júlia Manso Alves and Ana Maria Da Costa Ferreira
Antiparasitic Activity of Oxindolimine–Metal Complexes against Chagas Disease
Reprinted from: *Inorganics* **2023**, *11*, 420, doi:10.3390/inorganics11110420 . . . . **153**

Glenn Blade, Andrew J. Wessel, Karna Terpstra and Liviu M. Mirica
Pentadentate and Hexadentate Pyridinophane Ligands Support Reversible Cu(II)/Cu(I) Redox Couples
Reprinted from: *Inorganics* **2023**, *11*, 446, doi:10.3390/inorganics11110446 . . . . **169**

Micaela Richezzi, Joaquín Ferreyra, Sharon Signorella, Claudia Palopoli, Gustavo Terrestre, Nora Pellegri, et al.
Effect of Metal Environment and Immobilization on the Catalytic Activity of a Cu Superoxide Dismutase Mimic
Reprinted from: *Inorganics* **2023**, *11*, 425, doi:10.3390/inorganics11110425 . . . . **183**

Victoria Karner, Attila Jancso and Lars Hemmingsen
Probing the Bioinorganic Chemistry of Cu(I) with [111]Ag Perturbed Angular Correlation (PAC) Spectroscopy
Reprinted from: *Inorganics* **2023**, *11*, 375, doi:10.3390/inorganics11100375 . . . . **202**

James A. Isaac, Gisèle Gellon, Florian Molton, Christian Philouze, Nicolas Le Poul, Catherine Belle and Aurore Thibon-Pourret
Symmetrical and Unsymmetrical Dicopper Complexes Based on Bis-Oxazoline Units: Synthesis, Spectroscopic Properties and Reactivity
Reprinted from: *Inorganics* **2023**, *11*, 332, doi:10.3390/inorganics11080332 . . . . **215**

About the Editors

Christelle Hureau

Christelle Hureau is a bioinorganic chemist and leader of the Alzheimer, Amyloids, and Bio-Inorganic Chemistry Group at the Coordination Chemistry Lab in Toulouse. She is mainly interested in the following topics: the impact of metallic ions in Alzheimer's disease and other amyloid-related diseases; the design and study of new drug candidates targeting copper ions; and inorganic compounds that tune the self-assembly of the amyloid-forming peptides. Her main field of expertise is the characterizations of metal–peptide interactions in link with health and diseases and bioinspired catalysis.

Ana Maria Da Costa Ferreira

Ana Maria da Costa Ferreira is a graduate chemist and doctor of chemistry at the Institute of Chemistry, University of São Paulo (São Paulo, Brazil). Currently she is a full professor in bioinorganic chemistry, teaching different disciplines in inorganic and biological inorganic chemistry to undergraduate and graduate students. She is the leader of a research laboratory, "Bioinorganic, Catalysis, and Pharmacological Chemistry", focusing on the reactivity of metal ions in biological systems, particularly aiming at the development of anticancer and anti-parasite metallodrugs, as well as inhibitors of selected proteins. New ligands and corresponding essential metal complexes are designed, characterized mainly by spectroscopic methods, and have their main biological targets and mechanisms of action investigated.

Gianella Facchin

Dra. Gianella Facchin Muñoz is a pharmaceutical chemist and doctor of chemistry at Facultad de Química (Udelar, Uruguay), where she is currently an associate professor of inorganic chemistry.

Her research focuses on the synthesis and characterization of metal complexes with antitumor properties, with a particular emphasis on copper complexes.

Dr. Facchin is also actively involved in academic leadership, serving as the representative of researchers on the directive board of the Program for the Development of Basic Sciences (PEDECIBA, Uruguay), where she also coordinates the Gender Committee. She was the former president of the Uruguayan Society of Biosciences (SUB) from 2020 to 2024.

Editorial

Bioinorganic Chemistry of Copper: From Biochemistry to Pharmacology

Ana Maria Da Costa Ferreira [1,*], Christelle Hureau [2,*] and Gianella Facchin [3,*]

1. Departamento de Química Fundamental, Instituto de Química, Universidade de São Paulo, São Paulo 05508-000, SP, Brazil
2. Laboratoire de Chimie de Coordination UPR 8241, Centre National de la Recherche Scientifique, 31400 Toulouse, France
3. Química Inorgánica, Departamento Estrella Campos, Facultad de Química, Universidad de la República, Montevideo 11800, Uruguay
* Correspondence: amdcferr@iq.usp.br (A.M.D.C.F.); christelle.hureau@lcc-toulouse.fr (C.H.); gfacchin@fq.edu.uy (G.F.)

Citation: Da Costa Ferreira, A.M.; Hureau, C.; Facchin, G. Bioinorganic Chemistry of Copper: From Biochemistry to Pharmacology. *Inorganics* 2024, 12, 97. https://doi.org/10.3390/inorganics12040097

Received: 14 March 2024
Accepted: 22 March 2024
Published: 28 March 2024

Copyright: © 2024 by the authors. Licensee MDPI, Basel, Switzerland. This article is an open access article distributed under the terms and conditions of the Creative Commons Attribution (CC BY) license (https://creativecommons.org/licenses/by/4.0/).

Copper is an essential trace element found ubiquitously in humans [1,2], plants [3–5], vertebrates and invertebrates [6], and is present in different active sites at innumerous proteins and enzymes [7–11]. In such biological systems, copper enzymes perform functions such as uptake and transport of oxygen; electron transfer in the respiratory chain; catalytic oxidation or reduction of many substrates; antioxidant action; uptake, transport and storage of metal ions, etc. [12,13]. Structurally, copper compounds appear in many configurations, coordinated with simple ligands or biomolecules, in a wide range of arrangements [14]. The two common oxidation states of copper, Cu^+ and Cu^{2+}, present in biological systems exhibit peculiar properties, with a range of reactivity and nuclearity, forming mono-, bi-, poly-nuclear, or even cluster species. The proteins of copper may have one or many metal ion centers with different spectroscopic signatures and dissimilar activity [15]. On the other hand, copper ions are also involved in neurodegenerative diseases, in which their redox properties play important roles [16–22]. Considering the varying biological roles of copper described above, the development of new copper-containing coordination complexes is an intense topic of research, involving exploration of their pharmacological properties, especially their anticancer activities [23–31].

Consequently, the Bioinorganic Chemistry of copper constitutes a rich and challenging field of investigation, attracting the attention and interest of research groups around the world, as demonstrated by the huge number of files found in literature searches by using copper in combination with a second keyword, such as antibacterial, anticancer, diseases, catalysts, mimics, proteins, spectroscopy, reactivity, etc.

This diversity is clearly demonstrated in this Special Issue of *Inorganics*, 'Bioinorganic Chemistry of Copper', which contains 14 published articles that explore topics such as antiproliferative studies, anticancer agents, anti-inflammatory compounds, potential radioactive imaging diagnosis agents, reactive species related to amyloid peptides, antiparasitic activity, catalytic oxidative activity, and protein mimics.

Potential anticancer agents were reported in most of the published articles. A review about mixed chelate homoleptic or heteroleptic copper(II) complexes, known as Casiopeínas® and already used in clinical tests, was provided by Ruiz-Azuara and co-workers (contribution 1), describing translational medicine criteria to establish a normative process for new drug development.

Batista and coll. (contribution 2) isolated and characterized a series of $Cu(I)/PPh_3/$naphtoquinone complexes with anticancer properties against diverse tumor cells. Their mode of action also involves reactive oxygen species (ROS) generation, both in the absence (peroxyl radicals) and presence of irradiation (hydroxyl radicals).

The cytotoxicity of phenylcarboxylate–copper(II) complexes with typical binuclear paddle-wheel arrangements was investigated by Fernandez et al. (contribution 3), who studied their lipophilicity, DNA binding, and cytotoxicity toward metastatic breast adenocarcinoma, lung epithelial carcinoma and cisplatin-resistant ovarian carcinoma cells.

A series of mononuclear copper(II) complexes with ligands containing phenolate and imine moieties was verified by Serre et al. (contribution 4), to act as efficient artificial nucleases, activated by reduction with ascorbate, toward cancer cell lines sensitive or resistant to cisplatin itself, with IC_{50} values much lower than those for cisplatin.

New isothiosemicarbazone–copper(II) complexes with varied structural features were isolated and characterized by different techniques, as reported by Graur et al. (contribution 5), showing antioxidant activity similar to trolox, used as an antioxidant agent in medicine, as well as high antiproliferative activity against cells sensitive to doxorubicin, a standard chemotherapy medication. Additionally, these compounds showed significant antibacterial and antifungal activities.

A strategic combination of bioactive ligands and metals that are already consolidated in the synthesis of metallopharmaceutical agents, allowed Corbi and coll. (contribution 6), to prepare and investigate naproxen (Nap)-based complexes of copper(II) and platinum(II) which showed cytostatic behavior over a set of tumor cells, but no bactericidal activity.

Complexes with other pharmacological activities were also presented. Copper(II) complexes with bi-, tetra-, or pentadentate ligands showing potential anti-inflammatory activity against Rheumatoid Arthritis (RA) were evaluated regarding their diffusion and membrane permeability, as described by Jackson and coll. (contribution 7). Chemical speciation was used to determine the predominant complex in solution at physiological pH. However, no correlation was found between partition coefficient and/or molecular weight and tissue permeability.

Since oxidative stress and metal (especially copper) dyshomeostasis are crucial factors in the pathogenesis of Alzheimer's disease (AD), involving ROS generation, Density Functional Theory (DFT) computations were used by L. Bertini and coll. (contribution 8), to verify a possible mechanism of oxidation through the OH radical propagation toward the phospholipidic membrane.

In another study, Valensin and co-workers (contribution 9) described an active alkaloid lycorine (LYC) capable of suppressing induced amyloid β (Aβ) toxicity in differentiated SH-SY5Y cell lines, likely by binding to the N-terminal region of Aβ via electrostatic interactions, which are favored in the presence of copper ions.

In the work of Portes et al. (contribution 10), copper(II) and zinc(II) compounds with oxindolimine ligands were shown to act as efficient trypanocidal agents against trypomastigote and amastigote forms of the parasites, through the generation of reactive oxygen species (ROS), inducing apoptosis, and probably involving the inhibition of selected parasite proteins. The determined IC_{50} values are lower and selective indexes (LC_{50}/IC_{50}) are higher, after 24 or 48 h incubation, modulated by the metal and the ligand, in comparison to traditional antiparasitic drugs used in clinics, or other metal-based compounds previously reported in the literature.

New penta- and hexadentate ligands containing pyridine moiety were prepared and verified to form stable Cu(I) and Cu(II) complexes, characterized by different methods, as reported by Mirica and coll. (contribution 11). After that, further experiments were performed to verify their potential use in vivo as ^{64}Cu PET imaging agents.

In addition, studies on structure–function relationships, methodologies, and catalysis were reported. Signorella and coll. (contribution 12) described the critical role of the flexibility or rigidity of the ligands in the redox cycle of copper superoxide dismutase (SOD) and therefore in the design of their mimics. A combination of ligand flexibility, total charge, and labile binding sites provided optimized catalytic properties for a *trans*-[Cu(II)N$_4$-Schiff base] complex in the dismutation of superoxide ions.

Applications of ^{111}Ag perturbed angular correlation (PAC) of γ-ray spectroscopy to elucidate the chemistry of Cu(I) in biological systems were reviewed by V. Karner et al.

(contribution 13). Since monovalent copper ion is isoelectronic with Ag(I) (both closed-shell d10), and both ions share ligand and coordination geometry preferences, the focused spectroscopy is appropriate to investigate the structural aspects of some small blue copper proteins, such as plastocyanin and azurin, involved in electron transport and transfer.

Finally, a catalytic action of copper compounds was reported by J. Isaac et al. (contribution 14) in the study of symmetrical and unsymmetrical dicopper(I) complexes with oxazolines or mixed pyridine–oxazoline coordination moieties that react with O_2 at low temperature to form $\mu\text{-}\eta^2\text{:}\eta^2$ $Cu_2\text{:}O_2$ peroxido species. These may result in C–C coupling products after reaction with a phenolate substrate, with the formation of an intermediary mixed-valence $Cu^{II}Cu^{III}$ species, as indicated by electrochemical and EPR results.

This Special Issue includes a range of examples of copper(I) and copper(II) compounds reactivity, reported by many researcher groups, using distinct strategies to illustrate different aspects of their bioinorganic chemistry.

Conflicts of Interest: The authors declare no conflict of interest.

List of Contributions

1. Aguilar-Jiménez, Z.; Espinoza-Guillén, A.; Resendiz-Acevedo, K.; Fuentes-Noriega, I.; Mejía, C.; Ruiz-Azuara, L. The Importance of Being Casiopeina as Polypharmacologycal Profile (Mixed Chelate-Copper (II) Complexes and Their In Vitro and In Vivo Activities). *Inorganics* **2023**, *11*, 394. https://doi.org/10.3390/inorganics11100394.
2. Leite, C.; Araujo-Neto, J.; Guedes, A.; Costa, A.; Demidoff, F.; Netto, C.; Castellano, E.; Nascimento, O.; Batista, A. Copper(I)/Triphenylphosphine Complexes Containing Naphthoquinone Ligands as Potential Anticancer Agents. *Inorganics* **2023**, *11*, 367. https://doi.org/10.3390/inorganics11090367.
3. Fernández, C.; Rocha, A.; Azam, M.; Alvarez, N.; Min, K.; Batista, A.; Costa-Filho, A.; Ellena, J.; Facchin, G. Synthesis, Characterization, DNA Binding and Cytotoxicity of Copper(II) Phenylcarboxylate Complexes. *Inorganics* **2023**, *11*, 398. https://doi.org/10.3390/inorganics11100398.
4. Serre, D.; Erbek, S.; Berthet, N.; Philouze, C.; Ronot, X.; Martel-Frachet, V.; Thomas, F. Anti-Proliferation and DNA Cleavage Activities of Copper(II) Complexes of N3O Tripodal Polyamine Ligands. *Inorganics* **2023**, *11*, 396. https://doi.org/10.3390/inorganics11100396.
5. Graur, V.; Usataia, I.; Graur, I.; Garbuz, O.; Bourosh, P.; Kravtsov, V.; Lozan-Tirsu, C.; Balan, G.; Fala, V.; Gulea, A. Novel Copper(II) Complexes with N4,S-Diallylisothiosemicarbazones as Potential Antibacterial/Anticancer Drugs. *Inorganics* **2023**, *11*, 195. https://doi.org/10.3390/inorganics11050195.
6. Silva, A.; Frajácomo, S.; Cruz, Á.; Buglio, K.; Affonso, D.; Portes, M.; Ruiz, A.; de Carvalho, J.; Lustri, W.; Pereira, D.; da Costa Ferreira, A.; Corbi, P. Copper(II) and Platinum(II) Naproxenates: Insights on Synthesis, Characterization and Evaluation of Their Antiproliferative Activities. *Inorganics* **2023**, *11*, 331. https://doi.org/10.3390/inorganics11080331.
7. Umba-Tsumbu, E.; Hammouda, A.; Jackson, G. Evaluation of Membrane Permeability of Copper-Based Drugs. *Inorganics* **2023**, *11*, 179. https://doi.org/10.3390/inorganics11050179.
8. Rovetta, A.; Carosella, L.; Arrigoni, F.; Vertemara, J.; De Gioia, L.; Zampella, G.; Bertini, L. Oxidation of Phospholipids by OH Radical Coordinated to Copper Amyloid-beta; Peptide: A Density Functional Theory Modeling. *Inorganics* **2023**, *11*, 227. https://doi.org/10.3390/inorganics11060227.
9. Kola, A.; Vigni, G.; Valensin, D. Exploration of Lycorine and Copper(II)'s Association with the N-Terminal Domain of Amyloid β. *Inorganics* **2023**, *11*, 43. https://doi.org/10.3390/inorganics11110443.

10. Portes, M.; Ribeiro, G.; Sabino, G.; De Couto, R.; Vieira, L.; Alves, M.; Da Costa Ferreira, A.M. Antiparasitic Activity of Oxindolimine-Metal Complexes against Chagas Disease. *Inorganics* **2023**, *11*, 420. https://doi.org/10.3390/inorganics11110420.
11. Blade, G.; Wessel, A.; Terpstra, K.; Mirica, L. Pentadentate and Hexadentate Pyridinophane Ligands Support Reversible Cu(II)/Cu(I) Redox Couples. *Inorganics* **2023**, *11*, 446. https://doi.org/10.3390/inorganics11110446.
12. Richezzi, M.; Ferreyra, J.; Signorella, S.; Palopoli, C.; Terrestre, G.; Pellegri, N.; Hureau, C.; Signorella, S. Effect of Metal Environment and Immobilization on the Catalytic Activity of a Cu Superoxide Dismutase Mimic. *Inorganics* **2023**, *11*, 425. https://doi.org/10.3390/inorganics11110425.
13. Karner, V.; Jancso, A.; Hemmingsen, L. Probing the Bioinorganic Chemistry of Cu(I) with ^{111}Ag Perturbed Angular Correlation (PAC) Spectroscopy. *Inorganics* **2023**, *11*, 375. https://doi.org/10.3390/inorganics11100375.
14. Isaac, J.; Gellon, G.; Molton, F.; Philouze, C.; Le Poul, N.; Belle, C.; Thibon-Pourret, A. Symmetrical and Unsymmetrical Dicopper Complexes Based on Bis-Oxazoline Units: Synthesis, Spectroscopic Properties and Reactivity. *Inorganics* **2023**, *11*, 332. https://doi.org/10.3390/inorganics11080332.

References

1. Chen, J.; Jiang, Y.; Shi, H.; Peng, Y.; Fan, X.; Li, C. The molecular mechanisms of copper metabolism and its roles in human diseases. *Pflügers Arch.-Eur. J. Physiol.* **2020**, *472*, 1415–1429. [CrossRef] [PubMed]
2. Tapiero, H.; Townsend, D.M.; Tew, K.D. Trace elements in human physiology and pathology. Copper. *Biomed. Pharmacother.* **2003**, *57*, 386–398. [CrossRef] [PubMed]
3. Kumar, V.; Pandita, S.; Singh Sidhu, G.P.; Sharma, A.; Khanna, K.; Kaur, P.; Bali, A.S.; Setia, R. Copper bioavailability, uptake, toxicity and tolerance in plants: A comprehensive review. *Chemosphere* **2021**, *262*, 127810. [CrossRef] [PubMed]
4. Hay, R.W. Plant Metalloenzymes. In *Plants and the Chemical Elements*; VCH Verlagsgesellschaft mbH: Weinheim, Germany, 1994; pp. 107–148.
5. Baran, E.J. Copper in plants: An essential and multifunctional element. *Adv. Plant Physiol.* **2014**, *15*, 373–397.
6. Beeby, A. Toxic metal uptake and essential metal regulation in terrestrial invertebrates: A review. In *Metal Ecotoxicology Concepts and Applications*; CRC Press: Boca Raton, FL, USA, 2020; pp. 65–89.
7. Bertini, I.; Cavallaro, G.; McGreevy, K.S. Cellular copper management—A draft user's guide. *Coord. Chem. Rev.* **2010**, *254*, 506–524. [CrossRef]
8. Tsang, T.; Davis, C.I.; Brady, D.C. Copper biology. *Curr. Biol.* **2021**, *31*, R421–R427. [CrossRef] [PubMed]
9. Gray, H.B.; Malmström, B.G.; Williams, R.J.P. Copper coordination in blue proteins. *JBIC J. Biol. Inorg. Chem.* **2000**, *5*, 551–559. [CrossRef] [PubMed]
10. Pretzler, M.; Rompel, A. What causes the different functionality in type-III-copper enzymes? A state of the art perspective. *Inorg. Chim. Acta* **2018**, *481*, 25–31. [CrossRef]
11. Solomon, E.I.; Hadt, R.G. Recent advances in understanding blue copper proteins. *Coord. Chem. Rev.* **2011**, *255*, 774–789. [CrossRef]
12. Farver, O. Electron transfer. In *Protein Electron Transfer*, 1st ed.; Bendall, D., Ed.; Garland Science: New York, NY, USA, 1996; p. 249.
13. Festa, R.A.; Thiele, D.J. Copper: An essential metal in biology. *Curr. Biol.* **2011**, *21*, R877–R883. [CrossRef]
14. Boal, A.K.; Rosenzweig, A.C. Structural Biology of Copper Trafficking. *Chem. Rev.* **2009**, *109*, 4760–4779. [CrossRef]
15. Adman, E.T. Copper Protein Structures. In *Advances in Protein Chemistry*; Anfinsen, C.B., Edsall, J.T., Richards, F.M., Eisenberg, D.S., Eds.; Academic Press: Cambridge, MA, USA, 1991; Volume 42, pp. 145–197.
16. Gaggelli, E.; Kozlowski, H.; Valensin, D.; Valensin, G. Copper Homeostasis and Neurodegenerative Disorders (Alzheimer's, Prion, and Parkinson's Diseases and Amyotrophic Lateral Sclerosis). *Chem. Rev.* **2006**, *106*, 1995–2044. [CrossRef]
17. Cheignon, C.; Tomas, M.; Bonnefont-Rousselot, D.; Faller, P.; Hureau, C.; Collin, F. Oxidative stress and the amyloid beta peptide in Alzheimer's disease. *Redox Biol.* **2018**, *14*, 450–464. [CrossRef]
18. Acevedo, K.; Masaldan, S.; Opazo, C.M.; Bush, A.I. Redox active metals in neurodegenerative diseases. *JBIC J. Biol. Inorg. Chem.* **2019**, *24*, 1141–1157. [CrossRef] [PubMed]
19. Leal, M.F.C.; Catarino, R.I.L.; Pimenta, A.M.; Souto, M.R.S. Roles of Metal Microelements in Neurodegenerative Diseases. *Neurophysiology* **2020**, *52*, 80–88. [CrossRef]
20. Bisaglia, M.; Bubacco, L. Copper Ions and Parkinson's Disease: Why Is Homeostasis So Relevant? *Biomolecules* **2020**, *10*, 195. [CrossRef]
21. Liu, Y.; Nguyen, M.; Robert, A.; Meunier, B. Metal Ions in Alzheimer's Disease: A Key Role or Not? *Acc. Chem. Res.* **2019**, *52*, 2026–2035. [CrossRef] [PubMed]

22. Fasae, K.D.; Abolaji, A.O.; Faloye, T.R.; Odunsi, A.Y.; Oyetayo, B.O.; Enya, J.I.; Rotimi, J.A.; Akinyemi, R.O.; Whitworth, A.J.; Aschner, M. Metallobiology and therapeutic chelation of biometals (copper, zinc and iron) in Alzheimer's disease: Limitations, and current and future perspectives. *J. Trace Elem. Med. Biol.* **2021**, *67*, 126779. [CrossRef]
23. Tisato, F.; Marzano, C.; Porchia, M.; Pellei, M.; Santini, C. Copper in Diseases and Treatments, and Copper-Based Anticancer Strategies. *Med. Res. Rev.* **2010**, *30*, 708–749. [CrossRef]
24. Santini, C.; Pellei, M.; Gandin, V.; Porchia, M.; Tisato, F.; Marzano, C. Advances in Copper Complexes as Anticancer Agents. *Chem. Rev.* **2014**, *114*, 815–862. [CrossRef]
25. Gandin, V.; Ceresa, C.; Esposito, G.; Indraccolo, S.; Porchia, M.; Tisato, F.; Santini, C.; Pellei, M.; Marzano, C. Therapeutic potential of the phosphino Cu(I) complex (HydroCuP) in the treatment of solid tumors. *Sci. Rep.* **2017**, *7*, 13936. [CrossRef] [PubMed]
26. Balsa, L.M.; Baran, E.J.; León, I.E. Copper Complexes as Antitumor Agents: In vitro and In vivo Evidence. *Curr. Med. Chem.* **2023**, *30*, 510–557. [CrossRef] [PubMed]
27. Oliveri, V. Biomedical applications of copper ionophores. *Coord. Chem. Rev.* **2020**, *422*, 213474. [CrossRef]
28. Krasnovskaya, O.; Naumov, A.; Guk, D.; Gorelkin, P.; Erofeev, A.; Beloglazkina, E.; Majouga, A. Copper Coordination Compounds as Biologically Active Agents. *Int. J. Mol. Sci.* **2020**, *21*, 3965. [CrossRef] [PubMed]
29. Kellett, A.; Molphy, Z.; McKee, V.; Slator, C. Recent Advances in Anticancer Copper Compounds. In *Metal-Based Anticancer Agents*; Royal Society of Chemistry: London, UK, 2019; pp. 91–119.
30. Shobha Devi, C.; Thulasiram, B.; Aerva, R.R.; Nagababu, P. Recent Advances in Copper Intercalators as Anticancer Agents. *J. Fluoresc.* **2018**, *28*, 1195–1205. [CrossRef]
31. da Silva, D.A.; De Luca, A.; Squitti, R.; Rongioletti, M.; Rossi, L.; Machado, C.M.L.; Cerchiaro, G. Copper in tumors and the use of copper-based compounds in cancer treatment. *J. Inorg. Biochem.* **2022**, *226*, 111634. [CrossRef]

Disclaimer/Publisher's Note: The statements, opinions and data contained in all publications are solely those of the individual author(s) and contributor(s) and not of MDPI and/or the editor(s). MDPI and/or the editor(s) disclaim responsibility for any injury to people or property resulting from any ideas, methods, instructions or products referred to in the content.

Review

The Importance of Being Casiopeina as Polypharmacologycal Profile (Mixed Chelate–Copper (II) Complexes and Their In Vitro and In Vivo Activities)

Zenayda Aguilar-Jiménez [1], Adrián Espinoza-Guillén [1], Karen Resendiz-Acevedo [1], Inés Fuentes-Noriega [2], Carmen Mejía [3] and Lena Ruiz-Azuara [1,*]

1. Inorganic Chemistry Department, Faculty of Chemistry, UNAM, Av. Universidad 3000, Circuito Exterior s/n, CU, Mexico City 04510, Mexico; zenayda_aj@hotmail.com (Z.A.-J.); adrianeg24@gmail.com (A.E.-G.); rakfq031293@gmail.com (K.R.-A.)
2. Pharmacy Departament, UNAM, Av. Universidad 3000, Circuito Exterior s/n, CU, Mexico City 04510, Mexico; fuentesines16@gmail.com
3. Interdisciplinary Biomedicine Laboraty, Faculty of Natural Sciences, University of Queretaro, Juriquilla, Queretaro 76230, Mexico; maria.c.mejia@uv.es
* Correspondence: lenar701@gmail.com

Abstract: In this review, we present a timeline that shows the origin of mixed chelate copper (II) complexes, registered as Mark Title Casiopeínas®, as the first copper (II) compounds proposed as anticancer drugs in 1988 and 1992. In the late twentieth century, the use of essential metals as anticancer agents was not even considered, except for their antifungal or antibacterial effects; also, copper, as gold salts, was used for arthritis problems. The use of essential metals as anticancer drugs to diminish the secondary toxic effects of Cisplatin was our driving force: to find less toxic and even more economical compounds under the rational design of metal chelate complexes. Due to their chemical properties, copper compounds were the choice to continue anticancer drug development. In this order of ideas, the rational designs of mixed chelate–copper (II) complexes (Casiopeínas, (Cas) homoleptic or heteroleptic, depending on the nature of the secondary ligand) were synthesized and fully characterized. In the search for new, more effective, and less toxic drugs, Casiopeína® (Cas) emerged as a family of approximately 100 compounds synthesized from coordinated Cu(II) complexes with proven antineoplastic potential through cytotoxic action. The Cas have the general formula $[Cu(N\text{--}N)(N\text{--}O)]NO_3$ and $[Cu(N\text{--}N)(O\text{--}O)]NO_3$, where N–N is an aromatic substituted diimine (1,10-phenanthroline or 2,2′-bipyridine), and the oxygen donor (O–O) is acetylacetonate or salicylaldehyde. Lately, some similar compounds have been developed by other research groups considering a similar hypothesis after Casiopeína's discoveries had been published, as described herein. As an example of translational medicine criteria, we have covered each step of the established normative process for drug development, and consequently, one of the molecules (Casiopeína III ia (CasIIIia)) has reached the clinical phase I. For these copper compounds, other activities, such as antibacterial, antiparasitic and antiviral, have been discovered.

Keywords: cancer; copper; antiparasitic; antiviral; drug delivery; Casiopeínas®

1. Introduction

Metals such as copper, silver and gold, and even minerals containing those metal ions, have been used since early civilizations. Egyptians, Greeks and Romans used them for curative applications and for purifying water [1–3]; however, the use of inorganic compounds for medicinal purposes had not occurred until the beginning of the Twentieth Century, when Paul Ehrlich developed Salvarsan (Arsfenamina, $C_{12}H_{13}N_2ClO_2As_2$) for the treatment of syphilis, and coined the term chemotherapy [4] for the use of a compound to treat some type of disease [5,6]. Anticancer drugs were mainly derived from organic molecules until some years later, when Barnett Rosenberg discovered the cytotoxic activity of the pure inorganic molecule Cisplatin cis-[Pt(NH$_3$)$_2$Cl$_2$], which was approved by the FDA in 1978 as an anticancer drug [7–10]. The success of Cisplatin as the pioneering anticancer metallodrug is due to its activity mainly against esophageal, neck, lung, and ovarian cancer [11–13]. Due to its secondary toxic effects on the kidneys (ototoxicity), some analogues were developed—carboplatin, oxaliplatin, among others—with the purpose of reducing those toxic effects. Despite being useful in therapies, this compound presents several secondary toxic effects on the kidneys, as well as myelotoxicity and neuropathy [14–18]. Another very serious limitation of using Cisplatin in cancer chemotherapy is drug resistance [19,20]. After Cisplatin was accepted for use in anticancer therapy, the research turned to focus on platinum metal compounds, and several reviews have been dedicated to such compounds [1,2,5–7].

To overcome the side effects and drug resistance of Cisplatin and its derivatives, some new approaches have been considered, including the use of elements from the platinum family, such as palladium [21], ruthenium [22], and iridium [23]. A more innovative approach was designing drugs based on endogenous trace metals. Among the first transitional metal species, copper is a trace element involved in a series of fundamental biological processes associated with metalloproteins and -enzymes, as well as several related to mitochondrial metabolism [24,25] and acting towards cellular oxidant species protection [26,27]. Even essential metals, metal compounds, or organic compounds at higher doses can present side effects; therefore, copper is essential in healthy cells, and there are homeostatic natural control processes acting to stabilize the level of copper in organisms, but which at higher concentrations can produce tumor growth, angiogenesis, and metastasis [28–31]. In accordance with these characteristics of copper, it has been considered as a promising component of metallodrugs used in cancer therapy [32], and in the late 1980′s, new mixed chelates–copper (II) complexes were designed and then patented a few years later [33]. Afterwards, copper complexes were synthesized with ligands containing principally N, O, S and P as donor atoms (phenanthroline, disulfiram, and thiosemicarbazones) [34–36]. These complexes have been shown to be multitargeting and multifunctional agents, and their anticancer activity involves several mechanisms. The presence of ligands leads to a minor concentration of the metal, and reduces the growth of malignant cells and angiogenesis. Compounds containing different types of ligands interact with proteasomes, inhibiting their function [37]. The Cu–esclemol complex can interact with the ferredoxin function, producing cuproptosis, a mode of independent cell death that occurs in pathways parallel to apoptosis [38].

The use of essential metals as anticancer agents in the late Twentieth Century was not even considered, except for their antifungal and antibacterial purposes. Also, copper in the form of gold salts was used for arthritis problems [39,40]. In the search for new anticancer drugs with fewer side effects, Casiopeínas®, comprising mixed chelate–copper complexes, was synthesized, reported, and patented [33,41,42]. The use of essential metals as anticancer drugs to diminish the secondary toxic effects of Cisplatin was the driving force of this research, as we sought less toxic and more economical compounds, incorporating the rational design of metal chelate complexes. The chemical properties of copper compounds were favored in continuing anticancer drug development. In this order of ideas, the rational design of mixed chelate–copper (II) complexes, such as Casiopeínas, as either homoleptic or heteroleptic depending on the nature of the secondary ligand, was synthesized and fully characterized. In the search for new, more effective, and less toxic drugs, Casiopeína® (Cas) emerged [43]. A family of approximately 100 compounds has been synthesized from coordinated Cu(II) complexes with antineoplastic potential proven through cytotoxic action [44–46] (Figure 1).

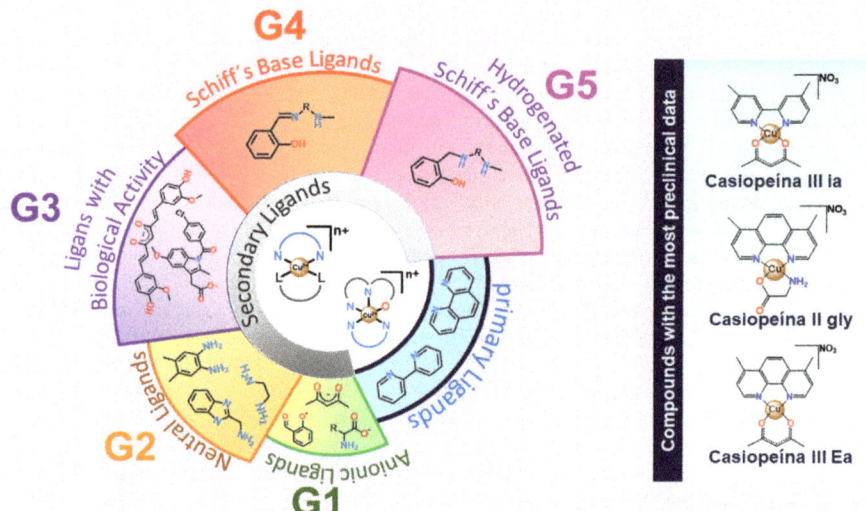

Figure 1. Circle of the five generations of Casiopeínas. The structures of the chelate–copper mixes comprise a primary ligand (substituted aromatic diamine) and a secondary ligand. Each generation is defined by the chemical characteristics of the secondary ligand. The first generation contains anionic ligands (e.g., amino acids, peptides, acetylacetonate and substituted salicylaldehydes); the second generation comprises neutral ligands (e.g., 2-aminomethyl benzimidazole, ethylendiamine and 4,5-dimethyl-o-phenanthroline); the third generation includes ligands with biological activity (e.g., indomethacin and Curcumine and its derivatives); the fourth generations includes Schiff's base ligands and the fifth generation involves the hydrogenation of the Schiff's base ligands (Adapted from Casiopeínas papers published until now).

Casiopeínas® are well-known copper compounds with proven potent anticancer activity. Due to the solubility of this compound in polar solvents, several X-ray structures have been reported [47–49] (Figure 2). Their general formula is $[Cu(N-N)(L-L)]^{n+}$ ($n = 1, 2$), where N–N = 4,7-dimethyl-1,10-phenanthroline or 4,4′-dimethyl-2,2′-bipyridine and L-L = different secondary ligands, that is, different bidentate chelate ligands (Table 1). One such ligand (CasIIIia:$[Cu(44'dmbipy)(acac)]^+$) is now being tested in clinical trials.

Table 1. Common name of some Casiopeínas of the first generation and Cisplatin.

Common Name	Formula	Short Common Name
Casiopeína I gly	[Cu(4,7-diphenyl-1,10-phenanthroline)(glycinate)]NO$_3$	CasIgly
Casiopeína II gly	[Cu(4,7-dimethyl-1,10-phenanthroline)(glycinate)]NO$_3$	CasIIgly
Casiopeína III ia	[Cu(4,4′ Dimethyl 2,2′ dipyridyl)(acac)]NO$_3$	CasIIIia
Casiopeína III Ja	[Cu(3,4,7,8-Tetramethyl-1,10-phenanthroline)(acac)]NO$_3$	CasIIIJa
Casiopeína III Ea	[Cu(4,7-dimethyl-1,10-phenanthroline)(acac)]NO$_3$	CasIIIEa
Casiopeína III La	[Cu(5,6-dimethyl-1,10-phenanthroline)(acac)]NO$_3$	CasIIILa
Casiopeína IV gly	[Cu(4,4′ Dimethyl 2,2′ dipyridyl)(glycinate)]NO$_3$	CasIVgly
Casiopeína VIII gly	[Cu(3,4,7,8-Tetramethyl-1,10-phenanthroline)(glycinate)]NO$_3$	CasVIIIgly
Cisplatin	cis-[Pt(NH$_3$)$_2$Cl$_2$]	CDDP

The second generation of Casiopeínas possess a neutral L-L ligand, such as ethylenediamine and substituted benzimidazoles, as well as 1,2-dianilines, with a general formula [Cu(N–N)(N1–N2)](NO$_3$)$_2$. The third generation of Casiopeínas have a formula of [Cu(N–N)(O1–O1)]NO$_3$, where the secondary ligand is a more hydrophilic monocharged molecule, and they have been previously used as commercial drugs, such as indometacinate and several curcuminates [48,50]. The fourth generation was conceived using tetra- and tridentated Shiff bases derived from salen ligand types. Tetradentate Shiff bases coordinated towards metal (II) ions produce neutral compounds that are generally insoluble in polar solvents such as water. Therefore, some tridentated ligands have been designed, synthesized and fully characterized. The mixed chelate compounds have the formula [Cu(NNO)(N–N)]$^+$, where NNO corresponds to asymmetric salen ligands (E)-2-((2-(methylamino)(ethylimino)methyl)phenolate (L1) and (E)-3-((2-(methylamino)ethylimino)methyl)naphthalenolate (LN1). The fifth generation contain hydrogenated derivatives 2-((2-(methylamino)ethylamino)methyl)phenolate (LH1) and 3-((2-(methylamino)ethylamino)methyl)naphthalenolate (LNH1); N–N corresponds to 4,4′-dimethyl-2,2′-bipiridyne (dmbpy) or 1,10-phenanthroline (phen). The insertion of this type of secondary ligand may modulate and increase the anticancer activity, as well as the anti-inflammatory activity, leading to an enhanced antitumor activity. However, the main mechanism of action of the first generation of Casiopeínas would not be different in the third generation, because of their main weak oxidant properties, and their ROS generation and apoptosis induction are similarly enacted through the reduction of Cu(II) to Cu(I). The mechanism of action has been studied in more than 20 compounds, and ROS generation has been identified, leading towards the induction of apoptosis and interaction with DNA, as well as nuclease action [51,52]. These compounds represent a viable, attractive, and available alternative for cancer treatment, including lung, cervix, and breast cancer. Additionally, Casiopeínas® show low toxicity against normal cells, which suggests high selectivity [53]. It has also been demonstrated that Casiopeínas® possess a multitarget cytotoxicity mechanism that converges in cellular apoptosis induction [54,55]. In addition, in silico studies suggest that all three generations of Casiopeínas® can act as antiviral agents via the inhibition of the main protease of SARS-CoV-2 [56].

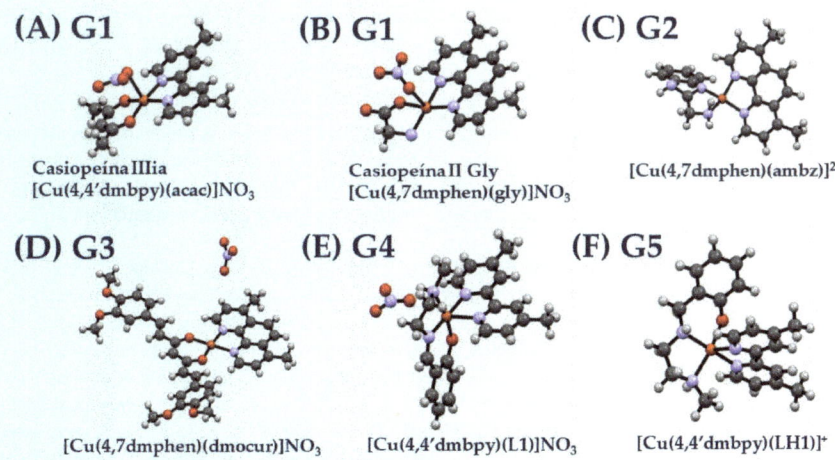

Figure 2. Structures of Casiopeínas and the generation to which they belong. X-ray structure of (**A**) CasIIIia, (**B**) CasIIgly, (**D**) [Cu(4,7dmphen)(dmethoxcur)]NO$_3$ and (**E**) [Cu(4,4′dmbpy)(L1)]NO$_3$, and the optimized M06/LanL2DZ structure of (**C**) [Cu(4,7dmphen)(ambz)]$^{2+}$ and (**F**) [Cu(4,4′dmbpy)(LH1)]$^+$ (Adapted from data from [47–49,53,56]).

2. Historical Timeline of Anticancer Drugs: Casiopeínas®

The use of metals in healthcare is an activity that has been described since ancient civilizations, who used metals such as copper, gold, or silver to disinfect water, food, or wounds [24,57,58].

In 1786, Thomas Fowler proposed a solution of As_2O_3 with potassium bicarbonate for treating periodic fever [59]. In subsequent years, arsenic trioxide was used to treat diseases such as psoriasis, malaria, asthma, and rheumatism [60]. However, because of its adverse effects and the emergence of new treatments, its use in treating these diseases has been avoided. Nonetheless, arsenic trioxide remains relevant, as in 2000, the FDA approved [61] this compound for use in the treatment of acute promyelocytic leukemia [34,62]. Paul Ehrlich, who is the "father of chemotherapy", having introduced the concepts of specific targets when developing new drugs, introduced Salvarsan in 1910 as an effective treatment for syphilis [35]. Salvarsan is an organic compound containing arsenic, which has led to the development of modern medicinal inorganic chemistry.

During the 1970s, the understanding of the molecular biology of cancer was limited, thus consequently constraining the identification of specific molecular targets when designing novel drugs [36,37]. Investigations of the mechanism of action of certain antitumor agents have piqued the interest in studying the characteristics of DNA when developing new anticancer molecules [38]. In 1978, the FDA approved Cisplatin, which was the first inorganic agent available for the treatment of cancer [11]. This spurred inorganic chemists to devise novel oncological medicines. As a result, some research groups involved in coordination chemistry have primarily focused on catalysis, and have shifted their focus to coordination compounds of the platinum metal group, which exhibit promising anticancer activities [2,41].

In the 1980s, Dr. Lena Ruiz Azuara, driven by the pursuit of novel cancer treatments, proposed a family of inorganic compounds with endogenous metals based on the following overarching criteria:

- Given DNA's status as a cancer target at the time, the synthesis of compounds that induce DNA damage was sought. Thus, molecules were designed with ligands situated in the equatorial plane of the metal's center to facilitate intercalation interactions with DNA;
- Molecules were devised to incorporate a copper atom, which, being an endogenous metal, was expected to mitigate compound toxicity. In addition, we explored whether the redox properties of the metal could be relevant in tumor cells;
- Small cationic molecules were proposed to enhance solubility in biological environments.

The research working group focused on the synthesis and characterization of the first copper ternary compounds to be designed, which were tested in 1986, whereby their potential anticancer activity was reported. With the patenting of the molecules in the early 1990s, the name Casiopeínas emerged, and preclinical studies of these compounds were initiated.

After more than 30 years since the design of Casiopeínas, there is solid evidence regarding the mechanism of action of these compounds. They are characterized as multitarget molecules that can be used for cancer treatment, inducing mitochondrial damage, increasing reactive oxygen species, and causing DNA damage. This cooperative action reinforces their effects on tumor cells. Their polypharmacological profiles ensure that the design of these molecules remains up to date as an alternative in the treatment of diseases like cancer. The relevance of polypharmaceutical activities of compounds, determined either by the nature of their ligands [63] or their oncological biomarkers, was later indicated [64].

The research group's expertise and preclinical results led to the selection of Casiopeína III ia (CasIIIia) for submission to the Mexican regulatory agency (COFEPRIS), marking the initiation of the first Phase I clinical trial of a copper-based anticancer compound in México. Recently, these molecules have garnered significant interest within the scientific community, due to their chemical characteristics and biological activities.

3. Activity and Properties

The anticancer activity of these compounds (Casiopeínas) depends mainly on the three descriptors obtained by QSAR study (Figure 3). Halfwave potential ($E_{1/2}$) is important in oxidant agents; aromaticity refers to the number of aromatic conjugated rings, and has a minor influence on the lopP of mono-charged cationic copper (II) complexes [42], as corroborated by analyses of the properties of the other compounds derived in the other generations [48,50], except for the second generation of Casiopeínas, wherein the mixed chelate complexes are bi-charged cationic complexes. These findings suggest that greater hydrophilicity in the cation complex decreases the membrane crossing efficiency in cancer cells. Structure–activity correlation analyses have been performed, and in vitro assays of IC_{50} have shown lower values against, mainly, $E_{1/2}$ and $Log D$ [42,48–50,65].

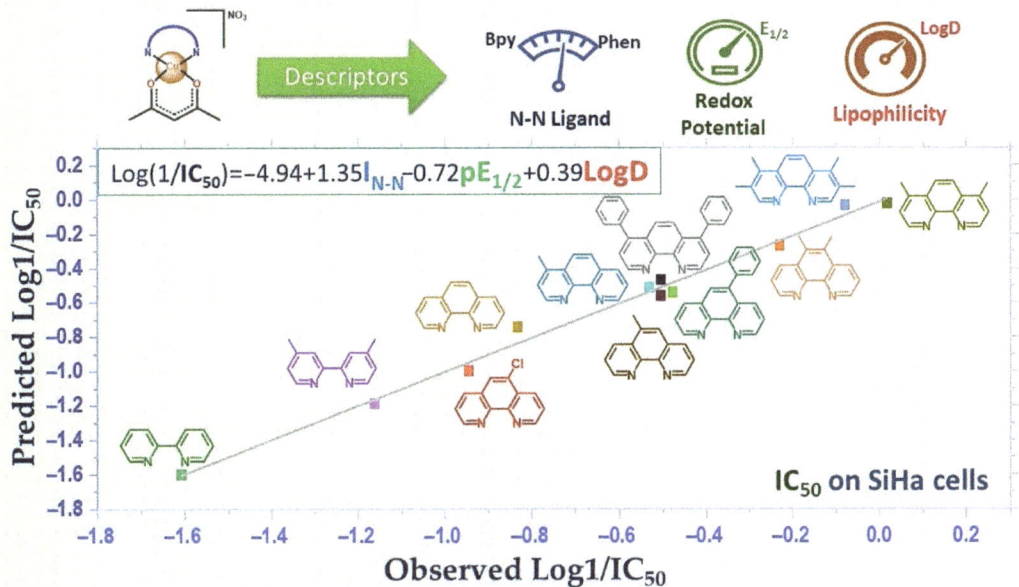

Figure 3. Plot of predicted vs. observed activity on SiHa (human cervical adenocarcinoma) cells, showing that the cytotoxicity of Casiopeínas may be successfully described by QSARs constructed with experimental values: I_{N-N}, $E_{1/2}$ and $Log D$ (adapted from data from [42]).

Casiopeinas and acetylacetonate analogues have been synthesized and studied by EPR and ENDOR spectroscopy. It has been found that small variations towards a square planar geometry can be detected by advanced EPR techniques [46,66].

It is important to mention that Casiopeínas of the first generation contain essential amino acids (aa); a large number of aa are used as secondary ligands, indicating that the amino acids produce very small variations in the activity compared with a change in the diimine; therefore, glycine was chosen to continue the studies [67].

Regarding the in vivo assays, the xenografting of several human cell lines into nu/nu mice models diminished the tumor volume, and even achieved an important increase in the activity towards the control drug Cisplatin. For example, when compared with the IC50 of cisplatin over HCT-15, CasIIIia has greater value, however, the *in vivo* value for the delay of specific growth as a function of the doses applied, the activity of CasIIIia is more effective than cisplatin [42,68] (Table 2, Figure 4).

Table 2. In vitro and in vivo metallodrugs' activities.

Compound	In Vitro IC$_{50}$ (µM) HCT-15	In Vivo Specific Growing Delay (Doses) HCT-15	In Vitro Lymphocyte (µM)
CasIIIia	40.5 [42]	1.4 (6 mg/kg) [68]	4700 [53]
CasIIgly	2 [42]	2 (1 mg/kg) [68]	1720 [53]
Cisplatin	21.8 [42]	0.1 (4 mg/kg) [68]	19 [53]

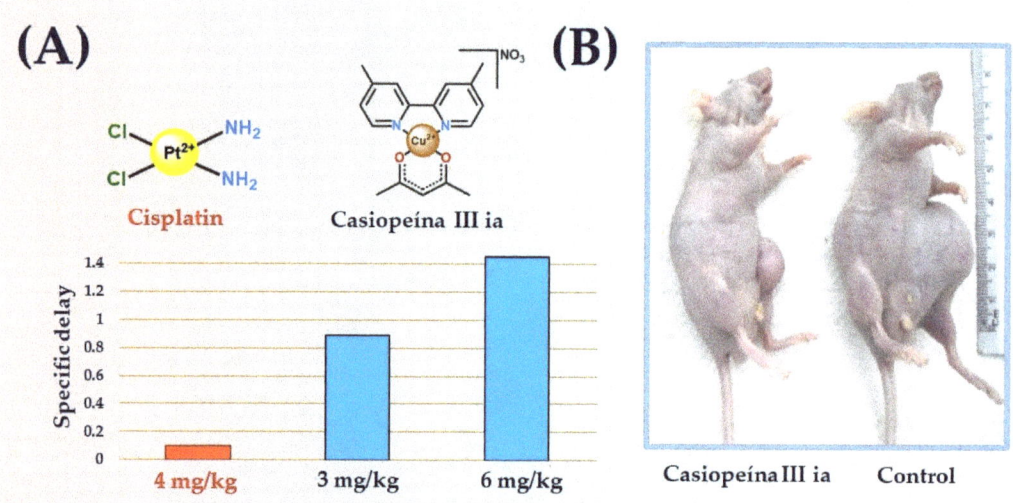

Figure 4. (**A**) Specific delay in growing under several treatment schemes of HCT-15 (human colon cancer) in a nu/nu model. (**B**) Left mouse treated with CasIII-ia 6 mg/kg (every 4 days for 6 ip doses), right mouse is the control mouse (adapted from data from [68]).

The tetramethylated phenanthroline analogues CasVIIIgly [Cu(3,4,7,8tm-phen)(gly)]NO$_3$ and CasIIIJa, [Cu(3,4,7,8tm-phen)(acac)]NO$_3$ were tested in an in vivo assay; when the HCT-15 (human colon adenocarcinoma) cell line was xenografted onto nu/nu mice, the acac analogue was more active than the IIgly one, which coheres with the prediction from the QSAR study. Then, the secondary ligand plays an important role in distribution in vivo. It is because of the above that there is a balance between two pharmacological properties—the tumor response of the drug and the access to the tumor—and this gives Casiopeínas its effect as a potent antitumor drug.

A lower cytotoxicity against non-cancer cell Lymphocytes, or stem cells, has been reported, and shows the safety of these compounds compared with commercial platinum drugs [69].

Pharmacological interactions between Casiopeínas analogues and Cisplatin were evaluated in HeLa cells (human cervical adenocarcinoma). An isobolographic analysis revealed that one of the combinations increased the antiproliferative activity from 50 to 77% when the cells were exposed to 4.59 and 9.70 µM of CasIIIia and Cisplatin, respectively. The results indicate that those compound analogous to tetraphenanthroline and Cisplatin had a synergistic effect with CasIIIia. The synergistic combinations assessed may be useful in future in vivo analyses with the aim of reducing the toxicity of Casiopeínas containing phenanthroline in their structure [70]. CasIIgly in combination with the 17 first-line antineoplastic drugs used at the National Institute of Cancer of México induces a synergistic response in HeLa cells (human cervical adenocarcinoma). The drugs combined with CasIIgly diminish the viability and proliferation rates at nanomolar doses, without any apparent effect on the normal proliferation of cells [71].

4. Toxicity

Lethal dose response was evaluated in twenty mongrel dogs. CasIIIia and CasIIgly were infused intravenously for 30 min. The reported LD_{99} was 10 mg/kg (200 mg/m^2) for CasIIIia and 8 mg/kg (160 mg/m^2) for CasIIgly. The therapeutic safety margin (TSM) and true therapeutic safety margin (TTSM) are observed in Figure 5. The authors recommend doses of 33.3 mg/m^2 for CasIIIia and 26.6 mg/m^2 for CasIIgly. At DL_{99}, the Casiopeínas caused acute dog death by pulmonary edema after a latency time of 30 to 50 min. Death due to pulmonary edema in dogs was confirmed by histopathological studies. In addition, it was found that the heart did not suffer direct cardiac toxicity; however, transmission electron microscopy studies revealed mitochondrial damage in myocardial cells [72]. CasIIgly and IIIEa were two and seven times more potent as inhibitors, respectively, than CasIIIia on respiratory activity. This security range was considered when proposing CasIIIia for use in clinical trial. Also, the effects of CasIIIia, CasIIgly, and CasIIIEa treatments on cardiac mitochondria-isolated cardiomyocytes were evaluated. It is proposed that phenanthroline, present in the structure of CasIIIEa and CasIIgly, may act as a transporter in the cellular uptake of copper compounds, meaning Casiopeínas provoke a loss of membrane potential, which increases the opening of the mitochondrial permeability transition pore (mPTP), and this could be associated with cardiotoxicity [73–75].

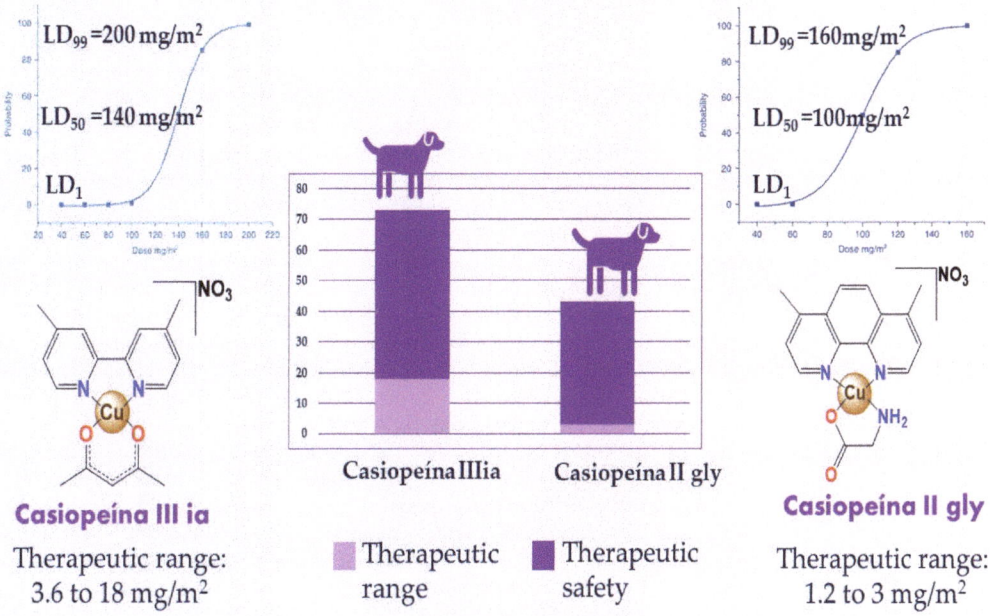

Figure 5. Therapeutic safety margin of Casiopeínas in mongrel dog study (adapted from data from [72]).

Hematoxicity

Hematoxicity assays of Casiopeínas were carried out in Wistar rats, via IV in singular doses. The results show that CasIIIia (3 mg/kg), CasIIgly (5 mg/kg), and CasIIIEa (4 mg/kg) caused hemolytic anemia at 24, 12, and 48 h, respectively. The restoration of hematic parameters was observed after 21 days (CasIIIia) and 15 days (CasIIgly). It seems that the evaluated Casiopeínas produce an early hematological effect with no delayed or permanent toxic effect. For CasIIIia and CasIIgly, hemolytic anemia was presented without any effect on biochemical parameters, but this was not the case for CasIIIEa, wherein alterations were observed in the concentrations of urea, bilirubin, albumin, and total proteins; however, recovery was observed at 96 h after the biochemical parameters were measured, whereat reversible renal damage was shown [76,77].

5. Pharmacokinetics

To determine the half-life ($t_{1/2}$) and distribution of Casiopeínas in a preclinical in vivo model, pharmacokinetic studies were carried out in species such as rats, dogs, and rabbits. A pharmacokinetic study was performed on CasIIgly in dog blood at two doses (1.5 and 3 mg/kg, n = 2 for each dose) given via intravenous infusion. The results obtained indicate that CasIIgly has a high elimination rate and a wide distribution for both tested doses. The reported half-life in dogs is slightly higher than that reported in male Wistar rats (8 mg/kg, n = 10). The results in dogs and rats suggest that CasIIgly is eliminated in a short time in animals [78]. Regarding CasIIIia, pharmacokinetic studies have been carried out in rats and rabbits. The volume of distribution and the half-life are both greater in rats than in rabbits, and the clearance is slower in rats than in rabbits. In addition, the half-life (3.9 h) and plasma concentration profile suggest that CasIIIia is distributed in several organs, with an apparent half-life of 13.4 µg/mL (administration dose was 4.5 mg/kg) [79].

Isomeric Casiopeínas have been synthesized. In order to study the differences in half-life, pharmacokinetic studies of CasIIILa (1 mg/kg) and CasIIIEa (4 mg/kg) in male Wistar rats (n = 6) were performed. The elimination of CasIIIEa was observed to be slower than that of CasIIILa. This compound had a greater distribution, even in organs with low blood perfusion. The differences found with respect to the elimination and distribution of both isomeric compounds are suggested to be related to the way they interact with cell membranes. When observing the structures of both isomeric compounds, it is possible to see that when the lipophilic character of the substituents increases, a decrease in the distribution and elimination rate of these compounds is also observed [69].

Plasma protein binding studies of CasIIIia have shown that the percentage of binding to plasma proteins is approximately 80% at the concentrations of 12.25 and 50 (µg/mL), which leads to the inference that CasIIIia is found in greater proportions within the blood circulation, which thus acts as a deposit at the concentrations tested. Correia et al. evaluated the binding of CasIIIia, CasIIgly, and CasIIIEa to human serum albumin (HSA). According to their results, the transport of Casiopeínas may occur through HSA, because they form 1:1 adducts within the physiological range of concentrations. The values were determined by circular dichroism, and the fluorescence emission spectra indicate that this binding takes place close to the Trp214 residue [80].

The data obtained from each study indicate that there is great interspecies variation, which is related to body weight and the physiological processes of each species. Therefore, further studies are required in these species in order to allow allometric scaling. Knowing the pharmacokinetic parameters in different animal species will help us in the future when designing dosing intervals, as well as in making an appropriate selection of dose levels for pharmacokinetic studies in future clinical stages.

6. Mechanism of Action

The biological activities of Casiopeínas have been evaluated both in vitro and in vivo, and they have demonstrated antiproliferative, cytotoxic and genotoxic effects, which has led to the clinical evaluation of several members of this family [68,80]. Research has focused on three main derivatives: CasIIgly, CasIIIia and CasIIIEa [53,54]. Various studies have been carried out on Casiopeínas in a wide variety of human and animal models, highlighting the antiproliferative influence in cell cultures of breast cancer [51], cervical cancer [81,82], lung cancer [83], rat C6 glioma [84], medulloblastoma and neuroblastoma [85]. On the other hand, non-tumor cells with accelerated growth, such as 3T3-L1 (healthy mice fibroblasts) treated with CasIIgly and CasIIIEa, presented low toxicity. As Casiopeínas have been shown to induce damage in healthy human peripheral lymphocytes, it is necessary to apply a higher mean inhibitory concentration (IC_{50}) (1 mM) with respect to CHP-212 neuroblastoma tumor cells (31.5 μM, 47.5 μM and 18.6 μM of CasIIgly, CasIIIa, and CasIIIEa, respectively) [53].

These experiments elucidate the tumor selectivity of Casiopeínas, as well as their low toxicity in healthy systems compared to most standard treatments used for different neoplasias.

In vivo, CasIIgly reduced tumor volume, mitosis, and cell proliferation, as well as promoting increased apoptosis in a model of C6 glioma cells xenografted in Wistar rats (doses of 0.4 and 0.8 mg/kg, injected intraperitoneally). The antineoplastic effect of CasIIgly occurred without leukopenia [86] or toxicity in the hepatobiliary or renal system, and furthermore, it did not induce animal mortality [84].

At high concentrations, CasIIIia underwent binding to proteins, specifically to the alpha-acid glycoprotein, thus intervening in binding to red blood cells [87]. On the other hand, the distribution of the whole blood assayed in Wistar rats showed a Cblood/Cplasma ratio > 2 (where C = concentration), which implies the binding of CasIIIia to erythrocytes. All the tests indicate a pattern in the mechanism of action of Casiopeínas based on the overproduction of reactive oxygen species (ROS) and their coupling to DNA, as well as the alteration of energy metabolism, thus promoting the induction of mitochondrial apoptosis [54,81–85].

From the energetic point of view, Casiopeínas can intervene in different levels of cellular respiration. Glycolysis is the main source of ATP in solid tumors, before even oxidative phosphorylation (OxPhos), known as the Warburg effect [88]. The protein responsible for initiating glycolysis is hexokinase (HK) [89], and so selective HK inhibitors such as 3-bromopyruvate (3-BrPyr) are important chemotherapeutic candidates [90,91]. The influence of CasIIgly on glycolysis was determined, whereby the inhibition of HK activity to a degree 1.3 to 21 times greater than that achieved with 3BrPy causes a drop in ATP production [92]. In a study in cardiac cells, it was determined that CasIIgy, CasIIIia, and CasIIIEa induce the loss of mitochondrial membrane potential ($\Delta\psi m$) due to increased ROS. In addition, the inhibition of the opening of the mitochondrial permeability transition pore (mPTP) by cyclosporin-A (CsA) prevents the depolarization of cells when treated with Casiopeínas, thus protecting cells from apoptosis [93], which suggests that the PTPm participates in the mechanism of action of Casiopeínas [73] (Figure 6).

In cancer cells, it has been reported that there is a high oxidative environment, which makes them susceptible to increased ROS [94]. In order to understand the participation of copper in oxidative stress processes, studies have described the oxidation of cysteine-containing peptides in the presence of copper [95]. Particularly, the redox pairs GSSG/GSH and Cystine/Cysteine have very similar oxidation-reduction potentials (−263 and 220 mV vs. NHE), and CasIIIia and CasIIgly show potentials of 62 mV and 90 mV vs. NHE, respectively. Therefore, the oxidation of cysteine and glutathione may be favored by the presence of Casiopeínas [83,96].

Cancer cells can adapt to these oxidant species by increasing concentrations of antioxidants, such as glutathione (GSH). Ref. [97] demonstrated that CasIIgly induced ROS overproduction by catalyzing the Fenton-type reaction and using GSH as an electron source in lung cancer cell lines (H157 and A549), inducing cell cytotoxicity via a rapid drop in GSH levels due to the delay in the cells' ability to initially control the levels of ROS. The hypothesis suggests that glutathione reacts with CasIIgly, which leads to the reduction of Cu(II) to Cu(I) and the formation of the glutaryl radical (GS•), which can react with another GS• product via the same reaction or via the superoxide (•-O_2)-mediated oxidation of GSH, which in turn causes GS• and •-O_2 to form oxidized glutathione (GSSG). GSSG is a compound composed of two glutathione molecules linked by a disulfide (S-S) bond. SOD (superoxide dismutase) catalyzes O_2 to form H_2O_2, which in turn reacts with Cu(I) to return to its oxidized Cu(II) state and produces the •OH radical, which initiates mitochondrial

DNA damage. This impairment promotes a drop in the expression of complex I proteins from the mitochondrial respiratory chain, causing their uncoupling, which is associated with the formation of O_2. The regression to oxidized CasIIgly restarts the GHS oxidation cycle, and consequently the subsequent reactions [83].

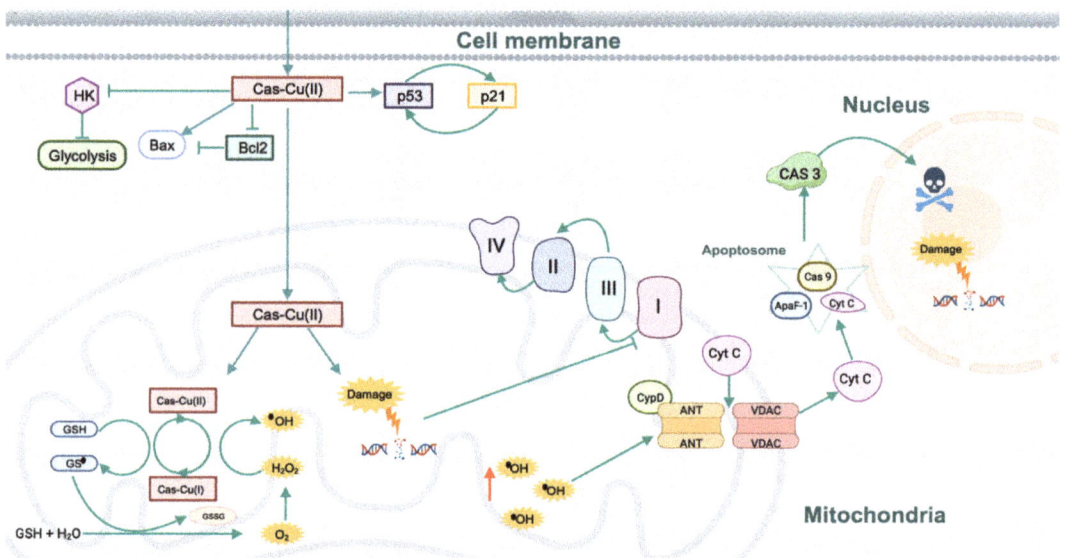

Figure 6. Joint theory on the mechanism of action of Casiopeínas. Casiopeína–Cu^{2+} (Cas-Cu(II)), Casiopeína–Cu^{1+} (Cas-Cu(I)), hydroxyl radical (•OH), hexokinase (HK), voltage-gated anion channel (VDAC), adenine nucleotide translocator (ANT), cyclophilin D (CyD), glutathione (GS), superoxide dismutase (SOD), reactive oxygen species (ROS) cytochrome c (Cyt c), apoptosis protease-activating factor-1 (Apaf-1), caspase-9 (cas-9) and caspase-3 (cas-3) (adapted from data from [53,54,73,82,83,92]).

On the other hand, bioinformatic analysis has shown that Casiopeínas have the characteristics required to intercalate between A and T bases, which redirects the bases towards the major groove of DNA. Experimental studies of the interaction of the same 21 Casiopeínas with pBR322 plasmid DNA via three different analytical techniques (circular dichroism, UV-Vis, and gel electrophoresis) have shown that, depending on the main diamine and the secondary ligand, Casiopeínas can be classified as intercalators with minor groove interaction, as well as with Ct-DNA/Cas and associated contents of the order of Kb (M−1) 105 [54].

In a complete transcriptome mapping study by Espinal-Enríquez et al. in 2016, it was determined that Casiopeínas block the cell cycle during the transition from the G1 to the S phase. Thus, in cervical cancer (HeLa) and neuroblastoma (CHP-212) cells treated with CasIIgly, the expressions of different apoptotic molecules (at 6 h and 2 h, respectively) were observed. The overexpression of the BAX, CIT-C (cytochrome-C), CAS9 (caspase-9), and CAS3 (caspase-3) genes, and the under-expression of the CAS8 (caspase-8) and BCL2 genes, identified the mitochondrial pathway as the preferred pathway by which Casiopeínas act. In turn, a high concentration of ROS was observed. Apoptosis was potentiated by the interruption of the cell cycle, which increased the expression of BAX, TP53, and P21 genes [82]

7. Metabolomics: CasIIgly and Its Effect on Triple-Negative Breast Cancer (TNBC) Metabolism

Metabolomics is an important tool that measures metabolic biology in response to systemic changes [98]. In cancer research, metabolomics provides information on complex changes in the molecular phenotype during malignant progression, whereby, in addition, the concentration of the metabolites can be associated with cancer state, and thus be used to complement other omics studies (genomic, proteomic, and transcriptomic). Thanks to this, the development of new treatments and patient stratification is possible [99–101].

The metabolomic studies that have been undertaken using coordination compounds are very few in number; therefore, they could be considered an emergent area. Until now, metabolomic studies in cancer have focused on Cisplatin, as well as some of its derivatives and coordination compounds, mainly ruthenium and palladium [102–104]. However, no reports have been found on this type of study using copper-based coordination compounds—except for the one on Casiopeínas®, wherein a triple-negative breast cancer (TNBC) cell line was used, which is a difficult type of cancer to treat [105].

TNBC represents 12 to 17% of all breast cancer cases, and it is characterized by its aggressiveness, poor prognosis, high incidence, and mortality [6,106]. A particular characteristic of TNBC is the absence of hormone receptors (progesterone and estrogen) and human epidermal growth factor receptor 2 (HER2); consequently, the available treatments are not very effective [6,107], and so it is necessary to explore new pharmacological targets and treatment alternatives for TNBC.

Cancer cell metabolism is of interest for TNBC treatment since it is modified in cancer cells to cover their nutrient and energy demands, thereby ensuring the cell's survival. Furthermore, this type of metabolic reprogramming is involved in metastatic and carcinogenesis processes [108–110].

Cisplatin is a chemotherapeutic commonly used agent many types of cancer, and its effect has been reported against TNBC metabolism when employing the MDA-MB-231 cell line. The principal pathways affected by Cisplatin treatment are phosphatidylcholine, phosphatidylethanolamine, phospholipid biosynthesis, and methionine metabolism [51]; besides this, it reduces the phosphocholine and betaine contents [111]. These processes are important for cancer cell growth and membrane cellular constitution; because MDA-MB-231 cells have a high concentration of lipids, this helps in the development of metastasis [112,113]. However, chemoresistance may occur when using Cisplatin in TNBC treatment, which could limit its use [114–116].

On the other hand, CasIIgly has a promising effect on TNBC metabolism, as it affects important biochemical pathways related to cancer cells within short treatment times (20 to 120 min), among which are included the Warburg effect, pyruvate metabolism, gluconeogenesis, glycolysis, the electron transport chain, β-oxidation, and the pentoses pathway [51]; all of these are important metabolic routes in cancer progression because they are related to how cancer cells undergo proliferation, metastasis, invasion and tumor growth [117–120].

These relevant alterations to the metabolism could cause cancer cells to undergo severe nutrient deprivation and eventually death. Besides this, further metabolomic studies could help us to understand and explore other types of pharmacological activities.

8. Encapsulation

Casiopeína nanoencapsulation has been carried out in colloidal systems (Figure 7). The aim is to increase the concentration in the blood of the drug, and reduce its inherent toxicity, as it is a compound used for cancer treatment [121,122]. It has been used in the encapsulation of CasIIIa within chitosan nanoparticles and niosomes formulations. Miranda in 2012 reported the encapsulation of CasIIIa in chitosan nanoparticles; the results show that the administration of CasIIIa within chitosan nanoparticles increased the survival time of CB6F1/Hsd mice transplanted with B16 melanoma tumor sixfold with respect to the administration of CasIIIa alone. Also, the CasIIIa within nanoparticles inhibited tumor growth more extensively than the free drug [123]. Niosome containing CasIIIa was formed using a Quality by Design tool [124] to ensure the prediction and optimization of a repeatable and reproducible formulation [125]. It was observed that encapsulated CasIIIia showed lower toxicity in female BALB/c mice with cells 4t1 (metastatic breast cancer model), compared with cis-Pt and nonencapsulated CasIIIia. When the weight of mice was evaluated at the end of the treatment, it was observed that only the groups treated with niosome with CasIIIia and niosome without CasIIIia recovered their initial weights [126].

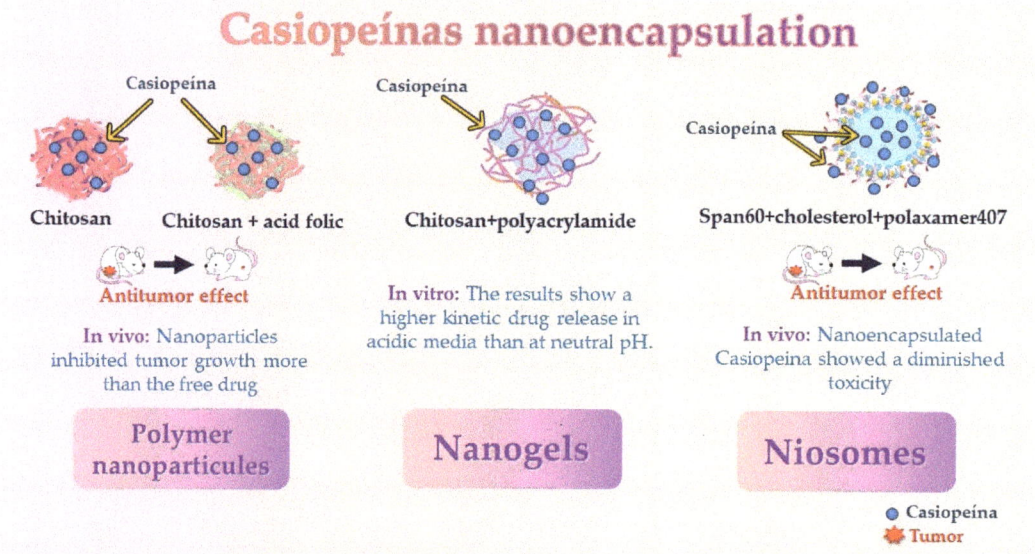

Figure 7. Copper-based drug encapsulation (adapted from data from [50,123,126]).

A third-generation compound of Casiopeínas, with the Indomethacin ligand, was synthesized and nanoencapsulated in chitosan–polyacrylamide nanogel nanoparticles. The nanogels were pH-dependent, and subjected to changes in structure, which opens up opportunities for pH-dependent chitosan nanogels to be used to encapsulate metal-based drugs [50]. In addition, a cytotoxic copper–phen aquaporin, inspired by Casiopeína, was synthesized and encapsulated in liposomes. In in vivo studies, male BALB/c mice showed nontoxicity after the parenteral administration of Cuphen liposomes in melanoma models [127] with Hag90. These preclinical in vivo studies show that the toxicity of copper-based compounds is reduced when encapsulated.

Also, a Cu(II) complex containing a b-diketone and phenanthroline has been synthesized, and the nanoparticles of this complex were prepared and evaluated in for their effects on the proliferation of MKN-45 cells. Cell proliferation was inhibited by all compounds and nanocompounds in a dose-dependent manner, but non-nanoparticle compounds were more active [128].

Although there are limited reports regarding encapsulated copper compounds, researchers have successfully synthesized nanoparticles employing copper as a carrier to augment the activity of these compounds [129]. Furthermore, poly(amidoamine) dendrimers complexed with copper(II) have been synthesized for use in radiotherapy and the treatment of metastatic cancer [130]. Copper chelates have also been formed with compounds possessing antitumor properties, such as disulfiram [131]. In addition, copper complexes with doxorubicin have been encapsulated within liposome [132]. A comprehensive review of metallodrugs used in the realm of cancer nanomedicine was conducted by Peña and collaborators [133].

9. Casiopeínas-like Compounds

Some reviews dedicated to copper(II) anticancer drugs have been published [43], including works related mono-, bis- and tris-chelates, and mainly mixed and non-mixed chelate compounds. Four years later, Santini et al. presented a number of mononuclear and binuclear Cu(II) and Cu(I) systems, describing their structures and later their mechanisms of action. In all cases, the induction of apoptosis was enabled by ROS production and DNA interaction. An interesting table is shown in Santini's review, which shows the exponential increase in the number of publications each year from 2000 (20) to 2012 (150). From 2014 to 2023, the number of publications increased exponentially; our search before that, from 1986 to 1999, found only 31 publications and four patents [134] (Figure 8).

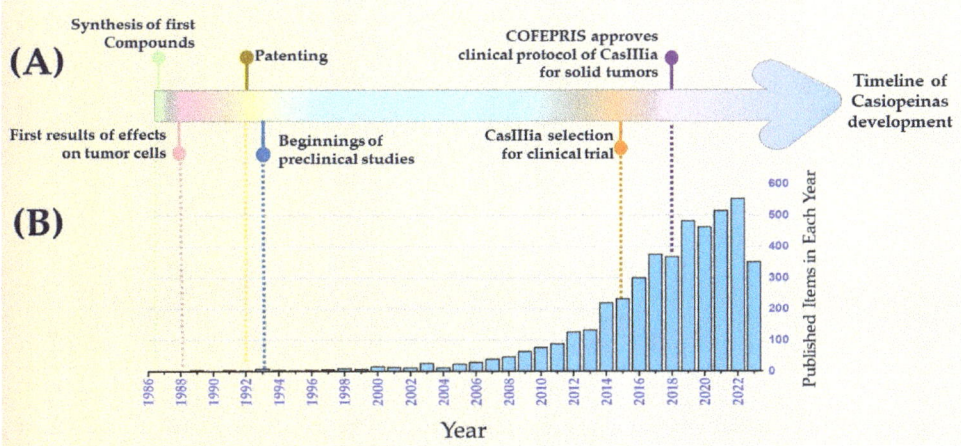

Figure 8. (**A**) Timeline of Casiopeínas development and (**B**) number of articles in Web of Science on the topic "copper and anticancer" from 1986 to 2023.

The principal characteristics of Casiopeína-like compounds include the presence of copper-containing diimines, such as 1,10-phenanthroline (phen) and 2,2-bipyridine (bipy), and eventually their substituted analogues, as well as the presence of secondary ligands to the ternary compounds, also called mixed chelate compounds. As precursors, we can mention the bischelate–phen complexes, which have shown nuclease activity; the strength of this property varies depending on the position and number of methyl substituents, such as Cu(II) or Cu(I) compounds [135,136]. Another approach to developing active metallodrugs is the use of bischelate non-diamine complexes, homoleptic biscurcuminates compounds synthesized with the purpose of improve the solubility of the ligands [137–140]. Similar examples of bis O–O chelates that are sterically bulky have been studied [141]. Homoleptic compounds are, in general, less active than ternary compounds. Also, some bis-functionalized phenanthroline complexes with carboxylic ligands have been prepared [Cu(RCOO)(1,10-phen)$_2$]+; their in vitro antiproliferative, antifungal, and antibacterial activities have been determined, as have their nuclease activity and albumin interaction [141]. Some bischelates, such as copper(II) square pyramide(sp) compounds with a cationic 2+ charge, [Cu(dmp)$_2$(CH$_3$CN)](ClO$_4$)$_2$, [Cu(phen)$_2$(CH$_3$CN)](ClO$_4$)$_2$, and tetrahedral copper(I) [Cu(dmp)$_2$](ClO$_4$)(th) where dmp = 2,9-dimethyl,1,10-phenantholine, have been studied. Their antiproliferative activities have been determined, and EPR studies have shown the redox process. The relevant conclusion here concerns the distortion of the square planar into a tetrahedral form and its relationship with the ability to modify the redox potential of copper [103].

In the present review, we are focusing on some articles concerning "Casiopeínas-like" that convey an innovative approach. Casiopeínas-like compounds contain Cu and either phen or bipy ligands, and the modification comes in the secondary ligand [66,142–144]. Some homoleptic compounds have been synthesized containing terpyridine derivatives such as [Cu(bitpy)$_2$](ClO$_4$)$_2$, and heteroleptic ones such as [Cu(bitpy)(phen)](NO$_3$)$_2$. For the second one, MTT assays revealed a significant oxidant challenge to both NIH3T3 and MG63 cells, which was observed upon treatment with a heteroleptic complex, even at a concentration as low as 5 mM. In support of the previously reported binding of DNA, we have found the following [43]: mixed chelates are more active than bis-chelates. A paper by I Correia et al. presents mixed chelate compounds containing phen and bipy, and tridented salgly or salala anions as the secondary ligand, giving five copper(II) complexes: [Cu(sal-Gly)(bipy)] (1), [Cu(sal-Gly)(phen)] (2), [Cu(sal-L-Ala)(phen)] (3), [Cu(sal-DAla)(phen)] (4), and [Cu(sal-L-Phe)(phen)] (5). MTT assays have been performed on four tumor lines, and DNA binding has been assessed by circular dichroism, as presented in [145]. Kordestani et al. reported a series of Cu(diimine)(X-sal)(NO$_3$) complexes, wherein the diimine is either 2,2′-bipyridine (bpy) or 1,10-phenanthroline (phen), and X-sal is a monoanionic halogenated salicylaldehyde (X = Cl, Br, I, or H). Their viability and antiproliferative activity were tested in vitro [146]. Several homoleptic bis-chelates have been reported with the following ligands: 2-(2-aminophenyl)-1H-benzimidazole (bm); N,N′-bis(salicylidene)ethylenediamine (salenH$_2$); 2-(aminomethyl)pyridine (amp); 4-nitrobenzene-1,2-diamine(nbda), with chloride as the counterion [147]. Tetrahedral neutral copper(I) complexes with 2,9-dmethyl,1,10-phenanthroline = dpm) [Cu(I)(dmp)(MPOH)] and [Cu(I)(dmp)(MPSG)] (MPOH = P(p-OCH$_3$-Ph)$_2$CH$_2$OH and MPSG = P(p-OCH$_3$-Ph)$_2$CH$_2$SarGly)) were prepared and characterized, and have shown cytotoxicity in cancer cell lines. The copper(I) complexes studied by fluorescence spectroscopy and cyclic voltamperometry have been proven to be able to generate reactive oxygen species as a result of redox processes [148]. Other examples containing 2,9dmphen and tridentated dipeptides, with the general formula [Cu(L-dipeptide)(neo)], and crystalline structures of [Cu(glyval)(neo)]·3H$_2$O, [Cu(gly-leu)(neo)]·H$_2$O, [Cu(ala-gly)(neo)]·4H$_2$O, [Cu(val-phe)(neo)]·4.5H$_2$O and [Cu(phephe)(neo)]·3H$_2$O, were determined by single-crystal X-ray diffraction, and their cytotoxicity was tested in several cancer cells (lung and breast) [149]. Bimetallic complexes have been developed with the formula [Cu(N,N′)(AA)]$_2$•(V$_4$O$_{12}$), where (N,N′) = 1,10-phenanthroline and 2,2′-bipyridine and (AA) = lysine and ornithine, and have shown innovative dodecavanadate counteranions with cytotoxic activity [150]. Compounds with diimines and monodentated ligands have been reported, such as [Cu(N–N))L$_2$], [Cu(N–N)$_2$L], [Cu(N–N))L$_2$] and [Cu$_2$(N–N)$_2$)L$_4$], where L = −4,5-dichloroisothiazole-3-carboxylic acid; they have shown antiproliferative effects, and the enzymes' cytochrome P450 dependence was studied [55].

The use of an active drug as a ligand, as in the third generation of Casiopeinas [50], has been reported, with the O–O donor plumbagin as a copper bis-chelate or mixed chelates with bipy. Both were tested in a series of cancer cells in vitro [151]. Other examples with O–O donors include mononuclear copper (II)ternary compounds with phen and 2Rsalal or diphosphate [152]. Indolacetate and phen compounds present nuclease activity [153,154].

Tridentated taurine Shiff bases and ONO donors with phen show octahedral geometries with a coordinated water molecule [155].

A review from 2015 has compiled the targets discovered for copper compounds related to their promising anticancer activities and lower toxicity in healthy cells, which are enhanced in the presence of copper [156].

A very recent review of copper's anticancer activity and its main targets concluded that the most relevant pathway is cancer cell death, such as via DNA oxidative cleavage, DNA intercalation, topoisomerase inhibition and proteasome inhibition [157].

10. Casiopeínas and Other Activities

The bacterial and so-called tropical neglected diseases (TND) are a cause of death in children and the adult population mainly in developing countries. Some results derived for Casiopeínas have been derived through assays using some of these bacteria or parasites. The antibacterial Inhibitory Fraction Index (IFI) values of CasIIIia, CasIIIEa and

CasIIgly, combined with INH, RIF, and EMB using a bidimensional checkerboard against susceptible and resistant clinical isolates of *M. tuberculosis* and *M. tuberculosis H37Rv*, have reported CasIIIEa/EMB as having a better synergic effect [158]. The cytotoxic effects of eight Cu (II) Casiopeínas against Giardia lamblia trophozoites, human peripheral blood lymphocytes (HPBL), and human peripheral blood macrophages (HPBM) were assessed, and their associated selectivity indexes were determined. The more active casiopeínas, with IC_{50} values of 36 μM, were the more lipophilic compounds—CasIgly [Cu(diphenyl,1-10-phen)(gly)]NO_3 and CasIIIHa [Cu(tm1,10-phen)(acac)]NO_3. TEM images showing the morphological changes in G. intestinalis trophozoites caused by 24 h of exposure to the IC_{50} doses evaluated compounds (Metronizadole, CasIgly, CasIIIia) compared with those without tratment [147].

Another protozoan parasite was evaluated using three Casiopeínas, and the IC_{50} (μM) values were determined for CasIIgly (3.9 ± 1.5), CaIIIEa (11.3 ± 3.8) and CasIIIia (6.9 ± 3.9). The compounds showed activity against *Trypanosoma cruzi* (*T. cruzi* epimastigotes (Dm28c strain)), an etiologic agent of the Chagas disease. The tested complexes showed in vitro anti-*T. cruzi* activity similar to that of the anti-trypanosomal reference drug Nifurtimox [159].

Halide derivatives [Cu(bipy)(acac)X] and [Cu(bipy)(acac)(H_2O)] X were derived, where X = Cl and X = Br, and the microbial activities were inferred [66].

Regarding the COVID-19 pandemic, the interaction between the active site of the SARS-CoV2 protein and Mpro was studied by docking using a DFT calculation and several casiopeínas. The study gave interesting results. The bond energies and bond constants of Casiopeínas/Mpro compared with Remdesivir show that mainly Casiopeínas from the first, second and third generations are the most promising for the treatment of this disease [56].

The several activities presented for these compounds can be explained by the different targets of activation, depending on their versatile ternary conformation and the variations on the descriptor, as derived from their behavior. This statement is supported by a recent review, wherein a mechanism of action approach is presented for mixed Cu(II) phenanthroline-based complexes, using a large number of secondary ligands as mono- or di-nuclear compounds. The authors summarized the overall targets derived from the mechanisms of action studied; those targets are (a) the nucleus (DNA, chromatine, nucleolar functions); (b) the mitochondria; (c) the rough endoplasmic reticulum and the (d) peroxisomas (ROS metabolism) [160].

11. Conclusions

The study of metallodrugs used in cancer therapy has been mainly focused on Cisplatin and its analogs; unfortunately, their secondary effects remain a problem. The search for more selective and less toxic molecules using bio-essential metals was the main goal of this review, aiming at enhancing the properties of copper compounds for pioneering Casiopeínas families and Casiopeínas-like compounds. We have presented a timeline of this development, showing the properties observed in chemical and preclinical models. As a result, we can conclude that copper metal drugs are safe, selective towards cancer cells, and produce low toxicity. Thus, they can be proposed as potent antineoplastic drugs that can be used in antineoplastic therapy. Furthermore, these compounds present a multiple-target mechanism of action, as evidenced by metabolomics and genomic studies; moreover, some of their other activities have been investigated, such as antibacterial, antivirus, and antiparasitic. From these findings, it is possible to infer that in the future, other activities can be investigated.

Author Contributions: All authors contributed to preparing and writing the text according to the review content, particularly as follows. Z.A.-J.: Encapsulation, writing—review and editing. A.E.-G.: Timeline of copper compounds development and figures preparation. K.R.-A.: Metabolomic part. I.F.-N.: Pharmacokinetic and toxicology parts. C.M.: Mechanism of action and figures. L.R.-A.: conceptualization, introduction, properties, cancer and other activities. All authors have read and agreed to the published version of the manuscript.

Funding: This research received no external funding.

Data Availability Statement: Not applicable.

Acknowledgments: The authors thank Cynthia Novoa-Ramirez, Yeshenia Figueroa-De-Paz, Nancy Vara-Gama, Luis-F. Hernandez and Miguel Reina, also Mauricio Gonzalez and Rogelio Hurtado for helpful discussions and information. Finally, Norma Hernandez is thanked for proof reading the manuscript. This review is dedicated to all the students and collaborators who have participated in the development of Casiopeínas and to those colleagues who have contributed to innovative idea of using copper compounds for cancer treatment.

Conflicts of Interest: The authors declare no conflict of interest.

References

1. Guo, Z.; Sadler, P.J. Metals in Medicine. *Angew. Chem. Int. Ed.* **1999**, *38*, 1512–1531. [CrossRef]
2. Mjos, K.D.; Orvig, C. Metallodrugs in Medicinal Inorganic Chemistry. *Chem. Rev.* **2014**, *114*, 4540–4563. [CrossRef]
3. Sadler, P.J. Inorganic Chemistry and Drug Design. In *Advances in Inorganic Chemistry*; Sykes, A.G., Ed.; Academic Press: Cambridge, MA, USA, 1991; Volume 36, pp. 1–48. ISBN 0898-8838.
4. Kaufmann, S.H.E. Paul Ehrlich: Founder of Chemotherapy. *Nat. Rev. Drug Discov.* **2008**, *7*, 373. [CrossRef]
5. Anthony, E.J.; Bolitho, E.M.; Bridgewater, H.E.; Carter, O.W.L.; Donnelly, J.M.; Imberti, C.; Lant, E.C.; Lermyte, F.; Needham, R.J.; Palau, M.; et al. Metallodrugs Are Unique: Opportunities and Challenges of Discovery and Development. *Chem. Sci.* **2020**, *11*, 12888–12917. [CrossRef]
6. Nayeem, N.; Contel, M. Exploring the Potential of Metallodrugs as Chemotherapeutics for Triple Negative Breast Cancer. *Chem. Eur. J.* **2021**, *27*, 8891–8917. [CrossRef]
7. Komeda, S.; Casini, A. Next-Generation Anticancer Metallodrugs. *Curr. Top. Med. Chem.* **2012**, *12*, 219–235. [CrossRef]
8. Lucaciu, R.L.; Hangan, A.C.; Sevastre, B.; Oprean, L.S. Metallo-Drugs in Cancer Therapy: Past, Present and Future. *Molecules* **2022**, *27*, 6485. [CrossRef]
9. Ndagi, U.; Mhlongo, N.; Soliman, M.E. Metal Complexes in Cancer Therapy—An Update from Drug Design Perspective. *Drug Des. Devel. Ther.* **2017**, *11*, 599–616. [CrossRef]
10. Wilson, J.J.; Johnstone, T.C. The Role of Metals in the next Generation of Anticancer Therapeutics. *Curr. Opin. Chem. Biol.* **2023**, *76*, 102363. [CrossRef]
11. Ghosh, S. Cisplatin: The First Metal Based Anticancer Drug. *Bioorg. Chem.* **2019**, *88*, 102925. [CrossRef]
12. Aldossary, S.A. Review on Pharmacology of Cisplatin: Clinical Use, Toxicity and Mechanism of Resistance of Cisplatin. *Biomed. Pharmacol. J.* **2019**, *12*, 7–15. [CrossRef]
13. Brown, A.; Kumar, S.; Tchounwou, P.B. Cisplatin-Based Chemotherapy of Human Cancers. *J. Cancer Sci. Ther.* **2019**, *11*, 97.
14. Florea, A.M.; Büsselberg, D. Cisplatin as an Anti-Tumor Drug: Cellular Mechanisms of Activity, Drug Resistance and Induced Side Effects. *Cancers* **2011**, *3*, 1351–1371. [CrossRef]
15. Tsvetkova, D.; Ivanova, S. Application of Approved Cisplatin Derivatives in Combination Therapy against Different Cancer Diseases. *Molecules* **2022**, *27*, 2466. [CrossRef]
16. Dasari, S.; Bernard Tchounwou, P. Cisplatin in Cancer Therapy: Molecular Mechanisms of Action. *Eur. J. Pharmacol.* **2014**, *740*, 364–378. [CrossRef]
17. Kopacz-Bednarska, A.; Król, T. Cisplatin—Properties and Clinical Application. *Oncol. Clin. Pract.* **2022**, *18*, 166–176. [CrossRef]
18. Barabas, K.; Milner, R.; Lurie, D.; Adin, C. Cisplatin: A Review of Toxicities and Therapeutic Applications. *Vet. Comp. Oncol.* **2008**, *6*, 1–18. [CrossRef]
19. Amable, L. Cisplatin Resistance and Opportunities for Precision Medicine. *Pharmacol. Res.* **2016**, *106*, 27–36. [CrossRef]
20. Galluzzi, L.; Senovilla, L.; Vitale, I.; Michels, J.; Martins, I.; Kepp, O.; Castedo, M.; Kroemer, G. Molecular Mechanisms of Cisplatin Resistance. *Oncogene* **2012**, *31*, 1869–1883. [CrossRef]
21. Czarnomysy, R.; Radomska, D.; Szewczyk, O.K.; Roszczenko, P.; Bielawski, K. Platinum and Palladium Complexes as Promising Sources for Antitumor Treatments. *Int. J. Mol. Sci.* **2021**, *22*, 8271. [CrossRef]
22. Sava, G. Ruthenium Compounds in Cancer Therapy. In *Metal Compounds in Cancer Therapy*; Fricker, S.P., Ed.; Springer: Dordrecht, The Netherlands, 1994; pp. 65–91. ISBN 978-94-011-1252-9.
23. Sharma, A.; Sudhindra, P.; Roy, N.; Paira, P. Advances in Novel Iridium (III) Based Complexes for Anticancer Applications: A Review. *Inorg. Chim. Acta* **2020**, *513*, 119925. [CrossRef]
24. Borkow, G.; Gabbay, J. Copper, an Ancient Remedy Returning to Fight Microbial, Fungal and Viral Infections. *Curr. Chem. Biol.* **2009**, *3*, 272–278. [CrossRef]
25. Kardos, J.; Héja, L.; Simon, Á.; Jablonkai, I.; Kovács, R.; Jemnitz, K. Copper Signalling: Causes and Consequences. *Cell Commun. Signal.* **2018**, *16*, 71. [CrossRef]
26. Ruiz, L.M.; Libedinsky, A.; Elorza, A.A. Role of Copper on Mitochondrial Function and Metabolism. *Front. Mol. Biosci.* **2021**, *8*, 711227. [CrossRef]

27. Klotz, L.-O.; Kröncke, K.-D.; Buchczyk, D.P.; Sies, H. Role of Copper, Zinc, Selenium and Tellurium in the Cellular Defense against Oxidative and Nitrosative Stress. *J. Nutr.* **2003**, *133*, 1448S–1451S. [CrossRef]
28. Peña, M.M.O.; Lee, J.; Thiele, D.J. Critical Review A Delicate Balance: Homeostatic Control of Copper Uptake and Distribution. *J. Nutr.* **1999**, *129*, 1251–1260. [CrossRef]
29. Tang, X.; Yan, Z.; Miao, Y.; Ha, W.; Li, Z.; Yang, L.; Mi, D. Copper in Cancer: From Limiting Nutrient to Therapeutic Target. *Front. Oncol.* **2023**, *13*, 1209156. [CrossRef]
30. Nasulewicz, A.; Mazur, A.; Opolski, A. Role of Copper in Tumour Angiogenesis—Clinical Implications. *J. Trace Elem. Med. Biol.* **2004**, *18*, 1–8. [CrossRef]
31. Alem, M.B.; Damena, T.; Desalegn, T.; Koobotse, M.; Eswaramoorthy, R.; Ngwira, K.J.; Ombito, J.O.; Zachariah, M.; Demissie, T.B. Cytotoxic Mixed-Ligand Complexes of Cu(II): A Combined Experimental and Computational Study. *Front. Chem.* **2022**, *10*, 1028957. [CrossRef]
32. Ji, P.; Wang, P.; Chen, H.; Xu, Y.; Ge, J.; Tian, Z.; Yan, Z. Potential of Copper and Copper Compounds for Anticancer Applications. *Pharmaceuticals* **2023**, *16*, 234. [CrossRef]
33. Ruiz-Azuara, L. Process to Obtain New Mixed Copper Aminoacidate from Methylate Phenathroline Complexes to Be Used as Anticancerigenic Agents. U.S. Patent 5,576,326, 20 December 1990.
34. Antman, K.H. Introduction: The History of Arsenic Trioxide in Cancer Therapy. *Oncologist* **2001**, *6*, 1–2. [CrossRef] [PubMed]
35. Valent, P.; Groner, B.; Schumacher, U.; Superti-Furga, G.; Busslinger, M.; Kralovics, R.; Zielinski, C.; Penninger, J.M.; Kerjaschki, D.; Stingl, G. Paul Ehrlich (1854–1915) and His Contributions to the Foundation and Birth of Translational Medicine. *J. Innate Immun.* **2016**, *8*, 111–120. [CrossRef] [PubMed]
36. De Vita, V.T., Jr.; Chu, E. A History of Cancer Chemotherapy. *Cancer Res.* **2008**, *68*, 8643–8653. [CrossRef] [PubMed]
37. Chabner, B.A.; Roberts, T.G., Jr. Chemotherapy and the War on Cancer. *Nat. Rev. Cancer* **2005**, *5*, 65–72. [CrossRef]
38. Hurley, L.H.; Boyd, F.L. DNA as a Target for Drug Action. *Trends Pharmacol. Sci.* **1988**, *9*, 402–407. [CrossRef]
39. Forestier, J. Comparative Results of Copper Salts and Gold Salts in Rheumatoid Arthritis. *Ann. Rheum. Dis.* **1949**, *8*, 132. [CrossRef] [PubMed]
40. Medici, S.; Peana, M.; Nurchi, V.M.; Lachowicz, J.I.; Crisponi, G.; Zoroddu, M.A. Noble Metals in Medicine: Latest Advances. *Coord. Chem. Rev.* **2015**, *284*, 329–350. [CrossRef]
41. Romero-Canelon, I.; Sadler, P.J. Next-Generation Metal Anticancer Complexes: Multitargeting via Redox Modulation. *Inorg. Chem.* **2013**, *52*, 12276–12291. [CrossRef]
42. Bravo-Gómez, M.E.; García-Ramos, J.C.; Gracia-Mora, I.; Ruiz-Azuara, L. Antiproliferative Activity and QSAR Study of Copper (II) Mixed Chelate [Cu (N − N)(Acetylacetonato)] NO_3 and [Cu (N − N)(Glycinato)] NO_3 Complexes, (Casiopeinas). *J. Inorg. Biochem.* **2009**, *103*, 299–309. [CrossRef]
43. Ruiz-Azuara, L.; Bravo-Gomez, M.E. Copper Compounds in Cancer Chemotherapy. *Curr. Med. Chem.* **2010**, *17*, 3606–3615. [CrossRef]
44. Rivero-Müller, A.; De Vizcaya-Ruiz, A.; Plant, N.; Ruiz, L.; Dobrota, M. Mixed Chelate Copper Complex, Casiopeina IIgly®, Binds and Degrades Nucleic Acids: A Mechanism of Cytotoxicity. *Chem. Biol. Interact.* **2007**, *165*, 189–199. [CrossRef] [PubMed]
45. Alemón-Medina, R.; Breña-Valle, M.; Muñoz-Sánchez, J.; Gracia-Mora, M.; Ruiz-Azuara, L. Induction of Oxidative Damage by Copper-Based Antineoplastic Drugs (Casiopeínas(R)). *Cancer Chemother. Pharmacol.* **2007**, *60*, 219–228. [CrossRef]
46. Folli, A.; Ritterskamp, N.; Richards, E.; Platts, J.A.; Murphy, D.M. Probing the Structure of Copper(II)-Casiopeina Type Coordination Complexes [Cu(O−O)(N−N)]+ by EPR and ENDOR Spectroscopy. *J. Catal.* **2021**, *394*, 220–227. [CrossRef]
47. Solans, X.; Ruíz-Ramírez, L.; Martínez, A.; Gasque, L.; Moreno-Esparza, R. Mixed Chelate Complexes. II. Structures of L-Alaninato (Aqua)(4, 7-Diphenyl-1, 10-Phenanthroline) Copper (II) Nitrite Monohydrate and Aqua (4, 7-Dimethyl-1, 10-Phenanthroline)(Glycinato)(Nitrato) Copper (II) Monohydrate. *Acta Crystallogr. C* **1993**, *49*, 890–893. [CrossRef]
48. Figueroa-Depaz, Y.; Pérez-Villanueva, J.; Soria-Arteche, O.; Martínez-Otero, D.; Gómez-Vidales, V.; Ortiz-Frade, L.; Ruiz-Azuara, L. Casiopeinas of Third Generations: Synthesis, Characterization, Cytotoxic Activity and Structure–Activity Relationships of Mixed Chelate Compounds with Bioactive Secondary Ligands. *Molecules* **2022**, *27*, 3504. [CrossRef]
49. Novoa-Ramírez, C.S.; Silva-Becerril, A.; González-Ballesteros, M.M.; Gomez-Vidal, V.; Flores-Álamo, M.; Ortiz-Frade, L.; Gracia-Mora, J.; Ruiz-Azuara, L. Biological Activity of Mixed Chelate Copper (II) Complexes, with Substituted Diimine and Tridentate Schiff Bases (NNO) and Their Hydrogenated Derivatives as Secondary Ligands: Casiopeína's Fourth Generation. *J. Inorg. Biochem.* **2023**, *242*, 112097. [CrossRef]
50. Godínez-Loyola, Y.; Gracia-Mora, J.; Rojas-Montoya, I.D.; Hernández-Ayala, L.F.; Reina, M.; Ortiz-Frade, L.A.; Rascón-Valenzuela, L.A.; Robles-Zepeda, R.E.; Gómez-Vidales, V.; Bernad-Bernad, M.J. Casiopeinas® Third Generation, with Indometachin: Synthesis, Characterization, DFT Studies, Antiproliferative Activity, and Nanoencapsulation. *RSC Adv.* **2022**, *12*, 21662–21673. [CrossRef] [PubMed]
51. Resendiz-Acevedo, K.; García-Aguilera, M.E.; Esturau-Escofet, N.; Ruiz-Azuara, L. ^{1}H-NMR Metabolomics Study of the Effect of Cisplatin and Casiopeina IIgly on MDA-MB-231 Breast Tumor Cells. *Front. Mol. Biosci.* **2021**, *8*, 742859. [CrossRef]
52. Rodrigues, J.H.V.; de Carvalho, A.B.; Silva, V.R.; de Santos, L.S.; Soares, M.B.P.; Bezerra, D.P.; Oliveira, K.M.; Corrêa, R.S. Copper(II)/Diiminic Complexes Based on 2-Hydroxybenzophenones: DNA- and BSA-Binding Studies and Antitumor Activity against HCT116 and HepG2 Tumor Cells. *Polyhedron* **2023**, *239*, 116431. [CrossRef]

53. García-Ramos, J.C.; Gutiérrez, A.G.; Vázquez-Aguirre, A.; Toledano-Magaña, Y.; Alonso-Sáenz, A.L.; Gómez-Vidales, V.; Flores-Alamo, M.; Mejía, C.; Ruiz-Azuara, L. The Mitochondrial Apoptotic Pathway Is Induced by Cu(II) Antineoplastic Compounds (Casiopeínas®) in SK-N-SH Neuroblastoma Cells after Short Exposure Times. *BioMetals* **2017**, *30*, 43–58. [CrossRef]
54. De Paz, F.-Y.; Resendiz-Acevedo, K.; Dávila-Manzanilla, S.G.; García-Ramos, J.C.; Ortiz-Frade, L.; Serment-Guerrero, J.; Ruiz-Azuara, L. DNA, a Target of Mixed Chelate Copper(II) Compounds (Casiopeinas®) Studied by Electrophoresis, UV–Vis and Circular Dichroism Techniques. *J. Inorg. Biochem.* **2022**, *231*, 111772. [CrossRef]
55. Eremina, J.A.; Lider, E.V.; Sukhikh, T.S.; Klyushova, L.S.; Perepechaeva, M.L.; Sheven, D.G.; Berezin, A.S.; Grishanova, A.Y.; Potkin, V.I. Water-Soluble Copper (II) Complexes with 4, 5-Dichloro-Isothiazole-3-Carboxylic Acid and Heterocyclic N-Donor Ligands: Synthesis, Crystal Structures, Cytotoxicity, and DNA Binding Study. *Inorg. Chim. Acta* **2020**, *510*, 119778. [CrossRef]
56. Reina, M.; Talavera-Contreras, L.G.; Figueroa-DePaz, Y.; Ruiz-Azuara, L.; Hernández-Ayala, L.F. Casiopeinas® as SARS-CoV-2 Main Protease (M pro) Inhibitors: A Combined DFT, Molecular Docking and ONIOM Approach. *New J. Chem.* **2022**, *46*, 12500–12511. [CrossRef]
57. Huaizhi, Z.; Yuantao, N. China's Ancient Gold Drugs. *Gold Bull.* **2000**, *46 34*, 24–29. [CrossRef]
58. Spear, M. Silver: An Age-Old Treatment Modality in Modern Times. *Plast. Aesthetic Nurs.* **2010**, *30*, 90–93. [CrossRef]
59. Kapp, R.W. Arsenic: Toxicology and Health Effects. In *Encyclopedia of Food and Health*; Caballero, B., Finglas, P.M., Toldrá, F., Eds.; Academic Press: Oxford, UK, 2016; pp. 256–265. ISBN 978-0-12-384953-3.
60. Gibaud, S.; Jaouen, G. Arsenic-Based Drugs: From Fowler's Solution to Modern Anticancer Chemotherapy. In *Medicinal Organometallic Chemistry*; Jaouen, G., Metzler-Nolte, N., Eds.; Springer: Berlin/Heidelberg, Germany, 2010; pp. 1–20. ISBN 978-3-642-13185-1.
61. Holman, J.; Kirkhart, B.; Maxon, M.S.; Oliva, C.; Earhart, R. Safety Experience with Trisenox® (Arsenic Trioxide) Injection. *Blood* **2004**, *104*, 4521. [CrossRef]
62. Hoonjan, M.; Jadhav, V.; Bhatt, P. Arsenic Trioxide: Insights into Its Evolution to an Anticancer Agent. *J. Biol. Inorg. Chem.* **2018**, *23*, 313–329. [CrossRef] [PubMed]
63. Proschak, E.; Stark, H.; Merk, D. Polypharmacology by Design: A Medicinal Chemist's Perspective on Multitargeting Compounds. *J. Med. Chem.* **2019**, *62*, 420–444. [CrossRef]
64. Jørgensen, J.T. The Importance of Predictive Biomarkers in Oncology Drug Development. *Expert. Rev. Mol. Diagn.* **2016**, *16*, 807–809. [CrossRef] [PubMed]
65. Reina, M.; Hernández-Ayala, L.F.; Bravo-Gómez, M.E.; Gómez, V.; Ruiz-Azuara, L. Second Generation of Casiopeinas®: A Joint Experimental and Theoretical Study. *Inorg. Chim. Acta* **2021**, *517*, 120201. [CrossRef]
66. Onawumi, O.O.E.; Odunola, O.A.; Suresh, E.; Paul, P. Synthesis, Structural Characterization and Microbial Activities of Mixed Ligand Copper (II) Complexes of 2,2′-Bipyridine and Acetylacetonate. *Inorg. Chem. Commun.* **2011**, *14*, 1626–1631. [CrossRef]
67. Bravo-Gómez, M.E.; Dávila-Manzanilla, S.; Flood-Garibay, J.; Muciño-Hernández, M.Á.; Mendoza, Á.; García-Ramos, J.C.; Moreno-Esparza, R.; Ruiz-Azuara, L. Secondary Ligand Effects on the Cytotoxicity of Several Casiopeína's Group II Compounds. *J. Mex. Chem. Soc.* **2012**, *56*, 85–92. [CrossRef]
68. Carvallo-Chaigneau, F.; Trejo-Solís, C.; Gómez-Ruiz, C.; Rodríguez-Aguilera, E.; MacÍas-Rosales, L.; Cortés-Barberena, E.; Cedillo-Peláez, C.; Gracia-Mora, I.; Ruiz-Azuara, L.; Madrid-Marina, V.; et al. Casiopeina III-Ia Induces Apoptosis in HCT-15 Cells in Vitro through Caspase-Dependent Mechanisms and Has Antitumor Effect in Vivo. *BioMetals* **2008**, *21*, 17–28. [CrossRef] [PubMed]
69. García-Ramos, J.C.; Vértiz-Serrano, G.; Macías-Rosales, L.; Galindo-Murillo, R.; Toledano-Magaña, Y.; Bernal, J.P.; Cortés-Guzmán, F.; Ruiz-Azuara, L. Isomeric Effect on the Pharmacokinetic Behavior of Anticancer CuII Mixed Chelate Complexes: Experimental and Theoretical Approach. *Eur. J. Inorg. Chem.* **2017**, *2017*, 1728–1736. [CrossRef]
70. Davila-Manzanilla, S.G.; Figueroa-de Paz, Y.; Mejia, C.; Ruiz-Azuara, L. Synergistic Effects between a Copper-Based Metal Casiopeína III-Ia and Cisplatin. *Eur. J. Med. Chem.* **2017**, *129*, 266–274. [CrossRef]
71. Rodríguez-Enríquez, S.; Gallardo-Pérez, J.C.; Hernández-Reséndiz, I.; Marín-Hernández, A.; Pacheco-Velázquez, S.C.; López-Ramírez, S.Y.; Rumjanek, F.D.; Moreno-Sánchez, R. Canonical and New Generation Anticancer Drugs Also Target Energy Metabolism. *Arch. Toxicol.* **2014**, *88*, 1327–1350. [CrossRef] [PubMed]
72. Leal-García, M.; García-Ortuño, L.; Ruiz-Azuara, L.; Gracia-Mora, I.; Luna-Delvillar, J.; Sumano, H. Assessment of Acute Respiratory and Cardiovascular Toxicity of Casiopeinas in Anaesthetized Dogs. *Basic Clin. Pharmacol. Toxicol.* **2007**, *101*, 151–158. [CrossRef] [PubMed]
73. Silva-Platas, C.; Guerrero-Beltrán, C.E.; Carrancá, M.; Castillo, E.C.; Bernal-Ramírez, J.; Oropeza-Almazán, Y.; González, L.N.; Rojo, R.; Martínez, L.E.; Valiente-Banuet, J. Antineoplastic Copper Coordinated Complexes (Casiopeinas) Uncouple Oxidative Phosphorylation and Induce Mitochondrial Permeability Transition in Cardiac Mitochondria and Cardiomyocytes. *J. Bioenerg. Biomembr.* **2016**, *48*, 43–54. [CrossRef]
74. García, N.; Martínez-Abundis, E.; Pavón, N.; Correa, F.; Chávez, E. Copper Induces Permeability Transition through Its Interaction with the Adenine Nucleotide Translocase. *Cell Biol. Int.* **2007**, *31*, 893–899. [CrossRef]
75. Zazueta, C.; Reyes-Vivas, H.; Zafra, G.; Sánchez, C.A.; Vera, G.; Chávez, E. Mitochondrial Permeability Transition as Induced by Cross-Linking of the Adenine Nucleotide Translocase. *Int. J. Biochem. Cell Biol.* **1998**, *30*, 517–527. [CrossRef]
76. De Vizcaya-Ruiz, A.; Rivero-Müller, A.; Ruiz-Ramirez, L.; Howarth, J.A.; Dobrota, M. Hematotoxicity Response in Rats by the Novel Copper-Based Anticancer Agent: Casiopeina II. *Toxicology* **2003**, *194*, 103–113. [CrossRef]

77. Vértiz, G.; García-Ortuño, L.E.; Bernal, J.P.; Bravo-Gómez, M.E.; Lounejeva, E.; Huerta, A.; Ruiz-Azuara, L. Pharmacokinetics and Hematotoxicity of a Novel Copper-Based Anticancer Agent: Casiopeina III-Ea, after a Single Intravenous Dose in Rats. *Fundam. Clin. Pharmacol.* **2014**, *28*, 78–87. [CrossRef] [PubMed]
78. Cañas-Alonso, R.C.; Fuentes-Noriega, I.; Ruiz-Azuara, L. Pharmacokinetics of Casiopeína IIgly in Beagle Dog: A Copper Based Compound with Antineoplastic Activity. *J. Bioanal. Biomed.* **2010**, *2*, 28–34. [CrossRef]
79. Fuentes-Noriega, I.; Ruiz-Ramırez, L.; Tovar, A.T.; Rico-Morales, H.; Gracia-Mora, I. Development and Validation of a Liquid Chromatographic Method for Casiopeina IIIi in Rat Plasma. *J. Chromatogr. B* **2002**, *772*, 115–121. [CrossRef] [PubMed]
80. Correia, I.; Borovic, S.; Cavaco, I.; Matos, C.P.; Roy, S.; Santos, H.M.; Fernandes, L.; Capelo, J.L.; Ruiz-Azuara, L.; Pessoa, J.C. Evaluation of the Binding of Four Anti-Tumor Casiopeínas® to Human Serum Albumin. *J. Inorg. Biochem.* **2017**, *175*, 284–297. [CrossRef]
81. Serment-Guerrero, J.; Cano-Sanchez, P.; Reyes-Perez, E.; Velazquez-Garcia, F.; Bravo-Gomez, M.E.; Ruiz-Azuara, L. Genotoxicity of the Copper Antineoplastic Coordination Complexes Casiopeinas. *Toxicol. Vitr.* **2011**, *25*, 1376–1384. [CrossRef]
82. Valencia-Cruz, A.I.; Uribe-Figueroa, L.I.; Galindo-Murillo, R.; Baca-Lopez, K.; Gutierrez, A.G.; Vazquez-Aguirre, A.; Ruiz-Azuara, L.; Hernandez-Lemus, E.; Mejía, C. Whole Genome Gene Expression Analysis Reveals Casiopeina-Induced Apoptosis Pathways. *PLoS ONE* **2013**, *8*, e54664. [CrossRef]
83. Kachadourian, R.; Brechbuhl, H.M.; Ruiz-Azuara, L.; Gracia-Mora, I.; Day, B.J. Casiopeína IIgly-Induced Oxidative Stress and Mitochondrial Dysfunction in Human Lung Cancer A549 and H157 Cells. *Toxicology* **2010**, *268*, 176–183. [CrossRef]
84. Trejo-Solís, C.; Palencia, G.; Zúniga, S.; Rodríguez-Ropon, A.; Osorio-Rico, L.; Luvia, S.T.; Gracia-Mora, I.; Marquez-Rosado, L.; Sánchez, A.; Moreno-García, M.E. Cas IIgly Induces Apoptosis in Glioma C6 Cells in Vitro and in Vivo through Caspase-Dependent and Caspase-Independent Mechanisms. *Neoplasia* **2005**, *7*, 563–574. [CrossRef]
85. Gutiérrez, A.G.; Vázquez-Aguirre, A.; García-Ramos, J.C.; Flores-Alamo, M.; Hernández-Lemus, E.; Ruiz-Azuara, L.; Mejía, C. Copper(Ii) Mixed Chelate Compounds Induce Apoptosis through Reactive Oxygen Species in Neuroblastoma Cell Line Chp-212. *J. Inorg. Biochem.* **2013**, *126*, 17–25. [CrossRef] [PubMed]
86. Su, Z.; Mao, Y.-P.; OuYang, P.-Y.; Tang, J.; Lan, X.-W.; Xie, F.-Y. Leucopenia and Treatment Efficacy in Advanced Nasopharyngeal Carcinoma. *BMC Cancer* **2015**, *15*, 429. [CrossRef] [PubMed]
87. Pitekova, B.; Uhlikova, E.; Kupcova, V.; Durfinova, M.; Mojto, V.; Turecky, L. Can Alpha-1-Acid Glycoprotein Affect the Outcome of Treatment in a Cancer Patient? *Bratisl. Med. J.* **2019**, *120*, 9–14. [CrossRef]
88. Liberti, M.V.; Locasale, J.W. The Warburg Effect: How Does It Benefit Cancer Cells? *Trends Biochem. Sci.* **2016**, *41*, 211–218. [CrossRef] [PubMed]
89. Roberts, D.J.; Miyamoto, S. Hexokinase II Integrates Energy Metabolism and Cellular Protection: Akting on Mitochondria and TORCing to Autophagy. *Cell Death Differ.* **2015**, *22*, 248–257. [CrossRef]
90. Silva, J.A.; Queirós, O.; Baltazar, F.; Ułaszewski, S.; Ko, Y.H.; Pedersen, P.L.; Preto, A.; Casal, M. The Anticancer Agent 3-Bromopyruvate: A Simple but Powerful Molecule Taken from the Lab to the Bedside. *J. Bioenerg. Biomembr.* **2016**, *48*, 349–362. [CrossRef]
91. Rai, Y.; Yadav, P.; Kumari, N.; Kalra, N.; Bhatt, A.N. Hexokinase II Inhibition by 3-Bromopyruvate Sensitizes Myeloid Leukemic Cells K-562 to Anti-Leukemic Drug, Daunorubicin. *Biosci. Rep.* **2019**, *39*, BSR20190880. [CrossRef]
92. Marín-Hernández, A.; Gallardo-Pérez, J.C.; López-Ramírez, S.Y.; García-García, J.D.; Rodríguez-Zavala, J.S.; Ruiz-Ramírez, L.; Gracia-Mora, I.; Zentella-Dehesa, A.; Sosa-Garrocho, M.; Macías-Silva, M. Casiopeina II-Gly and Bromo-Pyruvate Inhibition of Tumor Hexokinase, Glycolysis, and Oxidative Phosphorylation. *Arch. Toxicol.* **2012**, *86*, 753–766. [CrossRef] [PubMed]
93. Heusch, G.; Boengler, K.; Schulz, R. Inhibition of Mitochondrial Permeability Transition Pore Opening: The Holy Grail of Cardioprotection. *Basic Res. Cardiol.* **2010**, *105*, 151–154. [CrossRef] [PubMed]
94. Arfin, S.; Jha, N.K.; Jha, S.K.; Kesari, K.K.; Ruokolainen, J.; Roychoudhury, S.; Rathi, B.; Kumar, D. Oxidative Stress in Cancer Cell Metabolism. *Antioxidants* **2021**, *10*, 642. [CrossRef]
95. Giles, N.M.; Watts, A.B.; Giles, G.I.; Fry, F.H.; Littlechild, J.A.; Jacob, C. Metal and Redox Modulation of Cysteine Protein Function. *Chem. Biol.* **2003**, *10*, 677–693. [CrossRef]
96. Ramírez-Palma, L.G.; Espinoza-Guillén, A.; Nieto-Camacho, F.; López-Guerra, A.E.; Gómez-Vidales, V.; Cortés-Guzmán, F.; Ruiz-Azuara, L. Intermediate Detection in the Casiopeina–Cysteine Interaction Ending in the Disulfide Bond Formation and Copper Reduction. *Molecules* **2021**, *26*, 5729. [CrossRef] [PubMed]
97. Ong, W.K.; Jana, D.; Zhao, Y. A Glucose-Depleting Silica Nanosystem for Increasing Reactive Oxygen Species and Scavenging Glutathione in Cancer Therapy. *Chem. Commun.* **2019**, *55*, 13374–13377. [CrossRef] [PubMed]
98. Zhang, A.; Sun, H.; Xu, H.; Qiu, S.; Wang, X. Cell Metabolomics. *OMICS* **2013**, *17*, 495–501. [CrossRef] [PubMed]
99. Danzi, F.; Pacchiana, R.; Mafficini, A.; Scupoli, M.T.; Scarpa, A.; Donadelli, M.; Fiore, A. To Metabolomics and beyond: A Technological Portfolio to Investigate Cancer Metabolism. *Signal Transduct. Target. Ther.* **2023**, *8*, 137. [CrossRef] [PubMed]
100. Han, J.; Li, Q.; Chen, Y.; Yang, Y. Recent Metabolomics Analysis in Tumor Metabolism Reprogramming. *Front. Mol. Biosci.* **2021**, *8*, 763902. [CrossRef]
101. Schmidt, D.R.; Patel, R.; Kirsch, D.G.; Lewis, C.A.; Vander Heiden, M.G.; Locasale, J.W. Metabolomics in Cancer Research and Emerging Applications in Clinical Oncology. *CA Cancer J. Clin.* **2021**, *71*, 333–358. [CrossRef]

102. Vermathen, M.; Paul, L.E.H.; Diserens, G.; Vermathen, P.; Furrer, J. ^{1}H HR-MAS NMR Based Metabolic Profiling of Cells in Response to Treatment with a Hexacationic Ruthenium Metallaprism as Potential Anticancer Drug. *PLoS ONE* **2015**, *10*, e0128478. [CrossRef]
103. De Castro, F.; Stefàno, E.; De Luca, E.; Muscella, A.; Marsigliante, S.; Benedetti, M.; Fanizzi, F.P. A NMR-Based Metabolomic Approach to Investigate the Antitumor Effects of the Novel [Pt(H1-C$_2$H$_4$OMe)(DMSO)(Phen)] + (Phen = 1,10-Phenanthroline) Compound on Neuroblastoma Cancer Cells. *Bioinorg. Chem. Appl.* **2022**, *2022*, 8932137. [CrossRef]
104. De Castro, F.; Benedetti, M.; Del Coco, L.; Fanizzi, F.P. NMR-Based Metabolomics in Metal-Based Drug Research. *Molecules* **2019**, *24*, 2240. [CrossRef]
105. Yang, R.; Li, Y.; Wang, H.; Qin, T.; Yin, X.; Ma, X. Therapeutic Progress and Challenges for Triple Negative Breast Cancer: Targeted Therapy and Immunotherapy. *Mol. Biomed.* **2022**, *3*, 8. [CrossRef]
106. Almansour, N.M. Triple-Negative Breast Cancer: A Brief Review about Epidemiology, Risk Factors, Signaling Pathways, Treatment and Role of Artificial Intelligence. *Front. Mol. Biosci.* **2022**, *9*, 836417. [CrossRef]
107. Kumar, P.; Aggarwal, R. An Overview of Triple-Negative Breast Cancer. *Arch. Gynecol. Obstet.* **2016**, *293*, 247–269. [CrossRef]
108. Sun, X.; Wang, M.; Wang, M.; Yu, X.; Guo, J.; Sun, T.; Li, X.; Yao, L.; Dong, H.; Xu, Y. Metabolic Reprogramming in Triple-Negative Breast Cancer. *Front. Oncol.* **2020**, *10*, 428. [CrossRef]
109. Nong, S.; Han, X.; Xiang, Y.; Qian, Y.; Wei, Y.; Zhang, T.; Tian, K.; Shen, K.; Yang, J.; Ma, X. Metabolic Reprogramming in Cancer: Mechanisms and Therapeutics. *MedComm* **2023**, *4*, e218. [CrossRef]
110. Scatena, C.; Naccarato, A.G.; Ozsvari, B.; Scumaci, D. Metabolic Reprogramming in Breast Cancer. *Front. Oncol.* **2022**, *12*, 1081171. [CrossRef]
111. Maria, R.M.; Altei, W.F.; Selistre-de-Araujo, H.S.; Colnago, L.A. Impact of Chemotherapy on Metabolic Reprogramming: Characterization of the Metabolic Profile of Breast Cancer MDA-MB-231 Cells Using ^{1}H HR-MAS NMR Spectroscopy. *J. Pharm. Biomed. Anal.* **2017**, *146*, 324–328. [CrossRef]
112. Gupta, S.; Roy, A.; Dwarakanath, B.S. Metabolic Cooperation and Competition in the Tumor Microenvironment: Implications for Therapy. *Front. Oncol.* **2017**, *7*, 68. [CrossRef]
113. Tan, L.T.-H.; Chan, K.-G.; Pusparajah, P.; Lee, W.-L.; Chuah, L.-H.; Khan, T.M.; Lee, L.-H.; Goh, B.-H. Targeting Membrane Lipid a Potential Cancer Cure? *Front. Pharmacol.* **2017**, *8*, 12. [CrossRef]
114. Hill, D.P.; Harper, A.; Malcolm, J.; McAndrews, M.S.; Mockus, S.M.; Patterson, S.E.; Reynolds, T.; Baker, E.J.; Bult, C.J.; Chesler, E.J. Cisplatin-Resistant Triple-Negative Breast Cancer Subtypes: Multiple Mechanisms of Resistance. *BMC Cancer* **2019**, *19*, 1039. [CrossRef]
115. Sulaiman, A.; McGarry, S.; Chambers, J.; Al-Kadi, E.; Phan, A.; Li, L.; Mediratta, K.; Dimitroulakos, J.; Addison, C.; Li, X.; et al. Targeting Hypoxia Sensitizes TNBC to Cisplatin and Promotes Inhibition of Both Bulk and Cancer Stem Cells. *Int. J. Mol. Sci.* **2020**, *21*, 5788. [CrossRef]
116. Nedeljković, M.; Damjanović, A. Mechanisms of Chemotherapy Resistance in Triple-Negative Breast Cancer—How We Can Rise to the Challenge. *Cells* **2019**, *8*, 957. [CrossRef]
117. Raimondi, V.; Ciccarese, F.; Ciminale, V. Oncogenic Pathways and the Electron Transport Chain: A DangeROS Liaison. *Br. J. Cancer* **2020**, *122*, 168–181. [CrossRef]
118. Phan, L.M.; Yeung, S.-C.J.; Lee, M.-H. Cancer Metabolic Reprogramming: Importance, Main Features, and Potentials for Precise Targeted Anti-Cancer Therapies. *Cancer Biol. Med.* **2014**, *11*, 1.
119. Wang, Z.; Dong, C. Gluconeogenesis in Cancer: Function and Regulation of PEPCK, FBPase, and G6Pase. *Trends Cancer* **2019**, *5*, 30–45. [CrossRef]
120. Ma, Y.; Temkin, S.M.; Hawkridge, A.M.; Guo, C.; Wang, W.; Wang, X.-Y.; Fang, X. Fatty Acid Oxidation: An Emerging Facet of Metabolic Transformation in Cancer. *Cancer Lett.* **2018**, *435*, 92–100. [CrossRef]
121. Ali, I.; Rahis-Uddin, K.; Salim, K.; Rather, M.; Wani, W.; Haque, A. Advances in Nano Drugs for Cancer Chemotherapy. *Curr. Cancer Drug Targets* **2011**, *11*, 135–146. [CrossRef]
122. Blanco, E.; Shen, H.; Ferrari, M. Principles of Nanoparticle Design for Overcoming Biological Barriers to Drug Delivery. *Nat. Biotechnol.* **2015**, *33*, 941–951. [CrossRef]
123. Miranda-Calderón, J.E.; Macías-Rosales, L.; Gracia-Mora, I.; Ruiz-Azuara, L.; Faustino-Vega, A.; Gracia-Mora, J.; Bernad-Bernad, M.J. Effect of Casiopein III-Ia Loaded into Chitosan Nanoparticles on Tumor Growth Inhibition. *J. Drug Deliv. Sci. Technol.* **2018**, *48*, 1–8. [CrossRef]
124. Zagalo, D.M.; Sousa, J.; Simões, S. Quality by Design (QbD) Approach in Marketing Authorization Procedures of Non-Biological Complex Drugs: A Critical Evaluation. *Eur. J. Pharm. Biopharm.* **2022**, *178*, 1–24. [CrossRef]
125. Rawal, M.; Singh, A.; Amiji, M.M. Quality-by-Design Concepts to Improve Nanotechnology-Based Drug Development. *Pharm. Res.* **2019**, *36*, 153. [CrossRef]
126. Aguilar-Jiménez, Z.; González-Ballesteros, M.; Dávila-Manzanilla, S.G.; Espinoza-Guillén, A.; Ruiz-Azuara, L. Development and In Vitro and In Vivo Evaluation of an Antineoplastic Copper (II) Compound (Casiopeina III-Ia) Loaded in Nonionic Vesicles Using Quality by Design. *Int. J. Mol. Sci.* **2022**, *23*, 12756. [CrossRef]
127. Nave, M.; Castro, R.E.; Rodrigues, C.M.P.; Casini, A.; Soveral, G.; Gaspar, M.M. Nanoformulations of a Potent Copper-Based Aquaporin Inhibitor with Cytotoxic Effect against Cancer Cells. *Nanomedicine* **2016**, *11*, 1817–1830. [CrossRef] [PubMed]

128. Malekshah, R.E.; Salehi, M.; Kubicki, M.; Khaleghian, A. Synthesis, Structure, Computational Modeling and Biological Activity of Two New Casiopeínas® Complexes and Their Nanoparticles. *J. Coord. Chem.* **2019**, *72*, 2233–2250. [CrossRef]
129. Chen, M.; Huang, Z.; Xia, M.; Ding, Y.; Shan, T.; Guan, Z.; Dai, X.; Xu, X.; Huang, Y.; Huang, M.; et al. Glutathione-Responsive Copper-Disulfiram Nanoparticles for Enhanced Tumor Chemotherapy. *J. Control. Release* **2022**, *341*, 351–363. [CrossRef]
130. Fan, Y.; Zhang, J.; Shi, M.; Li, D.; Lu, C.; Cao, X.; Peng, C.; Mignani, S.; Majoral, J.-P.; Shi, X. Poly(Amidoamine) Dendrimer-Coordinated Copper(II) Complexes as a Theranostic Nanoplatform for the Radiotherapy-Enhanced Magnetic Resonance Imaging and Chemotherapy of Tumors and Tumor Metastasis. *Nano Lett.* **2019**, *19*, 1216–1226. [CrossRef] [PubMed]
131. Wu, W.; Yu, L.; Jiang, Q.; Huo, M.; Lin, H.; Wang, L.; Chen, Y.; Shi, J. Enhanced Tumor-Specific Disulfiram Chemotherapy by In Situ Cu^{2+} Chelation-Initiated Nontoxicity-to-Toxicity Transition. *J. Am. Chem. Soc.* **2019**, *141*, 11531–11539. [CrossRef]
132. Kheirolomoom, A.; Mahakian, L.M.; Lai, C.-Y.; Lindfors, H.A.; Seo, J.W.; Paoli, E.E.; Watson, K.D.; Haynam, E.M.; Ingham, E.S.; Xing, L.; et al. Copper–Doxorubicin as a Nanoparticle Cargo Retains Efficacy with Minimal Toxicity. *Mol. Pharm.* **2010**, *7*, 1948–1958. [CrossRef]
133. Peña, Q.; Wang, A.; Zaremba, O.; Shi, Y.; Scheeren, H.W.; Metselaar, J.M.; Kiessling, F.; Pallares, R.M.; Wuttke, S.; Lammers, T. Metallodrugs in Cancer Nanomedicine. *Chem. Soc. Rev.* **2022**, *51*, 2544–2582. [CrossRef]
134. Santini, C.; Pellei, M.; Gandin, V.; Porchia, M.; Tisato, F.; Marzano, C. Advances in Copper Complexes as Anticancer Agents. *Chem. Rev.* **2014**, *114*, 815–862. [CrossRef]
135. Gallagher, J.; Chen, C.B.; Pan, C.Q.; Perrin, D.M.; Cho, Y.-M.; Sigman, D.S. Optimizing the Targeted Chemical Nuclease Activity of 1, 10-Phenanthroline–Copper by Ligand Modification. *Bioconjug. Chem.* **1996**, *7*, 413–420. [CrossRef]
136. Hirohama, T.; Kuranuki, Y.; Ebina, E.; Sugizaki, T.; Arii, H.; Chikira, M.; Tamil Selvi, P.; Palaniandavar, M. Copper(II) Complexes of 1,10-Phenanthroline-Derived Ligands: Studies on DNA Binding Properties and Nuclease Activity. *J. Inorg. Biochem.* **2005**, *99*, 1205–1219. [CrossRef] [PubMed]
137. Lozada-García, M.C.; Enríquez, R.G.; Ramírez-Apán, T.O.; Nieto-Camacho, A.; Palacios-Espinosa, J.F.; Custodio-Galván, Z.; Soria-Arteche, O.; Pérez-Villanueva, J. Synthesis of Curcuminoids and Evaluation of Their Cytotoxic and Antioxidant Properties. *Molecules* **2017**, *22*, 633. [CrossRef]
138. Kunwar, A.; Simon, E.; Singh, U.; Chittela, R.K.; Sharma, D.; Sandur, S.K.; Priyadarsini, I.K. Interaction of a Curcumin Analogue Dimethoxycurcumin with DNA. *Chem. Biol. Drug Des.* **2011**, *77*, 281–287. [CrossRef] [PubMed]
139. Schneider, C.; Gordon, O.N.; Edwards, R.L.; Luis, P.B. Degradation of Curcumin: From Mechanism to Biological Implications. *J. Agric. Food Chem.* **2015**, *63*, 7606–7614. [CrossRef]
140. Meza-Morales, W.; Estévez-Carmona, M.M.; Alvarez-Ricardo, Y.; Obregón-Mendoza, M.A.; Cassani, J.; Ramírez-Apan, M.T.; Escobedo-Martínez, C.; Soriano-García, M.; Reynolds, W.F.; Enríquez, R.G. Full Structural Characterization of Homoleptic Complexes of Diacetylcurcumin with Mg, Zn, Cu, and Mn: Cisplatin-Level Cytotoxicity in Vitro with Minimal Acute Toxicity in Vivo. *Molecules* **2019**, *24*, 1598. [CrossRef]
141. Prisecaru, A.; McKee, V.; Howe, O.; Rochford, G.; McCann, M.; Colleran, J.; Pour, M.; Barron, N.; Gathergood, N.; Kellett, A. Regulating Bioactivity of Cu^{2+} Bis-1,10-Phenanthroline Artificial Metallonucleases with Sterically Functionalized Pendant Carboxylates. *J. Med. Chem.* **2013**, *56*, 8599–8615. [CrossRef] [PubMed]
142. Kubešová, K.; Dořičáková, A.; Trávníček, Z.; Dvořák, Z. Mixed-Ligand Copper (II) Complexes Activate Aryl Hydrocarbon Receptor AhR and Induce CYP1A Genes Expression in Human Hepatocytes and Human Cell Lines. *Toxicol. Lett.* **2016**, *255*, 24–35. [CrossRef] [PubMed]
143. Xu, Y.; Zhang, Q.; Lin, F.; Zhu, L.; Huang, F.; Zhao, L.; Ou, R. Casiopeina II-gly Acts on LncRNA MALAT1 by MiR-17-5p to Inhibit FZD2 Expression via the Wnt Signaling Pathway during the Treatment of Cervical Carcinoma. *Oncol. Rep.* **2019**, *42*, 1365–1379. [CrossRef] [PubMed]
144. Rajalakshmi, S.; Kiran, M.S.; Nair, B.U. DNA Condensation by Copper (II) Complexes and Their Anti-Proliferative Effect on Cancerous and Normal Fibroblast Cells. *Eur. J. Med. Chem.* **2014**, *80*, 393–406. [CrossRef]
145. Correia, I.; Roy, S.; Matos, C.P.; Borovic, S.; Butenko, N.; Cavaco, I.; Marques, F.; Lorenzo, J.; Rodríguez, A.; Moreno, V. Vanadium (IV) and Copper (II) Complexes of Salicylaldimines and Aromatic Heterocycles: Cytotoxicity, DNA Binding and DNA Cleavage Properties. *J. Inorg. Biochem.* **2015**, *147*, 134–146. [CrossRef]
146. Kordestani, N.; Rudbari, H.A.; Fernandes, A.R.; Raposo, L.R.; Baptista, P.V.; Ferreira, D.; Bruno, G.; Bella, G.; Scopelliti, R.; Braun, J.D. Antiproliferative Activities of Diimine-Based Mixed Ligand Copper (II) Complexes. *ACS Comb. Sci.* **2020**, *22*, 89–99. [CrossRef]
147. Zoroddu, M.A.; Aaseth, J.; Crisponi, G.; Medici, S.; Peana, M.; Nurchi, V.M. The Essential Metals for Humans: A Brief Overview. *J. Inorg. Biochem.* **2019**, *195*, 120–129. [CrossRef]
148. Komarnicka, U.K.; Kozieł, S.; Zabierowski, P.; Kruszyński, R.; Lesiow, M.K.; Tisato, F.; Porchia, M.; Kyzioł, A. Copper (I) Complexes with Phosphines P (p-OCH_3-Ph) $_2CH_2OH$ and P (p-OCH_3-Ph) $_2CH_2SarGly$. Synthesis, Multimodal DNA Interactions, and Prooxidative and in Vitro Antiproliferative Activity. *J. Inorg. Biochem.* **2020**, *203*, 110926. [CrossRef]
149. Alvarez, N.; Viña, D.; Leite, C.M.; Mendes, L.F.S.; Batista, A.A.; Ellena, J.; Costa-Filho, A.J.; Facchin, G. Synthesis and Structural Characterization of a Series of Ternary Copper (II)-L-Dipeptide-Neocuproine Complexes. Study of Their Cytotoxicity against Cancer Cells Including MDA-MB-231, Triple Negative Breast Cancer Cells. *J. Inorg. Biochem.* **2020**, *203*, 110930. [CrossRef]

150. Martínez-Valencia, B.; Corona-Motolinia, N.D.; Sánchez-Lara, E.; Noriega, L.; Sanchez-Gaytan, B.L.; Castro, M.E.; Melendez-Bustamante, F.; González-Vergara, E. Cyclo-Tetravanadate Bridged Copper Complexes as Potential Double Bullet pro-Metallodrugs for Cancer Treatment. *J. Inorg. Biochem.* **2020**, *208*, 111081. [CrossRef]
151. Chen, Z.-F.; Tan, M.-X.; Liu, L.-M.; Liu, Y.-C.; Wang, H.-S.; Yang, B.; Peng, Y.; Liu, H.-G.; Liang, H.; Orvig, C. Cytotoxicity of the Traditional Chinese Medicine (TCM) Plumbagin in Its Copper Chemistry. *Dalton Trans.* **2009**, *48*, 10824–10833. [CrossRef] [PubMed]
152. O'Connor, M.; Kellett, A.; McCann, M.; Rosair, G.; McNamara, M.; Howe, O.; Creaven, B.S.; McClean, S.; Foltyn-Arfa Kia, A.; O'Shea, D.; et al. Copper(II) Complexes of Salicylic Acid Combining Superoxide Dismutase Mimetic Properties with DNA Binding and Cleaving Capabilities Display Promising Chemotherapeutic Potential with Fast Acting in Vitro Cytotoxicity against Cisplatin Sensitive and Resista. *J. Med. Chem.* **2012**, *55*, 1957–1968. [CrossRef]
153. Zhang, Z.; Bi, C.; Schmitt, S.M.; Fan, Y.; Dong, L.; Zuo, J.; Dou, Q.P. 1,10-Phenanthroline Promotes Copper Complexes into Tumor Cells and Induces Apoptosis by Inhibiting the Proteasome Activity. *JBIC J. Biol. Inorg. Chem.* **2012**, *17*, 1257–1267. [CrossRef] [PubMed]
154. Levín, P.; Ruiz, M.C.; Romo, A.I.B.; Nascimento, O.R.; Di Virgilio, A.L.; Oliver, A.G.; Ayala, A.P.; Diógenes, I.C.N.; León, I.E.; Lemus, L. Water-Mediated Reduction of [Cu(Dmp)$_2$(CH$_3$CN)]$^{2+}$: Implications of the Structure of a Classical Complex on Its Activity as an Anticancer Drug. *Inorg. Chem. Front.* **2021**, *8*, 3238–3252. [CrossRef]
155. Itoh, S. Chemical Reactivity of Copper Active-Oxygen Complexes. In *Copper-Oxygen Chemistry*; John Wiley & Sons: Hoboken, NJ, USA, 2011; pp. 225–282. ISBN 9781118094365.
156. Denoyer, D.; Masaldan, S.; La Fontaine, S.; Cater, M.A. Targeting Copper in Cancer Therapy: "Copper That Cancer". *Metallomics* **2015**, *7*, 1459–1476. [CrossRef]
157. Balsa, L.M.; Baran, E.J.; León, I.E. Copper Complexes as Antitumor Agents: In Vitro and In Vivo Evidence. *Curr. Med. Chem.* **2023**, *30*, 510–557. [CrossRef] [PubMed]
158. Barbosa, A.R.; Caleffi-Ferracioli, K.R.; Leite, C.Q.F.; García-Ramos, J.C.; Toledano-Magaña, Y.; Ruiz-Azuara, L.; Siqueira, V.L.D.; Pavan, F.R.; Cardoso, R.F. Potential of Casiopeínas® Copper Complexes and Antituberculosis Drug Combination against Mycobacterium Tuberculosis. *Chemotherapy* **2016**, *61*, 249–255. [CrossRef] [PubMed]
159. Becco, L.; Rodríguez, A.; Bravo, M.E.; Prieto, M.J.; Ruiz-Azuara, L.; Garat, B.; Moreno, V.; Gambino, D. New Achievements on Biological Aspects of Copper Complexes Casiopeínas®: Interaction with DNA and Proteins and Anti-Trypanosoma Cruzi Activity. *J. Inorg. Biochem.* **2012**, *109*, 49–56. [CrossRef] [PubMed]
160. Masuri, S.; Vaňhara, P.; Cabiddu, M.G.; Moráň, L.; Havel, J.; Cadoni, E.; Pivetta, T. Copper(Ii) Phenanthroline-Based Complexes as Potential Anticancer Drugs: A Walkthrough on the Mechanisms of Action. *Molecules* **2022**, *27*, 49. [CrossRef]

Disclaimer/Publisher's Note: The statements, opinions and data contained in all publications are solely those of the individual author(s) and contributor(s) and not of MDPI and/or the editor(s). MDPI and/or the editor(s) disclaim responsibility for any injury to people or property resulting from any ideas, methods, instructions or products referred to in the content.

Article

Copper(I)/Triphenylphosphine Complexes Containing Naphthoquinone Ligands as Potential Anticancer Agents

Celisnolia M. Leite [1,2,*], João H. Araujo-Neto [3], Adriana P. M. Guedes [1], Analu R. Costa [2], Felipe C. Demidoff [4], Chaquip D. Netto [4], Eduardo E. Castellano [2], Otaciro R. Nascimento [2] and Alzir A. Batista [1,*]

1. Department of Chemistry, Federal University of São Carlos, São Carlos 13565-905, SP, Brazil; adriana_quimica@hotmail.com
2. São Carlos Institute of Physics, University of São Paulo, São Carlos 13560-970, SP, Brazil; anallucosta@gmail.com (A.R.C.); pino@ifsc.usp.br (E.E.C.); otaciro7f@gmail.com (O.R.N.)
3. Institute of Chemistry, University of São Paulo, São Paulo 27930-560, SP, Brazil; honoratoneto10@gmail.com
4. Multidisciplinary Institute of Chemistry, Federal University of Rio de Janeiro, Macaé 35400-000, RJ, Brazil; felipedemidoff@hotmail.com (F.C.D.); chaquip@gmail.com (C.D.N.)
* Correspondence: celisnolia@hotmail.com (C.M.L.); daab@ufscar.br (A.A.B.); Tel.: +55-16-3351-8285 (C.M.L. & A.A.B.)

Abstract: Four new Cu/PPh$_3$/naphtoquinone complexes were synthesized, characterized (IR, UV/visible, 1D/2D NMR, mass spectrometry, elemental analysis, and X-ray diffraction), and evaluated as anticancer agents. We also investigated the reactive oxygen species (ROS) generation capacity of complex **4**, considering the well-established photochemical property of naphtoquinones. Therefore, employing the electron paramagnetic resonance (EPR) "spin trap", 5,5-dimethyl-1-pyrroline N-oxide (DMPO) technique, we identified the formation of the characteristic •OOH species (hydroperoxyl radical) adduct even before irradiating the solution containing complex **4**. As the irradiation progressed, this radical species gradually diminished, primarily giving rise to a novel species known as •DMPO-OH (DMPO + •OH radical). These findings strongly suggest that Cu(I)/PPh$_3$/naphtoquinone complexes can generate ROS, even in the absence of irradiation, potentially intensifying their cytotoxic effect on tumor cells. Interpretation of the in vitro cytotoxicity data of the Cu(I) complexes considered their stability in cell culture medium. All of the complexes were cytotoxic to the lung (A549) and breast tumor cell lines (MDA-MB-231 and MCF-7). However, the higher toxicity for the lung (MRC5) and breast (MCF-10A) non-tumoral cells resulted in a low selectivity index. The morphological analysis of MDA-MB-231 cells treated with the complexes showed that they could cause decreased cell density, loss of cell morphology, and loss of cell adhesion, mainly with concentrations higher than the inhibitory concentration of 50% of cell viability (IC$_{50}$) values. Similarly, the clonogenic survivance of these cells was affected only with concentrations higher than the IC$_{50}$ values. An antimigratory effect was observed for complexes **1** and **4**, showing around 20–40% of inhibition of wound closure in the wound healing experiments.

Keywords: copper; triphenylphosphine; naphtoquinones; ROS; breast cancer cells

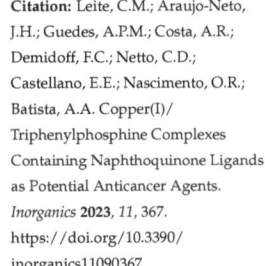

Citation: Leite, C.M.; Araujo-Neto, J.H.; Guedes, A.P.M.; Costa, A.R.; Demidoff, F.C.; Netto, C.D.; Castellano, E.E.; Nascimento, O.R.; Batista, A.A. Copper(I)/Triphenylphosphine Complexes Containing Naphthoquinone Ligands as Potential Anticancer Agents. *Inorganics* **2023**, *11*, 367. https://doi.org/10.3390/inorganics11090367

Academic Editors: Christelle Hureau and Eduardo Sola

Received: 25 July 2023
Revised: 1 September 2023
Accepted: 5 September 2023
Published: 9 September 2023

Copyright: © 2023 by the authors. Licensee MDPI, Basel, Switzerland. This article is an open access article distributed under the terms and conditions of the Creative Commons Attribution (CC BY) license (https://creativecommons.org/licenses/by/4.0/).

1. Introduction

Cancer is one of the leading causes of death, including premature death, in most countries [1]. The early detection and treatment of cancer at its advanced stages continues to pose immense challenges for the scientific community [2,3]. Among the diverse array of cancer types, breast cancer predominantly affects women [4,5]. These types of cancers exhibit substantial biological and clinical heterogeneity, resulting in varied responses to therapeutic agents and distinct prognoses [6,7]. While most breast tumors respond to hormone therapy, triple-negative (TN) breast cancer is an aggressive subtype without a targeted therapy, often requiring personalized treatment [8,9]. Similarly, lung cancer, characterized by its high aggressiveness, metastasis, and heterogeneity, can originate in various locations within the bronchial tree, leading to variable symptoms and signs depending on its anatomical site. Lung cancers are frequently diagnosed in the metastatic stage, limiting treatment options to palliative care [10–13].

The advancement of more efficient chemotherapeutic agents for treating these cancers aims to overcome limitations such as side effects and acquired resistance. Due to the success of cisplatin in treating various cancer types, researchers have extensively explored metal-based compounds in numerous studies as potential candidates for metallopharmaceuticals [14–16]. These compounds have shown promising antitumor activity through various mechanisms, including engaging in biomolecular interactions and generating reactive oxygen species [17,18]. Hence, different metals and ligands have been employed by researchers in the development of new compounds that can be used for the treatment of this disease. Copper, for example, an endogenous metal that plays a significant role in some cancer-related processes, has been widely employed to synthesize coordination compounds that target cancer cells [19–21]. Some of these copper compounds can efficiently generate reactive oxygen species (ROS) that can subsequently attack essential molecules such as proteins, lipids, and DNA, being a critical stimulus for apoptosis [22,23]. On the other hand, other complexes interfere with the cell cycle, potentially inducing regulated cell death [24,25]. Furthermore, some copper complexes possess outstanding photophysical characteristics, making them ideal candidates for photodynamic therapy (PDT) and photothermal therapy (PTT) [26,27].

Researchers commonly employ naphthoquinones as ligands in synthesizing new metal complexes. These ligands belong to the quinone group, organic molecules with diverse and important biological properties. In clinical medicine, several quinones, including daunorubicin, doxorubicin, idarubicin, mitomycin-C, and others, are utilized in cancer chemotherapy. One of the factors pointed out for the antitumor property of these molecules is due to the capacity that these molecules possess to induce oxidative stress by the intracellular generation of ROS, leading the cells to death by apoptosis [28]. Therefore, this study aims to discover new compounds as potential drug candidates through the synthesis and characterization of Cu–PPh$_3$ complexes with naphthoquinone ligands, as well as initial in vitro studies to ascertain the effects of the compounds on breast and lung cancer cells.

2. Results and Discussion

2.1. Syntheses of the Compounds

Complexes (**1–4**) were synthesized with good yields by reacting the precursor [Cu(NO$_3$)(PPh$_3$)$_2$] with the respective naphthoquinone (NQ) ligand in a 1:2 ratio in methanol/triethylamine (Et$_3$N) (Scheme 1). The use of a weak base was necessary to deprotonate the naphthoquinone for enhancing the exchange of the chloride atoms by the bidentate O-O ligand.

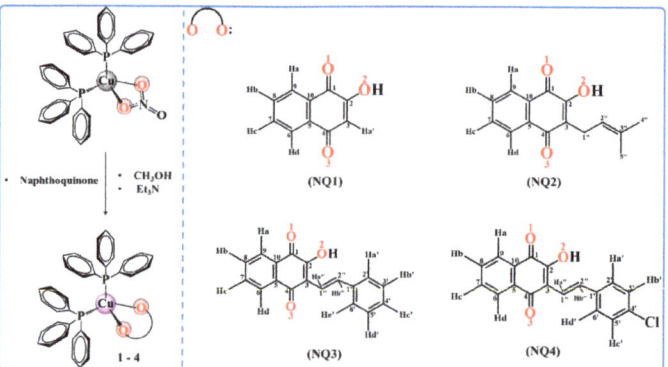

Scheme 1. Route of synthesis of complexes **1–4**.

2.2. Structural Studies

The crystal structures of complexes **1** and **2** showed the coordination of naphthoquinones to the metal center in a bidentate manner via phenolic and carbonyl oxygen atoms (Figure 1). The complexes exhibited a distorted tetrahedral geometry (Table 1). The bond angles of naphthoquinone to copper, O1-Cu-O2 (76.81(6) and 77.14(4)°), are similar to those found for the α-ketocarboxylate ligand in a Cu(I) complex, where the O-Cu-O angle is 79.18° [29]. The P1-Cu-P2 bond angles of 127.67(2)° (**1**) and 118.84(2)° (**2**) are more open. One factor that explains this deviation from the regular tetrahedron is the steric effects imposed by the two triphenylphosphine ligands, which tend to move apart due to their bulk. The bond distances concerning the Cu-O and Cu-P bonds are similar to values already reported in the literature for other copper complexes, including oxidation states I and II [30–35]. The selected bond distances and angles are summarized in Table 1.

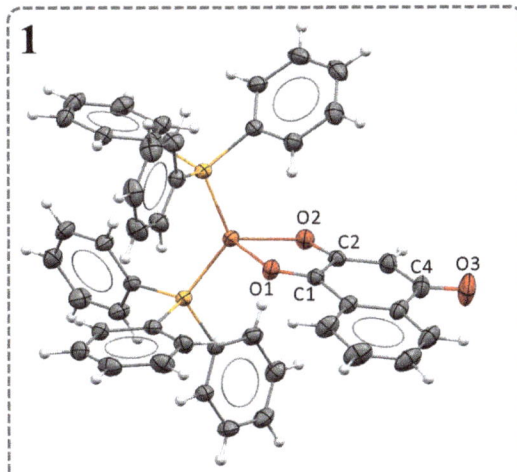
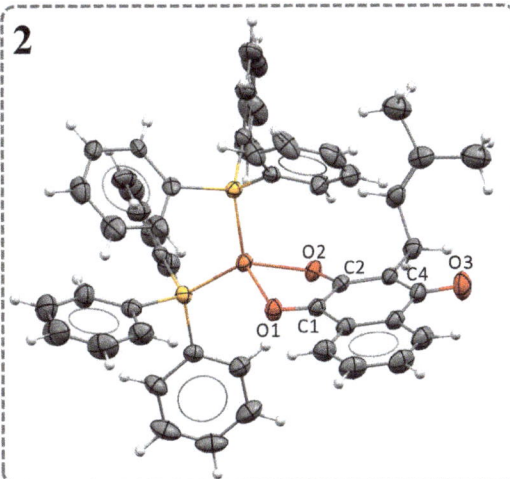

Figure 1. Crystal structures of complexes **1** and **2** (CCDC codes 2279524 and 2279525, respectively, thermal ellipsoids at 30%).

Table 1. Selected interatomic distances and bond angles for complexes **1** and **2**.

Bond Lengths (Å)	Ligands		Complexes		Bond Angles (°)	Complexes	
	NQ1 *	NQ2 **	1	2		1	2
Cu1–P1	-	-	2.243(4)	2.208(5)	P1–Cu1–P2	127.67(2)	118.84(2)
Cu1–P2	-	-	2.244(4)	2.259(5)	P1–Cu1–O1	102.92(4)	114.28(4)
Cu1–O1	-	-	2.227(1)	2.228(2)	O2–Cu1–P2	116.03(4)	109.92(5)
Cu1–O2	-	-	2.034(1)	2.015(1)	O2–Cu1–P1	114.65(4)	123.69(5)
C1–O1	1.217	1.226	1.224(2)	1.227(2)	O2–Cu1–O1	77.14(4)	76.81(6)
C2–O2	1.335	1.346	1.273(2)	1.278(2)	O1–Cu1–P2	100.04(4)	103.99(5)
C4–O3	1.226	1.225	1.230(2)	1.237(3)	-	-	-

* CCDC 1268837, ** CDCC 1189905.

Complexes **1** and **2** exhibited the same trends observed for the bond lengths of bidentate-coordinated naphthoquinone ligands in numerous metal complexes, whose most pronounced change is in the bond length corresponding to the distance of the C2-O2 bond compared to the free ligands (Table 1) [36–38]. Literature suggests that the decrease in the CO bond occurs due to the sharing of electrons in the σ bond between the metal and the ligand, making it stronger and shorter [39]. The crystal data collection and structural refinement parameters for the complexes are summarized in Table S1.

In the FTIR spectra of the free ligands, the ν(O-H) stretch is observed at around 3200 cm^{-1}. In the complexes, these stretching vibrations are absent, as illustrated in Figure 2a for complex **1**, indicating the anionic nature of the naphthoquinone ligands during coordination with the metal center [40]. The bidentate (O, O) and anionic coordination modes have limited literature for Cu(I) complexes [29], making this research an important report for this type of coordination to the metal center. However, for other metal centers, including Cu(II) complexes, the bidentate (O, O) coordination is commonly described in the literature [36–38,41–43]. The region between 1640 and 1680 cm^{-1} displays intense vibrational modes for naphthoquinone ligands, corresponding to the ν(C1=O1) and ν(C4=O3) stretching. Upon the complexation of the ligand to the metal, an increase in the electron density in the antiligand orbital of the carbonyl group leads to a shift of these bands towards lower frequencies in the IR spectra. These findings are consistent with the X-ray diffraction results, which reveal a slight elongation of these bonds after the coordination of the ligand to the metal. The vibrational mode of the C2–O2 bond, after the complexation of the ligand to the metal, shifts to the higher frequency region due to the electron sharing, between the Cu(I) and oxygen atom, which makes the C2–O2 bond stronger, corroborating with what was observed in the X-ray diffraction data. The assignment of the vibrational modes referring to the C1=O1, C2–O2, and C4=O3 bonds before and after coordination of the naphthoquinone ligand to Cu(I) are presented in Table S2.

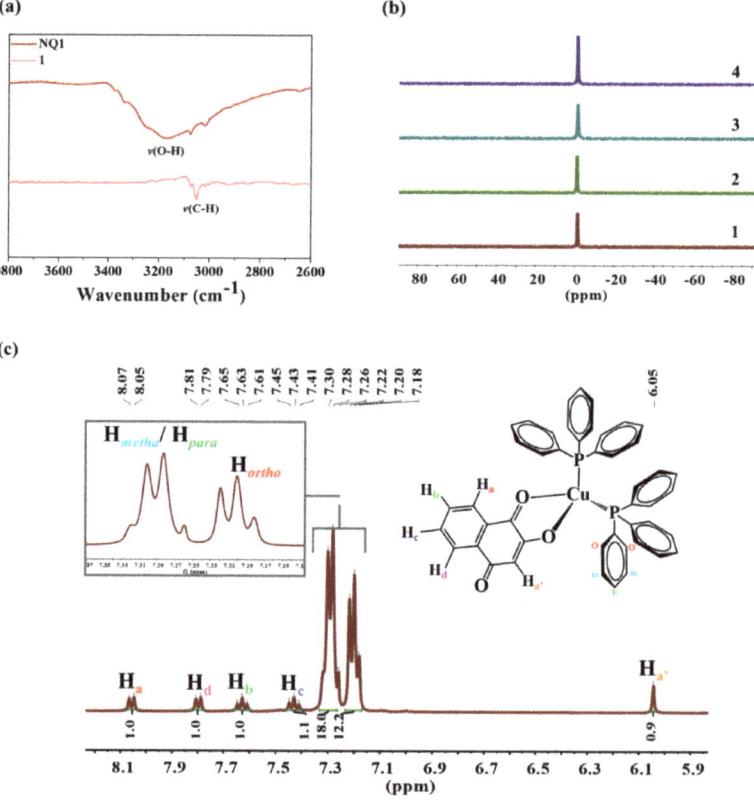

Figure 2. Characterization of the complexes. (**a**) FTIR spectra of complex **1** and its respective naphthoquinone ligand in the region of 3800–2600 cm^{-1}, (**b**) ^{31}P(^{1}H) NMR spectra for complexes **1** and **2** in CDCl$_3$ and **3** and **4** in (CD$_3$)$_2$OD, and (**c**) ^{1}H NMR spectrum of complex **1** in CDCl$_3$, and assignment of signals referring to the complex.

The ligands and complexes exhibit ν(C–H) bands at around 2950 cm^{-1}. The stretching of C=C bonds occurs in the range of 1600–1450 cm^{-1}, while the out-of-plane angular deformation of the =C–H bond can be observed between 900 and 690 cm^{-1}. Complexes **1–4** exhibit vibrational modes between 540 and 430 cm^{-1}, corresponding to the typical P–C$_{Aromatic}$ bonds of the triphenylphosphine ligand and Cu–P bonds [44–46]. Figures S1–S4 show the spectra of all complexes showing these vibrational modes.

The complexes were also analyzed via 1D and 2D NMR spectroscopy at different nuclei, such as phosphorus(^{31}P), carbon (^{13}C), and hydrogen (^{1}H). The ^{31}P(^{1}H) NMR spectra of the complexes exhibited only one signal, around −1.0 ppm (Figure 2b). This single signal indicates the equivalence of the phosphorus atoms of the triphenylphosphine ligands, as they are bound to the metal center in a tetrahedral arrangement confirmed by X-ray diffraction [30,47]. The proposed structures for the complexes are consistent with their ^{1}H NMR spectra. The analysis of the spectra confirms that the triphenylphosphine ligands are present in a 2:1 ratio concerning the naphthoquinone ligand, as illustrated in Figure 2c for complex **1**. The absence of the signal of the OH group from the naphthoquinones, observed for all four compounds, evidences the anionic nature of the ligands when coordinated with the copper, and corroborates with the results observed in the FTIR spectra of the complexes. The ^{13}C(^{1}H) NMR spectra of the complexes show the most deshielded signal, corresponding to the carbons of the C1=O1, C2–O2, and C4=O3 groups from the naphthoquinone ligand. Figures S5–S23 present the NMR spectra and correlation maps for the compounds.

Complexes **1** and **2** exhibit intense bands in their electronic spectra, with λ_{max} values of approximately 260 nm. These bands correspond to the transitions of IL: the π→π* nature of the ligands naphthoquinones (benzenoid and quinonoid systems) and PPh$_3$ and MLCT (Cu→π*naphthoquinone) [48,49]. Theoretical studies suggest that in Cu(I)–PPh$_3$ complexes, the triphenylphosphine ligand plays a crucial role in the stabilization of the Cu(I) center [50]. The spectra of complexes **1** and **2** also exhibit extended characteristic bands in the region between 300 and 550 nm, which can be attributed to the overlapping n→π*-type and intramolecular charge transfer (ICT) transitions from the substituent to the quinone ring, which is an electron acceptor [51].

The electronic spectra of complexes **3** and **4** display intense bands with an absorption maximum of approximately 260 nm, which are from the combined transitions of an IL nature, namely π→π* of the naphthoquinone and PPh$_3$ ligands, as well as MLCT transitions (Cu→π$_{naphthoquinone}$). The compounds also display a band with an absorption maximum of approximately 330 nm, referring to the π→π transitions of the styryl group of the naphthoquinones [52]. The band, which appears at approximately 370 nm, with strong molar absorptivity, can be attributed to the intramolecular charge transfer, from the substituent to the quinone ring, which acts as an electron acceptor. Finally, the low energy broad bands are observed in the visible regions, assignable to n→π* transitions of the carbonyl group of the quinone [53–55]. The electronic spectra for all complexes and the assignments of their transition bands are presented in Figures S24 and S25, and in Table S3.

We also examined the well-known photochemical properties of the naphthoquinone ligands, which generate reactive oxygen species that can potentially be involved in cell death mechanisms [56,57]. Representatively, we focused our analysis on complex **4**. In this case, in order to examine its photochemical behavior, we prepared solutions of complex **4** in dimethyl sulfoxide solvent (DMSO) and subjected them to irradiation with LED (375 nm). Subsequently, we analyzed the samples using the electron paramagnetic resonance (EPR) technique. Additionally, we assessed the photochemical stability of complex **4** using UV–Vis spectroscopy. Figure 3a shows the absorption spectrum of this complex after irradiation, revealing a decrease in absorption in all of its bands. Exhaustive photolysis of the complex led to an intense reduction of the absorption bands and the appearance of an isosbestic point, indicating the formation of more than one species (Figure 3b). Upon cessation of light exposure, the system did not return to its original state (Figure 3b). Furthermore, the photolysis also resulted in the loss of coloration of the solution (Figure 3c). These results indicate the photodegradation of complex **4**. Additionally, we noted that this process is oxygen-dependent, as no changes were observed in the spectrum of complex **4** in an argon atmosphere (Figure 3d).

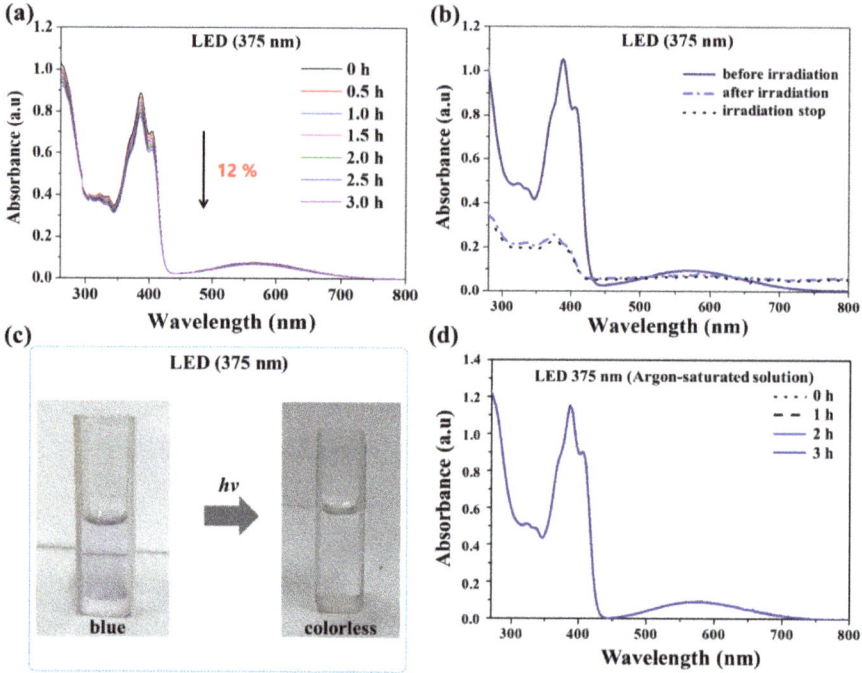

Figure 3. Photochemical stability of complex **4**. (**a**) UV–Vis spectra of a solution of complex **4** in DMSO in the presence of O_2 (open system) after irradiation with LED light (375 nm) at different times; (**b**) UV–Vis spectra of complex **4** in DMSO monitored before and after 19 h of irradiation with LED light (375 nm) and after irradiation ceased; (**c**) photographs of the cuvettes at the beginning and after 19 h of irradiation with LED light (375 nm); and (**d**) UV–Vis spectra of a solution of complex **4b** in DMSO irradiated at different times in the absence of O_2 (solution saturated with argon).

We used a "spin trap" 5,5-dimethyl-1-pyrroline N-oxide (DMPO) experiment to detect ROS via the EPR technique. Reactive oxygen species cannot be directly detected via EPR due to their brief lifetimes. Nevertheless, they rapidly react with spin trappers, such as DMPO, giving rise to stable spin adducts. These adducts can then be characterized through the EPR spectrum. We monitored a solution of complex **4** via EPR spectra in the presence of oxygen and LED (375 nm) irradiation at different time intervals, and the results are presented in Figure 4. It is interesting to note that the process of ROS formation was initiated even before irradiation (Figure 4a). We identified the characteristic adduct of the ·OOH species (hydroperoxyl radical) in DMSO. We identified this adduct by simulating the spectrum obtained experimentally (Figure 4b) and determining its hyperfine coupling constants as A_N = 1.297 mT (36.16 MHz), A_{H1} = 1.053 mT (29.37 MHz), and A_{H2} = 0.133 mT (3.71 MHz). These constants correspond to those reported in the literature for the DMPO–OOH adduct [58,59]. The ROS formation process was enhanced as the exposure time of the solution of complex **4** under irradiation increased, as observed in Figure 4c. We also observed a significant reduction in the DMPO–OOH adduct, while a predominantly new species attributed to the DMPO–OH adduct was formed. The EPR parameters obtained for the radical adducts are shown in Table S4.

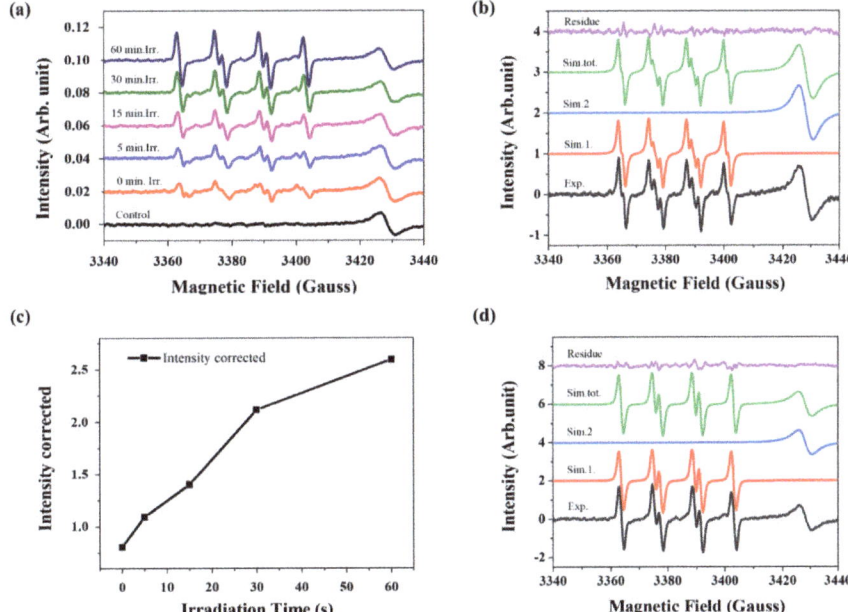

Figure 4. EPR analysis. (**a**) EPR spectra of a solution of complex **4** (10 µM) and DMPO (40 mM) in DMSO, obtained over 60 min of irradiation showing the formation of DMPO–OOH and DMPO-OH adducts; (**b**) experimental and simulated EPR spectra of a solution of complex **4** (10 µM) and DMPO (40 mM) in DMSO before irradiation, showing the formation of DMPO–OOH adduct; (**c**) graphic representation of the increase in the EPR signal intensity as a function of irradiation exposure time of the solution of complex **4**; and (**d**) experimental and simulated EPR spectra of a solution of complex **4** (10 µM) and DMPO (40 mM) in DMSO before irradiation, showing the formation of DMPO–OH adduct.

The DMPO–OH adduct is well-known for producing a distinctive EPR signal characterized by 1:2:2:1 line intensities that arise from nearly equal hyperfine coupling values of $A_N \approx A_H \approx 1.49$ mT in water-based solutions [60]. However, the spectral characteristics of the ·DMPO–OH adduct in DMSO showed a different EPR pattern, as depicted in Figure 4d, with hyperfine coupling constants $A_N = 1.395$ mT (38.89 MHz) and $A_H = 1.181$ mT (32.93 MHz). In a previous study, Zalibera et al. investigated the thermal generation of stable adducts with superhyperfine structures [61]. These researchers showed that a sample containing the DMPO–OH adduct in an aqueous solution displays a typical EPR pattern with constants $A_N = 1.493$ mT and $A_H = 1.474$ mT.

However, when the sample is diluted with DMSO in a 1:1 (v/v) ratio, the EPR pattern exhibits six lines resembling the radical adduct observed in the sample containing complex **4** and the DMPO. The authors attributed these differences to the lower dielectric permittivity of the DMSO/water mixture compared to pure water, which influences the hyperfine coupling constants. As a result, the observed signals deviate from the classical 1:2:2:1 line intensity pattern.

Although further studies are necessary to comprehend the mechanism of ROS formation in this system, we identified four spectral components in the EPR spectra measured at 77K (liquid N_2) and room temperature (Figures S26–S28 and Table S5). These findings indicate the formation of Cu (II) species. These preliminary results show that Cu(I)–PPh$_3$–naphthoquinone complexes have the potential to generate reactive oxygen species even before irradiation, and may have their cytotoxic action on tumor cells enhanced by the formation of these species. Additionally, studies to improve the stability of the complexes under biological conditions, such as changing the PPh$_3$ by bidentate phosphines or encapsulation of the complexes, may enable the use of these compounds as photosensitizers for photodynamic therapy targeting skin cancer.

3. Biological Studies

The in vitro cytotoxicity of the complexes against tumor cell lines (MCF-7, MDA-MB-231 and A549) and non-tumor cell lines (MCF-10A and MRC-5) was evaluated using the MTT method. The results were expressed as IC$_{50}$ values (inhibitory concentration of 50% of cell viability), and are described in Table 2. The concentration–response curves are shown in Figure S29. The cytotoxic effects presented by the complexes were similar in all cell lines. These results may be related to the instability of these compounds in the culture medium, as shown by the results of ^{31}P{^{1}H} NMR (Figure S30) and UV–Vis spectrophotometry (Figure S31), which may result in the same active species. All of the complexes were more toxic than cisplatin in the tested cells, although the high toxicity to the non-tumoral cells resulted in low values of the selectivity index. As the complexes displayed similar cytotoxicity in all of the cell lines tested, we chose only complexes **1** and **4** to undergo further biological investigations in the MDA-MB-231 cell line. Therefore, we evaluated the ability of complexes **1** and **4** to alter cell morphology, impede colony formation, and inhibit cell migration using the wound healing assay.

Table 2. In vitro cytotoxicity of complexes **1**–**4** against the MDA-MB-231, MCF7, and A549 tumor cell lines, as well as the MCF-10A and MRC-5 non-tumor cell lines, after 48 h of incubation.

	Inhibitory Concentration of 50% of Cell Viability, IC$_{50}$ (µM)							
	MCF7	MDA-MB-231	MCF-10A	A549	MRC-5	SI$_1$ *	SI$_2$ *	SI$_3$ *
1	15 ± 3	5.5 ± 0.3	5.9 ± 0.2	3.6 ± 0.3	2.7 ± 0.2	0.4	1.1	0.8
2	9 ± 1	6.5 ± 0.3	4.4 ± 0.8	6.1 ± 0.4	3.3 ± 0.2	0.5	0.7	0.5
3	11 ± 3	7.2 ± 0.3	5.4 ± 0.3	4.4 ± 0.1	2.8 ± 0.1	0.5	0.8	0.6
4	7 ± 2	7.9 ± 0.1	4.0 ± 0.2	4.5 ± 0.1	3.4 ± 0.2	0.6	0.5	0.8
Cu(NO$_3$)$_2$·3H$_2$O	>25	>25	>25	>25	>25	-	-	-
PPh$_3$	>25	>25	>25	>25	>25	-	-	-
NQ	>25	>25	>25	>25	>25	-	-	-
Cisplatin	13.9 ± 2.0	10.2 ± 0.2	23.9 ± 0.7	14.4 ± 1.4	29.9 ± 0.8	1.7	2.3	2.1

* Selectivity index for breast (SI$_1$ = IC$_{50}$ MCF-10A/MCF7 and SI$_2$ = IC$_{50}$ MCF-10A/MDA-MB-231) and lung (SI$_3$ = IC$_{50}$ MRC-5/ IC$_{50}$ A549) cell lines. PPh$_3$: triphenylphosphine; NQ: naphthoquinone.

The cytotoxicities of complexes **1** and **4** at the 24 h time point on the MDA-MB-231 cell line [7 ± 1 µM (**1**) and 6.4 ± 0.2 (**4**), respectively] were also evaluated. The IC$_{50}$ values of the complexes at this time were similar to those obtained at 48 h, showing that their cytotoxic effect is not time-dependent.

The morphological analysis of MDA-MB-231 cells treated with complexes **1** and **4** at different concentrations showed that the complexes could cause decreased cell density, loss of cell morphology (appearance of rounded cells), and loss of cell adhesion, especially at concentrations higher than the IC$_{50}$ (Figure 5). The exposure time of the cells to the complexes did not intensify these effects, which supports the observations made in the cell viability assay. Therefore, these changes indicate that complexes **1** and **4** induce cell death in MDA-MB-231 cells, and this effect is not dependent on time. The similarity of the results displayed by complexes **1** and **4** shows that the variation in the substituents of the naphthoquinone ligand does not influence the toxicity against analyzed cells.

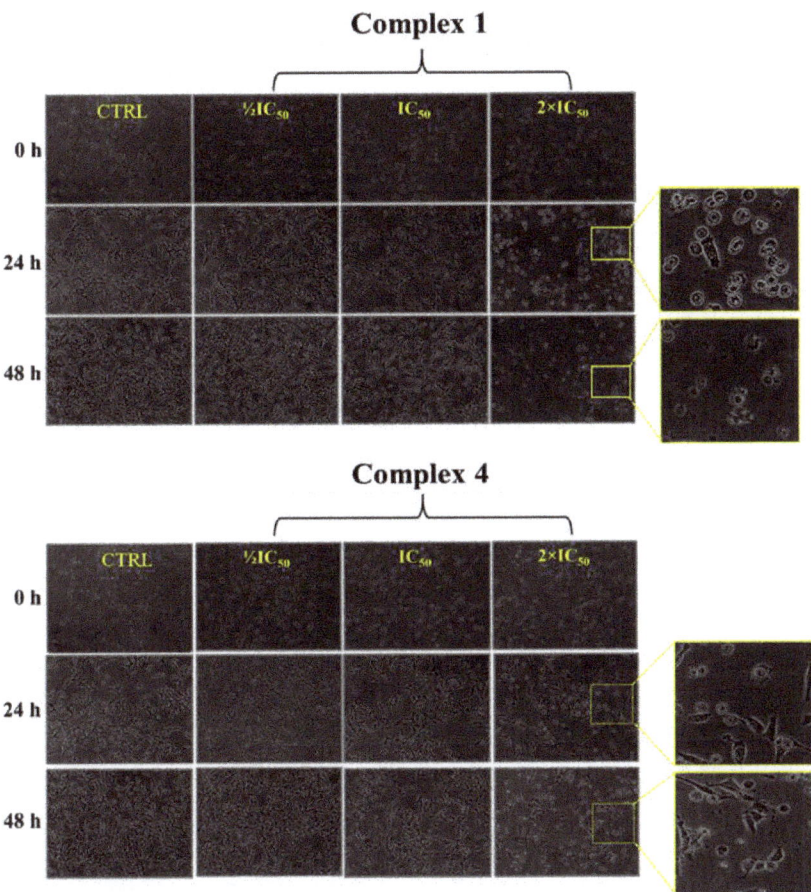

Figure 5. MDA-MB-231 cell morphology 48 h after treatment at $1/2IC_{50}$, IC_{50}, and $2 \times IC_{50}$ concentrations of complexes **1** and **4**.

The colony formation assay results demonstrated that complexes 1 and 4 inhibited the area and intensity of colonies at the $2 \times IC_{50}$ concentration (Figures 6 and S32). These results suggest that complexes **1** and **4** can interfere with the growth, development, and proliferation of MDA-MB-231 cells and exhibit cytotoxicity only at concentrations above the IC_{50}.

Cell migration is a process related to metastasis, which is one of the main causes of death in cancer [62,63]. Thus, the development of drugs that are capable of inhibiting cell migration is an important strategy for cancer therapy. For this reason, we investigated the ability of complexes **1** and **4** to inhibit cell migration using the wound healing assay [64,65]. This assay involves scratching a cell monolayer with approximately 90% of confluence, followed by treatment with solutions of the complexes. In order to observe the effect of inhibition of cell migration, the concentrations of the complexes used were below their IC_{50} concentrations. Mitomycin C, an antiproliferative agent, was used to prevent cell proliferation during the evaluation of wound closure [66]. The results of these experiments showed that complex **1** inhibited cell migration after 48 h of treatment.

The presence of the complex prevented complete closure of the scratch wound, resulting in an inhibition of around 20% wound closure compared to the negative control

(Figure 7). On the other hand, complex **4** was more effective, with an inhibition of around 40% after 24 h of treatment at a concentration of $1/2IC_{50}$, and this inhibition persisted after 48 h (Figure 7).

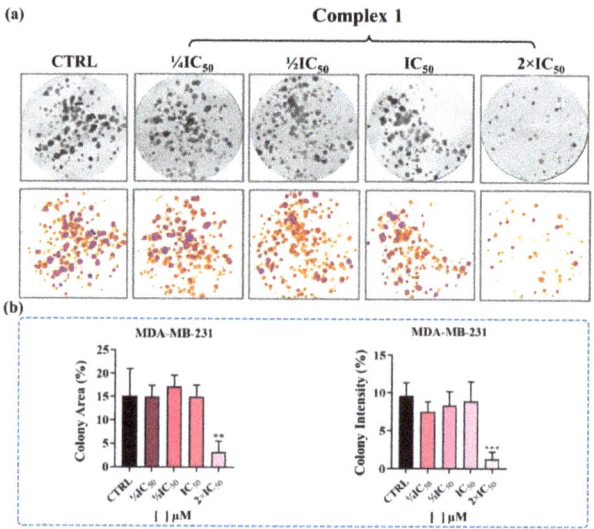

Figure 6. Clonogenic survival of MDA-MB-231 cells treated with different concentrations of complex **1** for 48 h. (**a**) Representation of wells and thresholds for the experiment, and (**b**) graphical quantifications of colony areas and intensities. Data represent the mean ± SD of assays in triplicate. Significance at ** $p < 0.01$, and *** $p < 0.001$ levels using ANOVA and Dunnet's test.

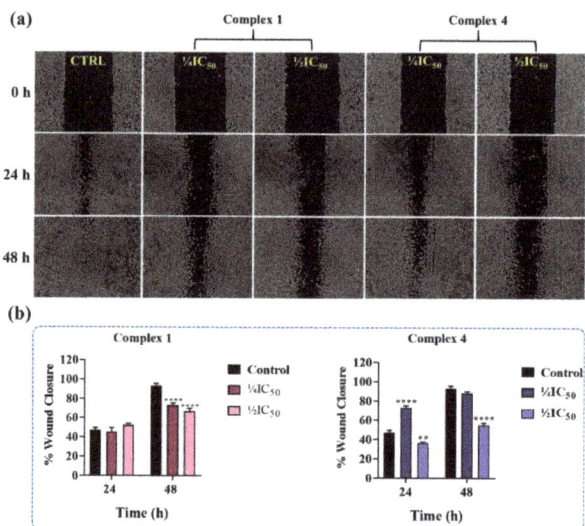

Figure 7. Complex **4** effects on MDA-MB-231 cell migration. (**a**) Representative images obtained in the wound healing assay, in which the effects of complexes **1** and **4** on MDA-MB-231 cell migration were evaluated. The objective used in the experiment had a magnification of 4×. (**b**) Graphical representation of the percentages of wound closure after 24 and 48 h of incubation with the complexes. Data represent the mean ± SD of triplicate assays. Significance at ** $p < 0.01$, and **** $p < 0.0001$ levels using ANOVA and Dunnet's test.

4. Experimental Section

4.1. Materials

The precursor [Cu(NO$_3$)(PPh$_3$)$_2$] was synthesized according to previous related studies in the literature [67]. The CuNO$_3 \cdot$3H$_2$O, triphenylphosphine (PPh$_3$), triethylamine (Et$_3$N) and lawsone (NQ1) were used as received from Sigma-Aldrich. The salts used for buffer preparation and MTT (3-(4,5-dimethylthiazol-2-yl)-2,5-diphenyltetrazolium bromide) were purchased from Sigma-Aldrich. All the solvents used in this study were purified using standard methods. The lapachol (NQ2) was kindly provided by Dr. Diogo Moreira from the Gonzalo Muniz Institute (IGM-FIOCRUZ—Salvador, Brazil) and the ligands NQ3 (3-styryl-lausone) and NQ4 (4-chloro-3-styryl-lausone) were synthesized by the group of Prof. Dr. Chaquip Daher Netto (UFRJ—Macaé, Brazil).

4.2. Physical Measurements

1D and 2D nuclear magnetic resonance (NMR) experiments were recorded on a Bruker DRX-400 spectrometer (9.4 T). 1H and 13C(1H) chemical shifts in chloroform (CD$_3$Cl) or acetone were referenced to the peak of residual nondeuterated solvent [(^{1}H) δ7.26, (^{13}C(^{1}H) δ 77.16 for CD$_3$Cl, and (^{1}H) δ2.09, and (^{13}C(^{1}H) δ 205.87 for (CD$_3$)$_2$CO]. The ^{31}P(^{1}H) NMR spectrometry was carried out in chloroform or acetone, and the chemical shifts were referenced to an external 85% H$_3$PO$_4$ standard at 0.00 ppm. Elemental analyses were performed on a FISIONS Instrument EA 1108 CHNS (Thermo Scientific, Waltham, MA, USA) elemental analyzer at the Analytical Laboratory at the Federal University of São Carlos, São Carlos (SP). Conductivity measurements in acetonitrile solutions (1.0 mM) of the complexes were carried out on a Meter Lab CDM2300 conductivity meter using a cell of constant 0.089 cm^{-1}. The UV–visible (UV–Vis) spectra were recorded on a Hewlett-Packard diode array −8452 A spectrophotometer in acetonitrile solutions (UV Cutoff of 190 nm) with a 1.0 cm quartz cell in the range of 200–800 nm. FTIR spectra in the range of 4000 and 200 cm^{-1} were recorded using as KBr pellets on a Bomem–Michelson FT-MB-102 instrument.

4.3. Syntheses of the Copper Complexes

In a two-mouth flask containing 5 mL of methanol previously deaerated for 1 h, the naphthoquinone ligand (0.30 mmol) and 50 µL of triethylamine were added. After 10 min of stirring, 0.10 g (0.15 mmol) of [Cu(NO$_3$)(PPh$_3$)$_2$] was added to the flask. The mixture was stirred at room temperature for 1 h, and a solid formed with shades ranging from purple (**1** and **2**) to dark blue (**3** and **4**). Then, the solid was filtered, washed with methanol, and dried under vacuum.

[Cu(NQ1)PPh$_3$)$_2$] (**1**). Purple solid. Yield (78%). Elemental analysis (%) calc. for: exp. (calc.) C, 72.36 (72.58); H, 4.94 (4.63). Molar conductance (dichloromethane): 0.20 S cm^2 mol^{-1}. IR (KBr, cm^{-1}): (υ(C-H)) 3053, (υ(C4=O3)) 1641, (υ(C1=O1)) 1587, (υ(C2-O2)) 1094, (υ(C-P)/(Cu-P)) 519, 507, 494. 31P{1H} NMR (162 MHz, CDCl$_3$, 298 K) [ppm, (multiplicity)]: −1.0 (s). ^{1}H NMR (400 MHz, CDCl$_3$, 298 K) [ppm, (multiplicity, integral, J (Hz), assignation)]: 6.05 (s, 1H, Ha' da NQ1); 7.16–7.24 (m, 12H, H$_{ortho}$ of PPh$_3$); 7.25–7.34 (m, overlapped signals: 12H, H$_{metha}$ of PPh$_3$ and 6H, H$_{para}$ of PPh$_3$); 7.43 (t, J = 7.5 Hz, 1H, Hc of NQ1); 7.63 (t, J = 7.5 Hz, 1H, Hb of NQ1); 7.80 (d, J = 7.6 Hz, 1H, Hd of NQ1); 8.06 (d, J = 7.6 Hz, 1H, Ha of NQ1). ^{13}C{^{1}H} NMR (100 MHz, CDCl$_3$, 298 K) [ppm, (multiplicity, J (Hz), assignation)]: 108.7 (C3 of NQ1); 125.8 (C6 of NQ1); 125.9 (C9 of NQ1); 128.8 (C$_{ortho}$ of PPh$_3$); 130.0 (CH of PPh$_3$); 130.4 (C7 of NQ1); 130.6 (C10 of NQ1); 132.5 (C$_{quaternary}$ of PPh$_3$); 133.8 (CH of PPh$_3$); 134.4 (C8 of NQ1); 135.4 (C5 of NQ1); 170.2 (C2 of NQ1); 184.5 (C1 of NQ1); 188.8 (C4 of NQ1).

[Cu(NQ2)PPh$_3$)$_2$] (**2**): Purple solid. Yield (80%). Elemental analysis (%) calc. for: exp. (calc.): C, 73.72 (73.86); H, 5.51 (5.23). Molar conductance (dichloromethane): 0.07 S cm^2 mol^{-1}. IR (KBr, cm^{-1}): (υ(C-H)) 3049, 2907; (υ(C$_4$=O$_3$)) 1628; (υ(C$_1$=O$_1$)) 1583; (υ(C$_2$-O$_2$)) 1095; (υ(C-P)/(Cu-P)) 514, 505, 490. ^{31}P{^{1}H} NMR (162 MHz, CDCl$_3$, 298 K) [ppm, (multiplicity)]: −1.4 (s). ^{1}H NMR (400 MHz, CDCl$_3$, 298 K) [ppm, (multiplicity, integral, J (Hz), assignation)]: 1.56 (s, 3H, CH$_3$ of NQ2); 1.74 (s, 3H, CH$_3$ of NQ2); 3.35 (d, J = 7.0 Hz, 2H, CH$_2$ of NQ2); 5.25–5.33 (m, 1H, CH of NQ2); 7.16–7.24 (m, 12H, H$_{ortho}$ of PPh$_3$); 7.27–7.39 (m, overlapped signals: 12H, H$_{metha}$ of PPh$_3$, 6H, H$_{para}$ of PPh$_3$ e 1H, Hb of NQ2); 7.57 (t, J = 7.5 Hz, 1H, Hc of NQ2); 7.72 (t, J = 7.6 Hz, 1H, Ha of NQ2); 8.04 (d, J = 7.6 Hz, 1H, Hd of NQ2). ^{13}C{^{1}H} NMR (100 MHz, CDCl$_3$, 298 K) [ppm, (multiplicity, J (Hz), assignation)]: 18.1 (CH$_3$ of NQ2); 23.1 (CH$_2$ da NQ2); 26.0 (CH$_3$ of NQ2); 120.9 (C$_3$ of NQ2); 123.9 (CH of NQ2); 125.3 (C9 of NQ2); 125.9 (C6 of NQ2); 128.7 (C$_{ortho}$ of PPh$_3$); 129.9 (CH of PPh$_3$); 130.3 (C$_{quaternary}$ of NQ2); 130.6 (C5 of NQ2); 132.7 (C8 of NQ2); 132.9 (C$_{quaternary}$ of NQ2); 133.9 (CH of PPh$_3$); 134.0 (C7 of NQ2); 135.5 (C10 of NQ2); 167.7 (C2 of NQ2); 182.5 (C4 of NQ2); 188.3 (C1 of NQ2).

[Cu(NQ3)PPh$_3$)$_2$] (3): Blue solid. Yield (79%). Elemental analysis (%) calc. for: exp. (calc.): C, 75.04 (75.12); H, 5.12 (4.79). Molar conductance (dichloromethane): 0.12 S cm^2 mol^{-1}. IR (KBr, cm^{-1}): (υ(C-H)) 3051, 3017; (υ(C$_4$=O$_3$)) 1628; (υ(C$_1$=O$_1$)) 1584; (υ(C$_2$-O$_2$)) 1094; (υ(C-P)/(Cu-P)) 513, 490. ^{31}P{^{1}H} NMR (162 MHz, (CD$_3$)$_2$CO, 298 K) [ppm, (multiplicity)]: −1.4 (s). ^{1}H NMR (400 MHz, (CD$_3$)$_2$CO, 298 K) [ppm, (multiplicity, integral, J (Hz), assignation)]: 7.15 (t, 1H, Hc' of NQ3); 7.27–7.35 (m, overlapped signals: 2H, Hb'/d' of NQ3 and 12H, H$_{ortho}$ of PPh$_3$); 7.37–7.45 (m, overlapped signals: 18H, H$_{metha/para}$ of PPh$_3$); 7.49 (t, J = 7.7 Hz, 2H, Ha'/e' of NQ3); 7.56 (t, J = 7.5 Hz, 1H, Hb of NQ3); 7.68–7.75 (m, overlapped signals: 2H, Ha'' and Hc da NQ3); 7.88 (d, J = 7.6 Hz, 1H, Ha of NQ3); 8.03 (d, J = 7.8 Hz, 1H, Hd of NQ3); 8.38 (d, J = 16.2 Hz, 1H, Hb'' of NQ3). ^{13}C{^{1}H} NMR (100 MHz, (CD$_3$)$_2$CO, 298 K) [ppm, (multiplicity, J (Hz), assignation)]: 117.8 (C1' of NQ3); 123.7 (CH of NQ3); 126.2 (C9 of NQ3); 126.8 (C2'/C6' of NQ3); 126.9 (C6 of NQ3); 127.0 (C4' of NQ3); 129.5 (C3'/C5' of NQ3); 129.8 (overlapped signals: CH of NQ3 e C$_{ortho}$ of PPh$_3$); 131.1 (CH of PPh$_3$); 131.8 (C5 of NQ3); 132.0 (C8 of NQ3); 133.5 (C$_{quaternary}$ of PPh$_3$); 134.6 (CH of PPh$_3$); 135.4 (C7 of NQ3); 135.7 (C10 of NQ3); 141.7 (C3 of NQ3); 168.3 (C2 of NQ3); 182.8 (C4 of NQ3); 188.5 (C1 of NQ3).

[Cu(NQ4)PPh$_3$)$_2$] (4): Blue solid. Yield (80%). Elemental analysis (%) calc. for: exp. (calc.): C, 72.12 (72.24); H, 4.74 (4.49). Molar conductance (dichloromethane): 0.13 S cm^2 mol^{-1}. IR (KBr, cm^{-1}): (υ(C-H)) 3049, 3013; (υ(C$_4$=O$_3$)) 1626; (υ(C$_1$=O$_1$)) 1582; (υ(C$_2$-O$_2$)) 1093; (υ(C-P)/(Cu-P)) 515, 507. ^{31}P{^{1}H} NMR (162 MHz, (CD$_3$)$_2$CO, 298 K) [ppm, (multiplicity)]: −1.2 (s). ^{1}H NMR (400 MHz, (CD$_3$)$_2$CO, 298 K) [ppm, (multiplicity, integral, J (Hz), assignation)]: 7.27–7.36 (m, overlapped signals: 2H, Hb'/c' of NQ4 e 12H, H$_{ortho}$ of PPh$_3$); 7.36–7.44 (m, overlapped signals: 18H, H$_{metha/para}$ of PPh$_3$); 7.47 (d, J = 8.2 Hz, 2H, Ha'/d' of NQ4); 7.56 (d, 21H, Hb of NQ4); 7.67–7.76 (m, 2H, Ha''/Hc of NQ4); 7.88 (d, J = 7.7 Hz, 1H, Ha of NQ4); 8.03 (d, J = 7.5 Hz, 1H, Hd of NQ4); 8.33 (d, J = 16.3 Hz, 1H, Hb'' of NQ4). ^{13}C{^{1}H} NMR (100 MHz, (CD$_3$)$_2$CO, 298 K) [ppm, (multiplicity, J (Hz), assignation)]: 117.5 (C1' of NQ4); 124.6 (CH of NQ4); 126.3 (C9 of NQ4); 126.9 (C6 of NQ4); 128.0 (CH of NQ4); 128.2 (C2'/C6' of NQ4); 129.5 (C3'/C5' of NQ4); 129.8 (C$_{ortho}$ of PPh$_3$); 131.2 (CH of PPh$_3$); 131.8 (C5 of NQ4); 131.9 (C4' of NQ4); 132.1 (C8 of NQ4); 133.5 (C$_{quaternary}$ of PPh$_3$); 134.6 (CH of PPh$_3$); 135.6 (C7 of NQ4); 135.7 (C10 of NQ4); 140.6 (C3 of NQ4); 168.6 (C2 of NQ4); 182.7 (C4 of NQ4); 188.5 (C1 of NQ4).

4.4. Stability of Complexes in Culture Medium

The stability of complexes 1–4 in Dulbecco's Modified Eagle's Medium (DMEM) without phenol red, in the presence of 10% fetal bovine serum (FBS), was assessed using UV–Vis spectroscopy and ^{31}P(^{1}H) NMR. Stock solutions of the complexes (0.5 mM) were prepared in DMSO, and then diluted with culture medium to obtain final solutions of 10 µM of the complexes at 0.5% DMSO (v/v) for UV–Vis analyses. For the ^{31}P(^{1}H) NMR spectroscopic assays, saturated solutions of the complexes were prepared with 90% DMSO and 10% culture medium. The samples were analyzed with the UV–Vis technique immediately after preparation of the solutions, and after 2, 4, 24, 48, and 72 h. For the ^{31}P(^{1}H) NMR experiments, the measurements were carried out immediately after preparation of the solutions, and after 24 and 48 h.

4.5. X-ray Crystallography

The complexes (**1**, CCDC code 2279524 and **2**, CCDC code 2279525) were crystallized in methanolic/dichloromethane solutions via slow evaporation of the solvents. The measurements of single crystals with X-ray diffraction were performed on a Rigaku XtaLAB mini II diffractometer (Rigaku Oxford Diffraction, distributed in Warriewood, Australia) with graphite monochromated Mo Kα radiation (λ = 0.71073 Å). Cell refinements were carried out using CrysAlisPro v.42 software, and the structures were obtained with the intrinsic phasing method using the SHELXT program. The Gaussian method was used for the absorption corrections. The tabular and structural representations were generated by OLEX2 and MERCURY, respectively.

4.6. EPR Measurements

The EPR spectroscopy experiments were performed at room temperature using a Varian E109 spectrometer operating in X-band (9.5 GHz). The compound, 5,5 dimethyl 1-pyrroline N-oxide (DMPO, 40 mmol L^{-1}), was used as a spin trap to verify the formation of ROS by complex **4** (10 μM solution) in DMSO. The solution of the complex was irradiated in a quartz cuvette with LED (375 nm) at certain time intervals (0, 5, 15, 30, and 60 min). Then, aliquots were collected and transferred to a quartz cuvette that contained the Cr (III) standard, which remained fixed on the outside of the cuvette throughout all of the measurements. A solution of DMPO (40 mM) in DMSO was measured as a control. The experimentally obtained spectra were simulated to obtain the EPR parameters such as the line width, absorption signal area, value of the hyperfine constants, and the g value, using the EasySpin program in the Matlab 7.5 (R2007b) environment. The experimental conditions were 0.25 G modulation, 20 mW power, time constant 0.128 s, and 10 scans.

4.7. Cell Culture

The MDA-MB-231 cell lines (triple-negative human breast tumor) were cultured in DMEM supplemented with 10% FBS. The MCF-7 cell line (human breast tumor cells) was cultured in Roswell Park Memorial Institute (RPMI) 1640 medium supplemented with 10% FBS. The MCF-10A cell line (non-tumor epithelial human breast cells) was cultured in Dulbecco's modified Eagle medium nutrient mixture F-12 (DMEM F-12) supplemented with 5% horse serum, EGF (20 ng mL^{-1}), hydrocortisone (0.5 μg mL^{-1}), insulin (0.01 mg mL^{-1}), and 1% penicillin/streptomycin. All of the cell lines were maintained in a humidified incubator at 37 °C and 5% CO_2.

4.8. Cell Viability Assay

For this experiment, 1.5×10^4 cells/well were seeded into 96-well plates and incubated at 37 °C in 5% CO_2 overnight to allow for cell adhesion. Copper complexes were dissolved in DMSO, and 0.75 μL was added to each well to achieve a final concentration of 0.5% DMSO/well. The concentration range for the copper complexes was 25–0.3 μM. The cells that were treated with 0.5% DMSO served as the negative control. The treated cells were then incubated for 48 h. After the treatment, MTT (50 μL, 1 mg/mL in culture medium) was added to each well, and the plate was incubated for an additional 3 h. Cell viability was detected by the reduction of MTT to purple formazan in live cells. The formazan crystals were solubilized in isopropanol (150 μL/well), and the optical density of each well was measured using a multi-scanner automatic reader at a wavelength of 540 nm. The IC_{50} value was obtained from the analysis of the absorbance data from three independent experiments conducted in triplicate.

4.9. Cell Morphology

MDA-MB-231 cells were seeded (0.5×10^5 cells/well) in a 12-well plate and incubated at 37 °C in 5% CO_2 overnight. The cell morphology was examined at 0, 24, and 48 h after treatment of the cells with complexes **1** and **4** at concentrations of $\frac{1}{2} \times IC_{50}$, IC_{50}, and $2 \times IC_{50}$ on an inverted microscope (Nikon, T5100) with a 10× objective.

4.10. Colony Formation

A density of 300 cells/well of the MDA-MB-231 cell line was seeded into a 6-well plate and incubated at 37 °C in 5% CO_2 for 24 h to cell adhesion. Then, the cells were treated with complexes **1** and **4** at concentrations of $\frac{1}{4} \times IC_{50}$, $\frac{1}{2} \times IC_{50}$, IC_{50}, and $2 \times IC_{50}$ for 48 h. The medium was replaced with fresh medium without any complex, and the cells were incubated for 10 days. Then, the cells were washed with PBS, fixed with methanol and acid acetic (3:1) for 5 min, and stained with a kit for fast differential staining in hematology Instant Prov (Newprov Products for Laboratóry Ltda, Pinhais, PR, Brazil). The relative survival was calculated using ImageJ software version 1.53t, using the plugin "ColonyArea" that measures the area and intensity of each colony in the selected image.

4.11. Scratch Assay (Wound Healing)

MDA-MB-231 cells were seeded at a density of 2×10^5 cells/well in 12-well plates and incubated in a humidified oven at 37 °C with 5% CO_2 until the culture reached about 90% confluence. Then, a stripe was made in the center of each well using a tip with a maximum volume of 200 µL and a ruler. Carefully, the wells were washed with PBS to remove cell fragments and detached cells from the scratched area. The cells were pre-treated with Mitomine C (10 µg/mL) for 2 h. When the Mitomycin C was removed, the cells were treated with complexes **1** and **4** at concentrations of $\frac{1}{4} \times IC_{50}$ and $\frac{1}{2} \times IC_{50}$. Images of the stripe from each well were captured at four different fields at 0 h, 24 h, and 48 h after treatment, using an inverted microscope (Nikon, T5100) coupled with a camera (Moticam 1000-1.3 Megapixels Live Resolution). The area of the stripe closure by cell migration was measured using ImageJ software, and the percentage of stripe closure was calculated using the following equation:

$$\%Wound\ Closure = [(A_{t=0h} - A_{t=\Delta h})] \times 100 \qquad (1)$$

where $A_{t=0h}$ is the measure of the scratched area at time 0 h, and $A_{t=\Delta h}$ is the measure of the scratched area at 24 or 48 h. The experiment was performed in triplicate. The cells treated with 0.5% DMSO were the negative control.

5. Conclusions

This study presented the synthesis, characterization, and biological evaluation of four Cu(I)/PPh_3/naphthoquinone complexes. The crystal structures of the complexes revealed the bidentate coordination of naphthoquinones to the metal center via phenolic and carbonyl oxygen atoms, resulting in distorted tetrahedral geometries. The FTIR and NMR spectroscopic analyses of the complexes confirmed the coordination modes of the ligands to the metal. The electronic spectra of the complexes displayed characteristic bands corresponding to ligand-to-metal charge transfer transitions and intramolecular charge transfer transitions. Furthermore, the photochemical properties of complex **4** were investigated, demonstrating the oxygen-dependent photodegradation and generation of reactive oxygen species (ROS) upon irradiation. The EPR analysis confirmed the formation of ·OOH and ·OH adducts, indicating the potential of these complexes to generate ROS and enhanced cytotoxicity. In terms of biological studies, the complexes exhibited cytotoxic effects against tumor cell lines (MCF-7, MDA-MB-231, and A549) and non-tumor cell lines (MCF-10A and MRC-5), with higher toxicity compared to cisplatin. However, the complexes also showed significant toxicity to non-tumoral cells, resulting in low selectivity indices. Complexes **1** and **4** were further investigated using the MDA-MB-231 cell line, showing that these complexes induced cell death, inhibited colony formation, and demonstrated their ability to inhibit cell migration. Overall, this study highlights the potential of Cu(I)–PPh_3–naphthoquinone complexes as cytotoxic agents with the ability to induce cell death and inhibit cell migration. Further studies are warranted to enhance the stability of these complexes and explore their potential as photosensitizers for photodynamic therapy.

Supplementary Materials: The following supporting information can be downloaded at: https://www.mdpi.com/article/10.3390/inorganics11090367/s1. Figure S1. FTIR spectra for the naphthoquinone ligand (NQ4) and complex **1**. Figure S2. FTIR spectra for the naphthoquinone ligand (NQ2) and complex **2**. Figure S3. FTIR spectra for the naphthoquinone ligand (NQ3) and complex **3**. Figure S4. FTIR spectra for the naphthoquinone ligand (NQ4) and complex **4**. Figure S5. 1H-1H COSY NMR of the aromatic region of complex **1**, in CDCl3. Figure S6. 13C(1H) NMR spectrum of complex **1**, in CDCl3. Figure S7. 1H–13C HMBC NMR of complex **1**, in CDCl3. Figure S8. 1H–13C HSQC NMR of complex **1**, in CDCl3. Figure S9. 1H NMR spectrum of complex **2**, in CDCl3. Figure S10. 1H–1H COSY NMR of complex **2**, in CDCl3. Figure S11. 13C(1H) NMR spectrum of complex **2**, in CDCl3. Figure S12. 1H–13C HMBC NMR of complex **2**, in CDCl3. Figure S13. 1H–13C HSQC NMR of complex **2**, in CDCl3. Figure S14. 1H NMR spectrum of complex **3**, in (CD3)2CO. Figure S15. 1H–1H COSY NMR of complex **3**, in (CD3)2CO. Figure S16. 13C(1H) NMR spectrum of complex **3**, in (CD3)2CO. Figure S17. 1H–13C HMBC NMR of complex **3**, in (CD3)2CO. Figure S18. 1H–13C HSQC NMR of complex **3**, in (CD3)2CO. Figure S19. 1H NMR spectrum of complex **4**, in (CD3)2CO. Figure S20. 1H–1H COSY NMR of complex **4**, in (CD3)2CO. Figure S21. 13C(1H) NMR spectrum of complex **4**, in (CD3)2CO. Figure S22. 1H–13C HMBC NMR of complex **4**, in (CD3)2CO. Figure S23. 1H–13C HSQC NMR of complex **4**, in (CD3)2CO. Figure S24. UV-vis absorbance for samples of the precursor complex, complexes **1** and **2**, and their respective naphthoquinone ligands, in acetonitrile. Figure S25. UV-vis absorbance for samples of the precursor complex, complexes **3** and **4**, and their respective naphthoquinone ligands, in acetonitrile. Figure S26. EPR spectra of compound **4** reacted with DMPO in DMSO solvent at 77K. Figure S27. EPR spectra of compound **4b** reacted with DMPO in DMSO solvent at room temperature. Figure S28. EPR spectra of Complex **4** reacted with DMPO in DMSO solvent at room temperature. Figure S29. Cytotoxicity of complexes. Figure S30. 31P(1H) NMR spectra in DMSO/Culture medium 9:1 of the complexes **1**, **2**, **3**, and **4** at different times. Figure S31. UV-vis spectra in DMSO/Culture medium 1:199 of the complexes **1**, **2**, **3**, and **4** at different times. Figure S32. Clonogenic survival of MDA-MB-231 cells treated with different concentrations of complex **4** for 48 h. Table S1. Crystal data and structure refinement parameters obtained for the complexes **1** and **2**. Table S2. Tentative assignment of the vibrational frequencies (cm^{-1}) of the ν(C1=O1 ν(C2-O2) and ν(C4=O3) stretches for the free and after coordinated naphthoquinone ligands, and the respective shifts (Δ) after coordination. Table S3. Maximum absorption wavelength (λ, nm), molar absorptivity (ε, mol^{-1}L cm^{-1}), and tentative assignment of the bands of the ligands NQ1 NQ4 and their respective complexes **1–4** in acetonitrile solution. Table S4. EPR parameters obtained from simulation of the experimental spectra of the DMPO-•OOH and DMPO-•OH adducts. Table S5. EPR parameters obtained from simulation of the spectrum measured at 77K temperature (liquid N2), with four spectral components.

Author Contributions: Conceptualization, A.A.B. and C.M.L.; methodology, C.M.L.; validation, O.R.N. and E.E.C.; investigation, A.P.M.G., F.C.D., C.M.L., J.H.A.-N., A.R.C., O.R.N., C.D.N., E.E.C., and A.A.B.; writing—original draft preparation, C.M.L., J.H.A.-N. and A.A.B.; supervision, A.A.B. All authors have read and agreed to the published version of the manuscript.

Funding: This research was funded by FAPESP (grants 2022/11924-8, 2020/14561-8 and 2017/15850-0), CNPq, and CAPES.

Data Availability Statement: Crystallographic data were deposited at the Cambridge Crystallographic Data Centre as a supplementary publication (CCDC 2279524 for complex **1** and CCDC 2279525 for complex **2**). Copies of the data can be obtained, free of charge, via www.ccdc.cam.ac.uk/conts/retrieving.html (or from the Cambridge Crystallographic Data Centre, CCDC, 12 Union Road, Cambridge CB2 1EZ, UK; fax: +44-1223-336033; or e-mail: deposit@ccdc.cam.ac.uk).

Acknowledgments: The authors thank Gianella Facchin, Facultad de Química, Facultad de Química, Univer sidad de la República, Montevideo, Uruguay, for the invitation.

Conflicts of Interest: The authors declare no conflict of interest.

References

1. Bray, F.; Laversanne, M.; Weiderpass, E.; Soerjomataram, I. The Ever-Increasing Importance of Cancer as a Leading Cause of Premature Death Worldwide. *Cancer* **2021**, *127*, 3029–3030. [CrossRef]
2. Crosby, D.; Bhatia, S.; Brindle, K.M.; Coussens, L.M.; Dive, C.; Emberton, M.; Esener, S.; Fitzgerald, R.C.; Gambhir, S.S.; Kuhn, P.; et al. Early Detection of Cancer. *Science* **2022**, *375*, eaay9040. [CrossRef] [PubMed]

3. Somarelli, J.A.; DeGregori, J.; Gerlinger, M.; Heng, H.H.; Marusyk, A.; Welch, D.R.; Laukien, F.H. Questions to Guide Cancer Evolution as a Framework for Furthering Progress in Cancer Research and Sustainable Patient Outcomes. *Med. Oncol.* **2022**, *39*, 137. [CrossRef] [PubMed]
4. Garreffa, E.; Arora, D. Breast Cancer in the Elderly, in Men and during Pregnancy. *Surgery* **2022**, *40*, 139–146. [CrossRef]
5. Smolarz, B.; Nowak, A.Z.; Romanowicz, H. Breast Cancer—Epidemiology, Classification, Pathogenesis and Treatment (Review of Literature). *Cancers* **2022**, *14*, 2569. [CrossRef]
6. Rossi, C.; Cicalini, I.; Cufaro, M.C.; Consalvo, A.; Upadhyaya, P.; Sala, G.; Antonucci, I.; Del Boccio, P.; Stuppia, L.; De Laurenzi, V. Breast Cancer in the Era of Integrating "Omics" Approaches. *Oncogenesis* **2022**, *11*, 17. [CrossRef]
7. Kavan, S.; Kruse, T.A.; Vogsen, M.; Hildebrandt, M.G.; Thomassen, M. Heterogeneity and Tumor Evolution Reflected in Liquid Biopsy in Metastatic Breast Cancer Patients: A Review. *Cancer Metastasis Rev.* **2022**, *41*, 433–446. [CrossRef]
8. Mehraj, U.; Mushtaq, U.; Mir, M.A.; Saleem, A.; Macha, M.A.; Lone, M.N.; Hamid, A.; Zargar, M.A.; Ahmad, S.M.; Wani, N.A. Chemokines in Triple-Negative Breast Cancer Heterogeneity: New Challenges for Clinical Implications. *Semin. Cancer Biol.* **2022**, *86*, 769–783. [CrossRef]
9. Fusco, N.; Sajjadi, E.; Venetis, K.; Ivanova, M.; Andaloro, S.; Guerini-Rocco, E.; Montagna, E.; Caldarella, P.; Veronesi, P.; Colleoni, M.; et al. Low-Risk Triple-Negative Breast Cancers: Clinico-Pathological and Molecular Features. *Crit. Rev. Oncol. Hematol.* **2022**, *172*, 103643. [CrossRef]
10. Herbst, R.S.; Morgensztern, D.; Boshoff, C. The Biology and Management of Non-Small Cell Lung Cancer. *Nature* **2018**, *553*, 446–454. [CrossRef]
11. Marino, F.Z.; Bianco, R.; Accardo, M.; Ronchi, A.; Cozzolino, I.; Morgillo, F.; Rossi, G.; Franco, R. Molecular Heterogeneity in Lung Cancer: From Mechanisms of Origin to Clinical Implications. *Int. J. Med. Sci.* **2019**, *16*, 981–989. [CrossRef] [PubMed]
12. De Sousa, V.M.L.; Carvalho, L. Heterogeneity in Lung Cancer. *Pathobiology* **2018**, *85*, 96–107. [CrossRef] [PubMed]
13. Testa, U.; Castelli, G.; Pelosi, E. Lung Cancers: Molecular Characterization, Clonal Heterogeneity and Evolution, and Cancer Stem Cells. *Cancers* **2018**, *10*, 248. [CrossRef] [PubMed]
14. Gamberi, T.; Hanif, M. Metal-Based Complexes in Cancer Treatment. *Biomedicines* **2022**, *10*, 2573. [CrossRef] [PubMed]
15. Ndagi, U.; Mhlongo, N.; Soliman, M.E. Metal Complexes in Cancer Therapy—An Update from Drug Design Perspective. *Drug Des. Devel. Ther.* **2017**, *11*, 599–616. [CrossRef]
16. Paprocka, R.; Wiese-Szadkowska, M.; Janciauskiene, S.; Kosmalski, T.; Kulik, M.; Helmin-Basa, A. Latest Developments in Metal Complexes as Anticancer Agents. *Coord. Chem. Rev.* **2022**, *452*, 85–94. [CrossRef]
17. Li, X.; Wang, Y.; Li, M.; Wang, H.; Dong, X. Metal Complexes or Chelators with ROS Regulation Capacity: Promising Candidates for Cancer Treatment. *Molecules* **2022**, *27*, 148. [CrossRef]
18. Neethu, K.S.; Sivaselvam, S.; Theetharappan, M.; Ranjitha, J.; Bhuvanesh, N.S.P.; Ponpandian, N.; Neelakantan, M.A.; Kaveri, M.V. In Vitro Evaluations of Biomolecular Interactions, Antioxidant and Anticancer Activities of Nickel(II) and Copper(II) Complexes with 1:2 Coordination of Anthracenyl Hydrazone Ligands. *Inorganica Chim. Acta* **2021**, *524*, 120419. [CrossRef]
19. Guan, D.; Zhao, L.; Shi, X.; Ma, X.; Chen, Z. Copper in Cancer: From Pathogenesis to Therapy. *Biomed. Pharmacother.* **2023**, *163*, 114791. [CrossRef]
20. Wang, C.; Yang, X.; Dong, C.; Chai, K.; Ruan, J.; Shi, S. Cu-Related Agents for Cancer Therapies. *Coord. Chem. Rev.* **2023**, *487*, 215156. [CrossRef]
21. Ji, P.; Wang, P.; Chen, H.; Xu, Y.; Ge, J.; Tian, Z.; Yan, Z. Potential of Copper and Copper Compounds for Anticancer Applications. *Pharmaceuticals* **2023**, *16*, 234. [CrossRef]
22. Singh, N.K.; Kumbhar, A.A.; Pokharel, Y.R.; Yadav, P.N. Anticancer Potency of Copper(II) Complexes of Thiosemicarbazones. *J. Inorg. Biochem.* **2020**, *210*, 111134. [CrossRef] [PubMed]
23. Castillo-Rodríguez, R.A.; Palencia, G.; Anaya-Rubio, I.; Gallardo-Pérez, J.C.; Jiménez-Farfán, D.; Escamilla-Ramírez, Á.; Zavala-Vega, S.; Cruz-Salgado, A.; Cervantes-Rebolledo, C.; Gracia-Mora, I.; et al. Anti-Proliferative, pro-Apoptotic and Anti-Invasive Effect of the Copper Coordination Compound Cas III-La through the Induction of Reactive Oxygen Species and Regulation of Wnt/β-Catenin Pathway in Glioma. *J. Cancer* **2021**, *12*, 5693–5711. [CrossRef] [PubMed]
24. da Silva, D.A.; De Luca, A.; Squitti, R.; Rongioletti, M.; Rossi, L.; Machado, C.M.L.; Cerchiaro, G. Copper in Tumors and the Use of Copper-Based Compounds in Cancer Treatment. *J. Inorg. Biochem.* **2022**, *226*, 111634. [CrossRef] [PubMed]
25. Chen, M.; Chen, X.; Huang, G.; Jiang, Y.; Gou, J. Synthesis, Anti-Tumour Activity, and Mechanism of Benzoyl Hydrazine Schiff Base-Copper Complexes. *J. Mol. Struct.* **2022**, *1268*, 133730. [CrossRef]
26. Jiang, Y.; Huo, Z.; Qi, X.; Zuo, T.; Wu, Z. Copper-Induced Tumor Cell Death Mechanisms and Antitumor Theragnostic Applications of Copper Complexes. *Nanomedicine* **2022**, *17*, 303–324. [CrossRef]
27. Jung, H.S.; Koo, S.; Won, M.; An, S.; Park, H.; Sessler, J.L.; Han, J.; Kim, J.S. Cu(ii)-BODIPY Photosensitizer for CAIX Overexpressed Cancer Stem Cell Therapy. *Chem. Sci.* **2023**, *14*, 1808–1819. [CrossRef]
28. Wellington, K.W. Understanding Cancer and the Anticancer Activities of Naphthoquinones-a Review. *RSC Adv.* **2015**, *5*, 20309–20338. [CrossRef]
29. Hong, S.; Huber, S.M.; Gagliardi, L.; Cramer, C.C.; Tolman, W.B. Copper(I)-α-Ketocarboxylate Complexes: Characterization and O$_2$ Reactions That Yield Copper-Oxygen Intermediates Capable of Hydroxylating Arenes. *J. Am. Chem. Soc.* **2007**, *129*, 14190–14192. [CrossRef]

30. Mohd Zubir, M.Z.; Jamaludin, N.S.; Abdul Halim, S.N. Hirshfeld Surface Analysis of Some New Heteroleptic Copper(I) Complexes. *J. Mol. Struct.* **2019**, *1193*, 141–150. [CrossRef]
31. Cabrera, A.R.; Gonzalez, I.A.; Cortés-Arriagada, D.; Natali, M.; Berke, H.; Daniliuc, C.G.; Camarada, M.B.; Toro-Labbé, A.; Rojas, R.S.; Salas, C.O. Synthesis of New Phosphorescent Imidoyl-Indazol and Phosphine Mixed Ligand Cu(i) Complexes-Structural Characterization and Photophysical Properties. *RSC Adv.* **2016**, *6*, 5141–5153. [CrossRef]
32. Villarreal, W.; Colina-Vegas, L.; Visbal, G.; Corona, O.; Corrêa, R.S.; Ellena, J.; Cominetti, M.R.; Batista, A.A.; Navarro, M. Copper(I)-Phosphine Polypyridyl Complexes: Synthesis, Characterization, DNA/HSA Binding Study, and Antiproliferative Activity. *Inorg. Chem.* **2017**, *56*, 3781–3793. [CrossRef] [PubMed]
33. Meza-Morales, W.; Machado-Rodriguez, J.C.; Alvarez-Ricardo, Y.; Obregón-Mendoza, M.A.; Nieto-Camacho, A.; Toscano, R.A.; Soriano-García, M.; Cassani, J.; Enríquez, R.G. A New Family of Homoleptic Copper Complexes of Curcuminoids: Synthesis, Characterization and Biological Properties. *Molecules* **2019**, *24*, 910. [CrossRef] [PubMed]
34. Małecki, J.G.; Maroń, A.; Palion, J.; Nycz, J.E.; Szala, M. A Copper(I) Phosphine Complex with 5,7-Dinitro-2-Methylquinolin-8-Ol as Co-Ligand. *Transit. Met. Chem.* **2014**, *39*, 755–762. [CrossRef]
35. Aliaga-Alcalde, N.; Marqués-Gallego, P.; Kraaijkamp, M.; Herranz-Lancho, C.; Den Dulk, H.; Görner, H.; Roubeau, O.; Teat, S.J.; Weyhermüller, T.; Reedijk, J. Copper Curcuminoids Containing Anthracene Groups: Fluorescent Molecules with Cytotoxic Activity. *Inorg. Chem.* **2010**, *49*, 9655–9663. [CrossRef] [PubMed]
36. Aguirrechu-Comerón, A.; Oramas-Royo, S.; Pérez-Acosta, R.; Hernández-Molina, R.; Gonzalez-Platas, J.; Estévez-Braun, A. Preparation of New Metallic Complexes from 2-Hydroxy-3-((5-Methylfuran-2-Yl)Methyl)-1,4-Naphthoquinone. *Polyhedron* **2020**, *177*, 114280. [CrossRef]
37. Oliveira, K.M.; Peterson, E.J.; Carroccia, M.C.; Cominetti, M.R.; Deflon, V.M.; Farrell, N.P.; Batista, A.A.; Correa, R.S. Ru(Ii)-Naphthoquinone Complexes with High Selectivity for Triple-Negative Breast Cancer. *Dalt. Trans.* **2020**, *49*, 16193–16203. [CrossRef]
38. Oliveira, K.M.; Liany, L.-D.; Corrêa, R.S.; Deflon, V.M.; Cominetti, M.R.; Batista, A.A. Selective Ru(II)/Lawsone Complexes Inhibiting Tumor Cell Growth by Apoptosis. *J. Inorg. Biochem.* **2017**, *176*, 66–76. [CrossRef]
39. Chaquin, P.; Canac, Y.; Lepetit, C.; Zargarian, D.; Chauvin, R. Estimating Local Bonding/Antibonding Character of Canonical Molecular Orbitals from Their Energy Derivatives. The Case of Coordinating Lone Pair Orbitals. *Int. J. Quantum Chem.* **2016**, *116*, 1285–1295. [CrossRef]
40. Salunke-Gawali, S.; Pereira, E.; Dar, U.A.; Bhand, S. Metal Complexes of Hydroxynaphthoquinones: Lawsone, Bis-Lawsone, Lapachol, Plumbagin and Juglone. *J. Mol. Struct.* **2017**, *1148*, 435–458. [CrossRef]
41. Majdi, C.; Duvauchelle, V.; Meffre, P.; Benfodda, Z. An Overview on the Antibacterial Properties of Juglone, Naphthazarin, Plumbagin and Lawsone Derivatives and Their Metal Complexes. *Biomed. Pharmacother.* **2023**, *162*, 114690. [CrossRef]
42. Kosiha, A.; Parthiban, C.; Ciattini, S.; Chelazzi, L.; Elango, K.P. Metal Complexes of Naphthoquinone Based Ligand: Synthesis, Characterization, Protein Binding, DNA Binding/Cleavage and Cytotoxicity Studies. *J. Biomol. Struct. Dyn.* **2018**, *36*, 4170–4181. [CrossRef] [PubMed]
43. Selvaraj, F.S.S.; Samuel, M.; Karuppiah, A.K.; Raman, N. Transition Metal Complexes Incorporating Lawsone: A Review. *J. Coord. Chem.* **2022**, *75*, 2509–2532. [CrossRef]
44. Oladipo, S.D.; Mocktar, C.; Omondi, B. In Vitro Biological Studies of Heteroleptic Ag(I) and Cu(I) Unsymmetrical N,N'-Diarylformamidine Dithiocarbamate Phosphine Complexes; the Effect of the Metal Center. *Arab. J. Chem.* **2020**, *13*, 6379–6394. [CrossRef]
45. Fan, W.W.; Li, Z.F.; Li, J.B.; Yang, Y.P.; Yuan, Y.; Tang, H.Q.; Gao, L.X.; Jin, Q.H.; Zhang, Z.W.; Zhang, C.L. Synthesis, Structure, Terahertz Spectroscopy and Luminescent Properties of Copper (I) Complexes with Bis(Diphenylphosphino)Methane and N-Donor Ligands. *J. Mol. Struct.* **2015**, *1099*, 351–358. [CrossRef]
46. Wu, Y.; Han, X.; Qu, Y.; Zhao, K.; Wang, C.; Huang, G.; Wu, H. Two Cu(I) Complexes Constructed by Different N-Heterocyclic Benzoxazole Ligands: Syntheses, Structures and Fluorescent Properties. *J. Mol. Struct.* **2019**, *1191*, 95–100. [CrossRef]
47. Leite, C.M.; Honorato, J.; Martin, A.C.B.M.; Silveira, R.G.; Colombari, F.M.; Amaral, J.C.; Costa, A.R.; Cominetti, M.R.; Plutin, A.M.; Aguiar, D.; et al. Experimental and Theoretical DFT Study of Cu(I)/N,N-Disubstituted-N'-acylthioureato Anticancer Complexes: Actin Cytoskeleton and Induction of Death by Apoptosis in Triple-Negative Breast Tumor Cells. *Inorg. Chem.* **2022**, *61*, 664–677. [CrossRef]
48. Gunasekaran, N.; Bhuvanesh, N.S.P.; Karvembu, R. Synthesis, Characterization and Catalytic Oxidation Property of Copper(I) Complexes Containing Monodentate Acylthiourea Ligands and Triphenylphosphine. *Polyhedron* **2017**, *122*, 39–45. [CrossRef]
49. Idriss, K.A.; Sedaira, H.; Hashem, E.Y.; Saleh, M.S.; Soliman, S.A. The Visible Absorbance Maximum of 2-Hydroxy-1,4-Naphthoquinone as a Novel Probe for the Hydrogen Bond Donor Abilities of Solvents and Solvent Mixtures. *Monatshefte Fur Chem.* **1996**, *127*, 29–42. [CrossRef]
50. McCormick, T.; Jia, W.L.; Wang, S. Phosphorescent Cu(I) Complexes of 2-(2'-Pyridylbenzimidazolyl) Benzene: Impact of Phosphine Ancillary Ligands on Electronic and Photophysical Properties of the Cu(I) Complexes. *Inorg. Chem.* **2006**, *45*, 147–155. [CrossRef] [PubMed]
51. Ibis, C.; Sahinler Ayla, S.; Babayeva, E. Reactions of Quinones with Some Amino Alcohols, Thiols and a UV-Vis Study. *Phosphorus Sulfur Silicon Relat. Elem.* **2020**, *195*, 474–480. [CrossRef]

52. Sørensen, T.; Nielsen, M. Synthesis, UV/Vis Spectra and Electrochemical Characterisation of Arylthio and Styryl Substituted Ferrocenes. *Open Chem.* **2011**, *9*, 610–618. [CrossRef]
53. Verma, S.K.; Singh, V.K. Synthesis, Electrochemical, Fluorescence and Antimicrobial Studies of 2-Chloro-3-Amino-1,4-Naphthoquinone Bearing Mononuclear Transition Metal Dithiocarbamate Complexes [M{κ^2S,S-S$_2$C–Piperazine–C$_2$H$_4$N(H)ClNQ}$_n$]. *RSC Adv.* **2015**, *5*, 53036–53046. [CrossRef]
54. Win, T.; Bittner, S. Novel 2-Amino-3-(2,4-Dinitrophenylamino) Derivatives of 1,4-Naphthoquinone. *Tetrahedron Lett.* **2005**, *46*, 3229–3231. [CrossRef]
55. Záliš, S.; Fiedler, J.; Pospíšil, L.; Fanelli, N.; Lanza, C.; Lampugnani, L. Electron Transfer in Donor–Acceptor Molecules of Substituted Naphtoquinones: Spectral and Redox Properties of Internal Charge Transfer Complexes. *Microchem. J.* **1996**, *54*, 478–486. [CrossRef] [PubMed]
56. Sutovsky, Y.; Likhtenshtein, G.I.; Bittner, S. Synthesis and Photochemical Behavior of Donor-Acceptor Systems Obtained from Chloro-1,4-Naphthoquinone Attached to Trans-Aminostilbenes. *Tetrahedron* **2003**, *59*, 2939–2945. [CrossRef]
57. Nowicka, B.; Walczak, J.; Kapsiak, M.; Barnaś, K.; Dziuba, J.; Suchoń, A. Impact of Cytotoxic Plant Naphthoquinones, Juglone, Plumbagin, Lawsone and 2-Methoxy-1,4-Naphthoquinone, on Chlamydomonas Reinhardtii Reveals the Biochemical Mechanism of Juglone Toxicity by Rapid Depletion of Plastoquinol. *Plant Physiol. Biochem.* **2023**, *197*, 107660. [CrossRef]
58. Pieta, P.; Petr, A.; Kutner, W.; Dunsch, L. In Situ ESR Spectroscopic Evidence of the Spin-Trapped Superoxide Radical, O2{radical Dot}-, Electrochemically Generated in DMSO at Room Temperature. *Electrochim. Acta* **2008**, *53*, 3412–3415. [CrossRef]
59. Clément, J.L.; Ferré, N.; Siri, D.; Karoui, H.; Rockenbauer, A.; Tordo, P. Assignment of the EPR Spectrum of 5,5-Dimethyl-1-Pyrroline N-Oxide (DMPO) Superoxide Spin Adduct. *J. Org. Chem.* **2005**, *70*, 1198–1203. [CrossRef] [PubMed]
60. Shoji, T.; Li, L.; Abe, Y.; Ogata, M.; Ishimoto, Y.; Gonda, R.; Mashino, T.; Mochizuki, M.; Uemoto, M.; Miyata, N. DMPO-OH Radical Formation from 5,5-Dimethyl-1-Pyrroline N-Oxide (DMPO) in Hot Water. *Anal. Sci.* **2007**, *23*, 219–221. [CrossRef]
61. Zalibera, M.; Rapta, P.; Staško, A.; Brindzová, L.; Brezová, V. Thermal Generation of Stable Spin Trap Adducts with Super-Hyperfine Structure in Their EPR Spectra: An Alternative EPR Spin Trapping Assay for Radical Scavenging Capacity Determination in Dimethylsulphoxide. *Free Radic. Res.* **2009**, *43*, 457–469. [CrossRef] [PubMed]
62. Massagué, J.; Obenauf, A.C. Metastatic Colonization by Circulating Tumour Cells. *Nature* **2016**, *529*, 298–306. [CrossRef] [PubMed]
63. Fares, J.; Fares, M.Y.; Khachfe, H.H.; Salhab, H.A.; Fares, Y. Molecular Principles of Metastasis: A Hallmark of Cancer Revisited. *Signal Transduct. Target. Ther.* **2020**, *5*, 28. [CrossRef] [PubMed]
64. Jonkman, J.E.N.; Cathcart, J.A.; Xu, F.; Bartolini, M.E.; Amon, J.E.; Stevens, K.M.; Colarusso, P. An Introduction to the Wound Healing Assay Using Live-Cell Microscopy. *Cell Adhes. Migr.* **2014**, *8*, 440–451. [CrossRef]
65. Stamm, A.; Reimers, K.; Strauß, S.; Vogt, P.; Scheper, T.; Pepelanova, I. In Vitro Wound Healing Assays—State of the Art. *BioNanoMaterials* **2016**, *17*, 79–87. [CrossRef]
66. Al Habash, A.; Aljasim, L.A.; Owaidhah, O.; Edward, D.P. A Review of the Efficacy of Mitomycin C in Glaucoma Filtration Surgery. *Clin. Ophthalmol.* **2015**, *9*, 1945–1951. [CrossRef]
67. Jardine, F.H.; Vohra, A.G.; Young, F.J. Copper(I) Nitrato and Nitrate Complexes. *J. Inorg. Nucl. Chem.* **1971**, *33*, 2941–2945. [CrossRef]

Disclaimer/Publisher's Note: The statements, opinions and data contained in all publications are solely those of the individual author(s) and contributor(s) and not of MDPI and/or the editor(s). MDPI and/or the editor(s) disclaim responsibility for any injury to people or property resulting from any ideas, methods, instructions or products referred to in the content.

Article

Synthesis, Characterization, DNA Binding and Cytotoxicity of Copper(II) Phenylcarboxylate Complexes

Carlos Y. Fernández [1,2], Analu Rocha [3], Mohammad Azam [4], Natalia Alvarez [1], Kim Min [5], Alzir A. Batista [3], Antonio J. Costa-Filho [6], Javier Ellena [7] and Gianella Facchin [1,*]

1. Química Inorgánica, Departamento Estrella Campos, Facultad de Química, Universidad de la República, Montevideo 11800, Uruguay; cyanez@fq.edu.uy (C.Y.F.)
2. Programa de Posgrados de la Facultad de Química, Facultad de Química, Universidad de la República, Gral. Flores 2124, Montevideo 11800, Uruguay
3. Departamento de Química, Federal University of São Carlos, CP 676, São Carlos 13565-905, SP, Brazil
4. Department of Chemistry, College of Sciences, King Saud University, P.O. Box 2455, Riyadh 11451, Saudi Arabia
5. Department of Safety Engineering, Dongguk University, 123 Dongdae-ro, Gyeongju 780714, Gyeongbuk, Republic of Korea
6. Physics Department, Ribeirão Preto School of Philosophy, Science and Literature, University of São Paulo, Av. Bandeirantes, Ribeirão Preto 14040-901, SP, Brazil
7. São Carlos Institute of Physics, University of São Paulo, Av. do Trabalhador São-Carlense 400, São Carlos 13566-590, SP, Brazil
* Correspondence: gfacchin@fq.edu.uy

Abstract: Coordination compounds of copper exhibit cytotoxic activity and are suitable for the search for novel drug candidates for cancer treatment. In this work, we synthesized three copper(II) carboxylate complexes, $[Cu_2(3\text{-}(4\text{-hydroxyphenyl})\text{propanoate})_4(H_2O)_2]\cdot 2H_2O$ (**C1**), $[Cu_2(\text{phenylpropanoate})_4(H_2O)_2]$ (**C2**) and $[Cu_2(\text{phenylacetate})_4]$ (**C3**), and characterized them by elemental analysis and spectroscopic methods. Single-crystal X-ray diffraction of **C1** showed the dinuclear paddle-wheel arrangement typical of Cu–carboxylate complexes in the crystal structure. In an aqueous solution, the complexes remain as dimeric units, as studied by UV-visible spectroscopy. The lipophilicity (partition coefficient) and the DNA binding (UV visible and viscosity) studies evidence that the complexes bind the DNA with low K_b constants. In vitro cytotoxicity studies on human cancer cell lines of metastatic breast adenocarcinoma (MDA-MB-231, MCF-7), lung epithelial carcinoma (A549) and cisplatin-resistant ovarian carcinoma (A2780cis), as well as a nontumoral lung cell line (MRC-5), indicate that the complexes are cytotoxic in cisplatin-resistant cells.

Keywords: copper complexes; phenyl-carboxylate; DNA interaction; cytotoxic activity

1. Introduction

Metal-based drugs play an important role in cancer treatment. Cisplatin and its congeners (carboplatin, oxalylplatin, heptaplatin and picoplatin) are successfully used against various cancer types, being curative in several cases [1]. Despite this, there is still a lack of effective treatment for all types of cancer. Furthermore, despite offering a variety of compounds and mechanisms of action, the development of new potential anticancer metallopharmaceuticals remains mainly academic, possibly due to the complexity of metal-coordination compounds' reactivity [2].

Copper-coordination compounds are an attractive class of compounds for the development of novel cancer treatments [2–5]. Different copper complexes with antitumor activity have been synthesized and characterized, with promising results, even presenting antimetastatic and antiangiogenic activities (in vitro assays) or being cytotoxic to cancer stem cells [3,4,6–13]. Cu(II) complexes of ligands with no appreciable cytotoxic activity are active, indicating that the metal itself plays a role in antitumor activity.

The mechanism of action of copper compounds may include various molecular processes, which have not been thoroughly characterized [3,4,13]. The lack of specificity against a single molecular target strengthens copper complexes' ability to fight a diverse cell population, such as those found in a tumor. DNA binding and producing reactive oxygen species (ROS), inducing redox stress, are commonly proposed as molecular events for most anticancer copper compounds [2,4,14–16].

As a part of our research of copper complexes with cytotoxic activity [17–25], we search for simple molecules, especially those already tested for their biological use, that can act as anion ligands. Phenylacetic acid is a compound used to treat high nitrogen levels in hepatic patients and, therefore, meets the safety regulations to be used as a drug [26]. In this work, we explored the chemical properties and cytotoxicity of copper complexes with phenylacetic acid, as well as two related compounds, phenylpropanoate and 3-(4-hydroxyphenyl)propanoate, in order to prepare complexes with varying lipophilicity and possibly other differences in chemical behavior.

The complexes were studied both in the solid state and aqueous solution, including a new crystal structure. The binding of the complexes to the DNA molecule was investigated. The cytotoxicity of the complexes was evaluated against MDA-MB-231, MCF-7 (human metastatic breast adenocarcinomas, the first triple negative), A549 (human lung epithelial carcinoma), A2780cis (cisplatin-resistant human ovarian carcinoma, SIGMA) and MRC-5 (human nontumoral lung epithelial cells), finding an interesting activity on cisplatin-resistant A2780cis cells.

2. Results

As described in the experimental section, three complexes were synthesized: [Cu$_2$(3-(4-hydroxyphenyl)propanoate)$_4$(H$_2$O)$_2$]·2H$_2$O (**C1**); [Cu$_2$(phenylpropanoate)$_4$(H$_2$O)$_2$] (**C2**); and [Cu$_2$(phenylacetate)$_4$] (**C3**).

2.1. Crystal Structures

The obtained complexes were recrystallized from water by slow evaporation at room temperature. Single crystals suitable for X-ray diffraction analysis were obtained only for **C1**, a new compound, and **C3**, which had two previously reported [27,28]. The most relevant structural features are described in this section. Table 1 summarizes crystallographic data and refinement details. A scheme of the complexes and the ligands is included in the supplementary material (Figure S1).

Table 1. Crystallographic data and refinement details for **C1** and **C3**.

Complex	C1	C3
Formula	C$_{36}$H$_{44}$Cu$_2$O$_{16}$	C$_{16}$H$_{14}$CuO$_4$
$D_{calc.}$/g cm^{-3}	1.594	1.618
μ/mm^{-1}	2.130	2.374
Formula Weight	859.832	333.81
Color	Blue	Blue
Shape	Prism	Plate
Size/mm^3	0.15 × 0.10 × 0.10	0.30 × 0.15 × 0.08
Crystal System	Triclinic	monoclinic
Space Group	$P\bar{1}$	$P2_1/c$
a/Å	8.6810(2)	5.17356(6)
b/Å	10.6746(3)	26.2143(3)
c/Å	11.3849(3)	10.20173(12)
α/°	66.930(3)	90
β/°	70.661(2)	97.8378(11)
γ/°	71.814(2)	90
V/Å^3	895.43(5)	1370.64(3)
Z	1	4

Table 1. Cont.

Complex	C1	C3
$\Theta_{min}/°$	4.347	3.372
$\Theta_{max}/°$	80.066	79.397
Measured Refl.	15,912	13,429
Independent Refl.	3875	2965
Reflections with I > 2σ(I)	3820	2743
R_{int}	0.0193	0.0451
Parameters	251	191
Restraints	0	0
Largest Peak	0.622	0.476
Deepest Hole	−0.737	−0.597
GooF	1.040	1.027
wR_2 (all data)	0.0708	0.0855
wR_2	0.0706	0.0835
R_1 (all data)	0.0278	0.0349
R_1	0.0275	0.0325
CCDC deposition number	2,288,430	2,288,436

2.1.1. [Cu$_2$(3-(4-Hydroxyphenyl)propanoate)$_4$(H$_2$O)$_2$]·2H$_2$O

[Cu$_2$(3-(4-hydroxyphenyl)propanoate)$_4$(H$_2$O)$_2$]·2H$_2$O, **C1**, crystallizes in the triclinic space group P$\bar{1}$ with one molecular formula per unit cell. Figure 1 presents both the asymmetric and cell unit of the structure, whereas Table 2 indicates bond lengths (Å) and angles (°) surrounding the coordination center. The copper ion presents a pentacoordinated environment where the equatorial donors are four carboxylate O atoms from four different ligands, and the apical position is occupied by an O atom from a water molecule. The carboxylate group acts as a bridging bidentate ligand, connecting the two copper(II) centers in the dimeric molecule. Figure 1b presents the molecular moiety where the dimeric paddle-well arrangement typical of dimeric Cu–carboxylate complexes can be observed. This motif is observed on several Cu(II) compounds with ligands containing carboxylate groups, such as acetate [29,30], propionate [31], dinitrobenzoates [32] and N-acetylglycinato [33], among others.

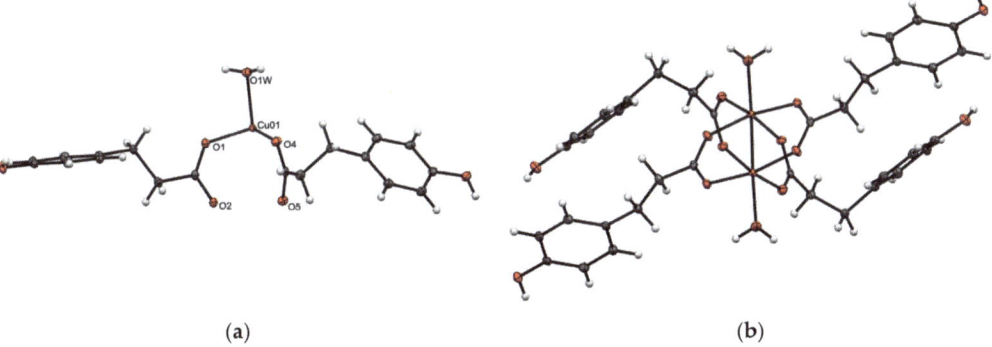

Figure 1. ORTEP representation at 50% probability of (**a**) the asymmetric unit and (**b**) molecular moiety of [Cu$_2$(3-(4-hydroxyphenyl)propanoate)$_4$(H$_2$O)$_2$]·2H$_2$O (**C1**). The hydration water molecule is omitted for clarity. Atom color code: Cu (orange), C (gray), O (red) and H (white).

Table 2. Selected bond lengths (Å) and angles (°) for C1.

Bond Lengths (Å)		Angles (°)	
Cu1-Cu2	2.6075(4)	O1-Cu1-O4	90.98(5)
Cu1-O4	1.9649(11)	O5'-Cu1-O4	169.16(4)
Cu1-O1	1.9604(10)	O5'-Cu1-O1	88.48(4)
Cu1-O5'	1.9751(10)	O2'-Cu1-O4	91.39(5)
Cu1-O2'	1.9628(11)	O2'-Cu1-O1	168.36(5)
		O5'-Cu1-O2'	87.05(5)

A crystallographic database search in the CSD [34] v2022.3.0, conducted using Conquest [35], found 786 related structures, which were analyzed in Mercury [36]. The bridging bidentate mode of coordination of the carboxylate group determines Cu···Cu distances in this dinuclear paddle-wheel type complexes. The distances in the analyzed structures range from 2.58 to 2.68 Å, including the 2.608 Å distance observed in C1. Other structures containing a 2.608 Å Cu···Cu distance include structures with acetate [37], propionate [31,38], benzoate [39,40] and paranitrobenzoate [41] as ligands.

The crystal packing is sustained primarily by strong classical H-bond interactions [42] involving the hydroxyl and carboxylate groups in the ligand and the coordinated and lattice water molecules. Each hydroxyl group acts as an H-bond acceptor with a coordinated water molecule in a contiguous complex molecule (H···O distance of 1.898 Å, O-H-O angle 172.4°) and donor with a lattice water molecule (H···O distance of 1.903 Å, O-H-O angle 172.5°). The lattice water molecule also acts as an H-bond donor to a carboxylate O atom with an H···O distance of 2.028 Å and an O-H-O angle of 153.0°. Nonclassical H-bonds are also observed in the C-H···π interactions between phenyl rings of ligands in contiguous molecules with a centroid to H distance of 2.658 Å and the angle between the phenyl rings of 47.45°.

2.1.2. [Cu$_2$(Phenylacetate)$_4$]·2H$_2$O and [Cu$_2$(Phenylpropanoate)$_4$(H$_2$O)$_2$]

The crystal structure of Cu$_2$(phenylacetate)$_4$]·2H$_2$O, C3, has been previously reported at 150 [27] and 298 [28] K. There are only slight differences in the cells' axis lengths and angles for these structures. We run the structure comparison tool available at the Bilbao Crystallographic Server [43] to compare the structure at 100 K reported in this article with the one obtained at room temperature, which presented the higher differences, finding a degree of lattice distortion of 0.0055 with a maximum difference of atomic positions of 0.1370 Å. C3 also exhibits a paddle-wheel coordination motif with the carboxylate group in a bis-chelate fashion. In the case of C1, each carboxylate O atom coordinates to one copper(II) center. Meanwhile, in C3, an O atom from the carboxylate group can be connected to one or two copper(II) centers. This coordination motif gives rise to the formation of a 1D chain along the *a* axis.

In C3, the Cu···Cu distance is 2.5787(5) Å, also contained in the expected range. The same intermetal distance was observed in the structures with hexanoate [44], benzoate [45,46] and 2,3-dihydro-1,4-benzodioxine-6-carboxylate [47]. C-H···π interactions can be observed between phenyl rings of ligands within the paddle wheel on the 1D chain contiguous molecules with a centroid to H distance of 3.062 Å and an angle between the phenyl rings of 71.00°. The infinite chains are sustained with each other through dispersive interactions involving the phenyl groups. No obvious hydrogen bonds or π-stacking interactions can be observed in the structure.

The structure of [Cu$_2$(phenylpropanoate)$_4$(H$_2$O)$_2$], C2, was also previously determined, showing a coordination scheme similar to that of C3 [48]. In spite of that, according to the molecular formula found, it is possible that, in the compound prepared by us, the structure is similar to that of C1.

2.2. Infrared Spectra

The studied ternary complexes present similar infrared spectra. Table 3 presents a tentative assignment of the bands related to coordinating groups. In particular, the values of the Δν (calculated as ν(COO)$_{as}$-ν(COO)$_s$) for **C1** = 157 cm^{-1}, **C2** = 157 cm^{-1} and **C3** = 176 cm^{-1} agree with a bidentate coordination of the carboxylate [49], as observed in the crystal structures of **C1** and **C3**. The spectra of the complexes and the ligands are included in the supplementary material (Figures S2–S7).

Table 3. Wavenumber (cm^{-1}) of common bands in the complexes, and their tentative assignment.

Compound	ν(O-H)	ν(C=O) + ν(COO)$_{as}$	ν(COO)$_s$	ν(Cu-O)
[Cu$_2$(3-(4-hydroxyphenyl)propanoate)$_4$(H$_2$O)$_2$]·2H$_2$O	3330 sh	1582 s, 1516 w	1425 m	532 w
[Cu$_2$(phenylpropanoate)$_4$(H$_2$O)$_2$]	3500–3200 sh	1588 s, 1516 w	1431 m	480 w
[Cu$_2$(phenylacetate)$_4$]	3500–3200 sh	1594 s, 1514 s	1438 m	532 w

2.3. Solution Studies

Major Species in Solution Characterization Using UV-Visible Spectra and Lipophilicity

The visible spectra of the complexes show an absorption band at around 710 nm (DMSO solution), as presented in Table 4, which, if compared with the wavelength of the maxima calculated according to the empiric correlation of Prenesti et al. [50,51], agrees with an equatorial coordination by four carboxylate oxygen atoms (calculated λ$_{max}$ 708 nm), as observed in the solid state. In relation to the dimeric structure, the occurrence of a band between 350 and 400 nm has been related to this species' existence in solution [52]. This band is present in the complexes' UV spectra but not in the ligand spectra. According to this analysis, in a DMSO solution, the complexes remain as dimers like the solid-state form of **C1**. The complexes are not soluble in H$_2$O, but, as an approach to studying their behavior in this solvent, spectra were also registered in a DMSO:water mixture (80:20), Table 4 presents the obtained results. The λ$_{max}$ shifts slightly, and the shape of the spectra-changed difference was accounted for by n (n = ε$_{850}$/ε$_{max}$ × 100), which is higher in an aqueous solution, suggesting a different degree of distortion of the coordination geometry depending on the solvent [53], as previously observed with other Cu complexes.

Table 4. Maximum absorption wavelength (λ$_{max}$, nm), molar absorptivity (ε, M^{-1}cm^{-1}) and n (ε$_{850}$/ε$_{max}$ × 100) of the spectra in DMSO and DMSO:H$_2$O (80:20) and partition coefficients (P) between n-octanol and physiologic solution.

Compound	λmax/ε *	n *	λmax/ε **	n **	P
[Cu$_2$(3-(4-hydroxyphenyl)propanoate)$_4$(H$_2$O)$_2$]·2H$_2$O	710/388	43	726/134	65	0.10
[Cu$_2$(phenylpropanoate)$_4$(H$_2$O)$_2$]	715/313	44	713/288	73	0.24
[Cu$_2$(phenylacetate)$_4$]	711/404	48	736/150	72	0.47

* DMSO, ** DMSO:H$_2$O (80:20), ε calculated per Cu mole.

The lipophilicity of the complexes is similar, with the hydroxyl group of **C1** giving rise to a slightly more hydrophilic compound, as expected.

2.4. Complex–DNA Binding Studies

2.4.1. K_b Determination (UV-Visible Spectra)

The intrinsic binding constants of the complexes to the DNA (K_b) were determined via UV-visible titration (Figure 2 and supplementary material Figures S8 and S9). Their values are presented in Table 5. The ligands produce nonappreciable DNA binding as studied via this technique.

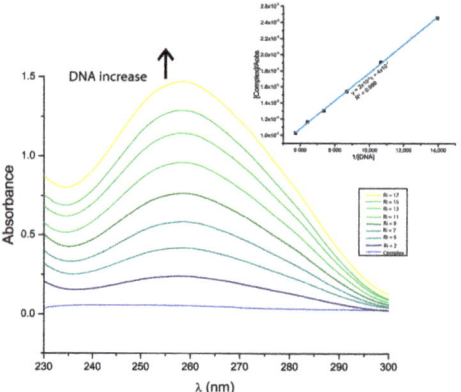

Figure 2. UV spectra of **C1** with increasing [DNA]/[complex] (*Ri*) ratio. Inset: [complex]/Aobs (i.e., the complex concentration/the measured absorbance) as a function of 1/[DNA] plot with regression parameters.

Table 5. DNA binding constants (K_b), as determined by the Benesi–Hildebrand method.

Compound	C1	C2	C3	L1	L2	L3
K_b (M-1)	5.2×10^2	2.0×10^2	8.7×10^2	ND *	ND *	ND *

* Not Determined.

The observed values of K_b are relatively low if compared with other Cu-carboxylate complexes. For instance, compounds [Cu$_2$(nitrofenilacetate)$_4$)(H$_2$O)$_2$] and [Cu$_2$(fenilbutanoate)$_4$]$_n$ present K_b values in the 10^3–10^4 range [54,55]. In particular, the binding of **C3** on salmon sperm DNA was already reported and determined by the same methodology, with $K_b = 1.4 \times 10^4$ M^{-1} nm on the used DNA being suggestive of intercalation in addition to binding by the grooves [27].

2.4.2. Mode of Binding (Relative Viscosity)

Relative viscosity is a highly sensitive method to detect changes in the overall length of the DNA caused by the interaction of small molecules [49]. Figure 3 presents the effect of the increasing concentration of the complexes on the relative viscosity of CT-DNA. Free ligands induce no appreciable change in DNA's relative viscosity, as detected by this technique. The complexes induce a slight relative viscosity decrease at the studied ratios. This suggests that the binding provokes bends in the DNA helix [56]. A small slope is observed, in agreement with the low K_b of the complexes, evidencing that the binding event is relatively minor compared to other Cu complexes and induces only small changes in DNA conformation.

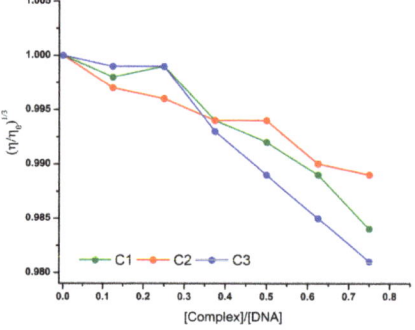

Figure 3. Effect of the increasing concentration of the complexes on the relative viscosity of CT-DNA.

2.5. Cytotoxicity of the Compounds

The cytotoxicity of the complexes and free ligands was evaluated on four tumor and one nontumor cell lines; Table 6 presents the results expressed by IC_{50}. The ligands L1–L3 present no detectable cytotoxicity up to 100 µM.

Table 6. Cytotoxic activity (expressed by IC_{50}) of the studied complexes after 48 h of incubation, against MCF-7, MDA-MB-231 (human metastatic breast adenocarcinomas, the latter triple negative), A549 (human lung epithelial carcinoma), A278cis (human ovarian cisplatin-resistant) and MRC-5 (lung nontumoral) cell lines.

Compound	Cytotoxicity, IC50 (µM)				
	MCF-7	MDA-MB-231	A549	A278cis	MRC-5
C1	>50	>50	>50	26.80 ± 4.50	>50
C2	20.20 ± 0.78	>50	>50	13.50 ± 0.57	>50
C3	>50	>50	>50	7.85 ± 0.86	>50
Cisplatin	8.91 ± 2.60	24.90 ± 3.40	14.40 ± 1.40	26.90 ± 0.60	29.09 ± 0.78

The complexes induce low cytotoxicity to four of the studied lines but are cytotoxic to the A278cis cell line and resistant to cisplatin, therefore showing no cross resistance. This activity can be classified as moderate compared to other Cu complexes [3]. There seems to be a correlation between the IC_{50} and lipophilicity (P). Both **C2** and **C3** are more cytotoxic than cisplatin on A278cis cells and are less toxic than cisplatin to the nontumor cell MRC-5, making both complexes **C2** and **C3** interesting complexes for further study of their activities on other tumor cells, especially those resistant to cisplatin.

3. Discussion

The compounds presented in this work are dimeric complexes in the solid state, with **C3** further extending into a polymeric structure. In a DMSO solution, the dimeric structure seems to be preserved. In the conditions of the biological assays, coordination may be altered, possibly including, in addition to carboxylate, other ligands such as residues from albumin. The biological activity of the compounds is different when compared with the free ligands, suggesting also that the ligands remain coordinated in the major species in these conditions.

The complexes bind the DNA with low K_b compared to other Cu complexes; therefore, this seems not to be part of the mechanism of the cytotoxicity of the complexes.

This work aimed to find new complexes with interesting cytotoxic activity, particularly with ligands that present no appreciable toxicity. The complexes were active only in one of the studied tumor cells, a cell line that is resistant to cisplatin. This opens an opportunity to further explore the activity of **C2** and **C3** on other tumor cell lines. To date, there are few Cu(II) complexes that have ligands with low toxicity and are cytotoxic to tumor cells.

4. Materials and Methods

4.1. Synthetic Procedures

All reagents were used as commercially available: copper(II) carbonate and copper(II) chloride (Fluka, SIGMA-Aldrich, St. Louis, MI, USA), carboxylic ligands (SIGMA-Aldrich, St. Louis, MI, USA) and calf thymus DNA (CT-DNA, SIGMA-Aldrich, St. Louis, MI, USA).

[Cu_2(phenylcarboxylate)$_4$] Complexes

An ethanolic solution of phenylcarboxylate (0,23 mmol, 5 mL) was added under constant stirring at room temperature to an aqueous solution of copper(II) chloride (0,23 mmol, 5 mL). The solution turned green instantly. It was allowed to slowly evaporate giving rise to green prismatic single crystals adequate for X-ray diffraction studies. [Cu_2(3-(4-hydroxyphenyl)propanoate)$_4$(H_2O)$_2$]·$2H_2O$ (**C1**) Calc. for $C_{36}H_{44}Cu_2O_{12}$/Found: %C: 50.29/

50.15 %H: 5.16/5.45; [Cu$_2$(phenylpropanoate)$_4$(H$_2$O)$_2$] (**C2**) Calc. for C$_{36}$H$_{40}$Cu$_2$O$_{10}$/Found: %C: 56.90/56.80 %H: 5.31/5.39; [Cu$_2$(phenylacetate)$_4$] (**C3**) Calc. for C$_{32}$H$_{28}$Cu$_2$O$_8$/Found: %C: 57.57/57.67 %H: 4.23/4.57.

4.2. Physical Methods

4.2.1. Characterization—General

Elemental analyses (C, N and H) of the samples were carried out on a Thermo Flash 2000 elemental analyzer (Thermo Fisher Scientific, USA). Infrared spectra were measured on a Shimadzu IR Prestige 21 (Shimadzu, Kyoto, Japan, 4000 to 400 cm^{-1}) as 1% KBr disks with a 4 cm^{-1} resolution. UV-visible spectra of 5 mM solutions in DMSO or DMSO H$_2$O (80:20) of the complexes were recorded on a ShimadzuUV1900 spectrophotometer (Shimadzu, Kyoto, Japan) in 1 cm path-length quartz cells.

4.2.2. Crystal Structure Determination

Suitable single crystals of **C1** and **C3** were obtained from recrystallization from DMSO aqueous solution slow evaporation. Samples were mounted, and their diffraction patterns were measured on a Rigaku XtaLAB Synergy-S diffractometer (Rigaku, USA) equipped with an Oxford Cryosystems Cryostream 800 PLUS. The crystals were kept at a steady $T = 100(2)$ K during data collection with a PhotonJet (CuKα = 1.54184 Å) X-ray Source. CrysAlisPro v 42.84a software (Rigaku) was used to evaluate the collection strategy, data reduction and scaling, as well as absorption correction. The structure was solved using direct methods with ShelXt [57] and refined using the atoms in the molecules model with ShelXL-2019/2 [58] using least squares minimization on F^2. Both ShelXt and ShelXL were used within Olex2 [59]. Hydrogen atoms were geometrically positioned and refined isotopically with the riding model. Molecular graphics were prepared using Mercury [36].

The nonhydrogen atoms were refined anisotropically. Then, all hydrogen atoms were located from electron-density difference maps and were positioned geometrically and refined using the riding model [Uiso(H) = 1.2 Ueq or 1.5 Ueq]. The Olex2 was also used for analysis and visualization of the structures and for graphic material preparation. Table 1 summarizes the X-ray diffraction data and refinement parameters obtained for the elucidated crystal structures. The CIF files of complexes **C1** and **C3** were deposited in the Cambridge Structural Data Base under the CCDC numbers 2,288,430 and 2,288,436, respectively. Copies of the data can be obtained, free of charge, via www.ccdc.cam.ac.uk.

4.3. Lipophility Assessment

Lipophilicity was studied by determining the partition coefficient of the complexes in n-octanol/physiological solution (0.9% NaCl in water). To 1 mL of n-octanol 0.2 mM solution of the complex, 1 mL of physiological solution was added. It was shaken for 1 h. Afterward, the samples were centrifuged, and the phases separated. UV-vis spectra were used to determine the concentration of the complex in each phase. The partition coefficient, P, was calculated as C n-octanol/Cwater.

4.4. DNA Interaction

A stock solution of Calf Thymus DNA (CT-DNA, 5 mg in 5 mL H$_2$O) was prepared by stirring overnight, stored at 4 °C and used within 3 days. Its concentration was determined spectroscopically at 260 nm (ε_{260} = 6600 M^{-1}cm^{-1}/base pair). The solution was free of protein, as determined by the A$_{260}$/A$_{280}$ ratio, which varied in the 1.8–1.9 range.

4.4.1. DNA Binding Constant: UV Absorption Titration Experiments

The DNA intrinsic binding constant (K_b) was determined by UV absorption measurements using the Benesi–Hildebrand model [60,61]. Solutions of the complexes 5 mM, in buffer Tris/HCl pH = 7.5 and 50 mM in NaCl were used, and their concentration was kept constant at 10–15 μM while adding CT-DNA to obtain concentrations in the 0–250 μM in the base pairs range. The Benesi–Hildebrand model can be described by the equation:

$$1/(\varepsilon a - \varepsilon f) = 1/(\varepsilon b - \varepsilon f) + 1/K_b[DNA]\,(\varepsilon b - \varepsilon f) \tag{1}$$

where [DNA] is the concentration of DNA, εa are the apparent absorption coefficients, εf and εb are the extinction coefficient for the free copper(II) complex and the extinction coefficient for the copper(II) complex in the fully bound form, respectively. In Equation (1), $1/(\varepsilon a - \varepsilon f)$ is equivalent to Aobserved/[Cu]. Therefore, according to this model, the K_b value equals the slope to the intercept ratio of the plot [complex]/Aobserved as a function of 1/[DNA].

4.4.2. DNA Binding Mode: Variation of Viscosity Experiments

Viscosity measurements were performed in an Ostwald-type viscosimeter (SIGMA-Aldrich, St. Louis, MI, USA) maintained at a temperature of 25.0 ± 0.1 °C in a thermostatic bath. Solutions of CT-DNA (150 μM base pairs) and complexes were prepared separately in Tris-HCl (5 mM, pH = 7.2, 50 mM NaCl) and thermostatized at 25 °C. Complex–DNA solutions (4 mL) were prepared just prior to running each experiment at different molar ratios ([complex]/[CT-DNA] = 0.125, 0.250, 0.375, 0.500, 0.625 and 0.750 (equivalent to [DNA]/[complex] ratio contained values of 8, 4, 2.7, 2 and 1.3). Solutions were equilibrated for 15 min at 25 °C, and, then, 5 flow times were registered. The relative viscosity of DNA in the absence (η_0) and presence (η) of complexes was calculated as $(\eta/\eta_0) = t - t_0/t_{DNA} - t_0$, where t_0 and t_{DNA} are the flow times of the buffer and DNA solution, respectively, and t is the flow time of the DNA solution in the presence of copper complexes. Data are presented as a plot of $(\eta/\eta_0)^{1/3}$ versus the ratio of [complex]/[DNA [62].

4.5. Cytotoxicity Studies

The cytotoxicity of the complexes was evaluated on human cancer cell lines: metastatic breast adenocarcinoma MDA-MB-231 (triple negative, ATCC: HTB-26), MCF-7 (ATCC: HTB-22), cisplatin-resistant ovarian carcinoma A2780cis (SIGMA), lung epithelial carcinoma A549 (ATCC: CCL-185) and on the nontumoral lung cell line MRC-5 (ATCC: CCL-171) using the MTT colorimetric assay. Cells were cultured in Dulbecco's Modified Eagle's Medium (DMEM) for MDA-MB-231, A549 and MRC-5, supplemented with 10% fetal bovine serum (FBS), Roswell Park Memorial Institute (RPMI) 1640 Medium for MCF-7 and A278cis, supplemented with 10% FBS, containing 1% penicillin and 1% streptomycin, at 310 K in a humidified 5% CO_2 atmosphere. In the assay, 1.5×10^4 cells/well were seeded in 150 μL of medium in 96-well plates and incubated at 310 K in 5% CO_2 for 24 h, to allow cell adhesion. Then cells were treated with copper complexes for 48 h. Cu complexes were dissolved in DMSO, and 0.75 μL of solution were added to each well with 150 μL of medium (final concentration of 0.5% DMSO/well). Cisplatin, used as a reference drug, was solubilized in DMF. Afterward, to detect cell viability, 3-(4,5-dimethylthiazol-2-yl)-2,5-diphenyltetrazolium bromide (MTT, 50 μL, 1 mg mL^{-1} in PBS) was added to each well, and the plate was further incubated for 4 h. Living cells reduce MTT to purple formazan. The formazan crystals were solubilized with isopropanol (150 μL/well), and each well was measured with a microplate spectrophotometer at a wavelength of 540 nm. The concentration to 50% (IC_{50}) of cell viability (Table) was obtained from the analysis of absorbance data from three independent experiments.

Supplementary Materials: The following supporting information can be downloaded at: https://www.mdpi.com/article/10.3390/inorganics11100398/s1, Figure S1: Scheme of complexes. Figures S2–S7: Infrared spectra of complexes **C1–C3** and ligands L1–L3; Figures S8 and S9: UV spectra of **C2** (S7) and **C3** (S8) with increasing amounts of DNA. Inset: [complex]/Aobs as a function of 1/[DNA] plot with regression parameters.

Author Contributions: Conceptualization, G.F.; methodology, G.F. and A.A.B.; validation, G.F.; formal analysis, N.A. and J.E.; investigation, C.Y.F. and A.R.; resources, G.F.; data curation, C.Y.F.

and N.A.; writing—original draft preparation, G.F.; writing—review and editing, G.F., N.A., M.A., K.M., J.E., A.A.B. and A.J.C.-F.; visualization, N.A.; supervision, G.F.; project administration, G.F.; funding acquisition, C.Y.F., G.F.,M.A., A.J.C.-F., J.E. and A.A.B. All authors have read and agreed to the published version of the manuscript.

Funding: This research was funded by the Comisión Sectorial de Investigación Científica and Comisión Sectorial de Posgrado (CSIC and CAP respectively, UdelaR, CSIC I+D Grant to G. Facchin, CAP Grant to C.Y.F.), Programa de Desarrollo de las Ciencias Básicas (PEDECIBA Química) and Agencia Nacional de Investigación e Innovación (ANII), Uruguay and Fundação de Amparo à Pesquisa do Estado de São Paulo (FAPESP, Grant no. 2015/50366-7 and 2020/15542-7) and Conselho Nacional de Desenvolvimento Científico e Tecnológico (CNPq, Grants no 306682/2018-4 and 312505/2021-3), Brazil and Researchers Supporting Project number (RSP2023R147), King Saud University, Riyadh, Saudi Arabia.

Institutional Review Board Statement: All the cell lines were obtained from Banco de Células do Rio de Janeiro, Parque Tecnológico de Xerém—Av. Nossa Sra. das Graças, 50—Vila Nossa Sra. das Gracas, Duque de Caxias—RJ, 25250-020, Brazil, which in turn were obtained from the commercial lines stated in the materials and methods section.

Data Availability Statement: The data presented in this study are available in the article and Supplementary Material.

Acknowledgments: The authors acknowledge all the participant institutions. M.A. acknowledges the financial support through the Researchers Supporting Project number (RSP2023R147), King Saud University, Riyadh, Saudi Arabia.

Conflicts of Interest: The authors declare no conflict of interest.

References

1. Romani, A.M.P. Cisplatin in cancer treatment. *Biochem. Pharmacol.* **2022**, *206*, 115323. [CrossRef]
2. Kellett, A.; Molphy, Z.; McKee, V.; Slator, C. Recent Advances in Anticancer Copper Compounds. In *Metal-Based Anticancer Agents*; Royal Society of Chemistry: London, UK, 2019; pp. 91–119.
3. Santini, C.; Pellei, M.; Gandin, V.; Porchia, M.; Tisato, F.; Marzano, C. Advances in Copper Complexes as Anticancer Agents. *Chem. Rev.* **2014**, *114*, 815–862. [CrossRef] [PubMed]
4. Zehra, S.; Tabassum, S.; Arjmand, F. Biochemical pathways of copper complexes: Progress over the past 5 years. *Drug Discov. Today* **2021**, *26*, 1086–1096. [CrossRef] [PubMed]
5. Krasnovskaya, O.; Naumov, A.; Guk, D.; Gorelkin, P.; Erofeev, A.; Beloglazkina, E.; Majouga, A. Copper Coordination Compounds as Biologically Active Agents. *Int. J. Mol. Sci.* **2020**, *21*, 3965. [CrossRef] [PubMed]
6. McGivern, T.; Afsharpour, S.; Marmion, C. Copper Complexes as Artificial DNA Metallonucleases: From Sigman's Reagent to Next Generation Anti-Cancer Agent? *Inorg. Chim. Acta* **2018**, *472*, 12–39. [CrossRef]
7. Shi, X.; Chen, Z.; Wang, Y.; Guo, Z.; Wang, X. Hypotoxic copper complexes with potent anti-metastatic and anti-angiogenic activities against cancer cells. *Dalton Trans.* **2018**, *47*, 5049–5054. [CrossRef] [PubMed]
8. Laws, K.; Suntharalingam, K. The next generation of anticancer metallopharmaceuticals: Cancer stem cell-active inorganics. *ChemBioChem* **2018**, *19*, 2246–2253. [CrossRef]
9. Kaur, P.; Johnson, A.; Northcote-Smith, J.; Lu, C.; Suntharalingam, K. Immunogenic Cell Death of Breast Cancer Stem Cells Induced by an Endoplasmic Reticulum-Targeting Copper(II) Complex. *ChemBioChem* **2020**, *21*, 3618–3624. [CrossRef]
10. Shi, X.; Fang, H.; Guo, Y.; Yuan, H.; Guo, Z.; Wang, X. Anticancer copper complex with nucleus, mitochondrion and cyclooxygenase-2 as multiple targets. *J. Inorg. Biochem.* **2019**, *190*, 38–44. [CrossRef]
11. Chang, M.R.; Rusanov, D.A.; Arakelyan, J.; Alshehri, M.; Asaturova, A.V.; Kireeva, G.S.; Babak, M.V.; Ang, W.H. Targeting emerging cancer hallmarks by transition metal complexes: Cancer stem cells and tumor microbiome. Part I. *Coord. Chem. Rev.* **2023**, *477*, 214923. [CrossRef]
12. Balsa, L.M.; Ruiz, M.C.; Santa Maria de la Parra, L.; Baran, E.J.; León, I.E. Anticancer and antimetastatic activity of copper(II)-tropolone complex against human breast cancer cells, breast multicellular spheroids and mammospheres. *J. Inorg. Biochem.* **2020**, *204*, 110975. [CrossRef] [PubMed]
13. da Silva, D.A.; De Luca, A.; Squitti, R.; Rongioletti, M.; Rossi, L.; Machado, C.M.L.; Cerchiaro, G. Copper in tumors and the use of copper-based compounds in cancer treatment. *J. Inorg. Biochem.* **2022**, *226*, 111634. [CrossRef] [PubMed]
14. Ruiz, M.C.; Perelmulter, K.; Levín, P.; Romo, A.I.B.; Lemus, L.; Fogolín, M.B.; León, I.E.; Di Virgilio, A.L. Antiproliferative activity of two copper (II) complexes on colorectal cancer cell models: Impact on ROS production, apoptosis induction and NF-κB inhibition. *Eur. J. Pharm. Sci.* **2022**, *169*, 106092. [CrossRef] [PubMed]
15. Figueroa-DePaz, Y.; Resendiz-Acevedo, K.; Davila-Manzanilla, S.G.; Garcia-Ramos, J.C.; Ortiz-Frade, L.; Serment-Guerrero, J.; Ruiz-Azuara, L. DNA, a target of mixed chelate copper(II) compounds (Casiopeinas(R)) studied by electrophoresis, UV-vis and circular dichroism techniques. *J. Inorg. Biochem.* **2022**, *231*, 111772. [CrossRef]

16. Peña, Q.; Sciortino, G.; Maréchal, J.-D.; Bertaina, S.; Simaan, A.J.; Lorenzo, J.; Capdevila, M.; Bayón, P.; Iranzo, O.; Palacios, Ò. Copper (II) N, N, O-Chelating Complexes as Potential Anticancer Agents. *Inorg. Chem.* **2021**, *60*, 2939–2952. [CrossRef]
17. Fernández, C.Y.; Alvarez, N.; Rocha, A.; Ellena, J.; Costa-Filho, A.J.; Batista, A.A.; Facchin, G. New Copper(II)-L-Dipeptide-Bathophenanthroline Complexes as Potential Anticancer Agents—Synthesis, Characterization and Cytotoxicity Studies—And Comparative DNA-Binding Study of Related Phen Complexes. *Molecules* **2023**, *28*, 896.
18. Alvarez, N.; Rocha, A.; Collazo, V.; Ellena, J.; Costa-Filho, A.J.; Batista, A.A.; Facchin, G. Development of Copper Complexes with Diimines and Dipicolinate as Anticancer Cytotoxic Agents. *Pharmaceutics* **2023**, *15*, 1345. [CrossRef]
19. Alvarez, N.; Leite, C.M.; Napoleone, A.; Mendes, L.F.S.; Fernandez, C.Y.; Ribeiro, R.R.; Ellena, J.; Batista, A.A.; Costa-Filho, A.J.; Facchin, G. Tetramethyl-phenanthroline copper complexes in the development of drugs to treat cancer: Synthesis, characterization and cytotoxicity studies of a series of copper(II)-L-dipeptide-3,4,7,8-tetramethyl-phenanthroline complexes. *J. Biol. Inorg. Chem.* **2022**, *27*, 431–441. [CrossRef]
20. Alvarez, N.; Viña, D.; Leite, C.M.; Mendes, L.F.; Batista, A.A.; Ellena, J.; Costa-Filho, A.J.; Facchin, G. Synthesis and structural characterization of a series of ternary copper (II)-L-dipeptide-neocuproine complexes. Study of their cytotoxicity against cancer cells including MDA-MB-231, triple negative breast cancer cells. *J. Inorg. Biochem.* **2020**, *203*, 110930. [CrossRef]
21. Alvarez, N.; Mendes, L.F.; Kramer, M.G.; Torre, M.H.; Costa-Filho, A.J.; Ellena, J.; Facchin, G. Development of Copper (II)-diimine-iminodiacetate mixed ligand complexes as potential antitumor agents. *Inorg. Chim. Acta* **2018**, *483*, 61–70. [CrossRef]
22. Alvarez, N.; Noble, C.; Torre, M.H.; Kremer, E.; Ellena, J.; Peres de Araujo, M.; Costa-Filho, A.J.; Mendes, L.F.; Kramer, M.G.; Facchin, G. Synthesis, structural characterization and cytotoxic activity against tumor cells of heteroleptic copper (I) complexes with aromatic diimines and phosphines. *Inorg. Chim. Acta* **2017**, *466*, 559–564. [CrossRef]
23. Iglesias, S.; Alvarez, N.; Kramer, G.; Torre, M.H.; Kremer, E.; Ellena, J.; Costa-Filho, A.J.; Facchin, G. Structural Characterization and Cytotoxic Activity of Heteroleptic Copper (II) Complexes with L-Dipeptides and 5-NO$_2$-Phenanthroline.: Crystal Structure of [Cu (Phe-Ala)(5-NO$_2$-Phen)]·4H$_2$O. *Struct. Chem. Crystallogr. Commun.* **2015**, *1*, 1–7.
24. Iglesias, S.; Alvarez, N.; Torre, M.H.; Kremer, E.; Ellena, J.; Ribeiro, R.R.; Barroso, R.P.; Costa-Filho, A.J.; Kramer, M.G.; Facchin, G. Synthesis, structural characterization and cytotoxic activity of ternary copper (II)–dipeptide–phenanthroline complexes. A step towards the development of new copper compounds for the treatment of cancer. *J. Inorg. Biochem.* **2014**, *139*, 117–123. [CrossRef] [PubMed]
25. Costa-Filho, A.J.; Nascimento, O.R.; Calvo, R. Electron paramagnetic resonance study of weak exchange interactions between metal ions in a model system: CuIIGly-Trp. *J. Phys. Chem. B* **2004**, *108*, 9549–9555. [CrossRef]
26. Balzano, T. Active Clinical Trials in Hepatic Encephalopathy: Something Old, Something New and Something Borrowed. *Neurochem. Res.* **2023**, *48*, 2309–2319. [CrossRef] [PubMed]
27. Iqbal, M.; Ali, S.; Tahir, M.N. Polymeric Copper(II) Paddlewheel Carboxylate: Structural Description, Electrochemistry, and DNA-binding Studies. *Z. Für Anorg. Und Allg. Chem.* **2018**, *644*, 172–179. [CrossRef]
28. Benslimane, M.; Redjel, Y.K.; Merazig, H.; Daran, J.-C. catena-Poly[bis([mu]3-2-phenylacetato-[kappa]3O,O′:O)bis([mu]2-2-phenylacetato-[kappa]2O:O′)dicopper(II)(Cu-Cu)]. *Acta Crystallogr. Sect. E* **2013**, *69*, m397. [CrossRef]
29. Rap, V.M.; Manohar, H. Synthesis and crystal structure of methanol and acetic acid adducts of copper acetate. Predominance of σ-interaction between the two copper atoms in the dimer. *Inorganica Chim. Acta* **1979**, *34*, L213–L214. [CrossRef]
30. Kanazawa, Y.; Mitsudome, T.; Takaya, H.; Hirano, M. Pd/Cu-Catalyzed Dehydrogenative Coupling of Dimethyl Phthalate: Synchrotron Radiation Sheds Light on the Cu Cycle Mechanism. *ACS Catal.* **2020**, *10*, 5909–5919. [CrossRef]
31. Kendin, M.; Nikiforov, A.; Svetogorov, R.; Degtyarenko, P.; Tsymbarenko, D. A 3D-Coordination Polymer Assembled from Copper Propionate Paddlewheels and Potassium Propionate 1D-Polymeric Rods Possessing a Temperature-Driven Single-Crystal-to-Single-Crystal Phase Transition. *Cryst. Growth Des.* **2021**, *21*, 6183–6194. [CrossRef]
32. Jassal, A.K.; Sharma, S.; Hundal, G.; Hundal, M.S. Structural Diversity, Thermal Studies, and Luminescent Properties of Metal Complexes of Dinitrobenzoates: A Single Crystal to Single Crystal Transformation from Dimeric to Polymeric Complex of Copper(II). *Cryst. Growth Des.* **2015**, *15*, 79–93. [CrossRef]
33. Udupa, M.R.; Krebs, B. Crystal and molecular structure of tetra-µ-N-acetylglycinatodiaquodicopper(II). *Inorganica Chim. Acta* **1979**, *37*, 1–4. [CrossRef]
34. Groom, C.R.; Bruno, I.J.; Lightfoot, M.P.; Ward, S.C. The Cambridge Structural Database. *Acta Crystallogr. Sect. B* **2016**, *72*, 171–179. [CrossRef] [PubMed]
35. Bruno, I.J.; Cole, J.C.; Edgington, P.R.; Kessler, M.; Macrae, C.F.; McCabe, P.; Pearson, J.; Taylor, R. New software for searching the Cambridge Structural Database and visualizing crystal structures. *Acta Crystallogr. Sect. B* **2002**, *58*, 389–397. [CrossRef]
36. Macrae, C.F.; Sovago, I.; Cottrell, S.J.; Galek, P.T.A.; McCabe, P.; Pidcock, E.; Platings, M.; Shields, G.P.; Stevens, J.S.; Towler, M.; et al. Mercury 4.0: From visualization to analysis, design and prediction. *J. Appl. Crystallogr.* **2020**, *53*, 226–235. [CrossRef]
37. Vaughan, G.B.M.; Schmidt, S.; Poulsen, H.F. Multicrystal approach to crystal structure solution and refinement. *Z. Für Krist. Cryst. Mater.* **2004**, *219*, 813–825. [CrossRef]
38. Wang, Y.Y.; Shi, Q.; Shi, Q.Z.; Gao, Y.C. Self-assembly of porous two-dimensional copper(II) α, β-unsaturated carboxylate complexes with trimethyl phosphate. *Transit. Met. Chem.* **2000**, *25*, 382–387. [CrossRef]
39. Pathak, S.; Biswas, N.; Jana, B.; Ghorai, T.K. Synthesis and characterization of a nano Cu2 cluster. *Adv. Mater. Proc.* **2017**, *2*, 275–279.
40. Li, J.-R.; Yu, J.; Lu, W.; Sun, L.-B.; Sculley, J.; Balbuena, P.B.; Zhou, H.-C. Porous materials with pre-designed single-molecule traps for CO2 selective adsorption. *Nat. Commun.* **2013**, *4*, 1538. [CrossRef]

41. Kristiansson, O.; Tergenius, L.-E. Structure and host–guest properties of the nanoporous diaquatetrakis(p-nitrobenzoato)dicopper(II) framework. *J. Chem. Soc. Dalton Trans.* **2001**, *9*, 1415–1420. [CrossRef]
42. Desiraju, G.R. A Bond by Any Other Name. *Angew. Chem. Int. Ed.* **2011**, *50*, 52–59. [CrossRef] [PubMed]
43. de la Flor, G.; Orobengoa, D.; Tasci, E.; Perez-Mato, J.M.; Aroyo, M.I. Comparison of structures applying the tools available at the Bilbao Crystallographic Server. *J. Appl. Crystallogr.* **2016**, *49*, 653–664. [CrossRef]
44. Doyle, A.; Felcman, J.; Gambardella, M.T.d.P.; Verani, C.N.; Tristão, M.L.B. Anhydrous copper(II) hexanoate from cuprous and cupric oxides. The crystal and molecular structure of $Cu_2(O_2CC_5H_{11})_4$. *Polyhedron* **2000**, *19*, 2621–2627. [CrossRef]
45. Katzsch, F.; Münch, A.S.; Mertens, F.O.R.L.; Weber, E. Copper(II) benzoate dimers coordinated by different linear alcohols—A systematic study of crystal structures. *J. Mol. Struct.* **2014**, *1064*, 122–129. [CrossRef]
46. Krause, L.; Herbst-Irmer, R.; Stalke, D. An empirical correction for the influence of low-energy contamination. *J. Appl. Crystallogr.* **2015**, *48*, 1907–1913. [CrossRef]
47. Sheng, G.-H.; Zhou, Q.-C.; Sun, J.; Cheng, X.-S.; Qian, S.-S.; Zhang, C.-Y.; You, Z.-L.; Zhu, H.-L. Synthesis, structure, and urease inhibitory activities of three binuclear copper(II) complexes with protocatechuic acid derivative. *J. Coord. Chem.* **2014**, *67*, 1265–1278. [CrossRef]
48. Massignani, S.; Scatena, R.; Lanza, A.; Monari, M.; Condello, F.; Nestola, F.; Pettinari, C.; Zorzi, F.; Pandolfo, L. Coordination polymers from mild condition reactions of copper(II) carboxylates with pyrazole (Hpz). Influence of carboxylate basicity on the self-assembly of the $[Cu_3(\mu_3\text{-OH})(\mu\text{-pz})_3]^{2+}$ secondary building unit. *Inorg. Chim. Acta* **2017**, *455*, 618–626. [CrossRef]
49. Nakamoto, K. *Infrared and Raman Spectra of Inorganic and Coordination Compounds, Applications in Coordination, Organometallic, and Bioinorganic Chemistry*, 6th ed.; Wiley-Interscience: Hoboken, NJ, USA, 2009.
50. Prenesti, E.; Daniele, P.; Prencipe, M.; Ostacoli, G. Spectrum–structure correlation for visible absorption spectra of copper (II) complexes in aqueous solution. *Polyhedron* **1999**, *18*, 3233–3241. [CrossRef]
51. Prenesti, E.; Daniele, P.G.; Berto, S.; Toso, S. Spectrum–structure correlation for visible absorption spectra of copper (II) complexes showing axial co-ordination in aqueous solution. *Polyhedron* **2006**, *25*, 2815–2823. [CrossRef]
52. Karaliota, A.; Kretsi, O.; Tzougraki, C. Synthesis and characterization of a binuclear coumarin-3-carboxylate copper(II) complex. *J. Inorg. Biochem.* **2001**, *84*, 33–37. [CrossRef]
53. Bhirud, R.G.; Srivastava, T.S. Synthesis, characterization and superoxide dismutase activity of some ternary copper (II) dipeptide-2, 2′-bipyridine, 1, 10-phenanthroline and 2, 9-dimethyl-1, 10-phenanthroline complexes. *Inorg. Chim. Acta* **1991**, *179*, 125–131. [CrossRef]
54. Muhammad, N.; Ikram, M.; Perveen, F.; Ibrahim, M.; Ibrahim, M.; Abel; Viola; Rehman, S.; Shujah, S.; Khan, W.; et al. Syntheses, crystal structures and DNA binding potential of copper(II) carboxylates. *J. Mol. Struct.* **2019**, *1196*, 771–782. [CrossRef]
55. Iqbal, M.; Ali, S.; Tahir, M.N.; Haleem, M.A.; Gulab, H.; Shah, N.A. A binary copper(II) complex having a stepped polymeric structure: Synthesis, characterization, DNA-binding and anti-fungal studies. *J. Serb. Chem. Soc.* **2020**, *85*, 203–214. [CrossRef]
56. Suh, D.; Chaires, J.B. Criteria for the mode of binding of DNA binding agents. *Bioorg. Med. Chem.* **1995**, *3*, 723–728. [CrossRef]
57. Sheldrick, G. SHELXT—Integrated space-group and crystal-structure determination. *Acta Crystallogr. Sect. A* **2015**, *71*, 3–8. [CrossRef]
58. Sheldrick, G. Crystal structure refinement with SHELXL. *Acta Crystallogr. Sect. C* **2015**, *71*, 3–8. [CrossRef] [PubMed]
59. Dolomanov, O.V.; Bourhis, L.J.; Gildea, R.J.; Howard, J.A.K.; Puschmann, H. OLEX2: A complete structure solution, refinement and analysis program. *J. Appl. Crystallogr.* **2009**, *42*, 339–341. [CrossRef]
60. Benesi, H.A.; Hildebrand, J.H. A Spectrophotometric Investigation of the Interaction of Iodine with Aromatic Hydrocarbons. *J. Am. Chem. Soc.* **2002**, *71*, 2703–2707. [CrossRef]
61. Sirajuddin, M.; Ali, S.; Badshah, A. DRUG-DNA Interactions and their study by UV-visible, fluorescence spectroscopies and cyclic voltametry. *J. Photochem. Photobiol. B Biol.* **2013**, *124*, 1–19. [CrossRef]
62. Scruggs, R.L.; Ross, P.D. Viscosity study of DNA. *Biopolymers* **1964**, *2*, 593–609. [CrossRef]

Disclaimer/Publisher's Note: The statements, opinions and data contained in all publications are solely those of the individual author(s) and contributor(s) and not of MDPI and/or the editor(s). MDPI and/or the editor(s) disclaim responsibility for any injury to people or property resulting from any ideas, methods, instructions or products referred to in the content.

Article

Anti-Proliferation and DNA Cleavage Activities of Copper(II) Complexes of N_3O Tripodal Polyamine Ligands

Doti Serre [1], Sule Erbek [2], Nathalie Berthet [1], Christian Philouze [1], Xavier Ronot [2], Véronique Martel-Frachet [2] and Fabrice Thomas [1,*]

[1] Département de Chimie Moléculaire—UMR CNRS-5250, Université Grenoble Alpes, 38041 Grenoble, France; nathalie.berthet@univ-grenoble-alpes.fr (N.B.)

[2] Institute for Advanced Biology—INSERM-UGA U823, Site Santé, 38700 La Tronche, France; veronique.frachet@univ-grenoble-alpes.fr (V.M.-F.)

* Correspondence: fabrice.thomas@univ-grenoble-alpes.fr

Abstract: Four ligands based on the 2-*tert*-butyl-4-X-6-{Bis[(6-methoxy-pyridin-2-ylmethyl)-amino]-methyl}-phenol unit are synthesized: X = CHO (HL^{CHO}), putrescine-pyrene (HL^{pyr}), putrescine (HL^{amine}), and 2-*tert*-butyl-4-putrescine-6-{Bis[(6-methoxy-pyridin-2-ylmethyl)-amino]-methyl}-phenol (H$_2$L^{bis}). Complexes **1**, **2**, **3**, and **4** are formed upon chelation to copper(II). The crystal structure of complex **1** shows a square pyramidal copper center with a very weakly bound methoxypridine moiety in the apical position. The pKa of the phenol moiety is determined spectrophotometrically at 2.82–4.39. All the complexes show a metal-centered reduction in their CV at $E_p^{c,red} = -0.45$ to -0.5 V vs. SCE. The copper complexes are efficient nucleases towards the ϕX174 DNA plasmid in the presence of ascorbate. The corresponding IC$_{50}$ value reaches 7 µM for **2**, with a nuclease activity that follows the trend: **2** > **3** > **1**. Strand scission is promoted by the hydroxyl radical. The cytotoxicity is evaluated on bladder cancer cell lines sensitive (RT112) or resistant to cisplatin (RT112 CP). The IC$_{50}$ of the most active complexes (**2** and **4**) is 1.2 and 1.0 µM, respectively, for the RT112 CP line, which is much lower than cisplatin (23.8 µM).

Keywords: copper; tripodal ligand; nuclease; DNA; phenol

Citation: Serre, D.; Erbek, S.; Berthet, N.; Philouze, C.; Ronot, X.; Martel-Frachet, V.; Thomas, F. Anti-Proliferation and DNA Cleavage Activities of Copper(II) Complexes of N$_3$O Tripodal Polyamine Ligands. *Inorganics* **2023**, *11*, 396. https://doi.org/10.3390/inorganics11100396

Academic Editors: Gianella Facchin and Vladimir Arion

Received: 28 August 2023
Revised: 18 September 2023
Accepted: 25 September 2023
Published: 9 October 2023

Copyright: © 2023 by the authors. Licensee MDPI, Basel, Switzerland. This article is an open access article distributed under the terms and conditions of the Creative Commons Attribution (CC BY) license (https://creativecommons.org/licenses/by/4.0/).

1. Introduction

Nucleases are important tools for manipulating genes and designing new chemotherapeutic agents [1]. Inorganic complexes are particularly well-suited for this purpose due to the intrinsic properties of the metal ion. Its acidity facilitates the activation of substrates, and hence favors hydrolytic processes [2,3]. On the other hand, redox-active metals may reduce molecular dioxygen into reactive oxygen species (ROS: O$_2^{2-}$, O$_2^{-}$ and OH), which subsequently cleave DNA in an oxidative pathway [2,3].

Numbers of artificial nucleases based on biologically relevant metal ions (manganese, copper, iron) have been developed [3,4]. Assuming that biological metals are less toxic than non-biological ones, these complexes may be used as therapeutic agents, in particular as an alternative to cisplatin in chemotherapy [4,5]. Amongst the biometals, copper has emerged as an attractive target for designing nucleases having an anti-cancer activity [4–10]. Indeed, copper is an essential cofactor for tumor angiogenesis [11], while correlations exist between the copper status and both malignant progression [12,13] and response to therapy in some human cancers [14]. Furthermore, copper(II) is capable of favoring nucleophilic attack at the DNA phosphates, while copper(I) reacts with dioxygen, affording ROS that is potent for strand scission in vitro [15].

The ligand platform plays a pivotal role in modulating the properties of the metal and its interaction (mode and strength) with DNA. Polypyridyl ligands have been widely used for designing copper nucleases because pyridines can intercalate into DNA and give rise to Cu^{II}/Cu^{I} redox potentials adequate for oxidative cleavage when engaged in coordination [16,17]. They are mostly based on bipyridine [18–20], phenanthroline [18,19,21–27], terpyridine [28], phenanthrene [1,29], and tripodal structures [17,30–33]. These moieties can be further engineered for adding functionalities [34,35]. On the other hand, phenol(ate) donors, especially sterically hindered ones, can be considered as phenoxyl radical precursors, which are efficient moieties for H-atom abstraction [36–38]. When combined with tripodal structures, these radicals may form and be involved in DNA strand scission through hydrogen atom transfer [39–42].

We previously designed copper nucleases from N$_3$O tripodal ligands featuring one sterically hindered phenol and two pyridines. It is striking to note that the nuclease activity is greatly improved when one pyridine has a methyl substituent in the α position [30]. This group is both electron-donating and sterically demanding. Although these properties are key points for complex activity, the relative influence of the two on nuclease activity is currently unknown. We have further demonstrated that both nuclease activity and anti-proliferative activity against bladder tumor cell lines can be enhanced by the addition of a putrescine chain and a pyrene moiety [43,44]. This functionalization pattern increases the affinity for DNA by both ionic interactions between the putrescine chain and the negatively charged phosphates, and intercalation of the pyrene moiety between consecutive base pairs. These encouraging results prompted us to investigate the role of the pyridine α-substituent in both the nuclease and anti-proliferative activities. We herein describe a new series of ligands derived from HL (Figure 1), in which the methyl substituent of the pyridine is replaced by a highly electron donating methoxy group. With this substitution pattern, we aim to increase the electron density at the metal, while decreasing the steric hindrance, and thus plan to decipher their respective importance in biological activity. This work is also a prelude to the insertion of PEG units on pyridine to further improve drug pharmacokinetics and bioavailability.

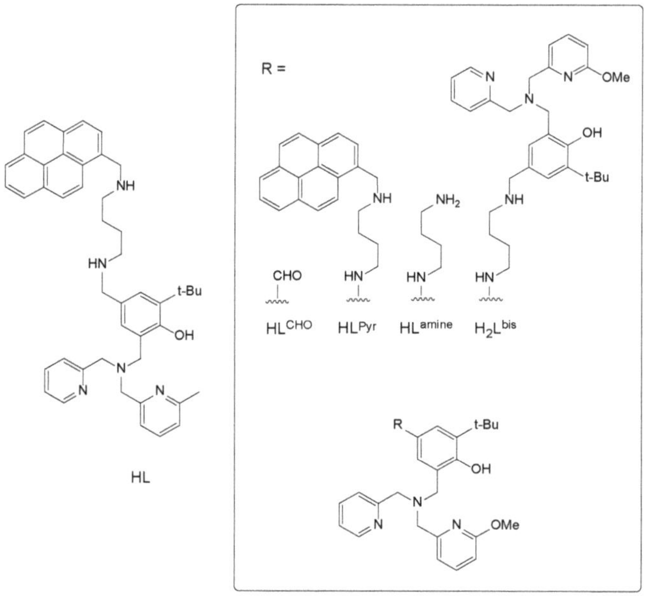

Figure 1. Nomenclature of the ligands under their neutral amino forms.

We present the synthesis of the ligands HL^{CHO}, HL^{Pyr}, HL^{amine}, and H$_2$L^{bis} (Figure 1), as well as the solution chemistry and spectroscopy of their copper complexes (**1**, **2**, **3**, and **4**, respectively). We investigated their nuclease activity and anti-proliferative activity towards bladder tumor cells.

2. Results and Discussion
2.1. Synthesis and Structure of the Complexes

The preparation of the series of ligands starts with the synthesis of the aldehyde precursor HL^{CHO} (Figure 1), which is next engaged in a reductive amination [43]. When N^1-(pyren-1-ylmethyl)butane-1,4-diamine was reacted with HL^{CHO}, the ligand HL^{pyr} was obtained. By using the *tert*-butyl-N-(4-aminobutyl)carbamate instead of N^1-(pyren-1-ylmethyl)butane-1,4-diamine, followed by acid deprotection, the ligand HL^{amine} was obtained instead. Finally, the reaction of neutral HL^{amine} with one molar equivalent of HL^{CHO} affords the binucleating ligands H$_2$L^{bis} (Figure 1). The mononuclear copper-phenolate complexes **1–3** were prepared in situ by mixing stoichiometric amounts of CuCl$_2$ 2H$_2$O, NEt$_3$, and the appropriate ligand in DMF (Scheme 1); the binuclear copper-phenolate complex **4** was prepared similarly by using two molar equivalents of metal salt instead of a single one. Both the nuclearity and nature of the complexes was established by mass spectrometry, UV-Vis, and EPR spectroscopies (see below). The mass spectra of the complexes are depicted in Figure S9. The spectrum of **1** shows a main peak corresponding to a monocation at m/z = 481.1421. It shows the expected pattern for a copper containing complex and corresponds to [**1**]$^+$ (calculated m/z = 481.1410). The pyrene-appended complex **2** exhibits a main peak at m/z = 384.1660 (dication), which is assigned to [**2** + H]$^{2+}$, while complex **3** shows a main peak at m/z = 277.1268 assigned to [**3** + H]$^{2+}$. For comparison, complex **4** demonstrates a peak at m/z = 1020.4, corresponding to [**4** + 2H]$^+$, without evidence for a monometalated species. The (2:1) (M:L) stoichiometry of **4** was further confirmed by the Jobs method (Figure S13).

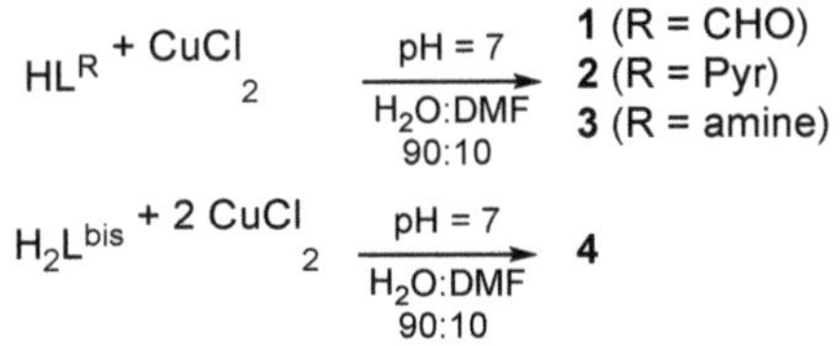

```
HL^R + CuCl_2    pH = 7         1 (R = CHO)
              ----------->      2 (R = Pyr)
               H_2O:DMF          3 (R = amine)
                 90:10

H_2L^bis + 2 CuCl_2   pH = 7
                    ----------->   4
                     H_2O:DMF
                       90:10
```

Scheme 1. Nomenclature of the complexes and their schematized synthesis. Complex **1** was isolated as single crystals; **2–4** were generated in situ for physical and biological characterizations.

The structure of **1** was substantiated by X-Ray diffraction (Figure 2). The other complexes proved to be difficult to isolate at the solid state, presumably because of the hygroscopic nature of the compounds (due to the presence of secondary amines) and the large flexibility of the putrescine arm. The structure of **1** shows a square pyramidal copper center where the tertiary amine nitrogen N3, the pyridine nitrogen N2, the phenolate oxygen O1, and the chloride Cl coordinate in the basal plane. The methoxypyridine is positioned apically, but this group is almost uncoordinated (Cu-N1 bond distance of 2.773(5) Å). This behavior is in sharp contrast with the methylpyridine analog of ^{Me}**1** that binds apically at a much shorter Cu-N1 bond distance of 2.303(2) Å [44]. The Cu-N3, Cu-N2, Cu-O1, and Cu-Cl bond distances are 2.034(3), 2.017(3), 1.902(3), and 2.251(1) Å, respectively. They are significantly shorter than in ^{Me}**1**, as a result of the lengthening of the axial Cu-N1 bond. The angle between the pyridine (equatorial) and substituted pyridine (axial) rings differs significantly between ^{Me}**1** and **1**; although 70° in ^{Me}**1**, it does not exceed 36° in the case of **1**. Furthermore, the distance between the centroids of these two rings reaches 4.21 Å for ^{Me}**1**, while it is only 3.80 Å for **1**. These structural features indicate that the more electron-rich methoxypyridine ring has a higher propensity to stack on the equatorial pyridine, weakening axial coordination.

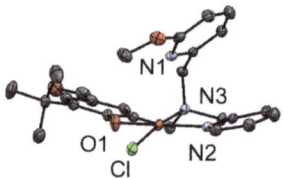

Figure 2. X-Ray crystal structure of **1**. The hydrogen atoms are omitted for clarity.

2.2. Spectroscopic Characterization

The complexes were formed in situ by mixing equimolar amounts of copper(II) chloride and the ligand (**1**, **2**, **3**) or two molar equivalents of the copper salt with the ligand (**4**). It is worth noting that a solution of single crystals of **1** shows the same spectroscopic signature than the in situ generated complex.

The EPR spectra of **1**–**4** show a clean axial ($S = \frac{1}{2}$) signal at pH 7 (Figures S11 and S12), disclosing the formation of a unique species in every case. Importantly, the intensity of the signal is identical for **1**–**3**. On the basis of the similarity of the donor set between HL, HL^{Pyr}, and HL^{amine} and given the resolution of the crystal structure of **1** (see above), we assume that the complexes formed in all three cases are mononuclear complexes. The spectrum of **4** is similar to that of **3**, although its intensity is twice as large. The magnetic interaction between the two copper centers is therefore negligible, which is not surprising given the length and flexibility of the linker connecting the tripodal units. Note that the shape of the spectra is similar for **2**–**4**, consistent with an identical N$_3$O coordination sphere, and combined with a similar electron-donating effect of the phenolic para-substituent.

The experimental spectra at arbitrary chosen pH values, which correspond to pKa + 1 and pKa − 1 (see below) where the acidobasic function is the phenolic hydroxyl, together with simulations are depicted in Figures 3 and S13–S15. A general trend is that $g_\perp$ is smaller than $g_{//}$, as expected for a d_{x2-y2} ground spin state. This ground state is consistent with the square pyramidal geometry of the copper(II) ion in the solid state structure. The spin Hamiltonian parameters differ significantly between pH = pKa−1 (1.80–3.34, for **1**–**4** see Table 1) and pH = pKa + 1 (3.80–5.35, for **1**–**4** see Table 1) for a given complex in the non–buffered medium. This suggests a pH-induced change in metal ion geometry and thus the existence of an acidobasic equilibrium involving a coordinating moiety, which we have attributed to the phenol/phenolate couple. On the other hand, the $g_{//}/A_{//}$ ratio has previously been correlated with copper ion distortion [45]. The $g_{//}/A_{//}$ ratio is 128 cm at low pH (phenol complex), corresponding to a main square planar geometry (very weak apical coordination), whereas it is 132–136 cm at high pH (phenolate). The copper environment is therefore more distorted in deprotonated complexes, which is likely a consequence of the stronger binding of the deprotonated phenolic moiety. It is worth noting that the EPR parameters diverge slightly between pH = pKa + 1 (unbuffered) and pH = 7 (buffered), which can be attributed to the binding of a different exogenous ion (chloride or water molecule depending on NaCl concentration.

In order to confirm the nature of the acidobasic residue and determine precisely the pKa, we conducted UV-Vis titrations of the four complexes over the pH range 2–7. The spectral evolution of **2** is depicted in Figure 4, while the other ones are shown in Figures S16–S18. Since the evolutions are similar, we will comment only on the spectra of **2**. Thus, the spectrum at pH = 6.94 is dominated by an intense band at 462 nm (Table 2), which is assigned to a phenolate-to-copper charge transfer (LMCT) transition (see TD-DFT section). As the pH decreases, the intensity of the LMCT decreases, indicating the protonation of the phenolate. At pH = 2.55, the 470 nm band has disappeared, while a low intensity band is still observed at 680 nm, corresponding to copper(II) d-d transitions.

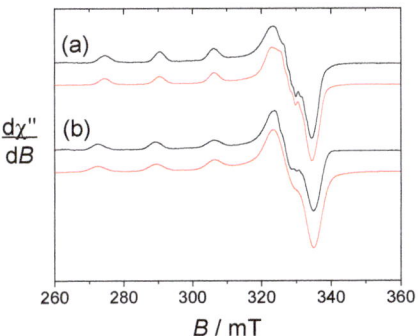

Figure 3. X-Band EPR spectra of 0.5 mM solutions of **2** in a (water:DMF) (90:10) medium at (**a**) pH = 5.15 and (**b**) pH = 3.11. Black: Experimental spectrum; Red: spectral simulation using the EASYSPIN 5.2.35 software and the parameters given in Table 1. [NaCl] = 0.1 M; T, 100 K; microwave frequency, 9.44 GHz; microwave power, 4 mW; mod. Freq., 100 KHz; mod. Amp. 0.4 mT.

Table 1. Spin Hamiltonian parameters for the copper complexes.

Complex	pH	$g_\perp$	$g_\parallel$	$A_\perp$	$A_\parallel$	g/A [c]
1	1.80 [a]	2.056	2.266	1.7	16.9	128
2	3.11 [a]	2.056	2.266	1.7	16.9	128
3	3.25 [a]	2.055	2.266	1.4	16.9	128
4	3.34 [a]	2.055	2.266	1.4	16.9	128
1	3.80 [a]	2.053	2.264	1.4	16.4	132
2	5.15 [a]	2.053	2.261	1.4	15.9	136
3	5.26 [a]	2.053	2.261	1.4	15.9	136
4	5.35 [a]	2.054	2.262	1.4	16.0	135
1	7 [b]	2.056	2.257	1.4	16.3	132
2	7 [b]	2.054	2.254	1.5	16.6	130
3	7 [b]	2.054	2.254	1.5	16.6	130
4	7 [b]	2.054	2.254	1.5	16.6	130

[a] In (H_2O:DMF) (90:10) frozen solution containing 0.1 M NaCl. $HClO_4$ was added to adjust the pH values to ca. pKa-1 and pKa + 1. [b] In (H_2O:DMF) (90:10) frozen solution. [Tris−HCl] = 0.05 M, [NaCl] = 0.02 M. [c] ratio expressed in cm.

Table 2. pKa values and electronic spectra of the copper complexes [a].

		Phenolate Form				Phenol Form [c]	
		Charge Transfer		d-d Band		d-d Band	
Complex	pKa	λ_{max}	ε	λ_{max}	ε	λ_{max}	ε
1	2.82 ± 0.02	482	662 [b]	680	170	680 [c]	<90 [c]
2	4.15 ± 0.01	462	1040	670	101	680	66
3	4.25 ± 0.01	461	780	670	232	687	147
4	8.77 ± 0.03 [d]	457	2005	670	328	687	147

[a] Fitted using the SPECFIT 3.0.37 software (from Biologic, Seyssinet-Pariset, France). The spectra were recorded in a (H_2O:DMF) (90:10) solution containing [NaCl] (0.1 M). $HClO_4$ or NaOH was added to adjust the pH. [b] Phenolate-to-copper charge transfer transition. [c] At the lowest pH investigated (1.8) the phenolate-to-copper charge transfer was still observable. [d] The value indicated corresponds to the $-\log\beta_2$ value associated to the equilibrium: $[(H_2L^{bis})Cu_2]^{4+} \rightleftharpoons 2H^+ + [(L^{bis})Cu_2]^{2+}$.

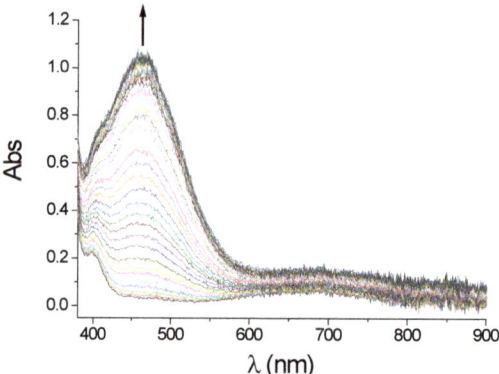

Figure 4. Electronic spectrum of **2** as a function of the pH. 1 mM (H$_2$O:DMF) (90:10) solution containing 0.1 M [NaCl], T = 298 K, l = cm. The pH is varied from 2.55 to 6.94, the arrow indicates spectral changes upon addition of base.

The phenol pKas were determined at 2.82 ± 0.02, 4.15 ± 0.01, 4.25 ± 0.01, and 8.77 ± 0.03 (Logβ for a two-proton process in the latter case) for **1**, **2**, **3**, and **4**, respectively. The lower value of **1** is explained by the electron-withdrawing properties of the aldehyde o-substituent. It is worth noting that the pKa values of **2**, **3**, and **4** (referred to one proton in the latter case, e.g., 4.38) are in a narrow range, consistent with a similar electronic effect of the phenol para substituent (same alkyl ammonium chain). It should be emphasized that these pKa values are 0.22 to 0.48 units higher than for analogous methyl complexes [44]. We interpret this increase in pKa by a decrease in Lewis acidity at the metal center due to the greater electron-donating capacity of the methoxy-substituted pyridine. This trend validates our working hypothesis that a change in the α-substituent of the pyridine would influence the electron density at the metal center. Importantly, all pKa values are much lower than those reported for complexes with axially bound phenolates [46,47], supporting an equatorial positioning of the phenolate in **1–4**. This interpretation is corroborated by the crystal structure of **1**.

2.3. Electrochemistry

The electrochemical behavior of the complexes has been investigated by cyclic voltammetry (CV) in a (H$_2$O:DMF) (90:10) solution containing 0.1 M NaCl as supporting electrolyte (Figures 5, S19 and S20). Redox potentials are summarized in Table 3. The CV curve for **1** demonstrates a cathodic peak at $E_p^{c,red}$ = −0.45 V, which is associated with an oxidation peak at $E_p^{a,red}$ = −0.14 V (red curve in Figure 5). This pair of semi-reversible peaks (ΔE = 0.31 V) is assigned to the Cu^{II}/Cu^{I} redox couple, in which significant structural rearrangements are associated with electron transfer. The apparent $E_{1/2}$ calculated as ($E_p^{c,red}$ + $E_p^{a,red}$)/2 is −0.30 V. An irreversible oxidation is also observed at $E_p^{a,ox}$ = +0.96 V and attributed to phenolate oxidation. The Cu^{II}/Cu^{I} redox couple in **2**, **3**, and **4** is again observed as a pair of separated peaks, similar to **1**. For **2**, the potentials values are $E_p^{c,red}$ = −0.51 V, $E_p^{a,red}$ = +0.16 V (ΔE = 0.67 V, apparent $E_{1/2}$ = −0.17 V), while for **4** they are $E_p^{c,red}$ = −0.47 V, $E_p^{a,red}$ = +0.07 V (ΔE = 0.54 V, apparent $E_{1/2}$ = −0.20 V). A broadening of the cathodic peak ($E_p^{c,red}$) is observed for **3**, presumably due to the presence of primary amines. This precludes a thorough comparison between **3** and the other complexes. The overall increase in ΔE in **2** and **4** compared with **1** reflects slower electron transfer, likely associated with greater structural rearrangements for the complexes appended by the alkylammonium chain. In addition, the significant anodic shift of $E_p^{a,red}$ (in comparison to **1**) together with small variations in $E_p^{c,red}$ suggest enhanced stability of the reduced form. Further comparison with previously reported methyl derivatives is unfortunately hampered by the broadening of the redox waves. Finally, two oxidation peaks are observed for **2**, **3**, and **4**. These are attributed to sequential oxidations of the phenolate unit [30,43,44,47–51] and an amine of the polyamine chain [43,44].

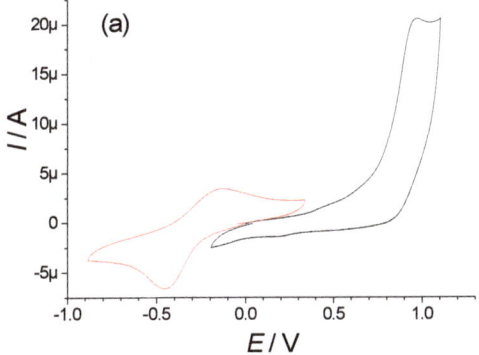

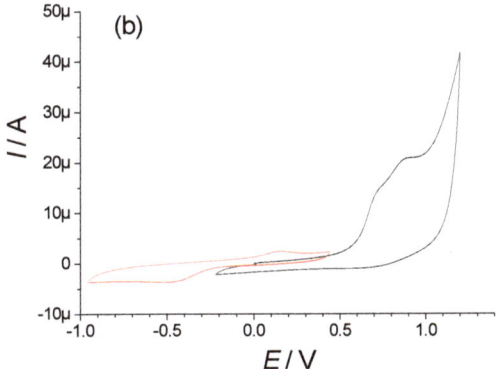

Figure 5. Cyclic voltammetry curves of (**a**) **1** and (**b**) **2** in a water:DMF 90:10 medium at pH = 7. [Tris−HCl] = 0.05 M, [NaCl] = 0.02 M; T, 298 K; scan rate, 0.1 V s^{-1}; The potentials are quoted relative to the SCE reference electrode. Red: scan in reduction; Black: scan in oxidation.

Table 3. Electrochemical data of the complexes at pH 7 [a].

Complex	$E_p^{c,red}$	$E_p^{a,red}$	$E_p^{a,ox}$
1	−0.45	−0.14	0.96
2	−0.51	+0.16	0.73, 0.88
3	−0.45	-[b]	0.69, 0.9 [b]
4	−0.47	+0.07	0.56, 0.73, 0.88
PyrPt [c]	-	-	0.86, 1.0 [b]

[a] In a (H$_2$O:DMF) (90:10) solution, [Tris−HCl] = 0.05 M, [NaCl] = 0.02 M. Reference, SCE; T, 298 K; Br: broad. [b] Ill-defined. [c] PyrPt: N^1-(pyren-1-ylmethyl)butane-1,4-diamine.

2.4. DFT and TD-DFT Calculations

In order to gain insight on the structures of the complexes and their UV-Vis signatures, we performed DFT calculations (Figure 6). In order to minimize computation time, we considered a model in which the para substituent of the phenol is a methyl group. We first optimized the structure of this model in its phenolate form with a chloride ligand. The methoxy group was arbitrary orientated either towards the N of the pyridine or away from the N. Strikingly, when orientated towards the N it creates significant steric hindrance that weakens the coordination of the methoxy pyridine (Cu-N distance of 2.80 Å). The almost uncoordinated methoxypyridine instead stacks up on the equatorial, unsubstituted pyridine. When the methoxy group is orientated away from the N, the steric hindrance is lower, allowing coordination of the methoxy pyridine (Cu-N bond distance of 2.29 Å), and no further stacking is observed. The most stable conformation is the first (by 3 kcal/mol), in agreement with the solid-state structure of **1** (with a chloride ligand). It is worth noting that a second conformation has been identified (1 kcal/mol less stable) in which the tertiary amine occupies the axial position. When the chloride ligand is replaced by a water molecule, a different behavior is observed, in which the methoxy oxygen establishes a H-bond with a hydrogen of the coordinated water molecule. As a result, the conformation in which the methoxy group points away from the pyridine nitrogen is favored. We have also optimized the structure of the phenol complexes, in which this latter group is located apically due to its lower coordinating capacity. Both orientations of the methoxy group lead to isoenergetic structures when the exogenous ligand is water, but the structure in which the methoxy points away from the N is more stabilized for the chloro complex (by 4.8 kcal/mol). For the lowest energy conformations (phenol and phenolate, chloro, and water adducts), we calculated the UV-Vis spectra by TD-DFT calculations. For the phenolate complexes, the calculations predict a relatively intense phenolate-to-copper charge transfer in the visible region, and the agreement between theory and experiment is excellent when the water adduct is taken in account (Table 4). For the phenol complex, a low-intensity d-d band is predicted in the low-energy range (Table 4), and once again agreement between theory and experiment is excellent when the water adduct is considered.

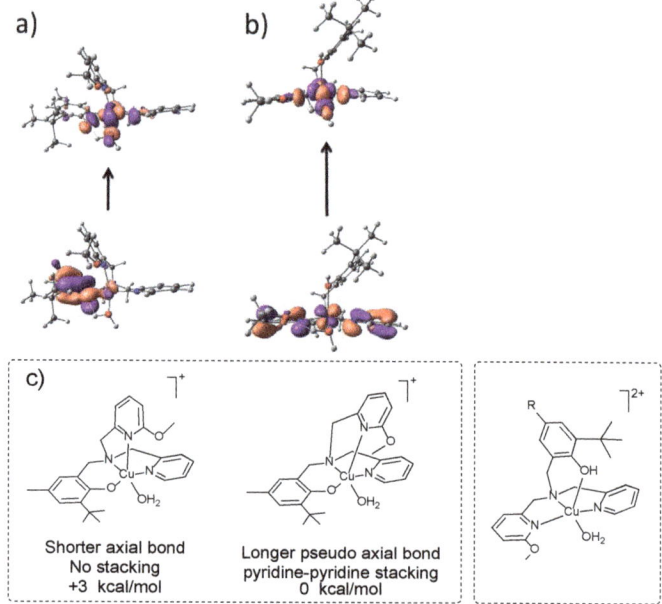

Figure 6. TD-DFT assignment of the main electronic excitations in the (**a**) phenolate (βHOMO → βLUMO) and (**b**) phenol (βHOMO-5 → βLUMO) complexes where the exogenous ligand is a water molecule. (**c**) Structures of the model complexes used for the calculations (B3LYP/6–31g*/SCRF, contour value: 0.040).

Table 4. TD-DFT assignment of the main visible band [a].

Protonation State	Exogenous Ligand	Main Contribution to the Electronic Excitation	λ_{cald} [nm] (f)
Phenolate	chloride	βHOMO → βLUMO (LMCT)	510 (0.0376)
Phenolate	water	βHOMO → βLUMO (LMCT)	563 (0.0646)
Phenol	chloride	βHOMO-7 → βLUMO (LMCT)	855 (0.0031)
Phenol	water	βHOMO-5 → βLUMO (dd)	661 (0.0012)

[a] B3LYP/6–31g*/SCRF (water). Most stable conformation for the phenolate complex.

2.5. DNA Cleavage

We investigated the nuclease activity of the complexes towards plasmidic supercoiled DNA. The reaction was monitored by gel electrophoresis in a medium constituted by a mixture (water:DMF) (90:10) in a phosphate buffer (10 mM) at pH = 7.2. Cleavage of the closed circular supercoiled (SC) plasmidic DNA affords either the nicked circular (NC) form (single-strand breakage) or the linear (L) form (double-strand breakage).

The DNA cleavage activity was first investigated in the absence of an exogenous agent (Figure 7, Table 5). Complex **1** cleaved 20% of DNA (single-strand cleavage) at the highest concentration investigated (50 μM, Figure 7a), and proved to be a better nuclease than its methyl congener (no cleavage at this concentration) [44]. Complex **2** and **4** (Figure 7b,d) behave drastically differently since the gel electrophoresis pattern shows the disappearance of the SC form as DNA concentration increases, with no appearance of the NC or L form. This behavior has already been observed for the methyl derivative as well as for polyamines [44,52]. It results from DNA condensation, which is favored by neutralization of the phosphate by protonated polyamines. This hypothesis is further supported by the observation of a delay in the migration of the SC form before its total disappearance. Under these conditions, it is not possible to conclude about the nuclease activity of **2** and **4**. Complex **3** behaves in a more "classical" manner, in that the NC form was visible in addition to the SC form. The NC/SC ratio is clearly in favor of the NC form (single-strand cleavage) on the gel electrophoresis at 50 μM complex concentration, reflecting significant cleavage activity. Since aggregation can occur, nuclease activity could be overestimated, so that only an upper limit of 65% cleavage can be calculated. However, complex **3** appears visually to be a much better nuclease than complex **1**, due to the incorporation of a putrescine chain.

Table 5. DNA cleavage activity at pH = 7.2 [a].

Complex	No Exogenous Agent	Mercaptan	Ascorbic Acid
1	>50	40	30
2	_[b]	13	7
3	>40	22.5	16
4	_[b]	_[b]	>18

[a] Expressed as the concentration of complex (in μM) that produces 50% of cleavage of φX174 supercoiled DNA. [DNA] = 20 mM base pairs; T = 37 °C; t = 1 h; [reductant] = 0.8 mM; phosphate buffer 10 mM; pH = 7.2; water:DMF (90:10). [b] The gel electrophoresis displays only the disappearance of the SC form, due to significant fragmentation or condensation of DNA (see the text). The IC_{50} value cannot be determined under these conditions.

The DNA cleavage activity was further investigated in the presence of two distinct reductants, namely mercaptoethanol and ascorbic acid. As depicted in Figure 8 and Table 5, the nuclease activity is increased when the reductant is present, suggesting a different cleavage mechanism (oxidative versus hydrolytic). Amongst the two reductants, ascorbic acid gave the best results. Since mercaptoethanol is a potential ligand due to its thiol functions, we will discuss only the results with ascorbic acid in the next section. It is worth noting that the cleavage can be now quantified for **1**, **2**, and **3** since the concentration of complex required to achieve 50% cleavage is lower in the presence of ascorbic acid than without. It is in these cases smaller than that which promotes DNA condensation. For **4**, multiple bands are observed at 15 µM, which disappear at higher concentrations, precluding an accurate determination of IC$_{50}$ (Figure 8d).

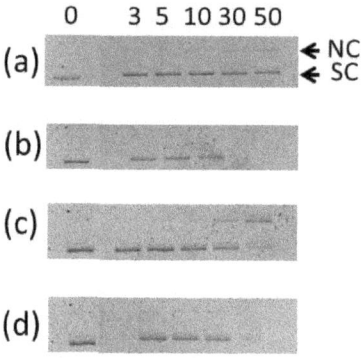

Figure 7. Cleavage of the ϕX174 supercoiled DNA by the copper complexes. The reaction is monitored by agarose gel electrophoresis in the absence of reductant at 37 °C for 1 h. [DNA] = 20 mM (base pairs); (water:DMF) (90:10) mixture; phosphate buffer (10 mM); pH = 7.2. Abbreviations: NC: nicked circular, SC: supercoiled. The concentrations of the complexes are expressed in µM. (**a**) **1**; (**b**) **2**; (**c**) **3** and (**d**) **4**.

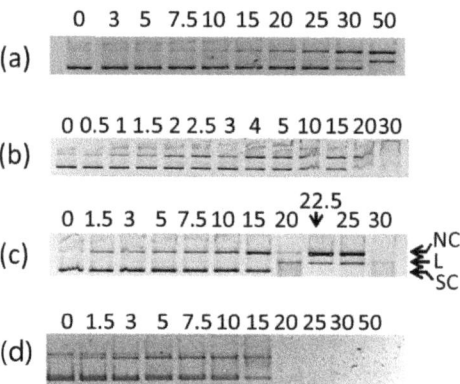

Figure 8. Cleavage of the ϕX174 supercoiled DNA by the copper complexes. The reaction is monitored by agarose gel electrophoresis in the presence of reductant at 37 °C for 1 h. [DNA] = 20 mM (base pairs); (water:DMF) (90:10) mixture; phosphate buffer (10 mM); [ascorbate] = 0.8 mM; pH = 7.2. Abbreviations: NC: nicked circular, SC: supercoiled, L: linear. The concentrations of the complexes are expressed in µM. (**a**) **1**; (**b**) **2**; (**c**) **3**; (**d**) **4**.

The most efficient nuclease is complex **2**, which cleaves 50% DNA at the concentration of 7 µM (Figure 8b). Complex **3**, which does not feature the pyrene moiety, is less active (16 µM for 50% of cleavage, Figure 8c), while complex **1** which features neither a pyrene, nor a putrescine chain as DNA anchoring function, is the poorest nuclease (30 µM for 50% of cleavage, Figure 8a). Interestingly, complex **3** catalyzes total DNA cleavage at a concentration of 25 µM, through both single and double-strand cleavage: The ratio is 60%

of double-strand cleavage and 40% of single-strand cleavage. Thus, the presence of a free terminal amine is important for the activity.

Important information about the cleavage mechanism was obtained from experiments in the presence of ascorbic acid and several scavengers at complex concentrations inducing significant cleavage. As illustrated in Figure 9 for **1** and **2** (and Figure S22 for **3**), the addition of superoxide dismutase (lane 4), NaN$_3$ (lane 5), or catalase (lane 7) does not induce changes in the cleavage profile, ruling out the involvement of superoxide, hydrogen peroxide, or singlet dioxygen in the reaction. In contrast, DMSO and to a lesser extend ethanol slightly inhibit the nuclease activity, pointing out the involvement of the hydroxyl radical in the reaction. Finally, EDTA, which is a strong copper chelator, has a significant inhibitory effect (lane 6). It can therefore be reasonably assumed that the reaction is initiated by the reduction of copper(II) to copper(I), which next activates dioxygen to form the reactive hydroxyl radical. The hydroxyl radical is the species ultimately involved in DNA strand scission [4,15].

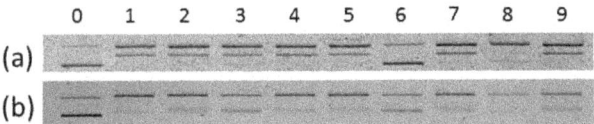

Figure 9. Cleavage of the ϕX174 supercoiled DNA by the copper complexes. The reaction is monitored by agarose gel electrophoresis in the presence of reductant at 37 °C for 1 h. [DNA] = 20 mM (base pairs); (water:DMF) (90:10) mixture; phosphate buffer (10 mM); [ascorbate] = 0.8 mM; pH = 7.2. (**a**) **1**, 50 µM; (**b**) **2**, 20 µM. Lane 0, DNA control; lane 1, DNA + complex (no scavenger); lanes 2–10, DNA + complex in the presence of various scavengers and agents: lane 2, ethanol; lane 3, DMSO (2 mL); lane 4, superoxide dismutase (0.5 unit); lane 5, NaN$_3$ (100 mM); lane 6, EDTA (10 mM); lane 7, catalase (0.1 unit); lane 8, Hoechst 33258 (100 mM); lane 9, NaCl (350 mM).

2.6. Mode of Binding of the Best Nuclease 2

In order to gain insight into the binding of the best nuclease, which is complex **2**, we investigated the nuclease activity in the presence of two binding agents: The minor groove binder Hoechst 33258 (Figure 9, lane 8) and NaCl (Figure 9, lane 9). The latter does not significantly alter the cleavage, but Hoechst 33258 proved to inhibit it significantly. This result suggests a minor groove binding of the putrescine chain. On the other hand, conjugated aromatic rings are known to intercalate into DNA. With the aim of confirming that the pyrene moiety of **2** is intercalated into DNA, we monitored its fluorescence at 468 nm upon titration with DNA. The pyrene units form excimers in solution, which give an intense fluorescence of around 470 nm. When DNA is added to **2**, the fluorescence at 468 nm is progressively quenched, showing that a process interferes with excimer formation, which is attributed to intercalation of the pyrene moiety [53]. The spectral changes were fitted by using a Scatchard–Von Hippel model (Figure 10) [54], giving a K value of 1.6×10^6 M^{-1} (n = 2). It is within the same order of magnitude than that measured for the methyl derivative [44], disclosing similar interactions with DNA. This further indicates that the main determinant for the interaction is not the nearby environment around the copper center, but ionic interactions mediated by the positively charged copper tripodal unit and the chain, as well as the intercalation of the pyrene group. Noteworthy, the K value is significantly higher than that of ethidium bromide (4.9×10^5) [55], confirming the high affinity of **2** for DNA.

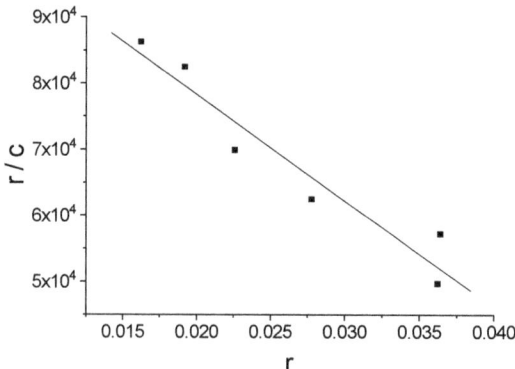

Figure 10. Fluorescence data fitted according to a Scatchard–Von Hippel model for the binding of **2** to CT DNA. The medium is a (water: DMF) (90:10) mixture containing 0.02 M NaCl. The pH is adjusted at 7 by a Tris–HCl buffer (0.05 M). T, 298 K.

2.7. Anti-Proliferative Activity

We investigated the anti-proliferative activity of complexes **1–4** on two bladder cancer cell lines by MTT assays: The RT112 cell lines are sensitive to cisplatin, whereas the RT112-CP are resistant to cisplatin. The results obtained after 48 h incubation are summarized in Table 6 and compared to those obtained with cisplatin.

Table 6. Anti-proliferative activity [a].

Complex	IC_{50} (µM)		RF
	RT112	RT112-CP	
Cisplatin	9.1 ± 1.2	23.8 ± 1.2	2.6
1	10.9 ± 0.2	13.2 ± 0.2	1.2
2	2.0 ± 0.2	1.2 ± 0.1	0.6
3	10.8 ± 0.2	10.5 ± 0.9	1.0
4	1.6 ± 0.2	1.0 ± 0.1	0.6

[a] The concentration resulting in 50% loss of cell viability relative to untreated cells (IC_{50}) was determined from dose-response curves. Results represent means ± SD of three independent experiments. RF is the ratio of IC_{50} against RT112-CP to RT112 cell lines. RT112-CP and RT112 are cisplatin-resistant and sensitive cell lines, respectively.

Complexes **2** and **4** exhibit the highest anti-proliferative activity, with remarkable IC_{50} values of 1–2 µM against both RT112 and cisplatin-resistant cells RT112-CP. The cytotoxicities of **2** and **4** are comparable to those of their methyl derivatives. They are much lower than those of cisplatin, which exhibits IC_{50} values of 9.1± 1.2 and 23.8± 1.2 µM against these two cell lines, respectively. Among the whole tripodal series of antiproliferative agents (incl. both the methylpyridine [44] and methoxypyridine derivatives), the best compounds proved to be **2** and **4** against the RT112-CP line (1.2 ± 0.1 and 1.0 ± 0.1 µM). Direct correlation between the nuclease activity and the cytotoxicity yet remains to be established, however, complex **2** proved to be both the best nuclease and the best anti-proliferative complex in the series. Finally, the resistance to cisplatin can be expressed through the resistance factor RF. It is calculated as the ratio of IC_{50} on RT112CP cell lines over IC_{50} on RT112 cell lines, the higher values being indicative of greater resistance. The RF is 2.6 for cisplatin but it is lower than 1.2 for the copper complexes, with remarkable values of 0.6 for complexes **2** and **4**. Hence, all the complexes appended by one putrescine chain overcome the resistance to cisplatin of RT112-CP cells.

3. Materials and Methods

3.1. Materials and Instruments

All chemicals were of reagent grade and were used without purification. NMR spectra were recorded on a Bruker AM 300 (^{1}H at 300 MHz) spectrometer (Bruker France SAS, Wissembourg, France). Chemical shifts are quoted relative to tetramethylsilane (TMS). Mass spectra were recorded on a Thermofinningen (EI/DCI) or a Bruker Esquire ESI-MS apparatus. For pKa determinations, UV/Vis spectra were recorded on a Cary Varian 50 spectrophotometer equipped with a Hellma immersion probe (1.000 cm path length, Hellma France, Paris, France). The temperature in the cell was controlled using a Lauda M3 circulating bath (Lauda France, Roissy-en-France, France) and the pH was monitored using a Methrom 716 DMS Titrino apparatus (Metrohm France, Villebon Courtaboeuf, France). A least square fit of the titration data was realized with the SPECFIT 3.0.37 software (from Biologic, Seyssinet-Pariset, France). The fluorescence spectra were recorded on a Cary Eclipse spectrometer. X-band EPR spectra were recorded on a Bruker EMX Plus spectrometer equipped with a Bruker nitrogen flow cryostat and a high sensitivity cavity (Bruker France SAS, Wissembourg, France). Spectra were simulated using the Easyspin 5.2.35 software [56]. Electrochemical measurements were carried out using a CHI 620 potentiostat. Experiments were performed in a standard three-electrode cell under argon atmosphere. A glassy carbon disc electrode (3 mm diameter), which was polished with 1 mm diamond paste, was used as the working electrode. The auxiliary electrode is a platinum wire, while a SCE was used as reference.

3.2. Synthesis

3-*tert*-butyl-4-hydroxy-benzaldehyde and N^1-(pyren-1-ylmethyl)butane-1,4-diamine were prepared according to literature procedures.

(6-methoxy-pyridin-2-ylmethyl)-pyridin-2-ylmethyl-amine.

6-methoxy-pyridin-2-carbaldehyde (2.74 g, 0.02 mol) and Pyridin-2-ylmethylamine (2.16 g, 0.02 mole) were each dissolved in methanol (15 mL). The two solutions were cooled down at 0 °C (ice bath) and mixed together. The ice bath was next removed, and the solution was stirred for 1 h at room temperature. The reaction mixture was cooled down at 0 °C and NaBH$_4$ (1 g, 0.026 mole) was slowly added in small portions (3 h). The solution was further stirred for 12 h at room temperature. Water was then added to the reaction mixture (30 mL), which was neutralized, extracted with CH$_2$Cl$_2$, and dried over anhydrous Na$_2$SO$_4$. To the residue were added CHCl$_3$ and pentane. The solution was cooled at −18 °C overnight. The supernatant was collected and concentrated under vacuum to give 3.66 g of an orange oil (yield: 80%). NMR ^{1}H (400 MHz, CDCl$_3$): δ (ppm) = 3.87 (s, 2H); 3.90 (s, 3H, OMe); 6.57 (d, 1H); 6.85 (d, 1H); 7.13 (m, 1H); 7.35 (d, 1H); 7.48 (d, 1H); 7.62 (td, 1H); 8.53 (d, 1H). NMR ^{13}C (Q.DEPT, 400 MHz, CDCl$_3$): δ (ppm) = 53.41 (CH$_3$, OMe); 54.50; 54.98 (2 CH$_2$); 108.85; 114.78; 122.05; 122.37; 136,57; 139.01; 149.47 (CH$_{aro}$); 157.49; 160.11; 164.03 (C$_{aro}$). HR-MS (Q-TOF): *m*/*z*, 230.1290; Calcd: 230.1293 for [M + H]$^+$.

HL^{CHO} (3-*tert*-Butyl-4-hydroxy-5-{[(6-methoxy-pyridin-2-ylmethyl)-pyridin-2-ylmethyl-amino]-methyl}-benzaldehyde). To a solution of 2-*tert*-butyl-4-hydroxybenzaldehyde (6.9 mmol) in ethanol (20 mL), 2 molar equivalents of paraformaldehyde and 1 equivalent of (6-methoxy-pyridin-2-ylmethyl)-pyridin-2-ylmethyl-amine diluted in ethanol (25 mL) were added. After 3 days under stirring at room temperature, the solvent was evaporated under vacuum and the reaction mixture was extracted with dichloromethane, washed with a saturated NaCl solution, and dried over Na$_2$SO$_4$. The solution was evaporated under vacuum, yielding a yellow-brown oil. The raw product was purified by column chromatography on silica gel (ethyl acetate/pentane (2/8) and 5% methanol to afford HL^{CHO} in a 46% yield. NMR ^{1}H (400 MHz, CDCl$_3$): δ (ppm) = 1.44 (s, 9H *t*-Bu); 3.85 (s, 2H); 3.99 (s, 2H); 4.02 (s, 2H); 4.03 (s, 3H, OCH$_3$); 6.62 (d, 1H, $^3J_{H-H}$ = 8.2 Hz, H$_7$ or H$_9$); 6.82 (d, 1H, $^3J_{H-H}$ = 7.1 Hz, H$_7$ or H$_9$); 7.28 (m, 1H, H$_4$); 7.49 (m, 3H, H$_2$-H$_6$-H$_8$); 7.72 (d, 1H, $^4J_{H-H}$ = 1.9 Hz, H$_1$); 7.75 (m, 1H, H$_5$); 8.56 (dd, 1H, $^3J_{H-H}$ = 5 Hz, $^4J_{H-H}$ = 0.85 Hz, H$_3$); 9.79 (s, 1H, CHO). NMR ^{13}C (Q.DEPT, 400 MHz, CDCl$_3$): δ (ppm) = 29.27 (3 CH$_3$, *t*-Bu); 34.91

(C, *t*-Bu); 53.51 (OCH$_3$); 57.46; 58.34; 59.09 (3 CH$_2$); 109.99; 116.22 (CH$_{aro}$); 122.79 (C$_{aro}$); 122.93; 124.14 (CH$_{aro}$); 127.96 (C$_{aro}$); 129.19; 129.92 (CH$_{aro}$); 137.65 (C$_{aro}$); 139.05 (CH$_{aro}$); 162.94; 164.18 (C$_{aro}$); 191.18 (CHO). MS (ESI): *m/z*, 420.3 [M + H]$^+$. Anal. Calcd for C$_{25}$H$_{29}$N$_3$O$_3$% C, 71.58; H, 6.97; N, 10.02. Found: C, 71.32; H, 6.84; N, 10.21%. HR-MS (Q-TOF): *m/z*, 420.2294; Calcd: 420.2287 for [M + H]$^+$.

HL^{pyr} ((2-*tert*-Butyl-6-{[(6-methoxy-pyridin-2-ylmethyl)-pyridin-2-ylmethyl-amino]-methyl}-4-({4-[(pyren-1-ylmethyl)-amino]-butylamino}-methyl)-phenol). N^1-(pyren-1-ylmethyl)butane-1,4-diamine (400 mg, 0.48 mmol) and 3-*tert*-butyl-4-hydroxy-5-{[(6-methoxy-pyridin-2-ylmethyl)-pyridin-2-ylmethyl-amino]-methyl}-benzaldehyde (145 mg, 0.48 mmol) were stirred overnight at room temperature in a MeOH/THF (3/1) mixture. Next, NaBH$_4$ (54 mg, 1.44 mmol) was added in small portions and the solution was stirred for two extra hours at room temperature. The reaction mixture was then extracted with AcOEt, washed with water and brine, and dried over anhydrous Na$_2$SO$_4$. The solution was evaporated under vacuum to afford yellow-brown oil. Column chromatography on silica gel with the eluent CH$_2$Cl$_2$/MeOH/Et$_3$N (92/5/3) afforded HL^{pyr} as a white solid in a 75% yield. NMR ^{1}H (400 MHz, CDCl$_3$): δ (ppm) = 1.37 (s, 9H, *t*-Bu); 1.64 (m, 4H, g and f); 2.64 (t, 2H, e or h); 2.85 (t, 2H, e or h); 3.57 (s, 2H, d); 3.59 (s, 2H, a); 3.63 (s, 2H, b); 3.71 (s, 2H, c); 4.00 (s, 3H, OCH$_3$); 4.50 (s, 2H, i); 6.57 (d, 1H, $^3J_{H-H}$ = 8.3 Hz, H$_7$); 6.72 (d, 1H, $^3J_{H-H}$ = 7.2 Hz, H$_9$); 7.01 (d, 1H, $^4J_{H-H}$ = 1.6 Hz, H$_2$); 7.11 (ddd, 1H, $^3J_{H-H}$ = 7.5 Hz; $^3J_{H-H}$ = 5.4 Hz; $^4J_{H-H}$ = 0.7 Hz, H$_4$); 7.42 (m, 2H, H$_6$ and H$_8$); 7.59 (td, 1H, $^3J_{H-H}$ = 7.5 Hz, $^4J_{H-H}$ = 1.7 Hz, H$_5$); 7.99–8.20 (m, 9H, 8H$_{pyrene}$ and H$_1$); 8.36 (d, 1H pyrene); 8,50 (d, 1H, $^3J_{H-H}$ = 5.4 Hz, H$_3$). NMR ^{13}C (Q.DEPT, 400 MHz, CDCl$_3$): δ (ppm) = 27.82; 28.33 (2 CH$_2$ f and g); 29.61 (3 CH$_3$ *t*-Bu); 34.79 (C, *t*-Bu); 48.89; 49.93; 51.81; 53.55 (4 CH$_2$); 53.58 (OCH$_3$); 57.88; 59.12; 59.44 (3 CH$_2$); 109.67; 116.17; 122.30 (CH$_{aro}$); 122.71 (C$_{aro}$); 123.19; 123.61; 125.00 (CH$_{aro}$); 125.09 (C$_{aro}$); 125.28; 125.37; 126.12; 126.69; 127.35; 127.41; 127.53; 127.64; 128.07 (CH$_{aro}$); 129.24; 130.95; 131.00; 131.50; 133.67 (C$_{aro}$); 136.57 (CH$_{aro}$); 136.59 (C$_{aro}$); 138.94; 149.08 (CH$_{aro}$); 155.28; 155.83; 158.21; 164.11 (C$_{aro}$). HR-MS (Q-TOF): *m/z*, 706.4128; Calcd: 706.4121 for [M + H]$^+$.

HL^{boc}. The ligand (3-*tert*-Butyl-4-hydroxy-5-{[(6-methoxy-pyridin-2-ylmethyl)-pyridin-2-ylmethyl-amino]-methyl}-benzaldehyde) (300 mg, 0.74 mmol) and *tert*-butyl-N-(4-aminobutyl)carbamate (140 mg, 0,74 mmol) were solubilized in MeOH (30 mL). The solution was stirred 12 h at room temperature. After that time, sodium borohydride (84 mg, 2.22 mmol) was added over 3 h by small fractions. The reaction mixture was further stirred for 2 h and next extracted with AcOEt, washed with water and brine, and finally dried over anhydrous Na$_2$SO$_4$. The organic phase was evaporated under vacuum, affording the raw product as a yellow-brown oil. Purification by column chromatography on silica gel with the eluent CH$_2$Cl$_2$/MeOH (95/5) using a gradient of triethylamine (0–5%) as the eluent afforded HL^{boc} as a colorless oil in a 64% yield. NMR ^{1}H (400 MHz, CDCl$_3$): δ (ppm) = 1.41 (s, 9H, *t*-Bu); 1.42 (s, 9H, *t*-Bu); 1.53 (m, 4H, f and g); 2.65 (t, 2H, e); 3.10 (m, 2H, h); 3.67 (s, 2H, b); 3.74 (s, 2H, d); 3.83 (s, 2H, a); 3.84 (s, 2H, c); 4.02 (s, 3H, OCH$_3$); 6.59 (d, 1H, $^3J_{H-H}$ = 8.2 Hz, H$_7$); 6.80 (d, 1H, $^3J_{H-H}$ = 7.2 Hz, H$_9$); 6.88 (d, 1H, $^4J_{H-H}$ = 1.8 Hz, H$_1$ or H$_2$); 7.09 (d, 1H, $^4J_{H-H}$ = 1.8 Hz, H$_1$ or H$_2$); 7.14 (ddd, 1H, $^3J_{H-H}$ = 7.5 Hz; $^3J_{H-H}$ = 5.0 Hz; $^4J_{H-H}$ = 0.7 Hz, H$_4$); 7.47 (m, 2H, H$_6$ and H$_8$); 7.63 (td, 1H, $^3J_{H-H}$ = 7.5 Hz, $^4J_{H-H}$ = 1.8 Hz, H$_5$); 8.52 (d, 1H, $^3J_{H-H}$ = 5.0 Hz, H$_3$), 10.83 (s, 1H, OH phenol). NMR ^{13}C (Q.DEPT, 400 MHz, CDCl$_3$): δ (ppm) = 27.13; 28.02 (2 CH$_2$, f and g); 28.59; 29.66 (6 CH$_3$, *t*-Bu); 34.90; 40.56 (2 C, *t*-Bu); 46.26; 48.93 (2 CH$_2$); 53.58 (OCH$_3$); 53.89; 58.19; 59.27; 59.60 (4 CH$_2$); 109.69; 116.26; 122.33 (CH$_{aro}$); 122.66 (C$_{aro}$); 123.68; 126.51; 127.63 (CH$_{aro}$); 128.98 (C$_{aro}$); 136.63; 138.95; 149.07 (CH$_{aro}$); 155.32; 155.79; 156.19; 158.24; 164.17 (C$_{aro}$). HR-MS (Q-TOF): *m/z*, 592.3873; Calcd: 592.3863 for [M + H]$^+$.

HL^{amine}. HL^{boc} (202 mg, mg, 0.35 mmol) was solubilized in an ethanol solution (20 mL), which was saturated with gaseous HCl beforehand. The solution was stirred during 1 h at room temperature. The solvent was next evaporated under vacuum, quantitatively affording HL^{amine} as a white solid. NMR ^{1}H (400 MHz, D$_2$O): δ (ppm) = 1.30 (s, 9H, *t*-Bu); 1.78 (m, 4H, f and g); 3.06 (m, 4H, e and h); 4.13 (s, 5H, a and OCH$_3$); 4.18 (s, 2H, b);

4.33 (s, 2H, d); 4.46 (s, 2H, c); 7.13 (d, 1H, $^3J_{H-H}$ = 8.3 Hz, H$_7$); 7.23 (m, 3H, H$_1$-H$_2$-H$_9$); 7.83 (m, 2H, H$_4$-H$_6$); 8.06 (t, 1H, $^3J_{H-H}$ = 8,3 Hz, H$_8$); 8.33 (t, 1H, $^3J_{H-H}$ = 7.9 Hz, H$_5$); 8.66 (d, 1H, $^3J_{H-H}$ = 5.6 Hz, H$_3$). NMR ^{13}C (Q.DEPT, 400 MHz, D$_2$O): δ (ppm) = 22.71; 23.99 (2 CH$_2$, f and g); 28.92 (3 CH$_3$, t-Bu); 34.08 (C, t-Bu); 38.81; 46.12; 50.48 (3 CH$_2$, a-e-h); 56.40 (OCH$_3$); 57.45; 57.82; 58.57 (3 CH$_2$, b-c-d); 109.56; 118.15 (CH$_{aro}$); 122.68; 123.56 (C$_{aro}$); 125.98; 127.07; 129.66; 130.42 (CH$_{aro}$); 139.07 (C$_{aro}$); 142.72; 145.20; 145.30 (CH$_{aro}$); 149.74; 151.22; 154.81; 162.63 (C$_{aro}$). HR-MS (Q-TOF): m/z, 492.3341; Calcd: 492.3339 for [M + H]$^+$.

H$_2$L^{bis}. (3-tert-Butyl-4-hydroxy-5-{[(6-methoxy-pyridin-2-ylmethyl)-pyridin-2-ylmethyl-amino]-methyl}-benzaldehyde) (500 mg, 1.24 mmol) and 1,4-butane diamine (55 mg, 0.62 mmol) were solubilized in MeOH (30 mL). After 10 min stirring, the reductant NaBH$_4$ (6 molar equivalents, 141 mg, 3.72 mmol) was progressively added to the solution (over 3 h) at room temperature. After 2 h of additional stirring, the solvent was evaporated under vacuum and the reaction mixture was extracted with AcOEt, washed with water and brine, and dried over anhydrous Na$_2$SO$_4$. The organic phase was evaporated under vacuum, giving a yellow-brown oil. The product was purified by column chromatography on silica gel with CH$_2$Cl$_2$/MeOH (95/5) with a gradient of triethylamine (0–5%) as the eluent, affording H$_2$L^{bis} as a white powder in a 50% yield. NMR ^{1}H (400 MHz, CDCl$_3$): δ (ppm) = 1.39 (s, 18H, 6 CH$_3$, t-Bu); 1.65 (m, 4H, f); 2.66 (m, 4H, e); 3.67 (s, 4H, b); 3.72 (s, 4H, d); 3.80 (s, 4H, a); 3.82 (s, 4H, c); 4.01 (s, 6H, 2 OCH$_3$); 6.58 (d, 2H, $^3J_{H-H}$ = 8.2 Hz, H$_9$); 6.78 (d, 2H, $^3J_{H-H}$ = 7,2 Hz, H$_7$); 6.87 (d, 2H, $^4J_{H-H}$ = 1.6 Hz, H$_2$); 7.04 (d, 2H, $^4J_{H-H}$ = 1.6 Hz, H$_1$); 7.13 (dd, 2H, $^3J_{H-H}$ = 4.9 Hz, $^3J_{H-H}$ = 7.4 Hz, H$_4$); 7.44 (m, 4H, H$_6$ et H$_8$); 7.61 (td, 2H, $^3J_{H-H}$ = 7.4 Hz, $^4J_{H-H}$ = 1.6 Hz, H$_5$); 8.50 (d, 2H, $^3J_{H-H}$ = 4.9 Hz, H$_3$); 10.94 (s, 2H, OHphenol). NMR ^{13}C (Q.DEPT, CDCl$_3$): δ (ppm) = 27.70 (CH$_2$, f); 29.65 (CH$_3$, t-Bu); 34.86 (C, t-Bu); 48.35; 52.98 (CH$_2$); 53.59 (OCH$_3$); 58.01; 59.20; 59.42 (CH$_2$); 109.74; 116.30; 122,38 (CH$_{aro}$); 122.92 (C$_{aro}$); 123.69; 126.79; 128.06; 136.67 (CH$_{aro}$); 136.86; 137.73 (C$_{aro}$); 139,03; 149.12 (CH$_{aro}$); 155.26; 156.22; 158.11; 164.16 (C$_{aro}$). HR-MS (Q-TOF): m/z, 895.5621; Calcd: 895.5598 for [M + H]$^+$.

Copper complex 1. The ligand HL^{CHO} (40 mg, 0.095 mmol) was dissolved in methanol (1 mL). The salt CuCl$_2$ • 2H$_2$O (16.2 mg, 0.095 mmol) was dissolved in methanol (1 mL) and this solution was added under stirring to the solution of the ligand. Triethylamine was next added (14 mL, 0.1 mmol) and the solution was concentrated under vacuum. The crude product was dissolved in acetonitrile (2 mL) and methyl acetate was slowly diffused into this solution, affording single crystals of complex 1 (39 mg, yield 80%). HR-MS (Q-TOF): m/z, 481.1410; Calcd m/z: 481.1421 for C$_{25}$H$_{28}$O$_3$N$_3$Cu ([M]$^+$). UV/Vis (H$_2$O:DMF 90:10, [NaCl] = 0.1 M, pH 7): 482 nm (ε = 662 M^{-1} cm^{-1}). EPR (X-band, H$_2$O:DMF 90:10, [Tris−HCl] = 0.05 M, [NaCl] = 0.02 M, pH = 7, 100 K): g$_\perp$ = 2.056, g$_{//}$ = 2.257, A$_\perp$ = 1.4 mT, A$_{//}$ = 16.3 mT (with Cu).

Other complexes. Complex 2 and 3 were generated in situ by reacting equimolar amounts of CuCl$_2$ • 2H$_2$O, base (triethylamine) and the appropriate ligand (HL^{pyr}, HL^{amine}, respectively) in DMF. Complex 4 was prepared in a similar way with H$_2$L^{bis}, except instead two molar equivalents of copper were used. The concentrated DMF solutions of the complexes were diluted in water in order to obtain a final medium of composition (DMF:H$_2$O) (10:90).

Complex 2. HR-MS (Q-TOF): m/z, 384.1660; Calcd m/z: 384.1664 for C$_{46}$H$_{51}$O$_2$N$_5$Cu ([M + H]$^{2+}$). UV/Vis (H$_2$O:DMF 90:10, [NaCl] = 0.1 M, pH 7): 462 nm (ε = 1040 M^{-1} cm^{-1}). EPR (X-band, H$_2$O:DMF 90:10, [Tris−HCl] = 0.05 M, [NaCl] = 0.02 M, 100 K): g$_\perp$ = 2.054, g$_{//}$ = 2.254, A$_\perp$ = 1.5 mT, A$_{//}$ = 16.6 mT (A$_{Cu}$).

Complex 3. HR-MS (Q-TOF): m/z, 277.1268; Calcd m/z: 277.1273 for C$_{29}$H$_{41}$O$_2$N$_5$Cu ([M + H]$^{2+}$). UV/Vis (H$_2$O:DMF 90:10, [NaCl] = 0.1 M, pH 7): 461 nm (ε = 780 M^{-1} cm^{-1}). EPR (X-band, H$_2$O:DMF 90:10, [Tris−HCl] = 0.05 M, [NaCl] = 0.02 M, 100 K): g$_\perp$ = 2.054, g$_{//}$ = 2.254, A$_\perp$ = 1.5 mT, A$_{//}$ = 16.6 mT (A$_{Cu}$).

Complex 4. HR-MS (ESI-MS): m/z, 1020.45; Calcd m/z: 1020.41 for C$_{54}$H$_{70}$Cu$_2$N$_8$O$_4$ [M + 2H]$^+$. UV/Vis (H$_2$O:DMF 90:10, [NaCl] = 0.1 M, pH 7): 457 nm (ε = 2005 M^{-1} cm^{-1}).

EPR (X-band, H$_2$O:DMF 90:10, [Tris—HCl] = 0.05 M, [NaCl] = 0.02 M, 100 K): g$_\perp$ = 2.054, g$_{//}$ = 2.254, A$_\perp$ = 1.5 mT, A$_{//}$ = 16.6 mT (A$_{Cu}$).

3.3. Crystal Structure Analysis

A single crystal of **1** was coated with a parafin mixture, picked up with nylon loops, and mounted in the nitrogen cold stream of the diffractometer. Mo-Kα radiation (λ = 0.71073Å) from a Mo-target X-ray micro-source equipped with INCOATEC Quazar mirror optics was used (INCOATEC, Geesthacht, Germany). Final cell constants were obtained from least squares fits of several thousand strong reflections. Intensity data were corrected for absorption using intensities of redundant reflections with the program SADABS [57]. The structures were readily solved by the charge flipping method. The OLEX2 1.2 software was used for the refinement [58]. All non-hydrogen atoms were anisotropically refined and hydrogen atoms were placed at calculated positions and refined as riding atoms with isotropic displacement parameters. CCDC-1479462 contains the crystallographic data for **1**; these data can be obtained free of charge via http://www.ccdc.cam.ac.uk/conts/retrieving.html (accessed on 25 September 2023).

3.4. Determination of the DNA Binding Constant

The binding constant towards CT-DNA was determined by fluorescence. CT-DNA (type I, fibers, from Sigma Aldrich) was purified beforehand, as reported [44]. Complex **2** (1 mM) was mixed with variable amounts of CT-DNA (typically 5–200 mM in base pairs) in a saline (H$_2$O:DMF) (90:10) solution ([NaCl] = 0.02 M), whose pH is buffered (pH = 7) by Tris—HCl (0.05 M). The fluorescence spectra were recorded after 10 min at 298 K. The DNA binding constant K was calculated by using a Scatchard–Von Hippel model from a duplicate experiment [54]: The ratio r/c was plotted against r, giving a straight line whose slope is K, where r represents the number of bound molecules per site and c corresponds to the concentration of free drug. The concentration of bound molecules was determined from $C_0 \times (f_0 - f)/(f_0 - f_b)$, where f_0 is the fluorescence of the free drug, f the fluorescence at any DNA concentration, f_b the fluorescence of the drug bound to DNA, and C_0 the total concentration of drug (calculated from the mass balance).

3.5. Procedure for DNA Cleavage Experiments

The plasmidic DNA φX174 RF1 was purchased from Fermentas. The experiments were performed in a (water:DMF) (90:10174) mixture containing 10 mM phosphate buffer (pH = 7.2). In a typical experiment, double-stranded supercoiled DNA was incubated with the complexes at 37 °C during the appropriate time. The reaction was next quenched by decreasing the temperature to −20 °C. A loading buffer was added (6× loading dye solution, Fermentas) and the solution was loaded on a 0.8% agarose gel in Tris-Boric acid-EDTA buffer, (pH = 8) (0.5 × TBE). The electrophoresis was performed at 70 V during about 3 h. Once DNA has migrated, the gels were stained by a 10 min incubation with an ethidium bromide solution (1 mg/mL) followed by washing with distilled water. The images of the gels were recorded by using the imager Typhoon 9400 (Cytiva France, Saint Germain en Laye, France). The fluorescence was quantified using the IQ Solutions v1.4 Software.

3.6. Cell Culture

Human bladder cancer cell line RT112 was obtained from Cell Lines Service (Eppelheim, Germany). Cisplatin-resistant RT112 cells (RT112-CP) were kindly provided by B. Köberle (Institute of Toxicology, Clinical Centre of University of Mainz, Mainz, Germany). RT112 and RT112-CP cells were cultured in RPMI 1640 medium supplemented with 10% (v/v) fetal calf serum (FCS) and 2 mM glutamine (Invitrogen Life Technologies, Paisley, UK). Cells were maintained at 37 °C in a 5% CO$_2$-humidified atmosphere and tested to ensure freedom from mycoplasma contamination.

3.7. Cell Proliferation Assay

Inhibition of cell proliferation by copper complexes was measured by MTT (3-(4,5-Dimethylthiazol-2-yl)-2,5-diphenyltetrazolium bromide) assays. RT112 and RT112-CP cells were seeded into 96-well plates (5×10^3 cells/well) in 100 µL of culture medium. After 24 h, cells were treated with cisplatin (Sigma-Aldrich, Lyon, France) or complexes at various concentrations. In parallel, a control with DMF (vehicle alone) at the same dilutions was performed. Following incubation for 48 h, 10 µL of MTT (Euromedex, Mundolsheim, France) stock solution in PBS at 5 mg/ml was added in each well and the plates were incubated at 37 °C for 3 h. Plates were then centrifuged 5 min at 1500 rpm before the medium was discarded and replaced with DMF (100 µL/well) to solubilize water-insoluble purple formazan crystals. After 15 min under shaking, absorbance was measured on an ELISA reader (Tecan, Männedorf, Switzerland) at a test wavelength of 570 nm and a reference wavelength of 650 nm. Absorbance obtained by cells treated with the same dilution of the vehicle alone (DMSO) was rated as 100% of cell survival. Each data point is the average triplicates of three independent experiments.

3.8. Computational Details

Full geometry optimizations were performed with the Gaussian 9 program [59], by using the B3LYP [60,61] functional. The 6–31g* basis set [62] was used for all the atoms. Frequency calculations were systematically performed in order to ensure that the optimized structure corresponds to a minimum and not a saddle point. Optical properties were computed by using time-dependent DFT (TD-DFT) [63], with the same basis set as for optimization. The solvent was taken into account by using a polarized continuum model (PCM) [64]. The 30 lowest energy excited states were calculated.

4. Conclusions

In summary, we prepared a series of copper tripodal complexes based on recently reported nucleases [44], in which the crucial α-methylpyridine moiety is replaced by an α-methoxypyridine. The methoxy group is both a stronger donor, and less sterically crowded. We establish both by X-Ray diffraction and DFT calculations that the copper ion geometry is significantly impacted by this substitution. In addition, the phenols' pKa are higher than for the "methyl" series [44], indicating weaker Lewis acidity at the metal center. Surprisingly, the Cu(II) reduction potential remains mostly unaffected ($E_p^{c,red}$ = −0.45 to −0.51 V). The DNA cleavage activity of the complexes was investigated: Without reductant, all the compounds featuring a putrescine chain promote DNA condensation, which hampers direct observation of strand cleavage. On the other hand, **1** cleaves 20% of DNA at the highest concentration investigated (50 µM), whereas its methyl congener did not promote any cleavage at this concentration [44]. This result cannot be explained in terms of Lewis acidity of the metal center, since an opposite trend would be expected. It may instead reflect structural effects, whereby the metal is less sterically crowded in the present series, and thus is more accessible for generating nucleophiles.

When ascorbic acid is present the nuclease activity is significantly enhanced, and switches to an oxidative pathway (through OH• formation). The IC$_{50}$ are 30, 7, 16, and >18 µM for complexes, **1**, **2**, **3**, and **4**, respectively. These values are in a narrow range, like the methyl derivatives (45, 1.7, 14 µM for the methyl derivatives of **1**, **2**, and **3**, respectively), with, however, two important differences: The activity of complex **1** is higher than that of the methyl congener, whereas that of complex **2** is lower. Thus, correlations between cleavage activity and nature of the pyridine α-substituent depend on more than a single parameter. The methoxy substitution herein appears more interesting for complexes that are not vectorized towards phosphate by the putrescine chain. Notably, the best nuclease in this series is complex **2** under reducing conditions. Although slightly less efficient than its methyl congener [44], it remains a very efficient nuclease. Its lower activity may be attributed to the slightly lower binding constant towards DNA (1.6×10^6 M^{-1} vs 2.6×10^6 M^{-1}). Complexes **2** and **4**, which are the best nucleases, inhibit the proliferation of bladder cancer cells much more efficiently than cisplatin (5 to 25-times better). The IC_{50} of complex **4** is slightly smaller than that of complex **2** (1–1.6 µM), but if the concentration is calculated on the basis of the copper content, complex **2** is the best agent. Finally, both overcome the resistance to cisplatin of RT112-CP cells.

In summary, the results on DNA cleavage demonstrate that a hydrolytic pathway can be favored when the steric bulk in α-position of the pyridine is decreased (methoxy vs. methyl substituent) [44]. The oxidative pathway is unfavored when a methoxy substituent replaces the methyl one, which is attributed to electronic rather than steric effects. These trends provide important insight into the strategy to be used for further functionalization and bioconjugation of the ligands, especially the incorporation of poly(O-alkyl) chains (PEG) instead of simple O-methyl groups, with the aim of enhancing the delivery of the copper nuclease once injected in an organism [65].

Supplementary Materials: The following supporting information can be downloaded at: https://www.mdpi.com/article/10.3390/inorganics11100396/s1, Figures S1–S4: ^{1}H NMR spectra of the ligands; Figures S5–S8: HR-MS of the ligands; Figure S9: Mass spectra of the complexes; Figure S10: Jobs' plot of **4**; Figures S11–S15: EPR spectra of the complexes; Figures S16–S18: Electronic spectra of the complexes; Figures S19–S24: CV curves of the complexes; Figures S25–S26: Agarose gel electrophoresis; Figure S27: summarizes the atom numbering used for assigning the ^{1}H NMR resonances. Output of geometry optimizations and optical properties.

Author Contributions: Conceptualization, F.T. and X.R.; funding acquisition, F.T.; investigation, D.S., S.E. and C.P.; methodology, V.M.-F. and N.B.; project administration, F.T.; writing—review & editing, F.T., V.M.-F. and N.B. All authors have read and agreed to the published version of the manuscript.

Funding: This research was funded by the French National Research Agency in the framework of the "Investissements d'avenir" program (ANR-15-IDEX-02) and Labex ARCANE and CBH-EUR-GS (ANR-17-EURE-0003).

Data Availability Statement: The data are given in Supplementary Materials.

Acknowledgments: The NanoBio-ICMG platforms (UAR 2607) are acknowledged for their technical support. The authors thank the CECIC cluster for providing computational resources.

Conflicts of Interest: The authors declare no conflict of interest.

References

1. Fantoni, N.Z.; Molphy, Z.; Slator, C.; Menounou, G.; Toniolo, G.; Mitrikas, G.; McKee, V.; Chatgilialoglu, C.; Kellett, A. Polypyridyl-Based Copper Phenanthrene Complexes: A New Type of Stabilized Artificial Chemical Nuclease. *Chem. Eur. J.* **2019**, *25*, 221–237. [CrossRef]
2. Desbouis, D.; Troitsky, I.P.; Belousoff, M.J.; Spiccia, L.; Graham, B. Copper(II), zinc(II) and nickel(II) complexes as nuclease mimetics. *Coord. Chem. Rev.* **2012**, *256*, 897–937. [CrossRef]
3. Mancin, F.; Scrimin, P.; Tecilla, P. Progress in artificial metallonucleases. *Chem. Commun.* **2012**, *48*, 5545–5559. [CrossRef]
4. Wende, C.; Lüdtke, C.; Kulak, N. Copper Complexes of N-Donor Ligands as Artificial Nucleases. *Eur. J. Inorg. Chem.* **2014**, *2014*, 2597–2612. [CrossRef]
5. Cristina, M.; Maura, P.; Francesco, T.; Carlo, S. Copper Complexes as Anticancer Agents. *Anti Cancer Agents Med. Chem.* **2009**, *9*, 185–211. [CrossRef]
6. Saverio, T.; Luciano, M. Copper Compounds in Anticancer Strategies. *Curr. Med. Chem.* **2009**, *16*, 1325–1348. [CrossRef]
7. Tisato, F.; Marzano, C.; Porchia, M.; Pellei, M.; Santini, C. Copper in diseases and treatments, and copper-based anticancer strategies. *Med. Res. Rev.* **2010**, *30*, 708–749. [CrossRef] [PubMed]
8. Duncan, C.; White, A.R. Copper complexes as therapeutic agents. *Metallomics* **2012**, *4*, 127–138. [CrossRef]
9. Santini, C.; Pellei, M.; Gandin, V.; Porchia, M.; Tisato, F.; Marzano, C. Advances in Copper Complexes as Anticancer Agents. *Chem. Rev.* **2014**, *114*, 815–862. [CrossRef]
10. Denoyer, D.; Masaldan, S.; La Fontaine, S.; Cater, M.A. Targeting copper in cancer therapy: 'Copper That Cancer'. *Metallomics* **2015**, *7*, 1459–1476. [CrossRef]
11. Huiqi, X.; Kang, Y.J. Role of Copper in Angiogenesis and Its Medicinal Implications. *Curr. Med. Chem.* **2009**, *16*, 1304–1314. [CrossRef]
12. Mazdak, H.; Yazdekhasti, F.; Movahedian, A.; Mirkheshti, N.; Shafieian, M. The comparative study of serum iron, copper, and zinc levels between bladder cancer patients and a control group. *Int. Urol. Nephrol.* **2010**, *42*, 89–93. [CrossRef] [PubMed]

13. Lelièvre, P.; Sancey, L.; Coll, J.-L.; Deniaud, A.; Busser, B. The Multifaceted Roles of Copper in Cancer: A Trace Metal Element with Dysregulated Metabolism, but Also a Target or a Bullet for Therapy. *Cancers* **2020**, *12*, 3594. [CrossRef] [PubMed]
14. Majumder, S.; Chatterjee, S.; Pal, S.; Biswas, J.; Efferth, T.; Choudhuri, S.K. The role of copper in drug-resistant murine and human tumors. *BioMetals* **2008**, *22*, 377–384. [CrossRef] [PubMed]
15. Pitié, M.; Pratviel, G. Activation of DNA Carbon−Hydrogen Bonds by Metal Complexes. *Chem. Rev.* **2010**, *110*, 1018–1059. [CrossRef]
16. Villarreal, W.; Colina-Vegas, L.; Visbal, G.; Corona, O.; Corrêa, R.S.; Ellena, J.; Cominetti, M.R.; Batista, A.A.; Navarro, M. Copper(I)–Phosphine Polypyridyl Complexes: Synthesis, Characterization, DNA/HSA Binding Study, and Antiproliferative Activity. *Inorg. Chem.* **2017**, *56*, 3781–3793. [CrossRef]
17. Banasiak, A.; Zuin Fantoni, N.; Kellett, A.; Colleran, J. Mapping the DNA Damaging Effects of Polypyridyl Copper Complexes with DNA Electrochemical Biosensors. *Molecules* **2022**, *27*, 645. [CrossRef]
18. Jaividhya, P.; Dhivya, R.; Akbarsha, M.A.; Palaniandavar, M. Efficient DNA cleavage mediated by mononuclear mixed ligand copper(II) phenolate complexes: The role of co-ligand planarity on DNA binding and cleavage and anticancer activity. *J. Inorg. Biochem.* **2012**, *114*, 94–105. [CrossRef]
19. Loganathan, R.; Ramakrishnan, S.; Suresh, E.; Riyasdeen, A.; Akbarsha, M.A.; Palaniandavar, M. Mixed Ligand Copper(II) Complexes of N,N-Bis(benzimidazol-2-ylmethyl)amine (BBA) with Diimine Co-Ligands: Efficient Chemical Nuclease and Protease Activities and Cytotoxicity. *Inorg. Chem.* **2012**, *51*, 5512–5532. [CrossRef]
20. Silva, P.P.; Guerra, W.; dos Santos, G.C.; Fernandes, N.G.; Silveira, J.N.; da Costa Ferreira, A.M.; Bortolotto, T.; Terenzi, H.; Bortoluzzi, A.J.; Neves, A.; et al. Correlation between DNA interactions and cytotoxic activity of four new ternary compounds of copper(II) with N-donor heterocyclic ligands. *J. Inorg. Biochem.* **2014**, *132*, 67–76. [CrossRef]
21. Bhat, S.S.; Kumbhar, A.A.; Heptullah, H.; Khan, A.A.; Gobre, V.V.; Gejji, S.P.; Puranik, V.G. Synthesis, Electronic Structure, DNA and Protein Binding, DNA Cleavage, and Anticancer Activity of Fluorophore-Labeled Copper(II) Complexes. *Inorg. Chem.* **2011**, *50*, 545–558. [CrossRef] [PubMed]
22. Silva, P.P.; Guerra, W.; Silveira, J.N.; Ferreira, A.M.d.C.; Bortolotto, T.; Fischer, F.L.; Terenzi, H.; Neves, A.; Pereira-Maia, E.C. Two New Ternary Complexes of Copper(II) with Tetracycline or Doxycycline and 1,10-Phenanthroline and Their Potential as Antitumoral: Cytotoxicity and DNA Cleavage. *Inorg. Chem.* **2011**, *50*, 6414–6424. [CrossRef] [PubMed]
23. Li, M.-J.; Lan, T.-Y.; Cao, X.-H.; Yang, H.-H.; Shi, Y.; Yi, C.; Chen, G.-N. Synthesis, characterization, DNA binding, cleavage activity and cytotoxicity of copper(ii) complexes. *Dalton Trans.* **2014**, *43*, 2789–2798. [CrossRef] [PubMed]
24. Leite, S.M.G.; Lima, L.M.P.; Gama, S.; Mendes, F.; Orio, M.; Bento, I.; Paulo, A.; Delgado, R.; Iranzo, O. Copper(II) Complexes of Phenanthroline and Histidine Containing Ligands: Synthesis, Characterization and Evaluation of their DNA Cleavage and Cytotoxic Activity. *Inorg. Chem.* **2016**, *55*, 11801–11814. [CrossRef]
25. Xiao, Y.; Wang, Q.; Huang, Y.; Ma, X.; Xiong, X.; Li, H. Synthesis, structure, and biological evaluation of a copper(ii) complex with fleroxacin and 1,10-phenanthroline. *Dalton Trans.* **2016**, *45*, 10928–10935. [CrossRef]
26. Bortolotto, T.; Silva, P.P.; Neves, A.; Pereira-Maia, E.C.; Terenzi, H. Photoinduced DNA Cleavage Promoted by Two Copper(II) Complexes of Tetracyclines and 1,10-Phenanthroline. *Inorg. Chem.* **2011**, *50*, 10519–10521. [CrossRef]
27. Lüdtke, C.; Sobottka, S.; Heinrich, J.; Liebing, P.; Wedepohl, S.; Sarkar, B.; Kulak, N. Forty Years after the Discovery of Its Nucleolytic Activity: [Cu(phen)2]2+ Shows Unattended DNA Cleavage Activity upon Fluorination. *Chem. Eur. J.* **2021**, *27*, 3273–3277. [CrossRef]
28. Zhou, W.; Wang, X.; Hu, M.; Guo, Z. Improving nuclease activity of copper(II)–terpyridine complex through solubilizing and charge effects of glycine. *J. Inorg. Biochem.* **2013**, *121*, 114–120. [CrossRef]
29. Molphy, Z.; Prisecaru, A.; Slator, C.; Barron, N.; McCann, M.; Colleran, J.; Chandran, D.; Gathergood, N.; Kellett, A. Copper Phenanthrene Oxidative Chemical Nucleases. *Inorg. Chem.* **2014**, *53*, 5392–5404. [CrossRef]
30. Berthet, N.; Martel-Frachet, V.; Michel, F.; Philouze, C.; Hamman, S.; Ronot, X.; Thomas, F. Nuclease and anti-proliferative activities of copper(ii) complexes of N3O tripodal ligands involving a sterically hindered phenolate. *Dalton Trans.* **2013**, *42*, 8468–8483. [CrossRef]
31. Xu, W.; Craft, J.; Fontenot, P.; Barens, M.; Knierim, K.; Albering, J.; Mautner, F.; Massoud, S. Effect of the central metal ion on the cleavage of DNA by [M(TPA)Cl]ClO4 complexes (M = CoII, CuII and ZnII, TPA = tris(2-pyridylmethyl)amine): An efficient artificial nuclease for DNA cleavage. *Inorg. Chim. Acta* **2011**, *373*, 159–166. [CrossRef]
32. Doniz Kettenmann, S.; Nossol, Y.; Louka, F.R.; Legrande, J.R.; Marine, E.; Fischer, R.C.; Mautner, F.A.; Hergl, V.; Kulak, N.; Massoud, S.S. Copper(II) Complexes with Tetradentate Piperazine-Based Ligands: DNA Cleavage and Cytotoxicity. *Inorganics* **2021**, *9*, 12. [CrossRef]
33. Castilho, N.; Gabriel, P.; Camargo, T.P.; Neves, A.; Terenzi, H. Targeting an Artificial Metal Nuclease to DNA by a Simple Chemical Modification and Its Drastic Effect on Catalysis. *ACS Med. Chem. Lett.* **2020**, *11*, 286–291. [CrossRef]
34. Levín, P.; Balsa, L.M.; Silva, C.P.; Herzog, A.E.; Vega, A.; Pavez, J.; León, I.E.; Lemus, L. Artificial Chemical Nuclease and Cytotoxic Activity of a Mononuclear Copper(I) Complex and a Related Binuclear Double-Stranded Helicate. *Eur. J. Inorg. Chem.* **2021**, *2021*, 4103–4112. [CrossRef]
35. McGorman, B.; Fantoni, N.Z.; O'Carroll, S.; Ziemele, A.; El-Sagheer, A.H.; Brown, T.; Kellett, A. Enzymatic Synthesis of Chemical Nuclease Triplex-Forming Oligonucleotides with Gene-Silencing Applications. *Nucleic Acids Res.* **2022**, *50*, 5467–5481. [CrossRef] [PubMed]

36. Philibert, A.; Thomas, F.; Philouze, C.; Hamman, S.; Saint-Aman, E.; Pierre, J.-L. Galactose Oxidase Models: Tuning the Properties of Cu^{II}–Phenoxyl Radicals. *Chem. Eur. J.* **2003**, *9*, 3803–3812. [CrossRef]
37. Thomas, F. Ten Years of a Biomimetic Approach to the Copper(II) Radical Site of Galactose Oxidase. *Eur. J. Inorg. Chem.* **2007**, *2007*, 2379–2404. [CrossRef]
38. Kunert, R.; Philouze, C.; Berthiol, F.; Jarjayes, O.; Storr, T.; Thomas, F. Distorted copper(ii) radicals with sterically hindered salens: Electronic structure and aerobic oxidation of alcohols. *Dalton Trans.* **2020**, *49*, 12990–13002. [CrossRef]
39. Maheswari, P.U.; Roy, S.; den Dulk, H.; Barends, S.; van Wezel, G.; Kozlevčar, B.; Gamez, P.; Reedijk, J. The Square-Planar Cytotoxic [CuII(pyrimol)Cl] Complex Acts as an Efficient DNA Cleaver without Reductant. *J. Am. Chem. Soc.* **2006**, *128*, 710–711. [CrossRef]
40. Ghosh, K.; Kumar, P.; Tyagi, N.; Singh, U.P.; Aggarwal, V.; Baratto, M.C. Synthesis and reactivity studies on new copper(II) complexes: DNA binding, generation of phenoxyl radical, SOD and nuclease activities. *Eur. J. Med. Chem.* **2010**, *45*, 3770–3779. [CrossRef]
41. Ghosh, K.; Kumar, P.; Mohan, V.; Singh, U.P.; Kasiri, S.; Mandal, S.S. Nuclease Activity via Self-Activation and Anticancer Activity of a Mononuclear Copper(II) Complex: Novel Role of the Tertiary Butyl Group in the Ligand Frame. *Inorg. Chem.* **2012**, *51*, 3343–3345. [CrossRef] [PubMed]
42. Cuzan, O.; Kochem, A.; Simaan, A.J.; Bertaina, S.; Faure, B.; Robert, V.; Shova, S.; Giorgi, M.; Maffei, M.; Réglier, M.; et al. Oxidative DNA Cleavage Promoted by a Phenoxyl-Radical Copper(II) Complex. *Eur. J. Inorg. Chem.* **2016**, *2016*, 5575–5584. [CrossRef]
43. Gentil, S.; Serre, D.; Philouze, C.; Holzinger, M.; Thomas, F.; Le Goff, A. Electrocatalytic O$_2$ Reduction at a Bio-inspired Mononuclear Copper Phenolato Complex Immobilized on a Carbon Nanotube Electrode. *Angew. Chem. Int. Ed.* **2016**, *55*, 2517–2520. [CrossRef]
44. Serre, D.; Erbek, S.; Berthet, N.; Ronot, X.; Martel-Frachet, V.; Thomas, F. Copper(II) complexes of N$_3$O tripodal ligands appended with pyrene and polyamine groups: Anti-proliferative and nuclease activities. *J. Inorg. Biochem.* **2018**, *179*, 121–134. [CrossRef]
45. Addison, A.W.; Rao, T.N.; Reedijk, J.; van Rijn, J.; Verschoor, G.C. Synthesis, structure, and spectroscopic properties of copper(II) compounds containing nitrogen-sulphur donor ligands; the crystal and molecular structure of aqua[1,7-bis(N-methylbenzimidazol-2[prime or minute]-yl)-2,6-dithiaheptane]copper(II) perchlorate. *J. Chem. Soc. Dalton Trans.* **1984**, 1349–1356. [CrossRef]
46. Vaidyanathan, M.; Viswanathan, R.; Palaniandavar, M.; Balasubramanian, T.; Prabhaharan, P.; Muthiah, P. Copper(II) Complexes with Unusual Axial Phenolate Coordination as Structural Models for the Active Site in Galactose Oxidase: X-ray Crystal Structures and Spectral and Redox Properties of [Cu(bpnp)X] Complexes. *Inorg. Chem.* **1998**, *37*, 6418–6427. [CrossRef]
47. Michel, F.; Thomas, F.; Hamman, S.; Philouze, C.; Saint-Aman, E.; Pierre, J.-L. Galactose Oxidase Models: Creation and Modification of Proton Transfer Coupled to Copper(II) Coordination Processes in Pro-Phenoxyl Ligands. *Eur. J. Inorg. Chem.* **2006**, *2006*, 3684–3696. [CrossRef]
48. Colomban, C.; Philouze, C.; Molton, F.; Leconte, N.; Thomas, F. Copper(II) Complexes of N3O Ligands as Models for Galactose Oxidase: Effect of Variation of Steric Bulk of Coordinated Phenoxyl Moiety on the Radical Stability and Spectroscopy. *Inorg. Chim. Acta* **2017**, *481*, 129–142. [CrossRef]
49. Michel, F.; Thomas, F.; Hamman, S.; Saint-Aman, E.; Bucher, C.; Pierre, J.-L. Galactose Oxidase Models: Solution Chemistry, and Phenoxyl Radical Generation Mediated by the Copper Status. *Chem. Eur. J.* **2004**, *10*, 4115–4125. [CrossRef]
50. Michel, F.; Hamman, S.; Thomas, F.; Philouze, C.; Gautier-Luneau, I.; Pierre, J.-L. Galactose Oxidase models: 19F NMR as a powerful tool to study the solution chemistry of tripodal ligands in the presence of copper(ii). *Chem. Commun.* **2006**, 4122–4124. [CrossRef]
51. Michel, F.; Hamman, S.; Philouze, C.; Del Valle, C.P.; Saint-Aman, E.; Thomas, F. Galactose oxidase models: Insights from 19F NMR spectroscopy. *Dalton Trans.* **2009**, 832–842. [CrossRef] [PubMed]
52. Iacomino, G.; Picariello, G.; D'Agostino, L. DNA and nuclear aggregates of polyamines. *Biochim. Biophys. Acta Mol. Cell Res.* **2012**, *1823*, 1745–1755. [CrossRef] [PubMed]
53. Xiong, Y.; Li, J.; Huang, G.; Yan, L.; Ma, J. Interacting mechanism of benzo(a)pyrene with free DNA in vitro. *Int. J. Biol. Macromol.* **2021**, *167*, 854–861. [CrossRef] [PubMed]
54. McGhee, J.D.; von Hippel, P.H. Theoretical aspects of DNA-protein interactions: Co-operative and non-co-operative binding of large ligands to a one-dimensional homogeneous lattice. *J. Mol. Biol.* **1974**, *86*, 469–489. [CrossRef]
55. Hinton, D.M.; Bode, V.C. Ethidium binding affinity of circular lambda deoxyribonucleic acid determined fluorometrically. *J. Biol. Chem.* **1975**, *250*, 1061–1067. [CrossRef]
56. Stoll, S.; Schweiger, A. EasySpin, a Comprehensive Software Package for Spectral Simulation and Analysis in EPR. *J. Magn. Reson.* **2006**, *178*, 42–55. [CrossRef]
57. Sheldrick, G.M. *SADABS, Bruker–Siemens Area Detector Absorption and Other Correction*; Version 2008/1; University of Göttingen: Göttingen, Germany, 2006.
58. Dolomanov, O.V.; Bourhis, L.J.; Gildea, R.J.; Howard, J.A.K.; Puschmann, H. Olex2: A Complete Structure Solution, Refinement and Analysis Program. *J. Appl. Cryst.* **2009**, *42*, 339–340. [CrossRef]
59. Frisch, M.J.; Trucks, G.W.; Schlegel, H.B.; Scuseria, G.E.; Robb, M.A.; Cheeseman, J.R.; Scalmani, G.; Barone, V.; Mennucci, B.; Petersson, G.A.; et al. *Gaussian 09, Revision D.01*; Gaussian, Inc.: Wallingford, CT, USA, 2009.

60. Lee, C.; Yang, W.; Parr, R.G. Development of the Colle-Salvetti Correlation-energy Formula into a Functional of the Electron Density. *Phys. Rev. B Condens. Matter Mater. Phys.* **1988**, *37*, 785–789. [CrossRef]
61. Becke, A.D. Density-functional Thermochemistry. III. The Role of Exact Exchange. *J. Chem. Phys.* **1993**, *98*, 5648–5652. [CrossRef]
62. Petersson, G.A.; Al-Laham, M.A. A complete basis set model chemistry. II. Open-shell systems and the total energies of the first-row atoms. *J. Chem. Phys.* **1991**, *94*, 6081–6090. [CrossRef]
63. Casida, M.E. *Recent Advances in Density Functional Methods*; Chong, D.P., Ed.; World Scientific: Singapore, 1995.
64. Miertuš, S.; Scrocco, E.; Tomasi, J. Electrostatic Interaction of a Solute with a Continuum. A Direct Utilizaion of Ab Initio Molecular Potentials for the Prevision of Solvent Effects. *Chem. Phys.* **1981**, *55*, 117–129. [CrossRef]
65. Kozlowski, A.; Charles, S.A.; Harris, J.M. Development of Pegylated Interferons for the Treatment of Chronic Hepatitis C. *BioDrugs* **2001**, *15*, 419–429. [CrossRef] [PubMed]

Disclaimer/Publisher's Note: The statements, opinions and data contained in all publications are solely those of the individual author(s) and contributor(s) and not of MDPI and/or the editor(s). MDPI and/or the editor(s) disclaim responsibility for any injury to people or property resulting from any ideas, methods, instructions or products referred to in the content.

Article

Novel Copper(II) Complexes with N^4,S-Diallylisothiosemicarbazones as Potential Antibacterial/Anticancer Drugs

Vasilii Graur [1,*], Irina Usataia [1], Ianina Graur [1], Olga Garbuz [1,2], Paulina Bourosh [3], Victor Kravtsov [3], Carolina Lozan-Tirsu [4], Greta Balan [4], Valeriu Fala [4] and Aurelian Gulea [1,*]

[1] Laboratory of Advanced Materials in Biofarmaceutics and Technics, Moldova State University, 60 Mateevici Street, MD-2009 Chisinau, Moldova
[2] Institute of Zoology, Moldova State University, 1 Academiei Street, MD-2028 Chisinau, Moldova
[3] Institute of Applied Physics, Moldova State University, 5 Academiei Street, MD-2028 Chisinau, Moldova
[4] Department of Preventive Medicine, State University of Medicine and Pharmacy "Nicolae Testemitanu", 165 Stefan cel Mare si Sfant Bd., MD-2004 Chisinau, Moldova
* Correspondence: vgraur@gmail.com (V.G.); guleaaurelian@gmail.com (A.G.)

Abstract: The six new copper(II) coordination compounds [Cu(HL^1)Cl$_2$] (**1**), [Cu(HL^1)Br$_2$] (**2**), [Cu(H$_2$O)(L^1)(CH$_3$COO)]·1.75H$_2$O (**3**), [Cu(HL^2)Cl$_2$] (**4**), [Cu(HL^2)Br$_2$] (**5**), [Cu(H$_2$O)(L^2)(CH$_3$COO)] (**6**) were synthesized with 2-formyl- and 2-acetylpyridine N^4,S-diallylisothiosemicarbazones (HL^1 and HL^2). The new isothiosemicarbazones were characterized by NMR, FTIR spectroscopy, and X-ray crystallography ([H$_2$L^2]I). All copper(II) coordination compounds were characterized by elemental analysis, FTIR spectroscopy, and molar conductivity of their 1mM methanol solutions. Furthermore, the crystal structure of complex **3** was determined using single-crystal X-ray diffraction analysis. The studied complexes manifest antibacterial and antifungal activities, that in many cases are close to the activity of medical drugs used in this area, and in some cases even exceed them. The complexes **4** and **5** showed the highest indexes of selectivity (280 and 154) and high antiproliferative activity against BxPC-3 cell lines that surpass the activity of Doxorubicin. The complexes **1–3** also manifest antioxidant activities against cation radicals ABTS$^{•+}$ that are close to that of trolox, the antioxidant agent used in medicine.

Keywords: isothiosemicarbazones; copper complexes; antiproliferative activity; antibacterial activity; antifungal activity; antiradical activity

---

1. Introduction

Copper is one of the crucial micronutrients that is located in different amounts in all human body tissues. The highest amount of copper is in the liver [1]. Various metalloproteins depend on copper as their active site, which makes it essential in a range of biochemical processes: electron transfer, oxidation, and oxygen transport. Copper also participates in cellular respiration, antioxidant protection, neurotransmission, connective tissue biosynthesis, and cellular iron metabolism [2]. Over the past few years, copper compounds have been studied as potential therapeutic agents for application as cancer medicine and as diagnostic drugs [3,4]. Many Cu(II) coordination compounds rapidly interact with glutathione in cells to form adducts and as a result the Cu(I) coordination compound is formed. This compound can generate a superoxide anion, which can induce ROS formation in a Fenton-like reaction [5]. However, antiproliferative action is not the only one for copper coordination compounds such as therapeutic agents, because of their high redox activity. For example, the copper(II) coordination compound with indomethacin is widely used as an anti-inflammatory drug in veterinary practice [6].

Cu(II) complexes of thiosemicarbazone are widely described in the literature because they are able to form stable complexes with different metal ions, which are lipophilic, and can easily permeate cell membranes. These complexes exhibit various types of biological activity: anticancer [7–12], antibacterial and antifungal [13–18], and antioxidant [19]. The antioxidant activity of copper(II) complexes is less studied.

There are many reasons why oxidative stress occurs: pollution, smoking, alcohol consumption, obesity etc. Antioxidants can protect us from free radicals that are produced in our body due to oxidative stress. Such free radicals can cause different diseases such as diabetes, cardiac diseases, cancer, and atherosclerosis [20].

In isothiosemicarbazones, alkylation of the sulfur fragment occurs, and they usually coordinate to the central metal atom through azomethine and thioamide nitrogen atoms. Therefore, in contrast to NS donor atoms of thiosemicarbazones, the isothiosemicarbazones have NN donor atoms. Due to the difference in coordination, it becomes possible to obtain coordination compounds of isothiosemicarbazones with a different structure, which will affect their chemical and biological properties. In some cases isothiosemicarbazones and their copper(II) coordination compounds outperform in activity the complexes of corresponding thiosemicarbazones [21]. Copper(II) complexes with isothiosemicarbazones are less often described in the literature [22–24] and there are several references to their biological activity, such as antibacterial [25,26] and anticancer [27,28].

Recently, we have synthesized 2-formylpyridine and 2-acetylpyridine 4-allyl-S-methylisothiosemicarbazones and their copper(II) coordination compounds [29,30]. Their biological activities such as anticancer, antibacterial, antifungal, and antioxidant have also been researched. These compounds showed promising results. In this paper we have replaced the S-methyl radical with the S-allyl one in the structure of isothiosemicarbazone to study how this will affect biological activity.

The aim of the present investigation is the synthesis, characterization, and study of antibacterial, antifungal, anticancer, and antioxidant activities of Cu(II) coordination compounds with 2-formylpyridine and 2-acetylpyridine N^4,S-diallylisothiosemicarbazones (HL^1 and HL^2, Figure 1).

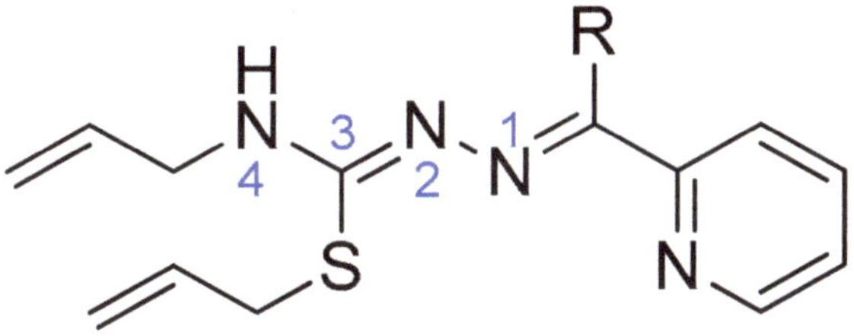

Figure 1. Structural formula of HL^1 (R = H) and HL^2 (R = CH$_3$).

2. Results and Discussion

In this work we have synthesized two new S-substituted isothiosemicarbazones, namely 2-formylpyridine N^4,S-diallylisothiosemicarbazone (HL^1) and 2-acetylpyridine N^4,S-diallylisothiosemicarbazone (HL^2), that were obtained by a three-step method starting with interaction between N^4-allylthiosemicarbazide with allyl iodide, then condensation with 2-formyl-/2-acetil-pyridine, and, finally, neutralization with sodium carbonate (Scheme 1).

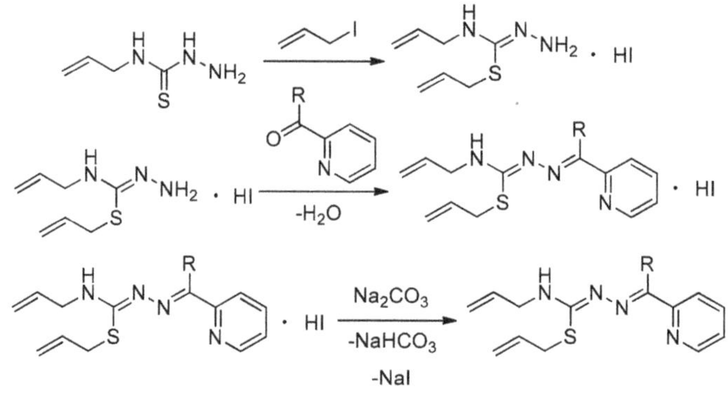

Scheme 1. Synthesis of N^4,S-diallylisothiosemicarbazones HL^1 and HL^2 (HL^1: R = H; HL^2: R = CH$_3$).

The structures of the HL^1 and HL^2 were confirmed using ^{1}H and ^{13}C NMR spectroscopy (Figures S1–S4). The NMR spectra of HL^1 contain peaks of three tautomeric forms that according to the literature [31] presumably are imino form and $cis(N^1$-$N^4)/trans(N^1$-$N^4)$ amino forms (Scheme 2). The NMR spectra of HL^2 contain peaks of two tautomeric forms. Only $cis(N^1$-$N^4)$ and $trans(N^1$-$N^4)$ amino forms of HL^2 can be observed in its spectra.

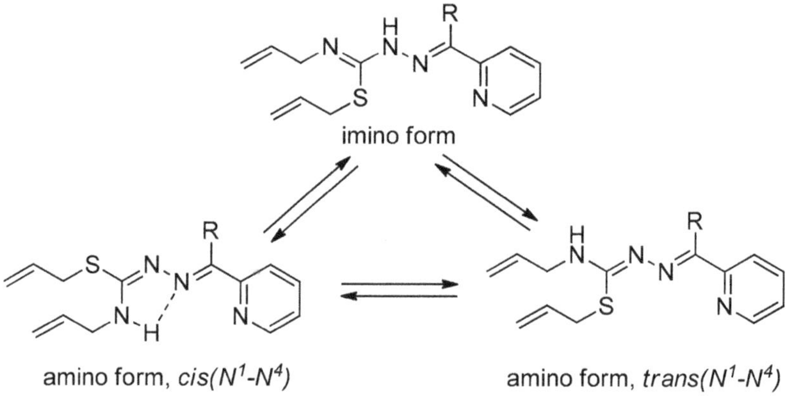

Scheme 2. The equilibrium in the solutions of HL^1 (R = H) and HL^2 (R = CH$_3$).

Furthermore, the single crystals of HL$^2 \cdot$HI were obtained by its recrystallization from methanol and their structure has been determined using single-crystal X-ray diffraction analysis. As a result, it was determined that this organic compound crystallizes in the triclinic space group P^-1 and represents an ionic compound [H$_2$L^2]I (Table 1, Figure 2a). The organic cation [H$_2$L^2]$^+$ forms upon the transfer of the proton from HI to HL^2.

The NNCN torsion angle of the isothiosemicarbazide fragment in this cation is 0.1°, which indicates its $cis(N^1$-$N^4)$ form (both terminal nitrogen atoms are on one side of the double C1=N2 bond). The C(1)–N(1) and C(1)–N(2) bonds equal 1.330(7) and 1.312(7) Å (Table 2). This indicates that the isothiosemicarbazide fragment is stabilized in the amino form [31]. The conformation of the [H$_2$L^2]$^+$ cation is favorable for formation of two intermolecular hydrogen bonds with the iodide anion (Table 3, Figure 2a) and for a tridentate co-ordination to the transition metal atoms. The survey of the Cambridge Structural Database (CSD) [32] revealed that non-coordinated isothiosemicarbazones are mainly stabilized in the amino form [30,33–36], but in the case of {2-[(2-oxyphenyl)methylidene]hydrazinyl}(methylsulfanyl)-N-(prop-2-en-1-yl)methaniminium iodide [37] the imino form is realized. The $cis(N^1$-$N^4)$ conformation similar to that in [H$_2$L^2]$^+$ cation was found in [30,37] with corresponding torsion angles in the range 0.56–2.31°, while in [33–36] these angles are in the range of 175.01–178.97°. In the crystal of [H$_2$L^2]I two intermolecular hydrogen bonds N–H···I link the organic cation to the iodide anion (Table 3). Two weak hydrogen C–H···I bonds unite charged components into chains (Figure 2b).

Table 1. Crystal and Structure Refinement Data for [H$_2$L^2]I and **3**.

Compound	[H$_2$L^2]I	3
Empirical formula	C$_{14}$H$_{19}$I$_1$N$_4$S$_1$	C$_{15}$H$_{23.5}$Cu$_1$N$_4$O$_{4.75}$S$_1$
Formula weight	402.29	431.48
Crystal system	Triclinic	Triclinic
Space group	$P\bar{1}$	$P\bar{1}$
Unit cell dimensions		
a (Å)	7.3553(8)	8.6225(5)
b (Å)	9.0535(9)	10.9536(5)
c (Å)	13.3945(18)	11.3493(8)
α (°)	103.136(10)	89.140(4)
β (°)	91.306(11)	69.700(6)
γ (°)	100.693(9)	81.612(4)
V (Å^3)	851.56(18)	993.85(11)
Z	2	2
ρ_{calc} (g cm^{-3})	1.569	1.442
μ_{Mo} (mm^{-1})	1.999	1.234
F(000)	400	449
Crystal size (mm)	0.60 × 0.12 × 0.08	0.48 × 0.40 × 0.21
θ Range (°)	3.12–25.05	3.39–25.25
Index range	$-8 \leq h \leq 8$, $-10 \leq k \leq 10$, $-15 \leq l \leq 15$	$-10 \leq h \leq 10$, $-12 \leq k \leq 13$, $-13 \leq l \leq 11$
Reflections collected/unique	6159/6159 (twin)	6114/3587 (R_{int} = 0.0238)
Completeness (%)	99.8 (θ = 25.05°)	99.6 (θ =25.25°)
Reflections with $I > 2\sigma(I)$	4518	3037
Number of refined parameters	184	240
Goodness-of-fit (GOF)	1.002	1.001
R (for $I > 2\sigma(I)$)	R_1 = 0.0437, wR_2 = 0.0954	R_1 = 0.0403, wR_2 = 0.1226
R (for all reflections)	R_1 = 0.0608, wR_2 = 0.0992	R_1 = 0.0496, wR_2 = 0.1296
$\Delta\rho_{max}/\Delta\rho_{min}$ (e·Å^{-3})	0.988/−0.521	0.687/−0.279

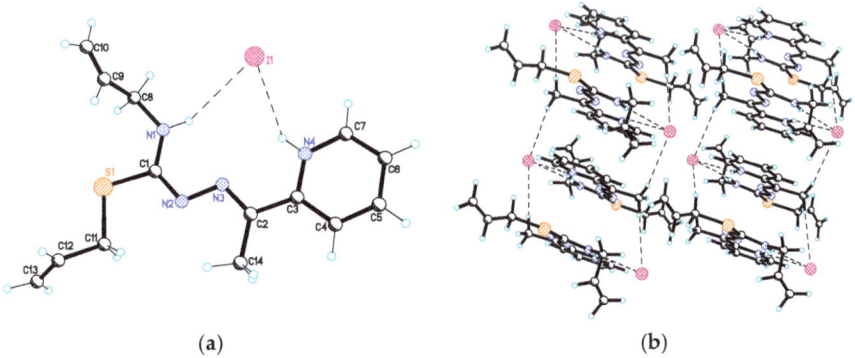

(a) (b)

Figure 2. (**a**) Molecular structure of [H$_2$L^2]I. (**b**) The formation of chains in the crystal of [H$_2$L^2]I.

Table 2. Selected Bond Lengths (Å) and Angles (deg) in fragments of isothiosemicarbazones in [H$_2$L^2]I and **3**.

Bonds	[H$_2$L^2]I	**3**
	(Å)	
N(3)–C(2)	1.292(7)	1.286(4)
N(3)–N(2)	1.374(6)	1.362(3)
C(1)–N(1)	1.330(7)	1.305(4)
C(1)–N(2)	1.312(7)	1.361(4)
C(1)–S(1)	1.760(6)	1.768(3)
S(1)–C(11)	1.821(6)	1.796(4)
N(1)–C(8)	1.463(7)	1.474(4)
Angles	(°)	
C(2)–N(3)–N(2)	112.8(5)	123.1(2)
N(3)–N(2)–C(1)	111.5(5)	107.0(2)
N(2)–C(1)–N(1)	127.1(6)	122.9(3)
N(2)–C(1)–S(1)	115.8(5)	117.0(2)
N(1)–C(1)–S(1)	117.1(5)	120.1(2)
C(1)–S(1)–C(11)	102.5(3)	104.4(2)
C(1)–N(1)–C(8)	126.6(5)	122.1(3)

Table 3. Hydrogen Bond Distances (Å) and Angles (deg) for [H$_2$L^2]I and **3**.

D–H⋯A	d(H⋯A)	d(D⋯A)	∠(DHA)	Symmetry Transformation for Acceptor
	[H$_2$L^2]I			
N(1)–H(1)⋯I(1)	2.84	3.622(5)	152	x, y, z
N(4)–H(2)⋯I(1)	2.75	3.490(5)	146	x, y, z
C(14)–H(2)⋯I(1)	3.31	4.241(6)	165	$-x+2, -y+1, -z+1$
C(14)–H(3)⋯I(1)	3.16	4.121(7)	175	$-x+1, -y+1, -z+1$
	3			
O(1W)–H(1)⋯O(3W)	1.88	2.761(4)	166	$-x, -y+1, -z+2$
O(1W)–H(2)⋯N(2)	1.94	2.835(3)	176	$-x, -y, -z+2$
O(2W)–H(1)⋯O(1W)	2.05	2.814(4)	151	x, y, z
O(2W)–H(2)⋯O(2)	1.95	2.759(4)	158	$x-1, y, z$
O(3W)–H(1)⋯O(2W)	1.92	2.735(5)	159	x, y, z
O(3W)–H(2)⋯O(1)	1.99	2.838(3)	174	x, y, z

Six new copper(II) complexes were obtained by the interaction of the corresponding copper(II) salts with isothiosemicarbazones HL^1 and HL^2 (Scheme 3). They have the following compositions: Cu(HL^1)Cl$_2$ (**1**), Cu(HL^1)Br$_2$ (**2**), Cu(L^1)(CH$_3$COO)·2.75H$_2$O (**3**), Cu(HL^2)Cl$_2$ (**4**), Cu(HL^2)Br$_2$ (**5**), Cu(L^2)(CH$_3$COO)·H$_2$O (**6**).

$$CuCl_2 \cdot 2H_2O + HL^{1,2} \longrightarrow Cu(HL^{1,2})Cl_2 + 2H_2O$$

$$CuBr_2 + HL^{1,2} \longrightarrow Cu(HL^{1,2})Br_2$$

$$Cu(CH_3COO)_2 \cdot H_2O + HL^{1,2} + (n-1) H_2O \longrightarrow Cu(L^{1,2})(CH_3COO) \cdot n(H_2O) + CH_3COOH$$

Scheme 3. Synthesis of complexes **1–6** (n = 1, 2.75).

Molar conductivity values of the complexes **1–2** and **4–5** in methanol are in the range of 169–192 $\Omega^{-1} \cdot cm^2 \cdot mol^{-1}$ which indicates that they behave like 1:2 electrolytes, while the molar conductivity values of complexes **3** and **6** are in the range of 82–85 $\Omega^{-1} \cdot cm^2 \cdot mol^{-1}$ which corresponds to the 1:1 type of electrolyte. The fact that the synthesized complexes **1–6** behave like electrolytes means that the anions of acid residues (Cl^-, Br^-, CH_3COO^-) from the inner sphere are readily substituted with solvent molecules while having been dissolved. It means that complexes **1–2** and **4–5** contain two anions of acid residue (Cl^-/Br^-) in their composition and that in the process of dissolution complex cations and two anions of acid residue are formed. In the case of complexes **3** and **6** only one anion acid residue is present in their composition.

The FTIR spectra of complexes **1–6** were compared with the spectra of corresponding isothiosemicarbazones (HL^1/HL^2) in order to determine the changes that occur during their formation (Figures S5–S12). It was observed that three donor nitrogen atoms of the isothiosemicarbazones HL^1 and HL^2 are involved in the coordination to the copper(II) central atoms. In the spectra of complexes **1–2** and **4–5** the $\nu(NH)$ stretching vibration band is shifted by 63–86 cm^{-1} towards lower wavenumbers. Meanwhile, this band disappears in the spectra of complexes **3** and **6**. It means that the NH group of isothiosemicarbazones is deprotonated in the process of coordination to the copper(II) ions in the presence of acetate ions that act like a weak base. The $\nu(C=N^1)$ and $\nu(C=N_{pyr})$ bands that are observed in the range of 1601–1558 cm^{-1} are shifted by 10–30 cm^{-1} suggesting the coordination of isothiosemicarbazones using azomethine and pyridine nitrogen atoms. Absorption bands of C–S bonds practically are not displaced in the spectra of complexes. Consequently, the sulfur atom is not involved in the coordination to the metal ion in these compounds. Furthermore, the characteristic bands of acetate ions are present in the FTIR spectra of complexes **3** (1620 and 1324 cm^{-1}) and **6** (1614 and 1312 cm^{-1}). According to the literature [38] the difference (Δ) between these two characteristic bands (Δ = 296 cm^{-3} for **3** and Δ = 302 cm^{-3} for **6**) corresponds to monodentate acetate ion in the inner sphere of the coordination compound.

Single crystals of complex **3** were obtained as a result of recrystallization from methanol and their structure was determined using single-crystal X-ray diffraction analysis. The complex **3** crystallizes in the triclinic space group $P\overline{1}$ (Table 1). Structural study determined that the formula of **3** is $[Cu(H_2O)(L^1)(CH_3COO)] \cdot 1.75H_2O$. The asymmetric part of the unit cell contains one molecular complex $[Cu(H_2O)(L^1)(CH_3COO)]$ (Figure 3) and 1.75 solvate water molecules. The Cu(II) in **3** is five-coordinated and the coordination polyhedron represents a square pyramid. The tridentate isothiosemicarbazone ligand is coordinated to the central atom in the monodeprotonated form $(L^1)^-$ using an N_3-set of donor atoms (Figure 3a) and forms two fused metallacycles. Such a coordination mode of similar ligands was found in the complexes of various transition metals [27,30,36,37,39]. Nevertheless, the sulfur atom of isothiosemicarbazones can also participate in coordination [30,35,40].

The basal plane of the Cu(II) polyhedron is formed by three donor atoms of the ligand $(L^1)^-$ and an oxygen atom of the acetate ion. The oxygen atom of the coordinated water molecule is at the apex of this polyhedron. The bond distances and angles in coordination surrounding are given in Table 4. The coordination of the $(L^1)^-$ to the Cu(II) ion did not lead to a change in its conformation, but affected the redistribution of bond lengths in the isothiosemicarbazide fragment: C–N interatomic distances, namely C(1)–N(1) and C(1)–N(2) values of 1.305(4) and 1.361(4) Å (Table 2) indicate the stabilization of the imino form.

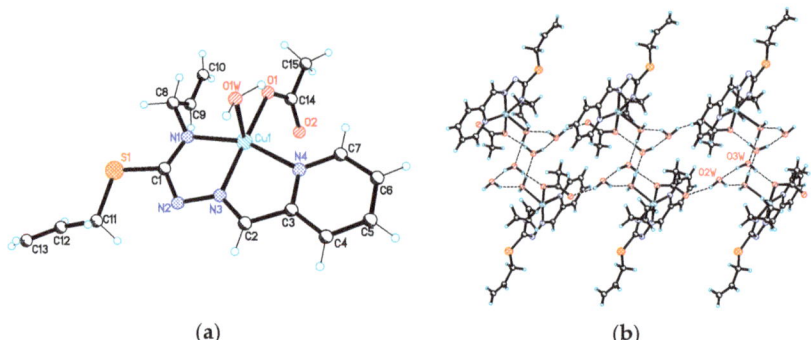

Figure 3. (a) The structure of the molecular complex [Cu(H$_2$O)(L^1)(CH$_3$COO)] in **3**. (b) The six-membered water cluster unites complexes in the chain in **3**.

Table 4. Bond Lengths (Å) and Angles (deg) in Coordination Metal Environment in **3**.

Bonds	Å
Cu(1)–N(1)	1.962(3)
Cu(1)–N(3)	1.948(2)
Cu(1)–N(4)	2.037(3)
Cu(1)–O(1)	1.942(2)
Cu(1)–O(1W)	2.353(2)
Angles	°
N(1)–Cu(1)–N(3)	78.61(10)
N(1)–Cu(1)–N(4)	158.36(11)
N(1)–Cu(1)–O(1)	99.28(10)
N(1)–Cu(1)–O(1W)	98.61(10)
N(3)–Cu(1)–N(4)	80.30(10)
N(3)–M(1)–O(1)	172.76(10)
N(3)–M(1)–O(1W)	99.94(9)
N(4)–M(1)–O(1)	101.04(10)
N(4)–M(1)–O(1W)	89.75(10)
O(1)–M(1)–O(1W)	87.21(9)

The components of the crystal are united in the chain by a system of hydrogen bonds in which two coordinated and four solvate water molecules from two formula units form a six-membered chair-like H-bonded cycle (Table 3, Figure 3b). These chains are associated in the layer parallel to (*ab*) crystallographic plane by intermolecular hydrogen bonds O(W)—H···O(acetate) and O(W)—H···N2.

In order to study the biological properties of the synthesized copper(II) complexes the antibacterial and antifungal properties of the complexes **1–6** were tested on Gram-positive (*S. aureus*, *B. cereus*) bacteria, Gram-negative (*E. coli*, *A. baumannii*) bacteria, and fungi (*C. albicans*). The obtained results in form of minimum inhibitory/bactericidal/fungicidal concentrations are shown in Table 5.

First of all, it is seen that copper(II) coordination compounds in most cases show higher activity than the corresponding N^4,S-diallylisothiosemicarbazones HL^1 and HL^2. The copper(II) complexes manifest higher antibacterial activity towards Gram-positive microorganisms. Among all synthesized copper(II) complexes, the least active ones were the complexes obtained from copper acetate (**3** and **6**). Other complexes showed approximately the same values of activities. So, the dependence between the activity and acid residue can be seen in these results. The activity decreases in the following order: Cl$^-$ ≈ Br$^-$ > CH$_3$COO$^-$. The ligand also affects the activity: copper(II) complexes with 2-acetylpyridine N^4,S-diallylisothiosemicarbazone (HL^2) are more active towards Gram-positive microorganisms and *A. baumanii* than the complexes with 2-formylpyridine N^4,S-diallylisothiosemicarbazone (HL^1). A group of antibiotics (Furacillinum [37,41] and Tetracycline [42–45]) and a group of antifungals (Nystatine [37] and Fluconazole [46]) were used in order to compare the antibacterial and antifungal activities of synthesized complexes with the corresponding activities of medicines. The synthesized complexes **1**, **2**, and **5** manifest greater activity than Furacillinum towards Gram-positive microorganisms and *E. coli*. Complexes **4** and **5** surpass 2–5 times the activity of Furacillinum in the case of *A. baumanii*. Furthermore, complex **5** approximately coincides with the activity of Tetracycline towards Gram-negative microorganism *E. coli*. All the studied copper(II) complexes surpass 4–20 times the activity of standard antifungals (Nystatine and Fluconazole).

Table 5. Minimal inhibitory, bactericidal, and fungicidal concentrations (μg mL^{-1}) of HL^1, HL^2, and copper(II) complexes **1–6**.

Compound	*Staphylococcus aureus* ATCC 25923		*Bacillus cereus* ATCC 11778		*Escherichia coli* ATCC 25922		*Acinetobacter baumannii* BAA-747		*Candida albicans* ATCC 10231	
	MIC	MBC	MIC	MBC	MIC	MBC	MIC	MBC	MIC	MFC
HL^1	125	250	31.3	62.5	>1000	>1000	-	-	15.6	31.3
1	0.977	1.95	0.977	1.95	15.6	31.3	15.6	31.3	7.81	15.6
2	0.977	1.95	1.95	3.91	15.6	31.3	15.6	31.3	3.91	7.81
3	31.3	62.5	31.3	62.5	250	500	-	-	31.3	62.5
HL^2	31.3	62.5	62.5	62.5	>1000	>1000	>1000	>1000	7.81	62.5
4	0.488	0.488	0.488	0.488	31.3	62.5	1.95	1.95	3.91	15.6
5	0.488	0.488	0.488	0.488	1.95	3.91	1.95	1.95	3.91	15.6
6	3.91	3.91	1.95	3.91	62.5	62.5	31.3	31.3	3.91	31.3
Furacillinum [37,41]	9.3	9.3	4.7	4.7	18.5	37.5	4.7	9.4	-	-
Tetracycline [42–45]	0.25	1.96	0.06	-	0.98	3.91	0.5	-	-	-
Nystatine [37]	-	-	-	-	-	-	-	-	80	80
Fluconazole [46]	-	-	-	-	-	-	-	-	15.6	31.3

Note: MIC—minimum inhibitory concentration; MBC—minimum bactericidal concentration; MFC—minimum fungicidal concentration; «-»—data not available.

The antibacterial activity of the synthesized copper(II) complexes can be compared with compounds with similar structures that were previously described in other articles: copper(II) coordination compounds with 2-formylpyridine and 2-acetylpyridine N^4-allyl-S-methylisothiosemicarbazones (S-MeT2FP and S-MeT2AP, correspondingly) [29,30]. Three types of microorganisms were taken for comparison: Gram-positive *S. aureus*, Gram-negative *E. coli* microorganisms, and fungus *C. albicans*. The copper(II) complexes with 2-formylpyridine N^4,S-diallylisothiosemicarbazone (**1**, **2**) showed more modest results towards *S. aureus* than their S-methyl substituted analogs (Figure 4a). While the copper(II) complexes with 2-acetylpyridine isothiosemicarbazone HL^2 obtained in this work surpass the activity of Cu(S-MeT2AP)Cl$_2$ and Cu(S-MeT2AP)Br$_2$ described in the literature. In the case of Gram-negative microorganisms *E. coli* complexes **1** and **2** are 4 times more active than the recently described copper(II) complexes (Figure 4b). The complex **5** manifests higher activity than coordination compounds with S-MeT2AP.

For comparison of antifungal properties, the activity against *C. albicans* was analyzed (Figure 5). All the synthesized complexes **1**, **2**, **4**, **5** exceed the activity of the corresponding coordination compounds with S-MeT2FP and S-MeT2AP.

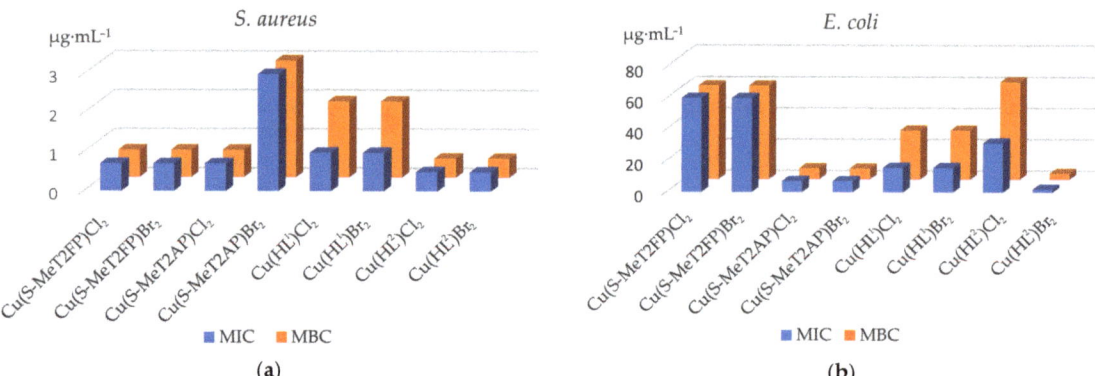

Figure 4. Comparison of the activity of studied complexes with their analogues against *S. aureus* (**a**) and *E. coli* (**b**).

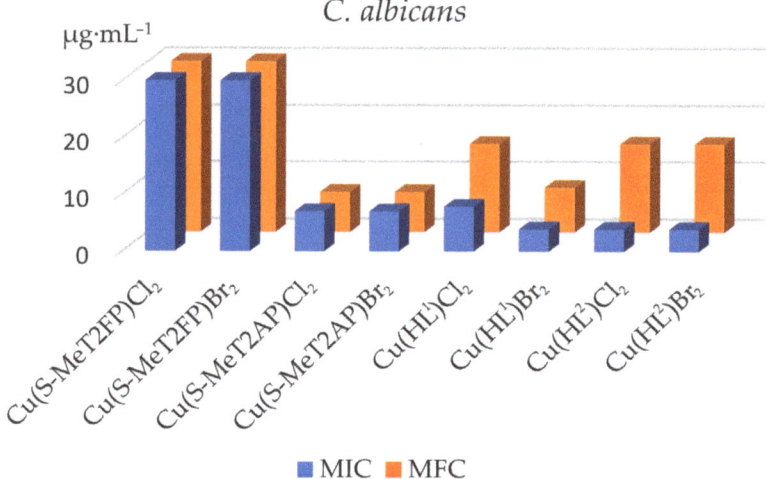

Figure 5. Comparison of the activity of studied complexes with their analogues against *C. albicans*.

Moreover, for the screening of the antiproliferative activity, HL^2 and copper(II) complexes **4** and **5** have been tested towards a series of cancer cell lines (HeLa, BxPC-3, RD) and a normal cell line (MDCK). The obtained results, in the form of semimaximal inhibitory concentrations (IC$_{50}$) and selectivity indexes (SI), are shown in Table 6 as well as the corresponding values of similar compounds, 2-acetylpyridine N^4-allyl-S-methylisothiosemicarbazone and its copper(II) complexes, that are described in [30].

While 2-acetylpyridine N^4,S-diallylisothiosemicarbazone (HL^2) does not manifest anticancer activity (only tested on HeLa and BxPC-3 cell lines), copper(II) complexes manifest a strongly marked antiproliferative activity. The complexes **4** and **5** manifest about the same level of activity. They showed the highest selectivity indexes, 280 and 154, towards BxPC-3 which is one of the most aggressive forms of neoplastic diseases [47]. Recently described copper(II) coordination compounds surpass the antiproliferative activity of studied complexes **4** and **5** towards MDCK and RD cell lines. Doxorubicin (DOXO) is a chemotherapy medication used to treat cancer that was used as a standard. Both synthesized complexes showed higher activity and selectivity than DOXO for all of the studied series of cancer cell lines.

Table 6. IC$_{50}$ values of HL^2 and complexes **4** and **5** towards non-cancerous cell line (MDCK), cancer cell lines (HeLa, BxPC-3, RD), and the corresponding selectivity indexes in comparison with doxorubicin and similar compounds described in [30].

Compound	MDCK IC$_{50}$, μM	HeLa IC$_{50}$, μM	SI	BxPC-3 IC$_{50}$, μM	SI	RD IC$_{50}$, μM	SI
DOXO	7.1 ± 0.3	10.0 ± 0.4	0.71	3.7 ± 0.3	1.9	16.2 ± 0.6	0.44
HL^2	-	>100	-	>100	-	-	-
4	1.4 ± 0.1	0.5 ± 0.1	2.80	0.005 ± 0.001	280	0.2 ± 0.1	7.00
5	1.23 ± 0.01	0.39 ± 0.01	3.15	0.008 ± 0.001	154	1.3 ± 0.4	0.95
S-MeT2AP	13.0 ± 1.3	47.6 ± 4.9	0.27	1.5 ± 0.5	8.7	>100	-
[Cu(S-MeT2AP)Cl$_2$]	1.00 ± 0.02	3.0 ± 1.2	0.33	0.09 ± 0.01	11	0.16 ± 0.01	6.3
[Cu(S-MeT2AP)Br$_2$]	0.35 ± 0.01	0.6 ± 0.2	0.58	0.02 ± 0.01	18	0.05 ± 0.01	7.0

Note: S-MeT2AP—2-acetylpyridine N^4-allyl-S-methylisothiosemicarbazone [30]; SI = IC$_{50}$(MDCK)/IC$_{50}$(cancer cell line)—selectivity index.

The antiradical activity against ABTS$^{•+}$ cation radicals was studied for HL^1, HL^2, and copper(II) complexes **1–6**. The obtained results in form of semimaximal inhibitory concentrations (IC$_{50}$) are shown in Table 7. The HL^1 and its copper(II) complexes **1–3** manifest the highest antiradical activity that is close to the activity of trolox, which is used in medicine as standard antioxidant agent. Complexes **4** and **5** are practically inactive towards ABTS$^{•+}$ cation radicals.

Table 7. Antiradical activity of complexes **1–6** against ABTS$^{•+}$.

Compound	IC$_{50}$, μM
HL^1	28.5 ± 4.0
1	28.9 ± 6.1
2	32.7 ± 0.9
3	30.1 ± 1.3
HL^2	80.8 ± 13.4
4	>100
5	>100
6	95.0 ± 7.3
Trolox	33.3 ± 0.2

3. Experimental Section

3.1. Materials and Instrumentation

All the reagents used were chemically pure. Copper(II) salts CuCl$_2$·2H$_2$O, CuBr$_2$, Cu(CH$_3$COO)$_2$·H$_2$O (Merck) were used as supplied. Allyl isothiocyanate, 50–60% (w/w) aqueous solution of hydrazine, allyl iodide, 2-formylpyridine, 2-acetylpyridine, and sodium carbonate were used as received (Sigma-Aldrich). N^4-Allyl-3-thiosemicarbazide was synthesized by reaction of fourfold excess of 50–60% (w/w) aqueous solution of hydrazine and allyl isothiocyanate [48]. The solvents were purified and dried according to standard procedures [49].

Bruker DRX-400 was used to record the ^{1}H and ^{13}C NMR spectra. Acetone-d$_6$ was used as a solvent to prepare probes for the NMR study. Bruker ALPHA FTIR spectrophotometer was used to record FTIR spectra of studied substances in the range of 4000–400 cm^{-1} at rt. The elemental analysis was performed similarly to the literature procedures [50] and on the automatic Perkin Elmer 2400 elemental analyzer. R-38 rheochord bridge was used to measure the resistance of 1 mM methanol solutions of complexes **1–6** at 20 °C.

3.2. Synthesis

3.2.1. Synthesis of N^4,S-Diallylisothiosemicarbazones

2-Formylpyridine N^4,S-Diallylisothiosemicarbazone (HL^1)

At the first step, the allyl iodide (1.68 g, 10.0 mmol) has been added to the solution of N^4-allylthiosemicarbazide (1.31 g, 10.0 mmol) in ethanol [51]. After 2 h of stirring at room temperature, 2-formylpyridine (1.07 g, 10.0 mmol) was added. The solution was stirred at 70 °C for 30 min. After cooling to room temperature, a yellow precipitate formed from the solution, which was filtered off, washed with ethanol, and dried in air. The obtained precipitate was dissolved in ethanol, and aqua solution of sodium carbonate was added dropwise to the obtained solution until the pH reached value 7–8. After that, the 2-formylpyridine N^4,S-diallylisothiosemicarbazone was extracted by chloroform and dried in vacuo.

Pale yellow solid. Yield: 75%; mp 62–63 °C. FW: 260.36 g/mol; Anal Calc. for $C_{13}H_{16}N_4S$: C, 59.97; H, 6.19; N, 21.52; S, 12.32; found: C, 60.28; H, 6.03; N, 21.48; S, 12.49%. FTIR data (cm^{-1}): ν(N-H) 3219; ν (C=N) 1599, 1575, 1560; ν(CH$_2$–S) 1096; ν (C–S) 766.

Form A (amino form, *cis(N^1-N^4)*). ^{1}H NMR (acetone-d$_6$): 8.59 (d, 1H, CH aromatic); 8.33 (s, 1H, CH=N); 8.04 (d, 1H, CH aromatic); 7.79 (t, 1H, CH aromatic); 7.33 (t, 1H, CH aromatic); 7.47 (br, 1H, NH); 6.12–5.88 (m, 2H, CH allyl); 5.44–4.96 (m, 4H, 2×CH$_2$=C); 3.96 (t, 2H, CH$_2$-N); 3.72 (d, 2H, CH$_2$-S). ^{13}C NMR (acetone-d$_6$): 163.61 (C-S); 154.91, 152.51, 135.38, 123.35, 119.92 (C aromatic); 149.46 (CH=N); 136.04, 134.17 (CH allyl); 117.24, 115.37, (CH$_2$=); 45.34 (CH$_2$-N); 32.24 (CH$_2$-S).

Form B (imino form). ^{1}H NMR (acetone-d$_6$): 8.57 (d, 1H, CH aromatic); 8.24 (s, 1H, CH=N); 8.13 (d, 1H, CH aromatic); 7.77 (t, 1H, CH aromatic); 7.31 (t, 1H, CH aromatic); 5.13 (br, 1H, NH); 6.12–5.88 (m, 2H, CH allyl); 5.44–4.96 (m, 4H, 2×CH$_2$=C); 4.09 (d, 2H, CH$_2$-N); 3.83 (d, 2H, CH$_2$-S). ^{13}C NMR (acetone-d$_6$): 163.56 (C-S); 155.21, 151.73, 134.91, 123.62, 120.46 (C aromatic); 149.37 (CH=N); 136.05, 133.75 (CH allyl); 117.83, 115.26, (CH$_2$=); 45.58 (CH$_2$-N); 32.49 (CH$_2$-S).

Form C (amino form, *trans(N^1-N^4)*). ^{1}H NMR (acetone-d$_6$): 8.69 (d, 1H, CH aromatic); 8.23 (s, 1H, CH=N); 8.24 (d, 1H, CH aromatic); 7.97 (t, 1H, CH aromatic); 7.48 (t, 1H, CH aromatic); 6.12–5.88 (m, 2H, CH allyl); 5.44–4.96 (m, 4H, 2×CH$_2$=C); 3.95 (t, 2H, CH$_2$-N); 3.93 (d, 2H, CH$_2$-S). ^{13}C NMR (acetone-d$_6$): 163.13 (C-S); 154.49, 152.53, 133.42, 124.18, 122.77 (C aromatic); 148.96 (CH=N); 137.25, 133.01 (CH allyl); 118.01, 116.87 (CH$_2$=); 47.50 (CH$_2$-N); 35.78 (CH$_2$-S).

2-Acetylpyridine N^4,S-Diallylisothiosemicarbazone (HL^2)

The isothiosemicarbazone HL^2 was synthesized similarly to HL^1 using 2-acetylpyridine (1.21 g, 10.0 mmol) instead of 2-formylpyridine.

Pale yellow solid. Yield: 80%; mp 96–97 °C. FW: 274.38 g/mol; Anal Calc. for $C_{14}H_{18}N_4S$: C, 61.28; H, 6.61; N, 20.42; S, 11.69; found: C, 61.07; H, 6.48; N, 20.37; S, 11.48%. FTIR data (cm^{-1}): ν(N-H) 3215; ν (C=N) 1601, 1583, 1558; ν(CH$_2$–S) 1044; ν (C–S) 743.

Form A (amino form, *cis(N^1-N^4)*). ^{1}H NMR (acetone-d$_6$): 8.58 (d, 1H, CH aromatic); 8.26 (d, 1H, CH aromatic); 7.71 (t, 1H, CH aromatic); 7.29 (t, 1H, CH aromatic); 7.27 (br, 1H, NH); 5.99 (m, 2H, CH allyl); 5.20 (m, 4H, 2×CH$_2$=C); 3.96 (t, 2H, CH$_2$-N); 3.86 (d, 2H, CH$_2$-S); 2.51 (s, 3H, CH$_3$). ^{13}C NMR (acetone-d$_6$): 161.53 (C-S); 157.19, 156.99, 135.66, 123.00, 120.16 (C aromatic); 148.46 (CH=N); 135.55, 134.41 (CH allyl); 116.98, 115.15, (CH$_2$=); 45.31 (CH$_2$-N); 32.35 (CH$_2$-S); 12.30 (CH$_3$).

Form B (amino form, *trans(N^1-N^4)*). ^{1}H NMR (acetone-d$_6$): 8.56 (d, 1H, CH aromatic); 8.20 (d, 1H, CH aromatic); 7.75 (т, 1H, CH aromatic); 7.31 (t, 1H, CH aromatic); 5.98 (m, 2H, CH allyl); 5.29 (m, 4H, 2×CH$_2$=C); 5.10 (br, 1H, NH); 4.11 (t, 2H, CH$_2$-N); 3.69 (d, 2H, CH$_2$-S); 2.43 (s, 3H, CH$_3$). ^{13}C NMR (acetone-d$_6$): 161.26 (C-S); 158.23, 156.79, 135.60, 123.25, 120.35 (C aromatic); 148.53 (C=N); 134.99, 133.99 (CH allyl); 117.67, 115.38 (CH$_2$=); 45.88 (CH$_2$-N); 32.28 (CH$_2$-S); 12.02 (CH$_3$).

3.2.2. Synthesis of Copper(II) Complexes

[Cu(HL^1)Cl$_2$] (1)

Copper(II) chloride dihydrate (CuCl$_2$·2H$_2$O) (0.170 g, 1 mmol) was added to a hot (55° C) ethanolic solution (25 mL) of 2-formylpyridine N^4,S-diallylisothiosemicarbazone HL^1 (0.260 g, 1 mmol). The mixture was stirred for 30 min at 55 °C. By cooling to room temperature, a green precipitate was obtained. It was filtered out, washed with cold ethanol, and dried in vacuo.

Green solid. Yield: 80%. Anal. Calc. for C$_{13}$H$_{16}$Cl$_2$CuN$_4$S (394.81 g mol^{-1}): C, 39.55; H, 4.08; Cl, 17.96; Cu, 16.10; N, 14.19; S, 8.12. Found: C, 39.38; H, 4.05; Cl, 17.91; Cu, 15.89; N, 14.02; S, 7.95. Main FTIR peaks (cm^{-1}): ν(NH) 3156, ν(C=N) 1591, 1567, 1534, ν(CH$_2$–S) 1095, ν(C–S) 768. χ(CH$_3$OH): 169 Ω^{-1} cm^{-2} mol^{-1}.

[Cu(HL^1)Br$_2$] (2)

The coordination compound **2** was synthesized similarly to compound **1** using CuBr$_2$ (0.223 g; 1 mmol) and HL^1 (0.260 g; 1 mmol).

Green solid. Yield: 85%. Anal. Calc. for C$_{13}$H$_{16}$Br$_2$CuN$_4$S (483.71 g mol^{-1}): C, 32.28; H, 3.33; Br, 33.04; Cu, 13.14; N, 11.58; S, 6.63. Found: C, 32.05; H, 3.20; Br, 33.17; Cu, 13.45; N, 11.71; S, 6.72. Main FTIR peaks (cm^{-1}): ν(NH) 3139, ν(C=N) 1593, 1567, 1538, ν(CH$_2$–S) 1098, ν(C–S) 765. χ(CH$_3$OH): 178 Ω^{-1} cm^{-2} mol^{-1}.

[Cu(H$_2$O)(L^1)(CH$_3$COO)]·1.75H$_2$O (3)

The coordination compound **3** was synthesized similarly to compound **1** using Cu(CH$_3$COO)$_2$·H$_2$O (0.200 g; 1 mmol) and HL^1 (0.260 g; 1 mmol).

Brown solid. Yield: 82%. Anal. Calc. for C$_{15}$H$_{23.5}$CuN$_4$O$_{4.75}$S (431.48 g mol^{-1}): C, 41.75; H, 5.49; Cu, 14.73; N, 12.98; S, 7.43. Found: C, 41.62; H, 5.58; Cu, 14.79; N, 12.81; S, 7.29. Main FTIR peaks (cm^{-1}): ν(C=O) 1620, ν(C=N) 1596, 1558, 1532, ν(C–O) 1324, ν(CH$_2$–S) 1091, ν(C–S) 766. χ(CH$_3$OH): 85 Ω^{-1} cm^{-2} mol^{-1}.

[Cu(HL^2)Cl$_2$] (4)

The coordination compound **4** was synthesized similarly to compound **1** using CuCl$_2$·2H$_2$O (0.170 g; 1 mmol) and HL^2 (0.274 g; 1 mmol).

Green solid. Yield: 78%. Anal. Calc. for C$_{14}$H$_{18}$Cl$_2$CuN$_4$S (408.84 g mol^{-1}): C, 41.13; H, 4.44; Cl, 17.34; Cu, 15.54; N, 13.70; S, 7.84. Found: C, 41.23; H, 4.56; Cl, 17.51; Cu, 15.72; N, 13.57; S, 7.93. Main FTIR peaks (cm^{-1}): ν(N–H) 3129, ν(C=N) 1591, 1571, 1544, ν(CH$_2$–S) 1044, ν(C–S) 746. χ(CH$_3$OH): 192 Ω^{-1} cm^{-2} mol^{-1}.

[Cu(HL^2)Br$_2$] (5)

The coordination compound **5** was synthesized similarly to compound **1** using CuBr$_2$ (0.223 g; 1 mmol) and HL^2 (0.274 g; 1 mmol).

Green solid. Yield: 72%. Anal. Calc. for C$_{14}$H$_{18}$Br$_2$CuN$_4$S (497.74 g mol^{-1}): C, 33.78; H, 3.65; Br, 32.11; Cu, 12.77; N, 11.26; S, 6.44. Found: C, 33.95; H, 3.82; Br, 32.29; Cu, 12.65; N, 11.10; S, 6.26. Main FTIR peaks (cm^{-1}): ν(NH) 3143, ν(C=N) 1591, 1569, 1542, ν(CH$_2$–S) 1043, ν(C–S) 747. χ(CH$_3$OH): 178 Ω^{-1} cm^{-2} mol^{-1}.

[Cu(H$_2$O)(L^2)(CH$_3$COO)] (6)

The coordination compound **6** was synthesized similarly to compound **1** using Cu(CH$_3$COO)$_2$·H$_2$O (0.200 g; 1 mmol) and HL^2 (0.274 g; 1 mmol).

Brown solid. Yield: 81%. Anal. Calc. for C$_{16}$H$_{22}$CuN$_4$O$_3$S (413.98 g mol^{-1}): C, 46.42; H, 5.36; Cu, 15.35; N, 13.53; S, 7.75. Found: C, 46.19; H, 5.42; Cu, 15.12; N, 13.59; S, 7.49. Main FTIR peaks (cm^{-1}): ν(C=O) 1614, ν(C=N) 1595, 1561, 1543, ν(C–O) 1312, ν(CH$_2$–S) 1041, ν(C–S) 741. χ(CH$_3$OH): 82 Ω^{-1} cm^{-2} mol^{-1}.

3.3. X-ray Crystallography

The single-crystal X-ray analysis of [H_2L^2]I and complex **3** were carried out at room temperature (293 K) on an Xcalibur E CCD diffractometer equipped with a CCD area detector and a graphite monochromator, Mo$K\alpha$ radiation (0.71073 Å). CrysAlis PRO software was used for data collection and reduction, and unit cell determination. The structures were solved and refined using the SHELXS97 and SHELXL2014 software packages [52,53]. The non-hydrogen atoms were treated anisotropically (full-matrix least squares method on F^2). The hydrogen atoms were placed in calculated positions and were treated using riding model approximations with $U_{iso}(H) = 1.2U_{eq}(C)$, while the oxygen-bounded H atoms were found from differential Fourier maps at an intermediate stage of the structure refinement. These hydrogen atoms were refined with the isotropic displacement parameter $U_{iso}(H) = 1.5U_{eq}(O)$.

The crystallographic data were deposited with the Cambridge Crystallographic Data Center, CCDC nos. 2253067 and 2253068 for [H_2L^2]I and **3**, respectively. Copies of this information may be obtained free of charge from the Director, CCDC, 12 Union Road, Cambridge CHB2 1EZ, UK (Fax: +44-1223-336033; e-mail: deposit@ccdc.cam.ac.uk or www.ccdc.cam.ac.uk (accessed on 26 April 2023)).

3.4. Antibacterial and Antifungal Activity

Antibacterial and antifungal activities of the isothiosemicarbazones HL^1, HL^2, and copper(II) coordination compounds **1–6** were studied on a series of standard strains: *Bacillus cereus* (ATCC 11778), *Staphylococcus aureus* (ATCC 25923), *Acinetobacter baumannii* (BAA-747), *Escherichia coli* (ATCC 25922), and *Candida albicans* (ATCC 10231). The minimum inhibitory concentrations (MICs, µg mL^{-1}), minimum bactericidal concentrations (MBCs, µg mL^{-1}), and minimum fungicidal concentrations (MFCs, µg mL^{-1}) were determined using the method of serial dilutions in liquid broth. The solutions of the tested substances were prepared in DMSO with a 10 mg mL^{-1} concentration. Subsequent dilutions were prepared by incorporating 2% peptonate bullion.

3.5. Antiproliferative Activity

3.5.1. Cell Cultures

The BxPC-3 (ATCC CRL-1687) cells were grown as a monolayer in Roswell Park Memorial Institute 1640 medium to which penicillin–streptomycin (final concentration of penicillin 100 U mL^{-1}; final concentration of streptomycin 100 µg mL^{-1}) was added. Furthermore, fetal bovine serum (FBS) was added to the medium at a concentration of 10% v/v.

The HeLa (ATCC CCL-2), RD (ATCC CCL-136), and MDCK (ATCC CCL-34) cell lines were grown in Dulbecco's modified essential medium. The medium contained glucose (4.5 g L^{-1}), L-glutamine (4 mM), HEPES buffer (20 mM), bovine albumin fraction (0.2% v/v), and penicillin-streptomycin (final concentration of penicillin 100 U mL^{-1}; final concentration of streptomycin 100 µg mL^{-1}). Moreover, the medium was supplemented with FBS at a concentration of 10% v/v.

The cells were cultured in 75-cm^2 dishes in a 5% humidified CO_2 environment at 37 °C.

3.5.2. Resazurin Test

The viability of cancer cells (BxPC-3, HeLa, RD) and normal cells (MDCK) was determined by using resazurin as a reagent.

Stock solutions (1×10^{-2} M) of the tested compounds (HL^1, HL^2, and complexes **1–6**) were prepared by dissolving 10^{-5} mol of each substance in 1 mL DMSO. These stock solutions were then used to prepare diluted solutions with final concentrations of 0.1, 1, 10, 100, and 1000 µM. Corresponding media were used for the dilution process.

To perform the assay, 90 µL of corresponding culture medium containing 1×10^4 cells were placed in the wells of a 96-well microtiter plate and incubated at 37 °C, 5% CO_2 for a 2–3 h period to allow the attachment of cells. Next, 10 µL of diluted solutions (0.1–1000 µM) of the tested compounds were added to the wells with culture medium. The incubation continued for 24 h, after which resazurin indicator solution (20 µL) was added to each well. After 4 h of incubation in presence of resazurin, the absorbance was measured at two wavelengths (570 nm and 600 nm).

3.6. Antiradical Activity

The ABTS$^{\bullet+}$ method [54] with modifications was used to study the antiradical activity of HL^1, HL^2, and complexes **1–6**.

The reaction of 2,20-azino-bis (3-ethylbenzothiazoline-6-sulphonic acid (ABTS, 7 nM) and potassium persulfate (140 mM) gave the ABTS$^{\bullet+}$ radical cations. The reaction was performed in the dark at 25 °C for 12 h. The acetate-buffered saline (0.02 M, pH 6.5) was used for dilution of the obtained solution up to a concentration at which its absorbance at 734 nm was 0.70 ± 0.01 AU.

Stock solutions (1×10^{-2} M) of the tested compounds (HL^1, HL^2, and complexes **1–6**) in DMSO were diluted to obtain final concentrations of 10, 100, and 1000 µM. After that, 180 µL of ABTS$^{\bullet+}$ working solution and 20 µL of each tested compound solution were mixed and homogenized in the wells of a 96-well microtiter plate. After 30 min of incubation at 25 °C, the absorbance of the solutions was measured at 734 nm. The experiment was conducted three times to ensure accuracy.

4. Conclusions

Two new N^4,S-diallylisothiosemicarbazones and six new copper(II) coordination compounds have been synthesized. The structure of isothiosemicarbazones HL^1 and HL^2 was determined using NMR spectroscopy. Isothiosemicarabzones exist in different tautomeric forms in the solution. Crystal structures of [H$_2$L^2]I and complex **3** ([Cu(H$_2$O)(L^1)(CH$_3$CO1O)]·1.75H$_2$O) were proved using X-ray diffraction analysis. The studied isothiosemicarbazones behave as tridentate ligands with N$_3$-set of donor atoms. All the studied complexes (**1–6**) are electrolytes, which indicates the process of substitution of acidic residues (Cl$^-$, Br$^-$, CH$_3$COO$^-$) by solvent molecules in the process of dissolution of these complexes.

Biological evaluation showed that the synthesized complexes manifest promising antibacterial, antifungal, and anticancer activity. Their antibacterial/antifungal activity in many cases is close to the activity of some drugs that are used in medicine for these purposes and, in some cases, surpass them. Complexes **4** and **5** selectively inhibit proliferation of BxPC-3 cancer cell line with IC$_{50}$ values 5–8 nM. Thus, these complexes exceed 400–700 times the corresponding activity of doxorubicin and 2.5–18 times the activity of the corresponding copper(II) complexes with 2-acetylpyridine N^4-allyl-S-methylisothiosemicarbazone. Moreover, their selectivity indexes are in the range of 150–280 which confirms their strongly marked selectivity.

In addition, HL^1 and complexes **1–3** exhibit antiradical activity that exceeds that of trolox. Therefore, copper(II) complexes with S-substituted N^4-allylisothiosemicarbazones manifest promising biological properties, which are also affected by the nature of S-substituent.

Supplementary Materials: The following supporting information can be downloaded at: https://www.mdpi.com/article/10.3390/inorganics11050195/s1, Figure S1: ^{1}H NMR spectrum of 2-formylpyridine N^4,S-diallylisothiosemicarbazone (HL^1); Figure S2: ^{13}C NMR spectrum of 2-formylpyridine N^4,S-diallylisothiosemicarbazone (HL^1); Figure S3: ^{1}H NMR spectrum of 2-acetylpyridine N^4,S-diallylisothiosemicarbazone (HL^2); Figure S4: ^{13}C NMR spectrum of 2-acetylpyridine N^4,S-diallylisothiosemicarbazone (HL^2); Figure S5: FTIR spectrum of HL^1; Figure S6: FTIR spectrum of **1**; Figure S7: FTIR spectrum of **2**; Figure S8: FTIR spectrum of **3**; Figure S9: FTIR spectrum of HL^2; Figure S10: FTIR spectrum of **4**; Figure S11: FTIR spectrum of **5**; Figure S12: FTIR spectrum of **6**.

Author Contributions: Conceptualization, A.G.; methodology, I.U. and O.G.; validation, V.G., O.G., V.K., G.B. and A.G.; formal analysis, O.G., P.B. and C.L.-T.; investigation, I.U., I.G., O.G., P.B., C.L.-T. and G.B.; resources, V.K., G.B., V.F. and A.G.; data curation, V.G., P.B., V.K. and A.G.; writing—original draft preparation, V.G., I.G., P.B.; writing—review and editing, V.G., V.K. and A.G.; visualization, V.G., P.B. and V.K.; supervision, A.G.; project administration, V.K. and A.G.; All authors have read and agreed to the published version of the manuscript.

Funding: This research was funded by ANCD, grant numbers 20.80009.5007.10 and 20.80009.5007.15.

Data Availability Statement: Data is contained within the article or supplementary material.

Acknowledgments: The authors are thankful to Olga Burduniuc, N. Testemitsanu State Medical and Pharmaceutical University for the assistance in conducting biological tests of synthesized substances.

Conflicts of Interest: The authors declare no conflict of interest.

References

1. Zatta, P.; Frank, A. Copper deficiency and neurological disorders in man and animals. *Brain Res.* **2007**, *54*, 19–33. [CrossRef] [PubMed]
2. Khalid, H.; Hanif, M.; Ali Hashmi, M.; Mahmood, T.; Ayub, K.; Monim-ul-Mehboob, M. Copper complexes of bioactive ligands with superoxide dismutase activity. *Mini-Rev. Med. Chem.* **2013**, *13*, 1944–1956. [CrossRef] [PubMed]
3. Paterson, B.M.; Donnelly, P.S. Copper complexes of bis(thiosemicarbazones): From chemotherapeutics to diagnostic and therapeutic radiopharmaceuticals. *Chem. Soc. Rev.* **2011**, *40*, 3005–3018. [CrossRef] [PubMed]
4. Barone, G.; Terenzi, A.; Lauria, A.; Almerico, A.M.; Leal, J.M.; Busto, N.; García, B. DNA-binding of nickel(II), copper(II) and zinc(II) complexes: Structure–affinity relationships. *Coord. Chem. Rev.* **2013**, *257*, 2848–2862. [CrossRef]
5. Chudal, L.; Pandey, N.K.; Phan, J.; Johnson, O.; Lin, L.; Yu, H.; Shu, Y.; Huang, Z.; Xing, M.; Liu, J.P.; et al. Copper-Cysteamine Nanoparticles as a Heterogeneous Fenton-Like Catalyst for Highly Selective Cancer Treatment. *ACS Appl. Bio Mater.* **2020**, *3*, 1804–1814. [CrossRef] [PubMed]
6. Weder, J.E.; Hambley, T.W.; Kennedy, B.J.; Lay, P.A.; MacLachlan, D.; Bramley, R.; Delfs, C.D.; Murray, K.S.; Moubaraki, B.; Warwick, B.; et al. Anti-Inflammatory Dinuclear Copper(II) Complexes with Indomethacin. Synthesis, Magnetism and EPR Spectroscopy. Crystal Structure of the N,N-Dimethylformamide Adduct. *Inorg. Chem.* **1999**, *38*, 1736–1744. [CrossRef] [PubMed]
7. Palanimuthu, D.; Shinde, S.V.; Somasundaram, K.; Samuelson, A.G. In vitro and in vivo anticancer activity of copper bis (thiosemicarbazone) complexes. *J. Med. Chem.* **2013**, *56*, 722–734. [CrossRef]
8. Fiadjoe, H.K.; Lambring, C.; Sankpal, U.T.; Alajroush, D.; Holder, A.A.; Basha, R. Anti-proliferative effect of two copper complexes against medulloblastoma cells. *Cancer Res.* **2023**, *83*, 6255. [CrossRef]
9. Mathews, N.A.; Kurup, M.P. Copper (II) complexes as novel anticancer drug: Synthesis, spectral studies, crystal structures, in silico molecular docking and cytotoxicity. *J. Mol. Struct.* **2022**, *1258*, 132672. [CrossRef]
10. Zheng, Y.; Li, B.; Ai, Y.; Chen, M.; Zheng, X.; Qi, J. Synthesis, crystal structures and anti-cancer mechanism of Cu (II) complex derived from 2-acetylpyrazine thiosemicarbazone. *J. Coord. Chem.* **2022**, *75*, 1325–1340. [CrossRef]
11. Paprocka, R.; Wiese-Szadkowska, M.; Janciauskiene, S.; Kosmalski, T.; Kulik, M.; Helmin-Basa, A. Latest developments in metal complexes as anticancer agents. *Coord. Chem. Rev.* **2022**, *452*, 214307. [CrossRef]
12. Adhikari, H.S.; Garai, A.; Yadav, P.N. Synthesis, characterization, and anticancer activity of chitosan functionalized isatin based thiosemicarbazones, and their copper (II) complexes. *Carbohydr. Res.* **2023**, *526*, 108796. [CrossRef]
13. Bajaj, K.; Buchanan, R.M.; Grapperhaus, C.A. Antifungal activity of thiosemicarbazones, bis (thiosemicarbazones), and their metal complexes. *J. Inorg. Biochem.* **2021**, *225*, 111620. [CrossRef] [PubMed]
14. Benns, B.G.; Gingras, B.A.; Bayley, C.H. Antifungal activity of some thiosemicarbazones and their copper complexes. *Appl. Microbiol.* **1960**, *8*, 353–356. [CrossRef] [PubMed]
15. Verma, K.K.; Nirwan, N.; Singh, R.; Bhojak, N. Microwave Assisted Synthesis, Characterisation and Biological Activities of Cu (II) Complexes of Few Thiosemicarbazones Ligands. *J. Sci. Res.* **2023**, *15*, 275–283. [CrossRef]
16. Dong, X.; Wang, H.; Zhang, H.; Li, M.; Huang, Z.; Wang, Q.; Li, X. Copper-thiosemicarbazone complexes conjugated-cellulose fibers: Biodegradable materials with antibacterial capacity. *Carbohydr. Polym.* **2022**, *294*, 119839. [CrossRef]
17. Petrasheuskaya, T.V.; Kovács, F.; Igaz, N.; Rónavári, A.; Hajdu, B.; Bereczki, L.; May, N.V.; Spengler, G.; Gyurcsik, B.; Kiricsi, M.; et al. Estradiol-Based Salicylaldehyde (Thio) Semicarbazones and Their Copper Complexes with Anticancer, Antibacterial and Antioxidant Activities. *Molecules* **2023**, *28*, 54. [CrossRef]
18. Nandaniya, B.; Das, S.; Jani, D. New thiosemicarbazone derivatives and their Mn (II), Ni (II), Cu (II) and Zn (II) complexes: Synthesis, characterization and in-vitro biological screening. *Curr. Chem. Lett.* **2023**, *12*, 289–296. [CrossRef]
19. Prathima, B.; Rao, Y.S.; Reddy, S.A.; Reddy, Y.P.; Reddy, A.V. Copper (II) and nickel (II) complexes of benzyloxybenzaldehyde-4-phenyl-3-thiosemicarbazone: Synthesis, characterization and biological activity. *Spectrochim. Acta Part A Mol. Biomol. Spectrosc.* **2010**, *77*, 248–252. [CrossRef]
20. Shah, S.S.; Shah, D.; Khan, I.; Ahmad, S.; Ali, U.; Rahman, A. Synthesis and antioxidant activities of Schiff bases and their complexes: An updated review. *Biointerface Res. Appl. Chem* **2020**, *10*, 6936–6963. [CrossRef]
21. Ohui, K.; Afanasenko, E.; Bacher, F.; Ting, R.L.X.; Zafar, A.; Blanco-Cabra, N.; Torrents, E.; Dömötör, O.; May, N.V.; Darvasiova, D.; et al. New water-soluble copper (II) complexes with morpholine–thiosemicarbazone hybrids: Insights into the anticancer and antibacterial mode of action. *J. Med. Chem.* **2018**, *62*, 512–530. [CrossRef] [PubMed]
22. Alizadeh, S.; Mague, J.T.; Takjoo, R. Structural, theoretical investigations and HSA-interaction studies of three new copper (II) isothiosemicarbazone complexes. *Polyhedron* **2022**, *224*, 115986. [CrossRef]

23. Takjoo, R.; Ramasami, P.; Rhyman, L.; Ahmadi, M.; Rudbari, H.A.; Bruno, G. Structural and theoretical studies of iron (III) and copper (II) complexes of dianion N1, N4-bis (salicylidene)-S-alkyl-thiosemicarbazide. *J. Mol. Struct.* **2022**, *1255*, 132388. [CrossRef]
24. Fasihizad, A.; Akbari, A.; Ahmadi, M.; Dusek, M.; Henriques, M.S.; Pojarova, M. Copper (II) and molybdenum (VI) complexes of a tridentate ONN donor isothiosemicarbazone: Synthesis, characterization, X-ray, TGA and DFT. *Polyhedron* **2016**, *115*, 297–305. [CrossRef]
25. Zalevskaya, O.A.; Gur'eva, Y.A. Recent Studies on the Antimicrobial Activity of Copper Complexes. *Russ. J. Coord. Chem.* **2021**, *47*, 861–880. [CrossRef]
26. Heinisch, L.; Fleck, W.F.; Jacob, H.E. Copper II complexes of N-heterocyclic formylisothiosemicarbazones with antimicrobial and beta-lactamase inhibitory activity. *Z. Allg. Mikrobiol.* **1980**, *20*, 619–626. [CrossRef]
27. Gulea, A.P.; Usataia, I.S.; Graur, V.O.; Chumakov, Y.M.; Petrenko, P.A.; Balan, G.G.; Burduniuc, O.S.; Tsapkov, V.I.; Rudic, V.F. Synthesis, Structure and Biological Activity of Coordination Compounds of Copper, Nickel, Cobalt, and Iron with Ethyl N′-(2-Hydroxybenzylidene)-N-prop-2-en-1-ylcarbamohydrazonothioate. *Russ. J. Gen. Chem.* **2020**, *90*, 630–639. [CrossRef]
28. Zaltariov, M.; Hammerstad, M.; Arabshahi, H.; Jovanović, K.; Richter, K.; Cazacu, M.; Shova, S.; Balan, M.; Andersen, N.; Radulović, S.; et al. New iminodiacetate–thiosemicarbazone hybrids and their copper (II) complexes are potential ribonucleotide reductase R2 inhibitors with high antiproliferative activity. *Inorg. Chem.* **2017**, *56*, 3532–3549. [CrossRef]
29. Balan, G.; Burduniuc, O.; Usataia, I.; Graur, V.; Chumakov, Y.; Petrenko, P.; Gudumac, V.; Gulea, A.; Pahontu, E. Novel 2-formylpyridine 4-allyl-S-methylisothiosemicarbazone and Zn (II), Cu (II), Ni (II) and Co (III) complexes: Synthesis, characterization, crystal structure, antioxidant, antimicrobial and antiproliferative activity. *Appl. Organomet. Chem.* **2020**, *34*, e5423. [CrossRef]
30. Graur, V.; Usataia, I.; Bourosh, P.; Kravtsov, V.; Garbuz, O.; Hureau, C.; Gulea, A. Synthesis, characterization, and biological activity of novel 3d metal coordination compounds with 2-acetylpyridine N4-allyl-S-methylisothiosemicarbazone. *Appl. Organomet. Chem.* **2021**, *35*, e6172. [CrossRef]
31. Yamazaki, C. The structure of isothiosemicarbazones. *Can. J. Chem.* **1975**, *53*, 610–615. [CrossRef]
32. Allen, F.H. The Cambridge Structural Database: A quarter of a million crystal structures and rising. *Acta. Crystallogr. B Struct. Sci. Cryst. Eng. Mater.* **2002**, *58*, 380–388. [CrossRef] [PubMed]
33. Arion, V.B.; Rapta, P.; Telser, J.; Shova, S.S.; Breza, M.; Lušpai, K.; Kozisek, J. Syntheses, electronic structures, and EPR/UV-Vis-NIR spectroelectrochemistry of nickel(II), copper(II), and zinc(II) complexes with a Tetradentate ligand based on S-methylisothiosemicarbazide. *Inorg. Chem.* **2011**, *50*, 2918–2931. [CrossRef] [PubMed]
34. Arion, V.B.; Platzer, S.; Rapta, P.; Machata, P.; Breza, M.; Vegh, D.; Dunsch, L.; Telser, J.; Shova, S.; Leod, T.C.O.; et al. Marked stabilization of redox states and enhanced catalytic activity in galactose oxidase models based on transition metal S-methylisothiosemicarbazonates with -SR group in ortho position to the phenolic oxygen. *Inorg. Chem.* **2013**, *52*, 7524–7540. [CrossRef] [PubMed]
35. Revenco, M.; Bulmaga, P.; Jora, E.; Palamarciuc, O.; Kravtsov, V.; Bourosh, P. Specificity of salicylaldehyde S-alkylisothiosemicarbazones coordination in palladium(II) complexes. *Polyhedron* **2014**, *80*, 250–255. [CrossRef]
36. Güveli, Ş.; Kılıç-Cıkla, I.; Ülküseven, B.; Yavuz, M.; Bal-Demirci, T. 5-Methyl-2-hydroxy-acetophenone-S-methyl-thiosemicarbazone and its nickel-PPh3 complex. Synthesis, characterization, and DFT calculations. *J. Mol. Struct.* **2018**, *1173*, 366–374. [CrossRef]
37. Pahontu, E.; Usataia, I.; Graur, V.; Chumakov, Y.; Petrenko, P.; Gudumac, V.; Gulea, A. Synthesis, characterization, crystal structure of novel Cu(II), Co(III), Fe(III) and Cr(III) complexes with 2-hydroxybenzaldehyde-4-allyl-S-methylisothiosemicarbazone: Antimicrobial, antioxidant and in vitro antiproliferative activity. *Appl. Organomet. Chem.* **2018**, *32*, e4544. [CrossRef]
38. Nakamoto, K. *Infrared and Raman Spectra of Inorganic and Coordination Compounds, Part B: Applications in Coordination, Organometallic, and Bioinorganic Chemistry*, 6th ed.; John Wiley & Sons: Hoboken, NJ, USA, 2009; pp. 288–290.
39. Güveli, Ş.; Agopcan Çınar, N.; Karahan, Ö.; Aviyente, V.; Ülküseven, B. Nickel (II)–PPh3 Complexes of S, N-Substituted Thiosemicarbazones–Structure, DFT Study, and Catalytic Efficiency. *Eur. J. Inorg. Chem.* **2016**, *2016*, 538–544. [CrossRef]
40. Revenco, M.D.; Simonov, Y.A.; Duca, G.G.; Bourosh, P.N.; Bulmaga, P.I.; Kukushkin, V.Y.; Zhora, E.I.; Gdaniec, M. Versatility and reactivity of salicylaldehyde S-methylisothiosemicarbazone in palladium (II) complexes. *Russ. J. Inorg. Chem.* **2009**, *54*, 698–707. [CrossRef]
41. Pahonţu, E.; Proks, M.; Shova, S.; Lupașcu, G.; Ilieș, D.C.; Bărbuceanu, Ș.F.; Socea, L.; Badea, M.; Păunescu, V.; Istrati, D.; et al. Synthesis, characterization, molecular docking studies and in vitro screening of new metal complexes with Schiff base as antimicrobial and antiproliferative agents. *Appl. Organomet. Chem.* **2019**, *33*, e5185. [CrossRef]
42. Masadeh, M.M.; Hussein, E.I.; Alzoubi, K.H.; Khabour, O.; Shakhatreh, M.A.K.; Gharaibeh, M. (2015). Identification, characterization and antibiotic resistance of bacterial isolates obtained from waterpipe device hoses. *Int. J. Environ. Res. Public Health* **2015**, *12*, 5108–5115. [CrossRef] [PubMed]
43. Khaledi, A.; Esmaeili, D.; Jamehdar, S.A.; Esmaeili, S.A.; Neshani, A.; Bahador, A. Expression of MFS efflux pumps among multidrug resistant *Acinetobacter baumannii* clinical isolates. *Pharm. Lett.* **2016**, *8*, 262–267.
44. Nikolić, M.; Vasić, S.; Đurđević, J.; Stefanović, O.; Čomić, L. Antibacterial and anti-biofilm activity of ginger (Zingiber officinale (Roscoe)) ethanolic extract. *Kragujev. J. Sci.* **2014**, *36*, 129–136. [CrossRef]
45. Sabo, V.A.; Gavric, D.; Pejic, J.; Knezevic, P. Acinetobacter calcoaceticus-A. baumannii complex: Isolation, identification and characterisation of environmental and clinical strains. *Biol. Serb.* **2022**, *44*, 3–17. [CrossRef]

46. Borcea, A.M.; Marc, G.; Ionuț, I.; Vodnar, D.C.; Vlase, L.; Gligor, F.; Pricopie, A.; Pîrnău, A.; Tiperciuc, B.; Oniga, O. A novel series of acylhydrazones as potential anti-Candida agents: Design, synthesis, biological evaluation and in silico studies. *Molecules* **2019**, *24*, 184. [CrossRef]
47. Tan, M.H.; Nowak, N.J.; Loor, R.; Ochi, H.; Sandberg, A.A.; Lopez, C.; Pickren, J.W.; Berjian, R.; Douglass, H.O.; Chu, T.M. Characterization of a new primary human pancreatic tumor line. *Cancer Investig.* **1986**, *4*, 15–23. [CrossRef]
48. Zhao, W.; Zhao, M. Synthesis and characterization of some multi-substituted thiosemicarbazones as the multi-dental ligands of metal ions. *Chin. J. Org. Chem.* **2001**, *21*, 681–684.
49. Perrin, D.D.; Armarego, W.L.; Perrin, D.R. *Purification of Laboratory Chemicals*, 4th ed.; Butterworth-Heinemann, Pergamon Press: Oxford, UK, 1966.
50. Fries, J.; Getrost, H.; Merck, D.E. *Organic Reagents Trace Analysis*; E. Merck: Darmstadt, Germany, 1977.
51. Graur, V.; Mardari, A.; Bourosh, P.; Kravtsov, V.; Usataia, I.; Ulchina, I.; Garbuz, O.; Gulea, A. Novel Antioxidants Based on Selected 3*d* Metal Coordination Compounds with 2-Hydroxybenzaldehyde 4,S-Diallylisothiosemicarbazone. *Acta Chim. Slov.* **2023**, *70*, 122–130. [CrossRef]
52. Sheldrick, G.M. A short history of SHELX. *Acta Crystallogr. Sect. A Found. Crystallogr.* **2008**, *64*, 112–122. [CrossRef]
53. Sheldrick, G.M. Crystal structure refinement with SHELXL. *Acta Crystallogr. Sect. C Struct. Chem.* **2015**, *71*, 3–8. [CrossRef]
54. Re, R.; Pellegrini, N.; Proteggente, A.; Pannala, A.; Yang, M.; Rice-Evans, C. Antioxidant activity applying an improved ABTS radical cation decolorization assay. *Free Radic. Biol. Med.* **1999**, *26*, 1231–1237. [CrossRef] [PubMed]

Disclaimer/Publisher's Note: The statements, opinions and data contained in all publications are solely those of the individual author(s) and contributor(s) and not of MDPI and/or the editor(s). MDPI and/or the editor(s) disclaim responsibility for any injury to people or property resulting from any ideas, methods, instructions or products referred to in the content.

Article

Copper(II) and Platinum(II) Naproxenates: Insights on Synthesis, Characterization and Evaluation of Their Antiproliferative Activities

Amanda A. Silva [1], Silmara C. L. Frajácomo [2], Állefe B. Cruz [3], Kaio Eduardo Buglio [4], Daniele Daiane Affonso [4], Marcelo Cecconi Portes [5], Ana Lúcia T. G. Ruiz [4], João Ernesto de Carvalho [4], Wilton R. Lustri [2], Douglas H. Pereira [3], Ana M. da Costa Ferreira [5] and Pedro P. Corbi [1],*

1. Institute of Chemistry, University of Campinas—UNICAMP, Campinas 13083-970, SP, Brazil; a265988@dac.unicamp.br
2. Department of Biological and Health Sciences, University of Araraquara—UNIARA, Araraquara 14801-320, SP, Brazil; scfrajacomo@uniara.edu.br (S.C.L.F.); wrlustri@uniara.edu.br (W.R.L.)
3. Chemistry Collegiate, Federal University of Tocantins—UFT, Gurupi 77402-970, TO, Brazil; allefe.cruz@uft.edu.br (Á.B.C.); doug@mail.uft.edu.br (D.H.P.)
4. Faculty of Pharmaceutical Sciences, University of Campinas-UNICAMP, Campinas 13083-871, SP, Brazil; kaiobuglio@gmail.com (K.E.B.); ana.ruiz@fcf.unicamp.br (A.L.T.G.R.); carvalho@fcf.unicamp.br (J.E.d.C.)
5. Department of Fundamental Chemistry, Institute of Chemistry, University of São Paulo—USP, São Paulo 05508-000, SP, Brazil; marcelo_cecconi@hotmail.com (M.C.P.); amdcferr@iq.usp.br (A.M.d.C.F.)
* Correspondence: ppcorbi@unicamp.br

Abstract: The growth of antibiotic resistance is a matter of worldwide concern. In parallel, cancer remains one of the main causes of death. In the search for new and improved antiproliferative agents, one of the strategies is the combination of bioactive ligands and metals that are already consolidated in the synthesis of metallopharmaceutical agents. Thus, this work deals with the synthesis, characterization, and study of naproxen (Nap)-based complexes of copper(II) and platinum(II) as antiproliferative agents. The copper complex (Cu–Nap) presents a binuclear paddle-wheel structure in a 1 Cu:2 Nap:1 H_2O molar composition, in which Cu(II) is bonded to the carboxylate oxygens from naproxenate in a bidentate bridging mode. The platinum complex (Pt–Nap) was identified as the square planar cis-$[Pt(Nap)_2(DMSO)_2]$ isomer, in which Pt(II) is bonded to the carboxylate oxygen atom of Nap in a monodentate fashion. Both complexes were inactive against the Gram-positive and Gram-negative bacterial strains assessed. Pt–Nap presented low cytostatic behavior over a set of tumor cells, but good viability for normal cells, while Cu–Nap was cytotoxic against all cells, with a cytocidal activity against glioma tumor cells.

Keywords: NSAIDs; copper(II); platinum(II); naproxen; antibacterial agents; antiproliferative activities

1. Introduction

An intense search for new broad-spectrum antimicrobial agents active against multidrug-resistant bacterial strains has been carried out worldwide. The misuse of antibiotics, not only in humans but also in livestock and agriculture, either to prevent or cure diseases led to microbial resistance and chronic infections. This situation not only caused the deaths of millions of people around the world, but also impacted the financial systems of many countries [1]. Furthermore, the search for new pharmacologically active agents in tumor cells, with improved activities and reduced adverse reactions, is necessary. By 2025, more than 20 million new cases of cancer are expected each year, requiring the development of new drug treatments and personalized therapies [1,2].

To overcome these two major challenges in current medicine, the preparation of bioactive coordination compounds with metals is considered a promising strategy. For a review, see Štarha and Trávníček [3]. In the 1960s, a new era of development of drugs containing metals began, with the discovery of the inhibitory activities of cisplatin on the proliferation of *Escherichia coli* strains and different tumor cell lines [2,4]. Today, cisplatin and its derivatives, carboplatin and oxaliplatin, are widely used in clinics as chemotherapeutics for ovarian, testicular, and melanoma cancers, among others [5]. However, resistance, combined with poor selectivity, has become an issue in therapy with platinum drugs. Additionally, high doses of the drugs became necessary to ensure that the exact concentration of Pt(II) reached the intracellular medium. This practice can lead to adverse effects, such as nephrotoxicity, neurotoxicity, ototoxicity, and gastrointestinal reactions, which can, in turn, lead to interruption of cancer treatment [5–7].

Thus, the search for alternative metals, combined with bioactive ligands, has risen in recent decades. A typical example is silver sulfadiazine, which has been used since the 1970s in the treatment of bacterial infections [8]. Metal complexes have unique advantages, such as their redox properties and their own reaction mechanisms [9,10], and their biological properties [11–13]. A wide variety of ligands can be used, and some drugs already used in clinics can be associated with metals. Both cancer and bacterial infections promote an inflammatory response in the body. Thus, associating anti-inflammatories with metals with known biological activities seems to be an effective proposal. The combination of antibiotics with anti-inflammatories is already a widespread practice (polytherapy). Inhibitors of COXs, and possibly of MMPs as well—such as non-steroidal anti-inflammatory drugs (NSAIDs)—have been studied for the treatment of cancer [6,14,15].

Several mononuclear-, binuclear, and trinuclear Cu(II) complexes, in combination with NSAIDs such as mefenamic acid, diclofenac, diflunisal, indomethacin, aspirin, and naproxen, among others, are reported in the literature [9,16,17]. Some of these complexes showed better biological activities in vitro than their parental molecules. A recently synthesized Cu(II)-nimesulide complex, for example, presented a minimum inhibitory concentration (MIC) of 3.0 mmol L^{-1} against *Staphylococcus aureus* and 1.5 mmol L^{-1} against *Escherichia coli* and *Pseudomonas aeruginosa*, while nimesulide alone did not have antibacterial potential [18]. In addition, several platinum(IV) prodrugs with NSAIDs were reported by Spector et al. [19], Tolan et al. [20], and Chen et al. [15], and many of them presented higher antiproliferative activity in vitro than that of pure cisplatin.

Naproxen (Nap), (+)-(S)-2-(6-methoxinaphtalen-2-yl)propanoic acid, is an antipyretic and analgesic NSAID. It is widely used in the treatment of arthritis, spondylitis, bursitis, gout, and chronic pain [12,21]. Compared to other NSAIDs, it has shown less unwanted cardiovascular and nephron side effects [15]. It is a COX 1 and COX 2 inhibitor, and has shown inhibitory activity over MMPs, especially when associated with Pt(IV). Naproxen reduced carcinogenesis and proliferation of prostate, colon, and bladder tumor cells [15,17].

A great number of articles on naproxenate-based complexes, proving that these compounds may present useful biological activities, can be found in the literature. Studies related to the synthesis of metal complexes and their anti-inflammatory, antioxidant, and antiproliferative activities were reported [15,22]. In the literature, zinc, cobalt, copper, and iron complexes with naproxen have also been reported [23–25]. In particular, three binuclear copper(II) naproxenate complexes call attention to a paddle-wheel type coordination. The structures of [Cu$_2$L$_4$(H$_2$O)$_2$] complexes were proposed for the ligands (L) as the NSAIDs diclofenac, ibuprofen, and naproxen, based on infrared (IR) and Raman spectroscopic analysis and the crystal structure for the diclofenac complex [26]. In another study, the [Cu$_2$(Nap)$_4$(H$_2$O)$_2$] presented good binding affinity to bovine serum albumin (BSA) and, especially, to human serum albumin (HSA) proteins, which take part in the transport of the metal complexes through the blood stream [27]. A [Cu$_2$(Nap)$_4$(DMSO)$_2$] complex that presented antitumor properties loaded to chitosan beads was characterized, and this system showed a controlled release in gastric and intestinal pH [28]. Additionally, the monomeric Cu(II) complex [Cu(Nap)$_2$(H$_2$O)$_3$](H$_2$O) presented a higher anti-inflammatory response and a lower gastric ulcer than those of free naproxen, acting as urease inhibitors [23].

Still, most reports of naproxenate-based complexes are compounds with various 3d transition metals (e.g., Co [21], Cu [23,29], and Zn [30]) associated to nitrogen donor heterocyclic ligands (1,10-phenanthroline, pyridine, and 2,2′-bipyridine, among others). They have been reported as antibacterial [12] or antitumoral agents [31], maintaining the anti-inflammatory and antioxidant properties.

In this context, other metals such as Pt(II) [7,32], Pd(II) [7], and Ru(II) [31] have also been combined with naproxen and nitrogen-donor ligands. Pt(II) naproxenate complexes with ethylenediamine and diaminocyclohexane presented selective antiproliferative activity against epithelial (MCF$_7$) and human cervical carcinoma cells (HeLa), respectively [32,33].

This work expands upon the studies on the synthesis and spectroscopic characterizations of Cu(II) and Pt(II) complexes with naproxen, and also explores the antiproliferative activities of the complexes over pathogenic bacterial strains and tumor cells, seeking new drug candidates. A combination of experimental and theoretical (DFT) data were applied to propose the most probable structural formulae for the complexes.

2. Results

2.1. Infrared and Raman Spectroscopic Measurements

The main bands on the ATR–FTIR spectra of Cu–Nap, Pt–Nap and NaNap are presented in Figure 1. As a complement, the Raman spectra of Cu–Nap, Pt–Nap, NaNap, and cis-[PtCl$_2$(DMSO)$_2$] are presented in Figure 2. The far-IR spectra from 100–500 cm^{-1} of Cu–Nap and Pt–Nap are presented in Supplementary Figure S1 (Supplementary Materials). In the FTIR spectra, the main absorption bands of NaNap and attributions were at 3564–3143 cm^{-1} (νOH, broad band), 3057–2830 cm^{-1} (νCH), 1585 cm^{-1} (νCOO$_{asym}$), and 1390 cm^{-1} (νCOO$_{sym}$), which were in good agreement with the literature [26]. Meanwhile, the Cu–Nap FTIR spectrum presented a sharp strong band at 3601 cm^{-1} (νOH), instead of the broad band observed for NaNap. Other main signals (in cm^{-1}) for Cu–Nap were 3064–2800 (νCH), 1554 (νCOO$_{asym}$), and 1403 cm^{-1} (νCOO$_{sym}$). Thus, the difference between the νCOO asymmetric and symmetric stretchings Δ(νCOO$_{asym}$ − νCOO$_{sym}$) for the Cu–Nap was 151 cm^{-1}, whereas for the free ligand the difference was 195 cm^{-1}. The weak signals from 100–600 cm^{-1} occurred for Cu–O stretching [27,34]. The broad medium intensity band with a maximum at ~1670 cm^{-1} may be assigned to the bending mode δ (HOH) of the coordination water molecule in the complex.

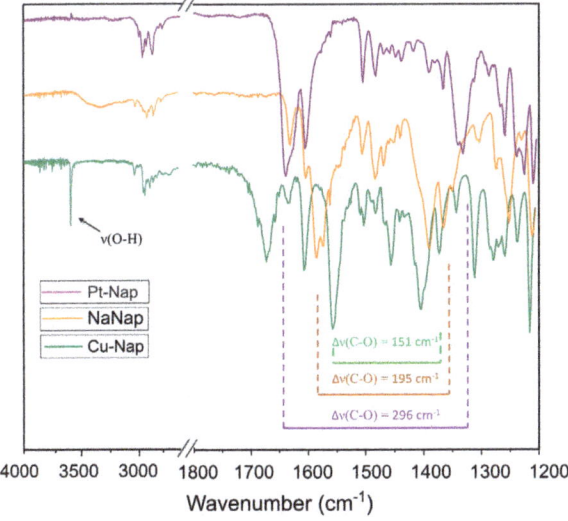

Figure 1. ATR–FTIR spectra of NaNap, Cu–Nap, and Pt–Nap (recorded from 4000 to 400 cm^{-1}) and identification of main bands.

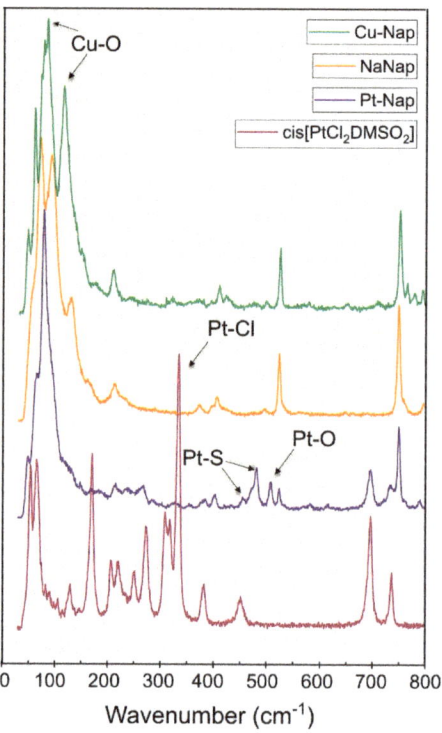

Figure 2. Raman spectra of the Cu–Nap and Pt–Nap complexes and their starting materials NaNap and cis-[PtCl$_2$(DMSO)$_2$].

Pt–Nap, for instance, showed FTIR bands for the respective wavenumbers (in cm^{-1}), as follows: 3064–2821 (νCH), 1637 (νCOO$_{asym}$), and 1341 (νCOO$_{sym}$). Thus, a Δ value of 296 cm^{-1} for Pt–Nap was observed. Additionally, a band and a shoulder were seen at 1144/1129 cm^{-1}, as shown in Supplementary Figure S2 in the Supplementary Materials. It corresponds to sulfoxide (νSO) stretching from DMSO. This result matches the one reported for cis-[PtCl$_2$(DMSO)$_2$] in the literature, attesting to the presence of DMSO in the structure in a cis-configuration [35]. In the Raman spectrum the Pt–S (Pt–DMSO), vibration was confirmed by the presence of a band and a shoulder at 480 cm^{-1} and 452 cm^{-1}, respectively, which also appeared for cis-[PtCl$_2$(DMSO)$_2$]. In the far-FTIR of Pt–Nap, it was also possible to see the Pt–S stretching at 453/430 (Figure S1) [35]. The Pt–O stretching at 523 cm^{-1} in Raman was also seen, as well as in the FTIR spectrum of Pt–Nap [36]. However, it was not seen in Pt–Nap the Pt–Cl stretch identified for the precursor in Raman at 349 cm^{-1}, proving the substitution of both chlorides. Furthermore, the absence of a broad band of around 3400–3600 cm^{-1} in the IR spectrum of Pt–Nap indicated the absence of hydration water in its composition.

2.2. UV–Vis Spectroscopic Analysis and Kinetic Studies for the Cu–Nap Complex

The UV–Vis spectrum of Cu–Nap exhibited three absorption bands at 240 nm (seen only in chloroform), 260 nm, and 320 nm (chloroform and dimethylsulfoxide, respetively), which may be attributed to intraligand transitions. For Cu–Nap, a band in the visible region with its maximum at 715 nm was also observed and was attributed to d–d transitions. This band was in accordance with the obtained results obtained by Dimiza et al. [27] for a paddle-wheel copper(II) complex with naproxenate, as previously described, which was obtained in this work by a different synthetic method. While Cu–Nap is reported to retain structure in either solid form or in solution [27], it was observed that, in DMSO, there was a minor change in the absorbance behavior throughout 24 h, indicating some solvent exchange. This is seen in Supplementary Figure S3, especially for the absorbance peaks of 260 nm and 275 nm during the first eight hours.

2.3. TGA Measurements

The TG/DTA thermograms of Cu–Nap and Pt–Nap are shown in Figure 3a,b, respectively. For Cu–Nap, an exothermic event was noticed in the range of 210 to 453 °C, which was responsible for two major mass losses, together totaling 79.32%. This was likely related to the oxidation of two naproxenate molecules (Calcd. 78.95%). The residue of 17.25% was attributed to the formation of copper oxide, CuO (Calcd. 14.74%). Additionally, a small loss of 3.43% slightly over 100 °C was observed, which indicated the loss of one water molecule (Calcd. 3.34%). The presence of one water molecule corroborated the FTIR results.

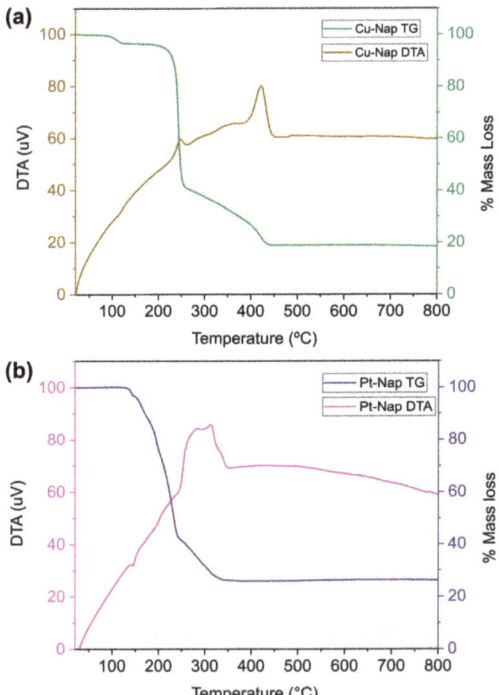

Figure 3. TG/DTA of Cu–Nap and Pt–Nap, from 20 °C to 800 °C, with an increase rate of 10 °C/min of (**a**) Cu–Nap and (**b**) Pt–Nap.

For Pt–Nap, TGA indicated a mass loss of 70.61%, from 160 to 350 °C, which likely corresponded to the oxidation of two naproxenate and two DMSO molecules (Calcd. 75.89%). The residue was 26.11%, which matched the composition of PtO (Calcd. 26.08%).

2.4. ESI Mass Spectrometric Analysis

Mass spectrometric analysis of Cu–Nap was conducted to confirm the most probable composition of the complex. The main signals identified in the positive mode (Figure 4a) were at m/z 179.0171 [(SO(CH$_3$)$_2$)$_2$Na]$^+$, seen as [DMSO$_2$ + Na]$^+$, at m/z 448.0435 [CuC$_{18}$H$_{25}$O$_5$S$_2$]$^+$ or [CuNap(DMSO)$_2$]$^+$, and at m/z 522.0665 [CuC$_{28}$H$_{26}$O$_6$ + H$^+$] or [Cu(Nap)$_2$ + H$^+$] for z = 1. Additionally, some important signals were identified in the negative mode (see Figure 4b): 229.0869 [C$_{14}$H$_{13}$O$_3$]$^-$ or Nap$^-$, 750.1894 [Cu(C$_{14}$H$_{13}$O$_3$)$_3$]$^-$

or [Cu(Nap)$_3$]$^-$. The signals at m/z 134.8654 and 185.0910 also appeared in the naproxen mass spectrum reported by Zayed et al. [37]. For Pt–Nap (Figure 4c), in the positive mode, the main signals identified were at m/z 231.1011 [(C$_{14}$H$_{14}$O$_3$) + H]$^+$ or [(NapH) + H$^+$], which is the protonated form of naproxen, and at m/z 253.0830 [(C$_{14}$H$_{13}$O$_3$Na) + H]$^+$ or [(NapNa) + H$^+$], corresponding to the protonated form of the NaNap. The [Pt(C$_{14}$H$_{13}$O$_3$)$_2$ + H]$^+$ or [Pt(Nap)$_2$ + H$^+$] was observed at m/z 653.1401, which corroborated the 1:2 Pt:Nap composition.

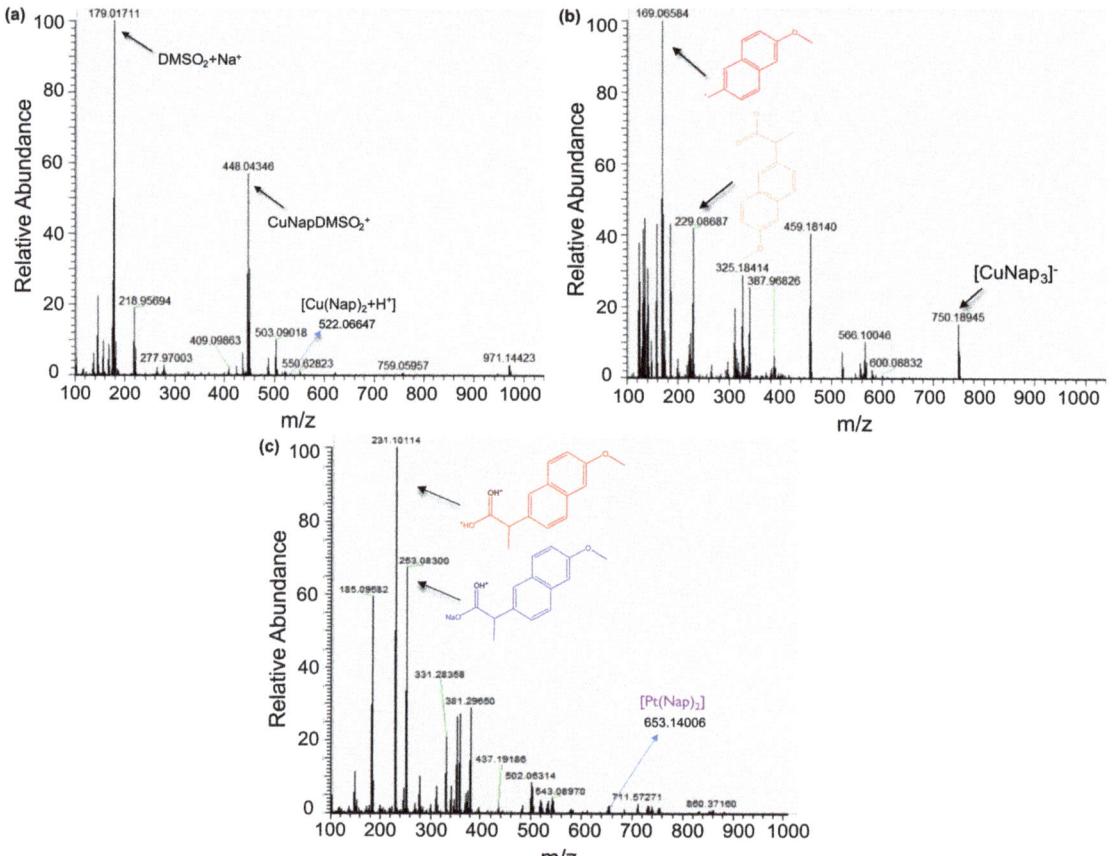

Figure 4. ESI–MS spectra of (**a**) Cu–Nap in the positive mode, (**b**) Cu–Nap in the negative mode, and (**c**) Pt–Nap in the positive mode.

2.5. NMR Measurements

To further evaluate the structure of the Pt–Nap, 1D and 2D NMR spectra were acquired for the complex and sodium naproxenate. The ^{1}H NMR spectrum of the complex is presented in Figure 5, together with the result for the kinetic study of the solvent exchange (DMSO), with measurements at t = 0, 12, and 24 h. The ^{13}C NMR spectra of the complex (DMSO-d$_6$) and the ligand (D$_2$O) are presented in Figures 6a and 6b, respectively. The shifts are numbered according to the attributions for the naproxenate portion corresponding to the schematized molecular structure. The most remarkable information from the ^{1}H NMR spectrum of the complex was the presence of two signals for DMSO. The signal at 2.50 ppm (quintet, 1H) was attributed to the residual peak of the solvent (CHD$_2$(SO)CD$_3$) used in the experiment. The other signal at 2.55 ppm was referred to as DMSO bonded to platinum (singlet, 6H, CH$_3$(SO)CH$_3$). To facilitate the visualization of both DMSO signals, Supplementary Figure S4 shows the ^{1}H signals. Moreover, when analyzing the ^{13}C spectra of the complex and the ligand, it was noted that the C1 (carboxylate group) in the complex shifted to a high field with a $\Delta\delta$ of -4.88. In addition, a small downshift was seen for C2. This observation led us to confirm the coordination of the Nap to Pt(II) by the oxygen atom of the carboxylate group, as will be discussed in the next session.

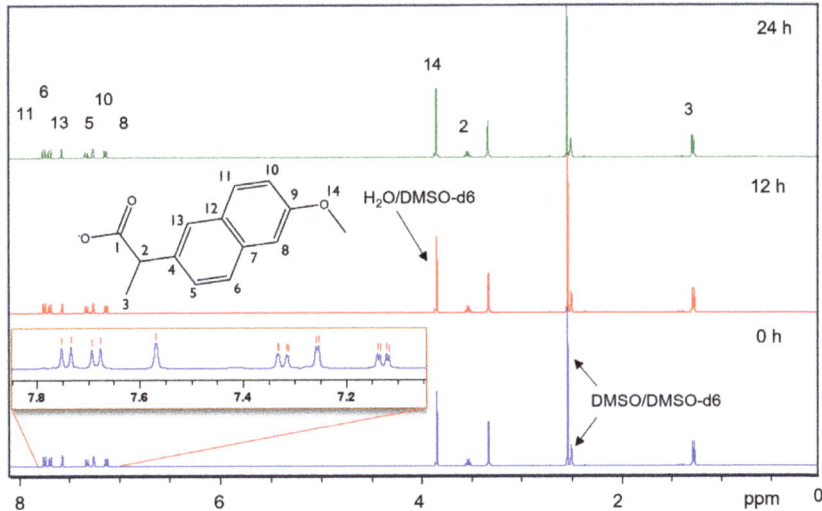

Figure 5. ^{1}H NMR (500 MHz, DMSO-d$_6$) spectrum of Pt–Nap in t = zero h, 12 h, and 24 h.

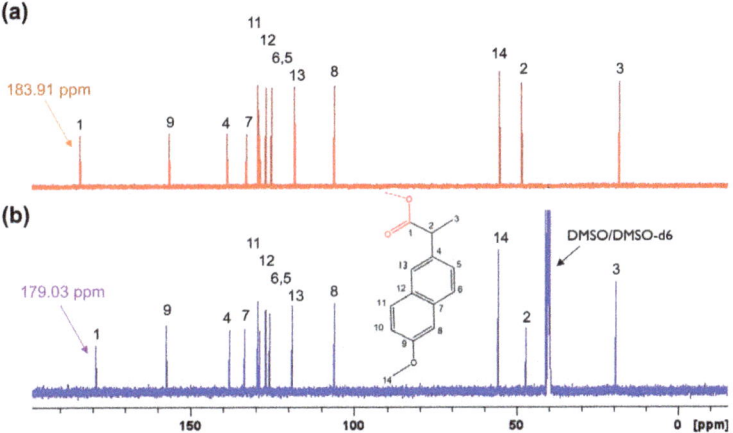

Figure 6. (a) ^{13}C NMR (500 MHz, D$_2$O) of NaNap; (b) ^{13}C NMR (500 MHz, DMSO-d$_6$) of Pt–Nap.

Moreover, ^{195}Pt NMR spectroscopic data were also acquired for Pt–Nap (Figure 7) and its starting Pt(II) complex, cis-[PtCl$_2$(DMSO)$_2$]. Potassium hexachloridoplatinate(IV), K$_2$PtCl$_6$ was used as a reference, being in accordance with the procedure reported by Quintanilha et al. in the literature [4]. For the precursor, a single peak at -3452.8 ppm was obtained, which was similar to the value of -3445.96 ppm found in the literature [4]. For Pt–Nap, a single peak at -3105.4 ppm was observed, with delta $\Delta\delta = -347.4$, in comparison to the single peak observed for its precursor.

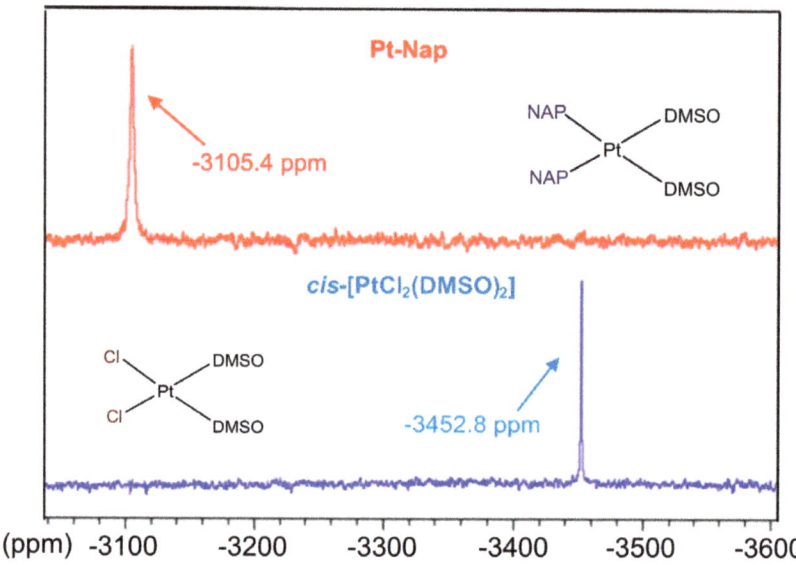

Figure 7. ^{195}Pt NMR spectra for Pt–Nap (**top**) and cis-[PtCl$_2$(DMSO)$_2$] (**bottom**).

2.6. EPR Spectra Analysis

The registered spectra for the complex Cu–Nap (see Figure 8) indicated a paddle-wheel structure, with two copper centers, bridged by four molecules of naproxenate, as observed in the literature [26,27]. The naproxenate molecules were coordinated by the carboxylate groups to both copper centers. At low temperature (77 K), only an isotropic signal was observed, with no hyperfine structure, but at room temperature (293 K), a typical paddle-wheel signal in a binuclear species was detected. Additionally, at room temperature (RT), a radical signal appeared at 350 mT. In the frozen DMSO solution, a central signal probably indicated the formation of a mononuclear species formed by the cleavage of the paddle-wheel structure through the coordination of solvent molecules, with spectroscopic parameters that were quite different from the original binuclear species, as shown in Table 1. The weak signals at ~150 mT and ~470 mT could be attributed to the binuclear species. However, a good fitting with experimental data was not achieved by simulation.

Table 1. EPR parameters determined for the complex Cu–Nap in a solid state and in a frozen DMSO solution.

Temperature	g_x	g_y	g_z	A_x	A_y	A_z
Solid at room temperature	1.9072	2.0430	2.5026	187 G	17 G	200 G
Solid at 77 K		g_{iso} 2.1780			A_{iso} 16 G	
In DMSO solution at 77 K	$g^\perp$ = 2.0355		$g_{//}$ = 2.3479		$A_{//}$ = 137 G	

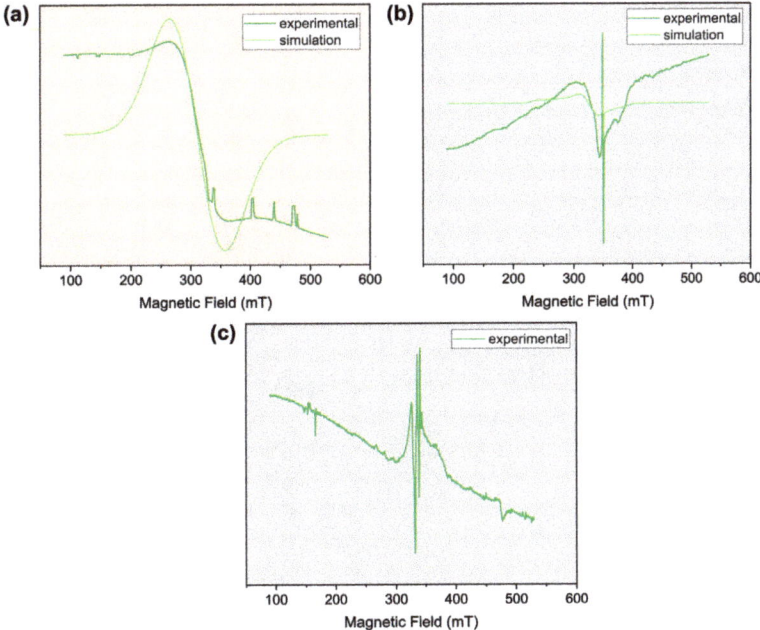

Figure 8. EPR spectra of complex Cu–Nap, registered in a solid state (**a**) at 77 K and (**b**) at RT (293 K) and (**c**) in a DMSO frozen solution at 77 K. In black, experimental results; in red, simulation data.

2.7. Evaluation of Biological Activities

In the disc diffusion assay, the respective masses of each compound added to the discs were calculated, based on the initial concentration of their solutions and molar masses. In these experimental conditions, none of the naproxenate complexes induced growth inhibition of Gram-positive (*S. aureus* and *B. cereus*) or Gram-negative (*E. coli* and *P. aeruginosa*) strains. Considering the starting materials, only copper(II) chloride showed inhibition zones against *S. aureus*, *B. cereus*, and *E. coli* (Table 2).

Table 2. Disc diffusion assay for the naproxenate complexes (Pt–Nap and Cu–Nap) and the starting materials (CuCl$_2$·2H$_2$O, NaNap and *cis*-[PtCl$_2$(DMSO)$_2$]).

Compounds	MW	Mass	Inhibition Zone (mm)			
			S. aureus	*B. cereus*	*E. coli*	*P. aeruginosa*
CuCl$_2$·2H$_2$O	170.48	1.549	11.0 (±0.5)	14.0 (±1.0)	11.0 (±0.3)	NI
NaNap	252.24	1.088	NI	NI	NI	NI
cis[PtCl$_2$(DMSO)$_2$]	422.25	1.353	NI	NI	NI	NI
Cu–Nap *	1080.12	0.965	NI	NI	NI	NI
Pt–Nap	809.85	1.042	NI	NI	NI	NI

MW = g/mol; Mass = in disc (mg); NI = no inhibition; * (binuclear).

When evaluated against a panel of human tumor cell lines, the copper(II) naproxenate complex Cu–Nap showed a cytostatic effect against glioblastoma (U251, GI$_{50}$ = 15 µg/mL, SI = 5.7), and multidrug resistant ovarian adenocarcinoma (NCI-ADR/Res, GI$_{50}$ = 44.3 µg/mL, SI = 1.9) (Figure 9e, Table 3), while platinum naproxenate complex Pt–Nap was inactive (GI$_{50}$ > 150 µg/mL). Both platinum chloride and sodium naproxenate were inactive (GI$_{50}$ > 150 µg/mL), while cooper(II) chloride showed an unspecific cytostatic activity against all tumor cell lines evaluated (GI$_{50}$ = 15 µg/mL, TGI = 105.9 µg/mL or 98.0 µmol/L, SI = 2.8) (Figure 9, Table 3).

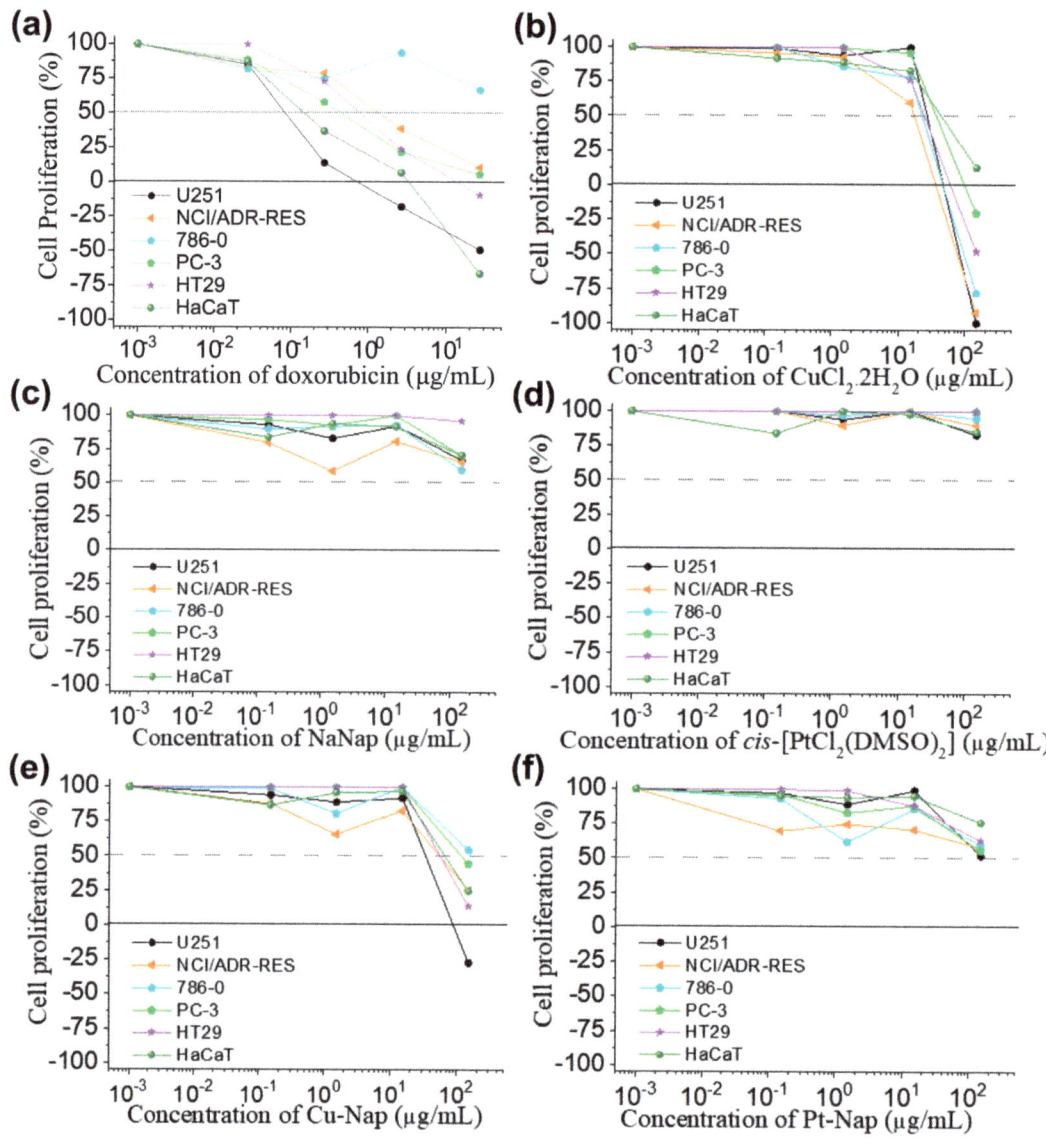

Figure 9. Anti-proliferative profile (%) of (**a**) doxorubicin (positive control); (**b**) $CuCl_2·2H_2O$; (**c**) *cis*-$[PtCl_2(DMSO)_2]$; (**d**) NaNap; (**e**) Cu–Nap; (**f**) Pt–Nap. Human tumor cell lines: U251 = glioblastoma; NCI-ADR/RES = multidrug resistant ovarian adenocarcinoma; 786-0 = adenocarcinoma of kidney; PC-3 = adenocarcinoma of prostate; HT-29 = adenocarcinoma colorectal. Human non-tumor cell line: HaCaT (immortalized keratinocyte). Sample concentration: 0.15–150 µg/mL (naproxenate complexes and starting material); 0.015–15 µg/mL (doxorubicin). Time exposure = 48 h.

Table 3. Anti-proliferative effect expressed as concentration required to inhibit 50% of growth (GI$_{50}$, in µg/mL) for NaNap, CuCl$_2 \cdot$2H$_2$O, Cu–Nap, cis-PtCl$_2$(DMSO)$_2$, Pt–Nap, and doxorubicin.

Cell Line	NaNap	CuCl$_2 \cdot$2H$_2$O	Cu–Nap	[PtCl$_2$(DMSO)$_2$]	Pt–Nap	Doxorubicin
U251	>150	15 *	15 *	>150	>150	0.05 ± 0.03
NCI/ADR-RES	>150	15 *	44.3 ± 62.7	>150	>150	0.8 ± 0.3
786-0	>150	15 *	>150	>150	>150	>15
PC-3	>150	15 *	147.2 ± 0.1	>150	>150	0.25 ± 0.01
HT29	>150	15 *	127.6 ± 0.1	>150	>150	0.4 ± 0.1
HaCaT	>150	42.3 ± 14.2	86.0 ± 37.8	>150	>150	0.092 ± 0.006

Results expressed as concentration required to elicit 50% (GI$_{50}$) in µg/mL followed by standard error, calculated by sigmoidal regression using Origin 8.0 software; * approximated effective concentration (standard error higher than calculated concentration). Human tumor cell lines: U251 = glioblastoma; NCI/ADR/RES = multidrug resistant ovarian adenocarcinoma; 786-0 = adenocarcinoma of kidney; PC-3 = adenocarcinoma of prostate; HT-29 = adenocarcinoma colorectal. Human non-tumor cell line: HaCaT (immortalized keratinocyte). Sample concentration: 0.15–150 µg/mL (naproxenate complexes and starting material); 0.015–15 µg/mL (doxorubicin). Time exposure = 48 h.

3. Discussion

3.1. Structure, Composition, and Coordination Sphere

Altogether, the results provide insights into the compositions and the coordination sites of naproxenate to the metal ions in the Cu–Nap and Pt–Nap complexes. It is worth mentioning that the formula suggested by elemental analyses and TGA for Cu–Nap was [Cu$_2$(Nap)$_4$(H$_2$O)$_2$], using either Cu(NO$_3$)$_2$ or CuCl$_2$ as the starting material. This indicates that the anion did not affect the composition of the complex in the synthetic procedure. The proportion of 1 Cu:2 Nap was also identified in the mass spectrum (Figure 4a), with a signal at m/z 522.0665, corresponding to [Cu(Nap)$_2$ + H]$^+$. Moreover, the mass spectrum showed signals with significant abundance in the negative mode and the presence of signals above m/z 520 in the positive mode. Even in low abundance, the signals could be evidence of polymeric or dimeric forms of Cu–Nap. This would explain the poor solubility of the complex in most of the solvents evaluated. The TGA (Figure 3a) indicated that the water molecule was lost at slightly over 100 °C, suggesting that it could be a coordination water rather than a hydration water. This hypothesis was reinforced in the FTIR (Figure 1), with the sharp signal at 3601 cm^{-1}, which was reported in the literature as a coordination water [27]. In addition, when comparing the FTIR spectra of Cu–Nap and NaNap, the $\Delta\nu$(COO) of 151 cm^{-1} for the complex was a bit lower than that of the ligand, $\Delta\nu$(COO) = 195 cm^{-1}. Such data suggests a bidentate bridging coordination mode of carboxylate to copper, just as proposed by Dimiza et al., where a $\Delta\nu$ value was of 170 cm^{-1} for the same complex [27]. This proposition of a paddle-wheel complex differs from the mononuclear copper(II) naproxenate complex [Cu(Nap)$_2$(H$_2$O)$_3$]·H$_2$O reported by Chu et al., in which Cu(II) was coordinated by two molecules of nap in the monodentate mode, and three water molecules plus a hydration water [23]. Finally, EPR studies (Figure 8) presented a typical pattern of paddle-wheel structure, confirming our hypothesis.

Furthermore, in the UV spectrum Cu–Nap preserved the naproxenate bands with no significant shifts. The broad low-intensity band that appeared for Cu–Nap at around 700 cm^{-1} corresponded to absorption in red, which matched the greenish color observed in the material. The UV–Vis kinetic study of the Cu–Nap complex in DMSO suggested little ligand exchange of a water molecule by DMSO throughout time (Supplementary Figure S3).

For instance, Pt–Nap presented higher solubility than Cu–Nap in various solvents. Its composition of PtC$_{32}$H$_{38}$O$_8$S$_2$, or [Pt(DMSO)$_2$(Nap)$_2$] determined by elemental analysis, was confirmed in TGA measurements (Figure 3b), with the loss of two Nap and two DMSO. The proportion of 1 Pt:2 Nap was also confirmed by mass spectrometric data with the presence of the [Pt(C$_{14}$H$_{13}$O$_3$)$_2$ + H]$^+$ ion or [Pt(Nap)$_2$+H]$^+$ (Figure 4c), at m/z 653.1401. The presence of DMSO in the complex was also reinforced by the FTIR (Figure 1) and Raman (Figure 2) spectroscopic data. Bands attributed to the ν(SO) at 1144/1129 cm^{-1} and ν(PtS) at 523 cm^{-1} stretching modes were identified. Due to the presence of a band and a shoulder for the stretching mode of the sulfoxide group, a cis configuration was proposed [35]. Finally, the presence of coordinated DMSO molecules to Pt(II) was confirmed by ^{1}H NMR spectroscopic measurements, as shown in Figure 5 and Figure S4.

The final evidence of the coordination sphere of Pt(II) in Pt–Nap comes from the FTIR (Figure 1, Figures S1 and S2) and Raman (Figure 2) stretchings. When analyzing the FTIR spectra of Pt–Nap, the Δ(COO) of 296 cm^{-1} was higher than that of the ligand (Δ195 cm^{-1}). This suggested a monodentate mode of coordination between the carboxylate and Pt(II), according to Dendrinou-Samara et al. [32]. Evidence of Pt(II) bonding to the carboxylate was seen in Raman and far-FTIR, with the presence of ν(PtO) stretching. The coordination of Pt(II) to the DMSO molecules occurred by the sulfur atom, which was confirmed by the presence of ν(PtS) stretchings on Raman spectra [35].

Additionally, the ^{13}C NMR spectrum of the Pt–Nap complex exhibited a significant shift of the carbon atom of the carboxylate group in Pt–Nap when analyzing both ligand (183.91 ppm, D$_2$O) and complex (179.03 ppm, DMSO-d$_6$) spectra (Figure 6). Thus, the former cis-[PtCl$_2$(DMSO)$_2$] used in the synthesis had the substitution of both its chlorides for naproxenate molecules. This was evidenced in ^{195}Pt NMR (Figure 7), with the shift of -347.4 ppm relative to the signal for cis-[PtCl$_2$(DMSO)$_2$]. In other words, there was less shielding effect and narrower electronic density around Pt(II) in Pt–Nap than in its starting complex. Nevertheless, the presence of a single ^{195}Pt peak for Pt–Nap proved that the sample consisted in only one of the geometric isomers.

3.2. Molecular Modeling

As discussed above, Cu–Nap was proposed in a dimeric/polymeric structure, with a bidentate bridging coordination mode of carboxylate to Cu(II). The paddle-wheel structure (Figure 10) was already suggested for Cu(II) complexes combined to NSAIDs, in studies by Trinchero et al. [26] and Mikuriya et al. [38]. The minimum molar composition of 1Cu(II):2Nap:1H$_2$O suggested by elemental analysis, TGA, and mass spectrometric data was considered in the construction of this structure. According to the literature, the water molecules should be placed in apical sites of binuclear Cu(II) paddle-wheel structures [39]. Thus, polymerization occurs by the water molecules, with a Cu–Ow interaction, where Ow refers to the oxygen of water molecules. A polymer with compositions [Cu$_2$Nap$_4$]$_n$ was previously reported by Abuhijleh et al., where the units were linked via Cu-carboxylate interactions from neighboring units [40]. DFT simulations for the dimeric form are presented in Figure 10b and the bond lengths are presented on Table 4. The results show that the distance between the copper atoms (Cu1–Cu2) was 2.870 Å and that the bonding distance between the oxygen atoms of the ligands and the Cu atoms (Cu-O) ranged from 1.872 Å to 2.556 Å. This was in accordance with the values reported in the literature for Cu(II) paddle-wheel complexes [17,34].

Table 4. Bond lengths for Cu–Nap complex formation.

Bond	Distance (Å)	Bond	Distance (Å)
Cu1-O1	2.295	Cu2-O2	1.875
Cu1-O3	2.124	Cu2-O4	1.888
Cu1-O6	2.134	Cu2-O5	1.895
Cu1-O8	2.556	Cu2-O7	1.872
Cu1-Ow1	2.158	Cu2-Ow2	2.343
Cu1-Cu2	2.870		

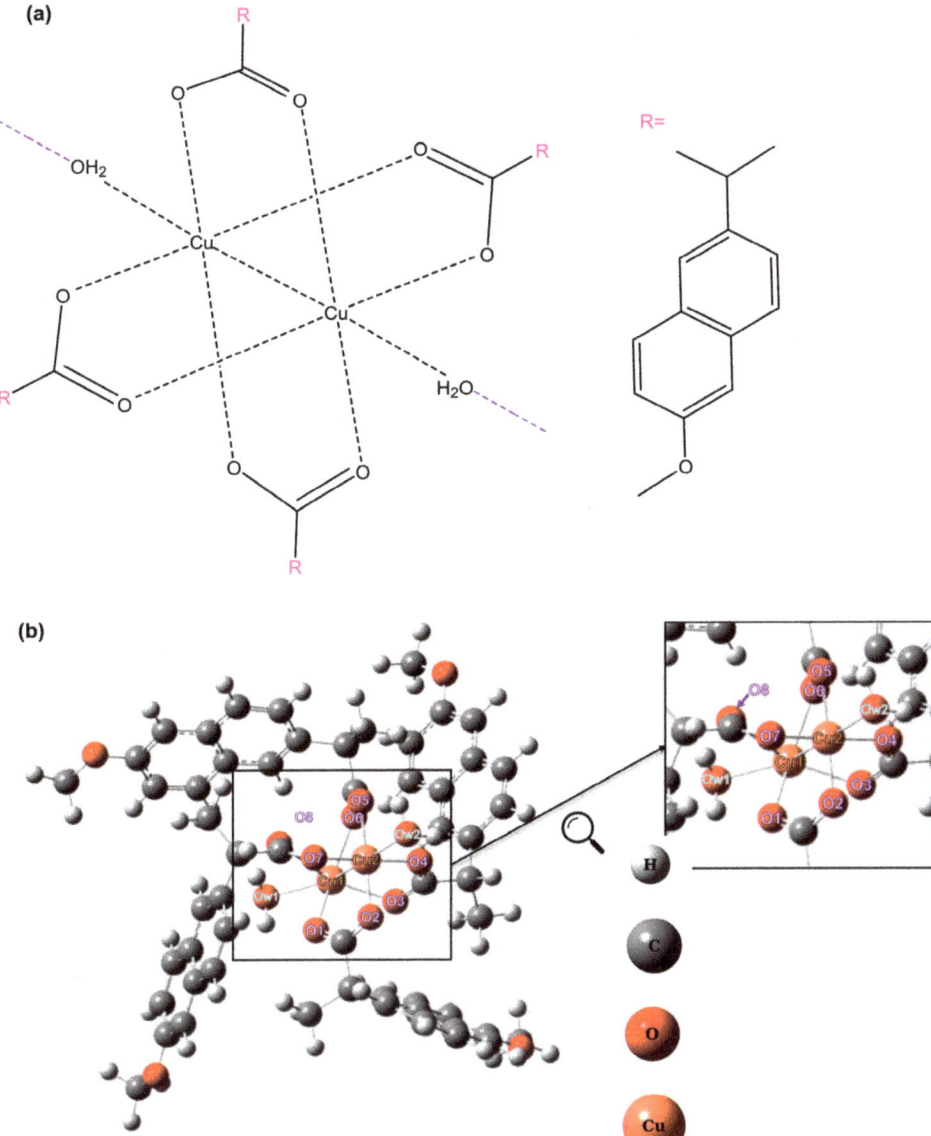

Figure 10. (a) Structural formula for the Cu–Nap complex; (b) molecular model for Cu–Nap.

Furthermore, as previously discussed, it was possible to suggest that Pt–Nap could be seen as [Pt(DMSO)$_2$(Nap)$_2$], with coordination in a monodentate mode of two nap molecules and two DMSO and with the absence of hydration water. Indeed, this coordination mode was also reinforced by the DFT studies. For the Pt–Nap complex, the possibilities of formation of *cis* and *trans* configurations were evaluated, and the structures are presented in Figure 11 with the bond lengths between the central atoms.

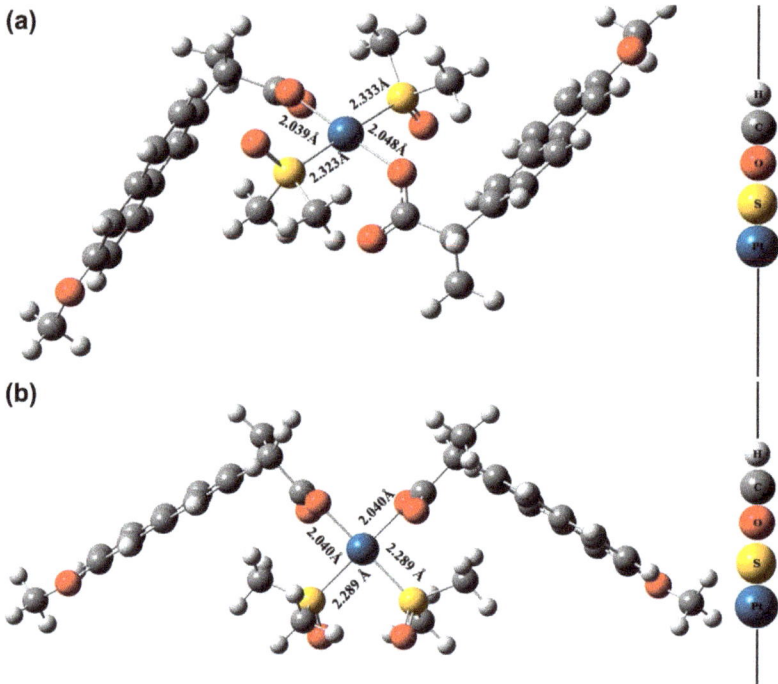

Figure 11. Structural parameters for the complexes: (**a**) *trans*-[Pt(DMSO)$_2$(Nap)$_2$]; (**b**) *cis*-[Pt(DMSO)$_2$(Nap)$_2$].

Analyzing the bond lengths for the *trans*-[Pt(DMSO)$_2$(Nap)$_2$] and *cis*-[Pt(DMSO)$_2$(Nap)$_2$] configurations, it was possible to observe that the Pt–O and Pt–S bonds length were very close between the structural configurations and the values were in accordance with the distances typically found in complexes in the literature [41].

The electronic energies values were calculated at the M06-2X/6-31 + G(d,p)/LANL2DZ level for the two conformations. It was possible to determine that the energy difference between the complexes *cis* and *trans* was 0.9 kcal mol^{-1}. This allowed us to infer that both isomers may occur. It should be noted that Pereira et al. [42] studied the formation of amantadine (atd), rimantadine (rtd), and memantine (mtn) complexes with platinum [PtCl$_2$(atd)(DMSO)], [PtCl$_2$(mtn)(DMSO)], and [PtCl$_2$(rtd)(DMSO)], and the crystallographic data showed the *cis* configuration for the complex [PtCl$_2$(mtn)(DMSO)] and *trans* for [PtCl$_2$(atd)(DMSO)] and [PtCl$_2$(rtd)(DMSO)]. Together with the X-ray data, the theoretical results showed that both *cis* and *trans* isomers are equally possible to occur, because the energy difference between the complexes was less than ±2.0 kcal·mol^{-1}, which corroborated with the results found in the present work. However, ^{195}Pt NMR showed a single peak and ^{1}H and ^{13}C NMR spectra did not show the existence of both isomers, suggesting that only one isomer was formed. With insight into Raman (Figure 2) and FTIR (Supplementary Figure S1) spectra, while comparing Pt–Nap to *cis*-[PtCl$_2$(DMSO)$_2$], also presented in literature [35,36], the bands corresponded to the *cis* isomer for Pt–Nap.

3.3. Biological Activity Assays

Antibacterial activities were not observed for the Pt–Nap and Cu–Nap complexes or for NaNap and *cis*-[PtCl$_2$(DMSO)$_2$] under the considered experimental conditions. As expected [43], CuCl$_2$·2H$_2$O showed antibacterial activity against *S. aureus*, *B. cereus*, and *E. coli*. Considering the stability assay results that showed that Cu–Nap appeared to exhibit only a small ligand exchange in DMSO, especially during the first 8 h (Supplementary Figure S3), the lack of effect can be partially explained by the impaired diffusion rate of Cu–Nap in agar medium [44].

Moreover, the unspecific cytostatic effect observed for copper chloride against tumor cell lines was modulated by the ligand naproxenate, as Cu–Nap selectively inhibited the proliferation of U251 (glioblastoma) and NCI–ADR/Res (multidrug resistant ovarian adenocarcinoma) cell lines (Figure 9). Copper ions have been described as inducing differently regulated cell death mechanisms, such as apoptosis and cuproptosis [45]. Despite the absence of an anti-proliferative effect, naproxen has been regarded as an efficient agent against different in vivo cancer models, by modulating the inflammatory microenvironment [46].

Furthermore, the promising cytostatic effects of Cu–Nap against the U251 and NCI–ADR/Res cell lines were accompanied by a weak anti-proliferative effect on the non-tumor cell line (Figure 9). As an on-target toxicological parameter [47], the selectivity index (SI > 2) indicated that Cu–Nap may inhibit tumor proliferation in preclinical models, without affecting normal tissues such as mucosa and bone marrow.

It is important to note that Pt–Nap showed stability in DMSO throughout 24 h (Figure 5). Thus, the lack of biological activities observed for Pt–Nap (Figure 9) could not be attributed to instability under experimental conditions. Indeed, no unwanted precipitation or color change was observed during the experimental procedures.

4. Materials and Methods

4.1. Materials and Equipment

Sodium naproxenate (NaNap) (98%), $CuCl_2 \cdot 2H_2O$ (99%), and cis-$[PtCl_2(DMSO)_2]$ (97%) were purchased from Sigma–Aldrich Laboratories (St. Louis, MO, USA). Dimethylsulfoxide (99.9%) was purchased from Synth (Diadema, Brazil). Elemental analyses were performed on a Perkin Elmer 2400 CHNS/O Analyzer (Shelton, CT, USA). Electronic absorption spectra in the range of 190 to 1100 nm were acquired using a 1.0 cm quartz cuvette in a diode array Hewlett Packard HP8453 UV/Vis absorption spectrophotometer (Shanghai, China). Infrared spectroscopic measurements (FTIR) were performed by using Agilent Cary 630 and 660 FTIR spectrometers (Penang, Malaysia) in the attenuated total reflectance (ATR) method. The FTIR spectra were recorded from 100 to 4000 cm^{-1}, with resolution of 2 cm^{-1}. Raman spectra from 0 to 1500 cm^{-1} were recorded using a Horiba Labram Soleil spectrometer (Loos, France). The thermal stability of the complexes was analyzed using a TGA/DTA simultaneous analyzer SEIKO EXSTAR 6000 (Oyama, Japan). The compounds were evaluated in oxidant atmosphere using an α-alumina crucible, at 25 to 900 °C with a heating rate of 10 °C/min. The ^1H, ^{13}C and the {$^{13}C,^1H$} HSQC and HMBC nuclear magnetic resonance (NMR) spectra of NaNap and Pt–Nap were recorded in a Bruker Avance III 500 MHz (11.7 T) spectrometer, while kinetic study of the Pt–Nap was conducted in a Bruker Avance III 400 MHz (9.4 T) NMR spectrometer (Wissembourg, France). Samples were analyzed at 298 K. NaNap was analyzed in D_2O, while the Pt–Nap was dissolved in DMSO-d_6, due to its solubility in this solvent. The chemical shifts for all solution measurements were provided, relative to tetramethylsilane (TMS). Electrospray ionization mass spectrometry (ESI-MS) measurements were conducted in a Thermo Q-Exactive using a Q-Orbitrap (Bremen, Germany). Samples were evaluated in the positive mode. Cu–Nap was also evaluated in the negative mode. Each solution was directly infused into the instrument's ESI source with capillary potential of 3.50 kV and a source temperature of 300 °C. The dilution used was of 50× with MeOH:H_2O 1:1 v/v, adding, when necessary, a second injection with 0.1% of formic acid. For Cu–Nap, direct preparation in methanol was not possible due to the low solubility of the material, so a solubilization in DMSO was carried out prior to the dilution in MeOH:H_2O. The prepared copper complex Cu–Nap was characterized by EPR spectroscopy, registering spectra in a CW–Bruker spectrometer model EMX (Ettlingen, Germany), working at X-band (9.5 GHz, 20.12 mW, 100 kHz modulation frequency) in solid state, and in frozen DMSO solutions. Samples were analyzed at RT (293 K) and at low temperature (77 K) in Wilmad (Vineland, NJ, USA) quartz tubes (4 mm internal diameter), using a quartz Dewar (Vineland, NJ, USA) with liquid nitrogen for the low temperature measurements. DPPH (α,α'-diphenyl-β-picrylhydrazyl) was used as the frequency calibrant, with g = 2.0036. Simulations of spectra were performed with EasySpin software (version 5.2.35) in the MATLAB environment [48].

4.2. Synthesis of Cu(II) Complex with Naproxenate (Cu–Nap)

To synthesize Cu–Nap, $CuCl_2 \cdot 2H_2O$ (0.25 × 10^{-3} mol) was first dissolved in 5.0 mL of deionized water. This solution was added dropwise to 5.0 mL of an aqueous solution of NaNap (0.50 × 10^{-3} mol) and maintained under constant stirring for 2 h at room temperature. The solid formed was filtered off under vacuum, washed with distilled water, and dried in a desiccator under P_4O_{10}. Anal. Calcd. for $[Cu_2(Nap)_4(H_2O)_2]$ (%): C, 62.27; H, 5.23. Found (%): C, 62.24; H, 5.42. Yield 79.04%. Nap refers to naproxenate. Additionally, the same procedure was performed by using $Cu(NO_3)_2 \cdot 3H_2O$ as the starting material and the elemental analysis was similar to the first method. Anal Calcd. for $[Cu_2(Nap)_4(H_2O)_2]$ (%): C, 62.27; H, 5.23. Found (%): C, 61.93; H, 5.40. Yield 71.32%. The complex is soluble in DMSO, and poorly soluble in methanol and chloroform. It is insoluble in water.

4.3. Synthesis of Pt(II) Complex with Naproxenate (Pt–Nap)

For the synthesis of Pt–Nap, cis-$[PtCl_2(DMSO)_2]$ (0.25 × 10^{-3} mol) was first dissolved in 5.0 mL of DMSO. This solution was added dropwise to 10 mL of an aqueous solution of NaNap (0.50 × 10^{-3} mol) and kept under constant stirring for 45 h at room temperature. The solid obtained was collected by filtration under vacuum, washed with distilled water, and dried under P_4O_{10}. Anal Calcd. for $PtC_{32}H_{38}O_8S_2$ or $[Pt(DMSO)_2(Nap)_2]$ (%): C, 46.33; H, 4.86. Found (%): C, 47.46; H, 4.73; Yield 72.61%. Nap refers to naproxenate, while DMSO is dimethylsulfoxide. H NMR(500 MHz, DMSO-d_6) δ (in ppm): 7.74 (d, J = 8.9 Hz, 1H), 7.68 (d, J = 8.5 Hz, 1H), 7.57 (s, 1H), 7.32 (dd, J = 8.6 Hz, 1H), 7.26 (s, 1H), 7.13 (dd, J = 6.3 Hz, 1H), 3.85 (s, 3H), 3.54 (q, J = 7.2 Hz, 1H), 2.55 (s, 6H), 1.28 (d, J = 7.2 Hz, 3H); ^{13}C NMR (500 MHz, DMSO-d_6) δ (in ppm): 179.03, 157.40, 137.94, 133.51, 129.54, 128.87, 127.15, 126.88, 125.80, 118.92, 106.28, 55.58, 47.05, 19.41. Although the percentage of carbon in the elemental analysis was outside the 0.4% range when compared to the expected results, the composition of the complex seemed to be the most reasonable one based on the hydrogen experimental percentage and the thermogravimetric and mass spectrometric data (Sections 2.3 and 2.4).

The complex is soluble in DMSO, dichloromethane, acetone, acetonitrile, and chloroform, and it is poorly soluble in methanol. The complex is insoluble in water.

4.4. Antibacterial Activity Assays In Vitro

Gram-positive *Staphylococcus aureus* (ATCC 25923) and *Bacillus cereus* (ATCC 14579), and Gram-negative *Escherichia coli* (ATCC 25922) and *Pseudomonas aeruginosa* (ATCC 27583) were acquired from the cell culture collection of Fundação André Tosello (Campinas, Brazil). The bacterial strains were inoculated in tubes containing BHI (Brain Heart Infusion FASVI) and incubated for 18 h at 35–37 °C. The disc diffusion method (antibiogram) was applied following the protocol described in the literature [49]. Sterile filter paper discs (Whatman 3 with 10.0 mm in diameter) were impregnated with 50 µL of Pt–Nap, Cu–Nap, and starting material solutions in triplicate. The amount of compound in each disc ranged from 0.96 to 1.55 mg. The impregnated discs were placed on the surface of the solid agar inoculated with a suspension, on the McFarland nephelometric scale 0.5 (~1.5 × 10^8 CFU/mL), of each strain, using a sterile cotton swab. The plates were incubated for 18 h at 35 °C. The sensitivity of the complexes was evaluated from the diameter of the zones of inhibition (in millimeters). $CuCl_2 \cdot 2H_2O$ was used as the positive control.

4.5. Antiproliferative Activity Assay over Tumor and Normal Cells

The anti-proliferative activities of the Pt–Nap and Cu–Nap complexes were evaluated against five human tumor cell lines (U251 = glioblastoma, NCI-ADR/RES = multidrug resistant ovarian adenocarcinoma, 786-0 = adenocarcinoma of kidney, PC-3 = adenocarcinoma of prostate, and HT-29 = adenocarcinoma colorectal) and one human non-tumor cell line (HaCaT, immortalized keratinocyte). The tumor cell lines were kindly provided by the Frederick Cancer Research & Development Center, National Cancer Institute, Frederick, MA, USA, while the non-tumor cell line was provided by Dr. Ricardo Della Coletta (UNICAMP).

Stock cultures were grown in RPMI-1640 (Vitrocell, Brazil) supplemented with 5% fetal bovine serum (Vitrocell, Brazil) (complete medium) at 37 °C in 5% CO_2. For the experiments, complete medium was supplemented with a 1% penicillin:streptomycin mixture (Nutricell, Brazil, 1000 U/mL:1000 mg/mL). Aliquots of $CuCl_2$, cis-$PtCl_2$, Pt–Nap, Cu-Nap, and NaNap were dissolved in DMSO (1.0 mg material: 10 µL DMSO). Afterwards, each sample solution (50 µL) was diluted in the complete medium (5000 µg/mL), followed by a serial dilution affording the final concentrations of 0.15, 1.5, 15, and 150 µg/mL. Doxorubicin (0.015, 0.15, 1.5, and 15 µg/mL) was used as the positive control. The DMSO final concentration ($\leq$0.15%) did not affect cell viability [50].

Cells in 96-well plates (100 µL cells/well, inoculation density: 3.5 to 6 × 10^4 cell/mL) were exposed to four concentrations of each sample and doxorubicin (100 µL/well), in triplicate, for 48 h at 37 °C and 5% of CO_2. Before (T_0 plate) and after (T_1 plates) the sample addition, the cells were fixed with 50% trichloroacetic acid (50 µL/well), and the cell proliferation was determined by the spectrophotometric quantification of cellular protein content using the sulforhodamine B assay [51]. The measurements were taken at 540 nm in a VersaMax Molecular Devices spectrometer (San Jose, CA, USA). The sample concentration that inhibits 50% cell growth (GI_{50}, cytostatic effect) was calculated via sigmoidal regression using Origin 9.0 software (OriginLab Corporation, Northampton, MA, USA) [52]. The selectivity index [47] was calculated as described in Equation (1).

$$SI = (GI_{50\ HaCaT}/GI_{50\ TCL}), \text{ where TCL = tumor cell line} \quad (1)$$

4.6. Computational Simulations

Gaussian 09 [53] and GaussView [54] programs were used for the computational simulation and to generate the structures of the compounds, respectively. Density functional theory (DFT) was used to evaluate the complexation of platinum and copper metal atoms with the sodium naproxenate ligand, Cu–Nap, and Pt–Nap complexes. For the simulations, the functional M06-2X [55] with 6-31 + G(d,p) [56–58] basis set for hydrogen, carbon, oxygen, and sulfur atoms and the LANL2DZ [59] effective core basis set for copper and platinum atoms were used. Frequency calculations were performed, and no imaginary frequency was found, which confirmed that the optimized structures were at minimum energy.

5. Conclusions

Our study described an efficient strategy to synthetize platinum(II) and copper(II) complexes, with naproxenate as the ligand. The complexes were successfully synthesized and characterized by chemical, spectroscopic, and molecular modeling analyses. Antiproliferative studies were conducted for bacteria and tumor cells, for evaluation of the biological activities of each complex.

Molecular formulae of the complexes were determined, by elemental and thermal analyses, as [$Cu_2(Nap)_4(H_2O)_2$] for Cu–Nap and [$Pt(DMSO)_2(Nap)_2$] for Pt–Nap. The coordination of Nap to Cu(II) occurred by the oxygen atoms of the carboxylate group in a bridging bidentate fashion. Mass spectrometric data, EPR, and molecular modeling optimization suggested the formation of a binuclear species for the Cu–Nap complex in a dimeric, or even polymeric, form, which explains its low solubility in most solvents. EPR spectra reinforced the existence of a binuclear species. In the DMSO solution, formation of mononuclear species occurred via the cleavage of the binuclear complex.

Conversely, infrared and NMR spectroscopic data permitted a proposal of a monodentate coordination of two Nap molecules to Pt(II). The square–planar coordination sphere was completed by two DMSO molecules, as indicated by elemental, infrared, and NMR spectroscopic analyses.

Preliminary antibacterial activity assays showed that the bacterial strains considered in the experiment were resistant to both complexes at the concentrations tested. However, while Pt–Nap did not present inhibition over 50% of the tumor cell lines' growth, Cu–Nap presented a significant cytostatic effect against HT-29 and PC-3 and a low but significant cytocidal effect over the U251 tumor cell line. The cell viability of the complexes was assessed against the HaCaT cell line, and Pt–Nap presented extremely low cytostatic effect at all concentrations assessed, as did its precursor NaNap. However, Cu–Nap presented a cytostatic effect of over 50%, with a GI$_{50}$ of 86 µg/mL (79.6 µmol/L). Still, it showed some selectivity over U251 tumor cells. Altogether, while the Pt(II) complex seems to be poorly effective, the obtained data concerning the biological activity of the Cu(II) complex with naproxen open new perspectives about its use as a cytotoxic agent over tumor cells. Further in vitro and in vivo studies are needed to confirm this hypothesis.

Supplementary Materials: The following supporting information can be downloaded at: https://www.mdpi.com/article/10.3390/inorganics11080331/s1, Figure S1: ATR–FTIR of Pt–Nap and NaNap from 1300 to 500 cm^{-1}. Figure S2: ATR–FTIR of Pt–Nap and NaNap from 1300 to 500 cm^{-1}. Figure S3: Kinetic UV–Vis of Cu–Nap in 24h at absorbances peaks 260, 275, 320, and 335 nm. Figure S4: ^{1}H NMR with integrals. Inset shows both DMSO signals, coordinated in Pt–Nap and solvent.

Author Contributions: A.A.S.: writing—original draft preparation, review and editing, conceptualization, visualization, methodology, investigation, formal analysis, and validation; S.C.L.F.: investigation; Á.B.C.: investigation; K.E.B.: investigation; D.D.A.: investigation; M.C.P.: investigation; A.L.T.G.R.: conceptualization, investigation, writing—review and editing, project administration, supervision, and funding acquisition; J.E.d.C.: investigation, project administration, supervision, and funding acquisition; W.R.L.: conceptualization, investigation, and funding acquisition; D.H.P.: methodology, investigation, funding acquisition, and writing—original draft preparation; A.M.d.C.F.: writing—original draft preparation, review and editing, methodology, and investigation; P.P.C.: conceptualization, visualization, investigation, writing—review and editing, formal analysis, project administration, supervision, and funding acquisition. All authors have read and agreed to the published version of the manuscript.

Funding: The authors thank CAPES (Financial code 001), FAPESP (2018/12062-4; 2021/07458-9, 2021/08717-8 and 2021/10265-8 Cancer Theranostics Innovation Center CancerThera, CEPID), FAEPEX (2432/22), and CNPq (407012/2018-4) for the financial support. Marcelo C. Portes and A.M. da Costa Ferreira thank CEPID-Redoxoma (FAPESP grant 2013/07937-8) for maintenance of EPR equipment. A.L.T.G. Ruiz thanks the National Council for Scientific and Technological Development (CNPq, grants #429463/2018-9 and #313440/2019-0), the São Paulo Research Foundation (FAPESP #2011/01114-4, #2014/23950-7), and the Fund for Support to Teaching, Research and Outreach Activities of the University of Campinas (FAEPEX/Unicamp, grants #2001/19, #2555/20, and #2235/21).

Data Availability Statement: Not applicable.

Conflicts of Interest: The authors declare no conflict of interest.

References

1. Gomes, F.R.; Addis, Y.; Tekamo, I.; Cavaco, I.; Campos, D.L.; Pavan, F.R.; Gomes, C.S.B.; Brito, V.; Santos, A.O.; Domingues, F.; et al. Antimicrobial and antitumor activity of *S*-methyl dithiocarbazate Schiff base zinc(II) complexes. *J. Inorg. Biochem.* **2021**, *216*, 111331. [CrossRef]
2. Manzano, C.M.; Bergamini, F.R.G.; Lustri, W.R.; Ruiz, A.L.T.G.; De Oliveira, E.C.S.; Ribeiro, M.A.; Formiga, A.L.B.; Corbi, P.P. Pt(II) and Pd(II) complexes with ibuprofen hydrazide: Characterization, theoretical calculations, antibacterial and antitumor assays and studies of interaction with CT-DNA. *J. Mol. Struct.* **2018**, *1154*, 469–479. [CrossRef]
3. Štarha, P.; Trávníček, Z. Non-platinum complexes containing releasable biologically active ligands. *Coord. Chem. Rev.* **2019**, *395*, 130–145. [CrossRef]

4. Quintanilha, M.M.; Schimitd, B.A.; Costa, A.M.F.; Nakahata, D.H.; Simoni, D.A.; Clavijo, J.C.T.; Pereira, D.H.; Massabni, A.C.; Lustri, W.R.; Corbi, P.P. A novel water-soluble platinum(II) complex with the amino acid deoxyalliin: Synthesis, crystal structure, theoretical studies and investigations about its antibacterial activity. *J. Mol. Struct.* **2021**, *1236*, 130316. [CrossRef]
5. Silconi, Z.B.; Benazic, S.; Milovanovic, J.; Jurisevic, M.; Djordjevic, D.; Nikolic, M.; Mijajlovic, M.; Ratkovic, Z.; Radić, G.; Radisavljevic, S.; et al. DNA binding and antitumor activities of platinum(IV) and zinc(II) complexes with some S-alkyl derivatives of thiosalicylic acid. *Transit. Metal Chem.* **2018**, *43*, 719–729. [CrossRef]
6. Ravera, M.; Zanellato, I.; Gabano, E.; Perin, E.; Rangone, B.; Coppola, M.; Osella, D. Antiproliferative activity of Pt(IV) conjugates containing the non-steroidal anti-inflammatory drugs (NSAIDs) ketoprofen and naproxen. *Int. J. Mol. Sci.* **2019**, *20*, 3074. [CrossRef]
7. Quirante, J.; Ruiz, D.; Gonzalez, A.; López, C.; Cascante, M.; Cortés, R.; Messeguer, R.; Calvis, C.; Baldomà, L.; Pascual, A.; et al. Platinum(II) and palladium(II) complexes with (N,N′) and (C,N,N′)—Ligands derived from pyrazole as anticancer and antimalarial agents: Synthesis, characterization and in vitro activities. *J. Inorg. Biochem.* **2011**, *105*, 1720–1728. [CrossRef] [PubMed]
8. Fox, C.L. Silver sulfadiazine—A new topical therapy for Pseudomonas in burns. Therapy of Pseudomonas infection in burns. *Arch. Surg.* **1968**, *96*, 184–188. [CrossRef] [PubMed]
9. Leung, C.H.; Lin, S.; Zhong, H.; Ma, D. Metal complexes as potential modulators of inflammatory and autoimmune responses. *Chem. Sci.* **2015**, *6*, 871. [CrossRef] [PubMed]
10. Xun-Zhong, Z.; An-Sheng, F.; Fu-Ran, Z.; Min-Cheng, L.; Yan-Zhi, L.; Meng, M.; Yu, L. Synthesis, Crystal Structures, and Antimicrobial and Antitumor Studies of Two Zinc(II) Complexes with Pyridine Thiazole Derivatives. *Bioinorg. Chem. Appl.* **2020**, *2020*, 8852470. [CrossRef]
11. Frei, A.; Zuegg, J.; Elliott, A.G.; Baker, M.; Braese, S.; Brown, C.; Chen, F.; Dowson, C.G.; Dujardin, G.; Jung, N.; et al. Metal complexes as a promising source for new Antibiotics. *Chem. Sci.* **2020**, *11*, 2627–2639. [CrossRef]
12. Ali, H.A.; Fares, H.; Darawsheh, M.; Rappocciolo, E.; Akkawi, M.; Jaber, S. Synthesis, characterization and biological activity of new mixed ligand complexes of Zn(II) naproxen with nitrogen-based ligands. *Eur. J. Med. Chem.* **2015**, *89*, 67–76. [CrossRef] [PubMed]
13. Oliveira, L.P.; Carneiro, Z.A.; Ribeiro, C.M.; Lima, M.F.; Paixão, D.A.; Pivatto, M.; Souza, M.V.N.; Teixeira, L.R.; Lopes, C.D.; Albuquerque, S.; et al. Three new platinum complexes containing fluoroquinolones and DMSO: Cytotoxicity and evaluation against drug-resistant tuberculosis. *J. Inorg. Biochem.* **2018**, *183*, 77–83. [CrossRef] [PubMed]
14. Juin, S.; Muhammad, N.; Sun, Y.; Tan, Y.; Yuan, H.; Song, D.; Guo, Z.; Wang, X. Multispecific Platinum(IV) Complex Deters Breast Cancer via Interposing Inflammation and Immunosuppression as an Inhibitor of COX-2 and PD-L1. *Angew. Chem. Int.* **2020**, *59*, 23313–23321. [CrossRef]
15. Chen, Y.; Wang, Q.; Li, Z.; Liu, Z.; Zhao, Y.; Zhang, J.; Liu, M.; Wang, Z.; Li, D.; Han, J. Naproxen platinum(IV) hybrids inhibiting cycloxygenases and matrix metalloproteinases and causing DNA damage: Synthesis and biological evaluation as antitumor agents in vitro and in vivo. *Dalton Trans.* **2020**, *49*, 5192–5204. [CrossRef] [PubMed]
16. Bhattacherjee, P.; Roy, M.; Naskar, A.; Tsai, H.; Ghosh, A.; Patra, N.; John, R.P. A trinuclear copper (II) complex of naproxen-appended salicylhydrazide: Synthesis, crystal structure, DNA binding and molecular docking study. *Appl. Organomet. Chem.* **2022**, *36*, 6459. [CrossRef]
17. Abuhijleh, A.L. Mononuclear and Binuclear Copper (II) Complexes of the Antinflammatory Drug Ibuprofen: Synthesis, Characterization, and Catecholase-Mimetic Activity. *J. Inorg. Biochem.* **1994**, *55*, 255–262. [CrossRef] [PubMed]
18. Nunes, J.H.B.; Paiva, R.E.F.; Cuin, A.; Ferreira, A.M.C.; Lustri, W.R.; Corbi, P.P. Synthesis, spectroscopic characterization, crystallographic studies and antibacterial assays of new copper(II) complexes with sulfathiazole and nimesulide. *J. Mol. Struc.* **2016**, *1112*, 14–20. [CrossRef]
19. Spector, D.; Krasnovskaya, O.; Pavlov, K.; Erofeev, A.; Gorelkin, P.; Beloglazkina, E.; Majouga, A. Pt(IV) Prodrugs with NSAIDs as Axial Ligands. *Int. J. Mol. Sci.* **2021**, *22*, 3817. [CrossRef] [PubMed]
20. Tolan, D.A.; Abdel-Monem, Y.K.; El-Nagar, M.A. Anti-tumor platinum (IV) complexes bearing the anti-inflammatory drug naproxen in the axial position. *Appl Organometal. Chem.* **2019**, *33*, 4763. [CrossRef]
21. Dimiza, F.; Papadopoulos, A.N.; Tangoulis, V.; Psycharis, V.; Raptopoulou, C.P.; Kessissoglou, D.P.; Psomas, G. Biological evaluation of cobalt(II) complexes with non-steroidal anti-inflammatory drug naproxen. *J. Inorg. Biochem.* **2012**, *107*, 54–64. [CrossRef]
22. Shaheen, M.A.; Feng, S.; Anthony, M.; Tahir, M.N.; Hassan, M.; Seo, S.; Ahmad, S.; Iqbal, M.; Saleem, M.; Lu, C. Metal-Based Scaffolds of Schiff Bases Derived from Naproxen: Synthesis, Antibacterial Activities, and Molecular Docking Studies. *Molecules* **2019**, *24*, 1237. [CrossRef]
23. Chu, Y.; Wang, T.; Ge, X.; Yang, P.; Li, W.; Zhao, J.; Zhu, H. Synthesis, characterization and biological evaluation of naproxen Cu(II) complexes. *J. Mol. Struct.* **2019**, *1178*, 564–569. [CrossRef]
24. Wang, T.; Tang, G.; Wan, W.; Wu, Y.; Tian, T.; Wang, J.; He, C.; Long, X.; Wang, J.; Ng, S.W. New homochiral ferroelectric supramolecular networks of complexes constructed by chiral S-naproxen ligand. *CrystEngComm* **2012**, *14*, 3802. [CrossRef]
25. Hasan, M.S.; Das, N. A detailed in vitro study of naproxen metal complexes in quest of new therapeutic possibilities. *Alex. J. Med.* **2017**, *53*, 157–165. [CrossRef]
26. Trinchero, A.; Bonora, S.; Tinti, A.; Fini, G. Spectroscopic Behavior of Copper Complexes of Nonsteroidal Anti-Inflammatory Drugs. *Biopolymers* **2004**, *74*, 120–124. [CrossRef]

27. Dimiza, F.; Perdih, F.; Tangoulis, V.; Turel, I.; Kessissoglou, D.P.; Psomas, G. Interaction of copper(II) with the non-steroidal anti-inflammatory drugs naproxen and diclofenac: Synthesis, structure, DNA- and albumin-binding. *J. Inorg. Biochem.* **2011**, *105*, 476–489. [CrossRef] [PubMed]
28. Martins, D.J.; Hanif-Ur-Rehman; Rico, S.A.; Costa, I.M.; Santos, A.C.P.; Szszudlowski, R.G.; Silva, D.O. Interaction of chitosan beads with a copper-naproxen metallodrug. *RSC Adv.* **2015**, *5*, 90184. [CrossRef]
29. Abuhijleh, A.L. Mononuclear copper (II) salicylate complexes with 1,2-dimethylimidazole and 2-methylimidazole: Synthesis, spectroscopic and crystal structure characterization and their superoxide scavenging activities. *J. Mol. Struct.* **2010**, *980*, 201–207. [CrossRef]
30. Sharma, J.; Singla, A.K.; Dhawan, S. Zinc–naproxen complex: Synthesis, physicochemical and biological evaluation. *Int. J. Pharm.* **2003**, *260*, 217–227. [CrossRef]
31. Srivastava, P.; Mishra, R.; Verma, M.; Sivakumar, S.; Patra, A.K. Cytotoxic ruthenium(II) polypyridyl complexes with naproxen as NSAID: Synthesis, biological interactions and antioxidant activity. *Polyhedron* **2019**, *172*, 132–140. [CrossRef]
32. Dendrinou-Samara, C.; Tsotsou, G.; Ekateriniadou, L.V.; Kortsaris, A.H.; Raptopoulou, C.P.; Terzis, A.; Kyriakidis, D.A.; Kessissoglou, D.P. Anti-inflammatory drugs interacting with Zn(II), Cd(II) and Pt(II) metal ions. *J. Inorg. Biochem.* **1998**, *71*, 171–179. [CrossRef]
33. Srivastava, P.; Singh, K.; Verma, M.; Sivakumar, S.; Patra, A.K. Photoactive platinum(II) complexes of nonsteroidal anti-inflammatory drug naproxen: Interaction with biological targets, antioxidant activity and cytotoxicity. *Eur. J. Med. Chem.* **2018**, *144*, 243–254. [CrossRef]
34. Terracina, A.; McHugh, L.N.; Todaro, M.; Agnello, S.; Wheatley, P.S.; Gelardi, F.M.; Morris, R.E.; Buscarino, G. Multitechnique Analysis of the Hydration in Three Different Copper Paddle-Wheel Metal-Organic Frameworks. *J. Phys. Chem. C* **2019**, *123*, 28219–28232. [CrossRef]
35. Price, J.H.; Williamson, A.N.; Schramm, R.F.; Wayland, B.B. Palladium(II) and platinum(II) alkyl sulfoxides complexes. Examples of sulfur-bonded, mixed sulfur- and oxygen-bonded, and totally oxygen-bonded complexes. *Inorg. Chem.* **1972**, *11*, 1280–1284. [CrossRef]
36. Zhang, J.; Li, Y.; Sun, J. Synthesis, characterization and cytotoxicity of amine/ethylamine platinum(II) complexes with carboxylates. *Eur. J. Med. Chem.* **2009**, *44*, 2758–2762. [CrossRef] [PubMed]
37. Zayed, M.A.; Hawash, M.F.; El-Desawy, M.; El-Gizouli, A.M.M. Investigation of naproxen drug using mass spectrometry, thermal analyses and semi-empirical molecular orbital calculation. *Arab. J. Chem.* **2017**, *10*, 351–359. [CrossRef]
38. Mikuriya, M.; Chihiro, Y.; Tanabe, K.; Nukita, N.; Amabe, Y.; Yoshioka, D.; Mitsuhashi, R.; Tatehata, R.; Tanaka, H.; Handa, M.; et al. Copper(II) Carboxylates with 2,3,4-Trimethoxybenzoate and 2,4,6-Trimethoxybenzoate: Dinuclear Cu(II) Cluster and μ-Aqua-Bridged Cu(II) Chain Molecule. *Magnetochemistry* **2021**, *7*, 35. [CrossRef]
39. Wang, Z.; Rodewald, K.; Medishetty, R.; Rieger, B.; Fischer, R.A. Control of Water Content for Enhancing the Quality of Copper Paddle-Wheel-Based Metal-Organic Framework Thin Films Grown by Layer-by-Layer Liquid-Phase Epitaxy. *Cryst. Growth Des.* **2018**, *18*, 7451–7459. [CrossRef]
40. Abuhijleh, A.L.; Khalaf, J. Copper (II) complexes of the anti-inflammatory drug naproxen and 3-pyridylmethanol as auxiliary ligand. Characterization, superoxide dismutase and catecholase-mimetic activities. *Eur. J. Med. Chem.* **2010**, *45*, 3811–3817. [CrossRef] [PubMed]
41. Vaz, R.H.; Silva, R.M.; Reibenspies, J.H.; Serra, O.A. Synthesis and X-ray Crystal Structure of a Stable *cis*-1,2-bis(diphenylphosphino)ethene Monodentate Thiolate Platinum Complex and TGA Studies of its Precursors. *J. Braz. Chem. Soc.* **2002**, *13*, 82–87. [CrossRef]
42. Pereira, A.K.S.; Manzano, C.M.; Nakahata, D.H.; Clavijo, J.C.T.; Pereira, D.H.; Lustri, W.R.; Corbi, P.P. Synthesis, crystal structures, DFT studies, antibacterial assays and interaction assessments with biomolecules of new platinum(II) complexes with adamantane derivatives. *New J. Chem.* **2020**, *44*, 11546–11556. [CrossRef]
43. Vincent, M.; Duval, R.E.; Hartemann, P.; Engels-Deutsch, M. Contact killing and antimicrobial properties of copper. *J. Appl. Microbiol.* **2018**, *124*, 1032–1046. [CrossRef]
44. Jorgensen, J.H.; Ferraro, M.J. Antimicrobial susceptibility testing: A review of general principles and contemporary practices. *Clin. Infect. Dis. Off. Publ. Infect. Dis. Soc. Am.* **2009**, *49*, 1749–1755. [CrossRef]
45. Xue, Q.; Kang, R.; Klionsky, D.J.; Tang, D.; Liu, J.; Chen, X. Copper metabolism in cell death and autophagy. *Autophagy* **2023**, *19*, 2175–2195. [CrossRef] [PubMed]
46. Han, M.İ.; Küçükgüzel, Ş.G. Anticancer and Antimicrobial Activities of Naproxen and Naproxen Derivatives. *Mini Rev. Med. Chem.* **2020**, *20*, 1300–1310. [CrossRef]
47. Muller, P.Y.; Milton, M.N. The determination and interpretation of the therapeutic index in drug development. *Nat. Rev. Drug Discov.* **2012**, *11*, 751–761. [CrossRef]
48. Stoll, S.; Schweiger, A. EasySpin, a comprehensive software package for spectral simulation and analysis in EPR. *J. Magn. Res.* **2006**, *178*, 42–55. [CrossRef]
49. CLSI. *Performance Standards for Antimicrobial Susceptibility Testing*, 32nd ed.; CLSI supplement M100; Clinical and Laboratory Standards Institute: Wayne, PA, USA, 2022.
50. Sobreiro, M.A.; Della Torre, A.; de Araújo, M.E.M.B.; Canella, P.R.B.C.; de Carvalho, J.E.; Carvalho, P.O.; Ruiz, A.L.T.G. Enzymatic hydrolysis of rutin: Evaluation of kinetic parameters and anti-proliferative, mutagenic and anti-mutagenic effects. *Life* **2023**, *13*, 549. [CrossRef]

51. Da Silva, G.G.; Della Torre, A.; Braga, L.E.O.; Bachiega, P.; Tinti, S.V.; de Carvalho, J.E.; Dionísio, A.P.; Ruiz, A.L.T.G. Yellow-colored extract from cashew byproduct—Nonclinical safety assessment. *Regul. Toxicol Pharmacol.* **2020**, *115*, 104699. [CrossRef]
52. Monks, A.; Scudiero, D.; Skehan, P.; Shoemaker, R.; Paull, K.; Vistica, D.; Hose, C.; Langley, J.; Cronise, P.; Vaigro-Wolff, A.; et al. Feasibility of a high-flux anticancer drug screen using a diverse panel of cultured human tumor cell lines. *JNCI J. Natl. Cancer Inst.* **1991**, *83*, 757–766. [CrossRef]
53. Frisch, J.; Trucks, G.W.; Schlegel, H.B.; Scuseria, G.E.; Robb, M.A.; Cheeseman, J.R.; Scalmani, G.; Barone, V.; Mennucci, B.; Petersson, G.A.; et al. *Gaussian09, Revision D.1*; Gaussian, Inc.: Wallingford, CT, USA, 2009.
54. Dennington, R.; Keith, T.; Millam, J. *Gauss View, Version 5*; Shawnee Mission. (n.d.); Semichem Inc.: Shawnee, KS, USA, 2009.
55. Zhao, Y.; Truhlar, D.G. The M06 suite of density functionals for main group thermochemistry, thermochemical kinetics, noncovalent interactions, excited states, and transition elements: Two new functionals and systematic testing of four M06-class functionals and 12 other functionals. *Theor. Chem. Acc.* **2008**, *120*, 215–241.
56. Ditchfield, R.; Hehre, W.J.; People, J.A. Self-consistent molecular-orbital methods IX: An extended Gaussian-type basis for molecular-orbital studies of organic molecules. *J. Chem. Phys.* **1971**, *54*, 724–728. [CrossRef]
57. Hehre, W.J.; Ditchfield, R.; Pople, J.A. Self-consistent molecular orbital methods XII: Further extensions of Gaussian-type basis sets for use in molecular orbital studies of organic molecules. *J. Chem. Phys.* **1972**, *56*, 2257–2261. [CrossRef]
58. Hariharan, P.C.; Pople, J.A. The influence of polarization functions on molecular orbital hydrogenation energies. *Theor. Chim. Acta* **1973**, *28*, 213–222. [CrossRef]
59. Hay, P.J.; Wadt, W.R. *Ab-initio* effective core potentials for molecular calculations: Potentials for the transition metal atoms Sc to Hg. *J. Chem. Phys.* **1985**, *82*, 270–283. [CrossRef]

Disclaimer/Publisher's Note: The statements, opinions and data contained in all publications are solely those of the individual author(s) and contributor(s) and not of MDPI and/or the editor(s). MDPI and/or the editor(s) disclaim responsibility for any injury to people or property resulting from any ideas, methods, instructions or products referred to in the content.

Communication

Evaluation of Membrane Permeability of Copper-Based Drugs

Evariste Umba-Tsumbu [1], Ahmed N. Hammouda [1,2] and Graham Ellis Jackson [1,*]

[1] Department of Chemistry, University of Cape Town, Cape Town 7701, South Africa
[2] Department of Chemistry, Faculty of Science, University of Benghazi, Benghazi 16063, Libya
* Correspondence: graham.jackson@uct.ac.za; Tel.: +61-490675225

Abstract: Membrane permeability of copper complexes with potential anti-inflammatory activity were measured using an artificial membrane in a modified Franz cell. Using $CuCl_2$ as the control, all the ligands tested enhanced the diffusion of copper, with enhancement factors ranging from 2 to 7. Octanol/water partition coefficients (log $K_{o/w}$) were measured and correlated with the permeability coefficients (K_p). In addition, chemical speciation was used to determine the predominant complex in solution at physiological pH. No correlation was found between the measured permeability coefficients and either molecular weight (MW) or log $K_{o/w}$.

Keywords: transdermal drug delivery; absorption enhancer; tissue partition; diffusion; permeability; Cerasome 9005

Citation: Umba-Tsumbu, E.; Hammouda, A.N.; Jackson, G.E. Evaluation of Membrane Permeability of Copper-Based Drugs. *Inorganics* **2023**, *11*, 179. https://doi.org/10.3390/inorganics11050179

Academic Editor: Bruno Therrien

Received: 27 February 2023
Revised: 10 April 2023
Accepted: 21 April 2023
Published: 23 April 2023

Copyright: © 2023 by the authors. Licensee MDPI, Basel, Switzerland. This article is an open access article distributed under the terms and conditions of the Creative Commons Attribution (CC BY) license (https://creativecommons.org/licenses/by/4.0/).

1. Introduction

Rheumatoid Arthritis (RA) is a chronic inflammatory disease, which mainly affects the articulate joints and in severe cases can lead to joint destruction [1]. In the early stages of the disease, serum copper levels are significantly elevated [2,3]. Whether this is causative or a response to the inflammation is unknown, but serum copper levels are a potential biomarker of disease activity in RA patients [4]. Sorenson [5] and Jackson et al. [6,7] have shown that the inflammation associated with RA can be reduced by the subcutaneous injection of Cu(II) complexes. Walker et al. observed that topical applications of an ethanolic Cu(II) salicylate-containing preparation (Alcusal) produces anti-inflammatory effects in human volunteers. A gel-based version is available in Australia [8]. Puranik et al. showed that the copper complexes of non-steroidal anti-inflammatory drugs inhibit acute inflammation in vivo [9].

Dermal absorption is the preferred route of administration for long-term drug therapy because it is slow, tolerable and less painful compared to injection or oral administration. This leads to improved patient compliance. For oral administration, there is the added complication that the stomach pH may affect the drug. The efficacy of the dermal absorption route, however, depends on the ability of the drug to pass through the skin. Hostynek and Maibach have reviewed the interaction between copper and the skin [10]. Preliminary dermal absorption studies have been performed on some of the copper complexes using Cu-67 and BALB/c mice [7,11–13]. However, these studies are expensive, time consuming and require ethical approval. For the rapid screening of different copper complexes, an efficient model of dermal absorption is needed. Since Flynn [14] proposed the use of physicochemical properties to predict skin permeation, several experimental and theoretical models have been proposed [15–18]. Much of this work have been based on immobilized artificial membranes [19]. Meanwhile, the skin is composed of three main layers with the outer layer—the epidermis—being the main barrier to dermal absorption. The difficulty is to find an artificial mimic of this layer. Cerasome 9005 is a mixture of lipids similar to those found in the stratum corneum. Krulikowska et al. [20] found a 95% correlation between the penetration coefficients of porcine skin and Cerasome 9005. For this reason, Cerasome has been used in this study.

2. Results and Discussion

The ligands used in this study were chosen because they have been investigated as potential anti-inflammatory copper(II) drugs [7,11–13,21]. Figure 1 shows the structure of some of the ligands. Figure 2 shows the results for the diffusion of several copper complexes, over 24 h, through an immobilized Cerasome membrane. For each complex, there was a slow induction period (8 h) before a steady state flux was obtained. From the slope of the curves, the steady state flux and the permeability coefficients, K_p, were calculated (Table 1). The use of K_p is now encouraged [22] although it has been criticized [23].

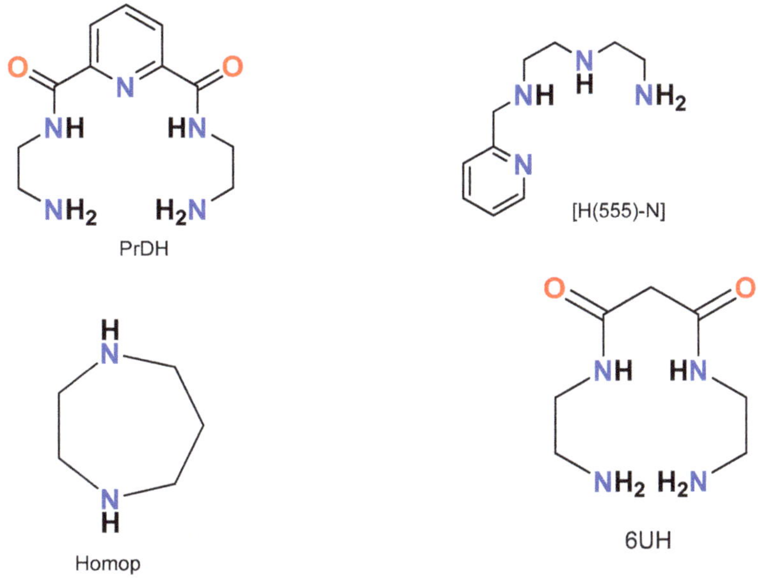

Figure 1. Schematic diagram of some of the ligands used in this study: N,N'-di(aminoethylene)-2,6-pyridine-dicarbonylamine (PrDH), N-[2-(2-aminoethylamino)ethyl]picolinamide (H(555)-N), homopiperazine (homop) and N,N'-bis[aminoethyl]propanediamide (6UH).

The increased uptake of Cu(II) in the presence of amino acids is in agreement with previous results obtained with Ehrlich ascites tumor, brain, liver and kidney cells [24–26]. The K_p of Cu(II) complexed to alanine (6.31 ± 0.01) × 10^{-6} cm/s and glycine (5.79 ± 0.04) × 10^{-6} cm/s are comparable to values found by Mazurowsky for the same complexes (alanine (1.90 ± 0.16) × 10^{-6} cm/s; glycine (1.62 ± 0.06) × 10^{-6} cm/s) [24–26]. Mazurowsky used liposomes as a model membrane, with potassium phosphate buffered at pH 7.4 as the acceptor phase.

CuCl$_2$, was included in the results as a control. The copper CuCl$_2$ experiments were performed at pH 4.23 since, at pH 7 and the concentrations used, Cu(OH)$_2$ would precipitate. Our results show that the ligands used were able to keep Cu(II) in solution at physiological pH and increase the rate of diffusion of copper through the membrane. From Equation (1), these resulted in an enhancement factor (EF) which can be calculated to provide the relative effect of the ligand upon Cu(II) diffusion through the membrane. The EF ranged from 2 for dtpa to 6.8 for H(555N) (Table 1).

$$EF = \frac{K_{p(Cu(II)-ligand)}}{K_{p(Cu^{2+})}} \quad (1)$$

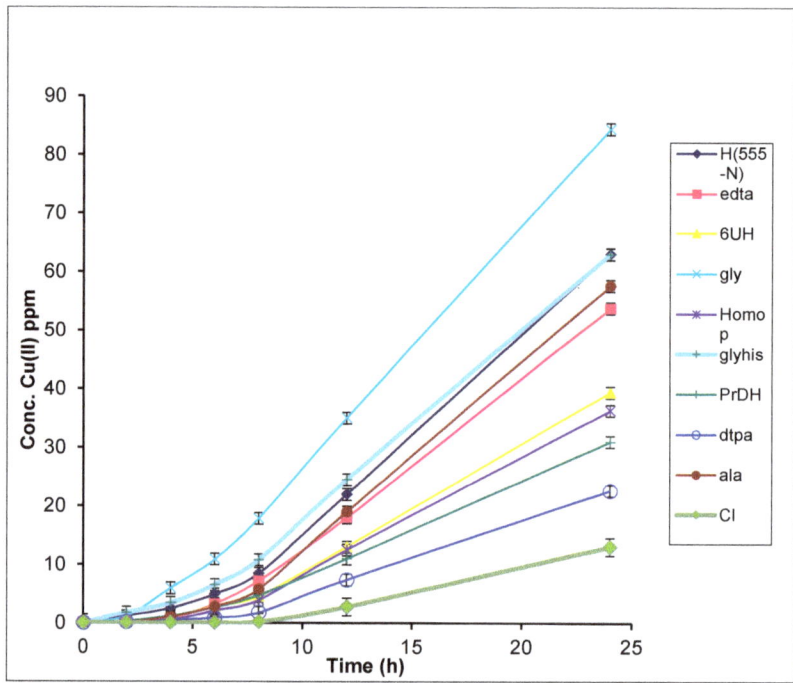

Figure 2. Concentration vs. time of Cu(II) in receiver phase after passing through a Cerasome 9005 membrane at 25 °C. The error bars represent the standard deviation in the concentration.

Table 1. Diffusion and distribution coefficients of different copper complex species present in solution at pH 7.0. Copper concentration was 20 mM.

Formula	Cu:L Ratio	$-\log(K_p)$	$-\log(K_{o/w})$	MW	EF
$[Cu(gly)_2(H_2O)_2]$	1:2	5.79	2.66	249.5	5.2
$[Cu(Homop)(H_2O)_4]$	1:2	5.63	3.48	300.2	5.1
$[Cu(6UH)(H_2O)_2]$	1:2	6.05	3.02	287.5	5.4
$[Cu(H(555N))(H_2O)_2]$	1:1	7.60	3.00	308.5	6.8
$[Cu(PrDH)(H_2O)_2]$	1:1	2.28	3.45	350.5	2.0
$[Cu(edta)]$	1:1	6.49	3.07	351.7	5.8
$[Cu(dtpa)]$	1:1	2.17	3.62	451.9	2.0
$[Cu(H_2O)_6]$	-	1.11	-	171.5	1

2.1. Partition Coefficient

Table 1 lists the partition coefficients (log $K_{o/w}$) of the different Cu(II) complexes, measured at room temperature and a physiological pH of 7.00. Cu-gly and Cu-H(555-N) have the highest lipophilicity to the other complexes studied, although none of the complexes are very lipophilic. Zvimba has suggested that, for reasonably absorption, a log $K_{o/w}$ of at least 0.6 is needed [27]. The negative values of log $K_{o/w}$ found in our study indicate that these complexes are largely hydrophilic.

2.2. Data Analysis

One aim of this study was to derive a simple relationship between permeability coefficient and another more easily measured physical parameter of copper complexes, such as $K_{o/w}$ or molecular weight (MW). Hence, it is necessary to know the MW of the complexes. For organic molecules, this is relatively easy, but for labile inorganic complexes such as copper, the speciation and, hence, MW changes with pH. In this study, we have used measured equilibrium constants [7–13,21] to calculate the speciation of the different complexes in solution. This is illustrated in Figure 3, which shows the concentration of the copper species as a function of pH. At pH 7, copper is 100% complexed to dtpa (Figure 3a), but with PrDH, the copper is distributed between two main complexes of different stoichiometry, [Cu(PrDH)(H$_2$O)$_3$] (12%) and [Cu(PrDH)(H$_2$O)$_3$H$_{-1}$] (60%) (Figure 3b). An added complication is that the number of coordinated water molecules is not specified by the equilibrium constant. However, in this study, we have assumed that the copper is 6-coordinate with any vacant sites occupied by water molecules. The resultant stoichiometries and MWs are given in Table 1.

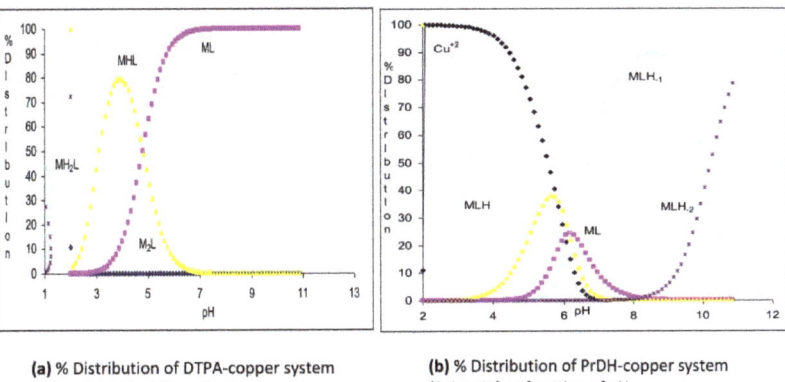

(a) % Distribution of DTPA-copper system (1:1 ratio) as function of pH

(b) % Distribution of PrDH-copper system (1:1 ratio) as function of pH

Figure 3. Species distribution curves for (a) Cu(II)/dtpa and (b) Cu(II)/PrDH. The stoichiometry of the complex is denoted by MHL, where M = Cu(II), L = dtpa or PrDH and H = H$^+$.

Linear regression analysis of log K_p against log $K_{o/w}$ presented an R^2 of 0.49, while log K_p versus MW presented an R^2 of 0.35. Consequently, for these compounds, there is no correlation between log K_p and either log $K_{o/w}$ or MW.

Potts and Guy [17] have proposed a quantitative structure–permeability relationship model (Equation (2)), which depends upon both the size of the drug (MW) and $K_{o/w}$.

$$\text{Log } K_p = \log (D^0/h) + f \log K_{o/w} - \beta' \text{ MW} \qquad (2)$$

where, h is the membrane thickness; D^0 is the diffusivity of a hypothetical molecule of zero molecular weight; f is a constant which accounts for the difference between the membrane lipids and octanol; β' converts molecular volume to molecular weight.

Multiple linear regression of log K_p, log $K_{o/w}$ and MW presented an R^2 of 0.53. Thus, these two parameters on their own are poor descriptors of log K_p and indicates that only 53% of log K_p is explained by log $K_{o/w}$ and MW. This is not surprising as the different copper complexes have different charges, which would affect their diffusivity. Even though the correlation was poor, values were obtained for β' (0.01), log (D^0/h) (−17.4) and f (−2.7). Flynn [14,22] found that the values of β', Log (D^0/h) and f for 90 drugs were 0.0061, −2.72 and 0.71, respectively. The diffusivity of these charged complexes would be expected to be much lower, as was found, but that f should be similar as the charge of the complex should have the same effect on partitioning for both octanol and the membrane. Thus, for labile metal complexes, factors other than size and partition coefficient are important.

3. Materials and Methods

CuCl$_2$.2H$_2$O, glycine (gly), ethylenediaminetetraacetic acid (edta), alanine (ala), diethylenetriaminepentaacetic acid (dtpa) and homopiperazine (homop) were obtained commercially. Ligands, N,N'-di(aminoethylene)-2,6-pyridine-dicarbonylamine (PrDH), N-[2-(2-

aminoethylamino)ethyl]picolinamide (H(555)-N) and N,N′-bis[aminoethyl]propanediamide (6UH) (Figure 1) were synthesized in our laboratory [7,12,21]. Cerasome (product name: Cerasome 9005) was kindly donated by Lipoid GMBH (Ludwigshafen, Germany). Cerasome 9005 is composed of hydrogenated lecithin, cholesterol, ceramides (NP and NS) and fatty acids (palmitic acid and oleic acid) in distilled water with ~10% ethanol as a preservative. The concentration of total lipids is 6.60 g/100 g. The particle size and pH value of Cerasome, offered by Lipoid GMBH, were 48.1 nm and 7.3, respectively. Cerasome was stored between 15 °C and 25 °C, as recommended in the product information sheet. A comparison of the structure of Cerasome 9005 and the stratum corneum is detailed in Figure 1 of [28].

10 mM or 5 mM solutions of the copper complexes were prepared from $CuCl_2 \cdot 2H_2O$ and the different ligands in MilliQ-water. The pH of the solutions was adjusted to 7.00 using concentrated NaOH or HCl. Different metal/ligand ratios were used contingent upon the ligand so as to avoid the formation of precipitate. In order to avoid precipitation of copper hydroxide, a pH of 4.23 was used for the $CuCl_2$ control. This was the highest pH that could be used and still avoid precipitation.

The artificial membrane was made using filter paper (Macherey-Nagel) of 3.2 cm^2 diameter and thickness 0.12 cm. The filter paper was submerged in the Cerasome 9005 lipid solution for a few minutes at 25 °C and then weighed. The amount of lipid absorbed was determined by mass difference and was typically 0.0131g with variability ≈ 7%.

Figure 4 shows a modified Franz cell consisting of two 50 cm^3 cylinders connected through a sintered glass membrane. This arrangement had the advantage that samples could be removed for analysis without disturbing the hydrostatic pressure. Each solution could be stirred and the whole apparatus was placed in a temperature-controlled environment. The disadvantage of this apparatus, relative to the Franz cell, is that the donor and receiver phases could not be independently thermostated. Each experiment was repeated 3 times.

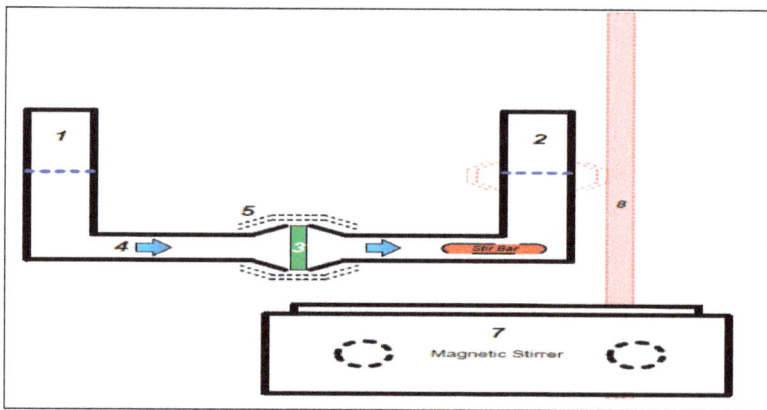

Figure 4. Modified Franz cell apparatus used in this study, where: (1) donor phase filled with 20 mL of Cu(II) complex; (2) receiver phase filled with blank solution (distilled/deionized water; (3) the artificial membrane; (4) passive diffusion direction; (5) clamp; (6) stirrer bar; (7) magnetic stirrer; and (8) burette stands with clamp.

The shake flask method was used to measure partition coefficients where the organic phase was 1-octanol pre-saturated with water [29]. An amount of 40 mL of 1-octanol was mixed with 10 mL of the aqueous complex solution and shaken for 5 min. Afterward, 1 mL of the aqueous layer and 38 mL of the organic layer were removed using micropipettes. The copper in the organic layer was back-extracted using 5% v/v HNO_3. The concentration of copper in the two layers was measured by atomic absorption spectroscopy (AAS) using a Varian AA-5 spectrometer. The instrument was calibrated in the 1–15 ppm range and sample concentrations were adjusted accordingly. The spectrometer settings used were as follows: 3 mA (copper lamp), 1.5 units/70 kPa (acetylene), 350 kPa (compress air), 324.7 nm (wavelength), 2 s (time), abs. exp. factor (1) and 0.03–10 µg/mL. The analytical standard deviation was found to be <1%.

Analysis of Data

Permeability coefficients (K_p) were determined from Fick's first law of diffusion (Equation (3)) [22,23]:

$$K_p = \frac{J}{C_i} \quad (3)$$

where C_i is the initial permeant concentration in the donor solution and J is the mass passing through unit area of the membrane per unit time and is given by:

$$J = -\frac{dM}{S.dt} \quad (4)$$

where J = flux in g/cm^{2}s; S = cross-section of barrier in cm^2; and dM/dt = rate of diffusion in g/s.

4. Conclusions

In this study, we have demonstrated that a simple artificial membrane, Cerasome, in a modified Franz cell, can be used to study the diffusion of copper complexes. In addition, we showed that simple ligands can be used to enhance the membrane permeability of copper. However, the partition coefficient and/or molecular weight cannot be used to predict tissue permeability.

Author Contributions: Conceptualization, G.E.J.; Methodology, G.E.J.; Writing—original draft, E.U.-T.; Writing—review and editing, G.E.J. and A.N.H. The project was conceptualized by G.E.J., executed by E.U.-T. and the data was interpreted by all of the authors. All authors have read and agreed to the published version of the manuscript.

Funding: This research was funded by the National Research Foundation of South Africa, grant number 93450 and 85466 and the University of Cape Town Research Committee. E.U. thanks the Eric Abraham Foundation for financial support.

Institutional Review Board Statement: Not applicable.

Informed Consent Statement: Not applicable.

Data Availability Statement: The data presented in this study are available in the article.

Acknowledgments: The authors wish to thank Lipoid GmbH Company (Germany) for supplying Cerasome 9005.

Conflicts of Interest: The authors declare no conflict of interest. The funders had no role in the design of the study; in the collection, analyses, or interpretation of data; in the writing of the manuscript, or in the decision to publish the results.

Abbreviations

K_p = permeability coefficient defined according to Fick's first law of diffusion; dtpa = diethylenetriamine; edta = ethylenediamine; gly = glycine; MW = molecular weight.

References

1. McInnes, I.B.; Schett, G. The pathogenesis of rheumatoid arthritis. *N. Engl. J. Med.* **2011**, *365*, 2205–2219. [CrossRef]
2. Strecker, D.; Mierzecki, A.; Radomska, K. Copper levels in patients with rheumatoid arthritis. *Ann. Agric. Environ. Med.* **2013**, *20*, 312–316.

3. Wang, H.; Zhang, R.; Shen, J.; Jin, Y.; Chang, C.; Hong, M.; Guo, S.; He, D. Circulating Level of Blood Iron and Copper Associated with Inflammation and Disease Activity of Rheumatoid Arthritis. *Biol. Trace Elem. Res.* **2023**, *201*, 90–97. [CrossRef]
4. Chakraborty, M.; Chutia, H.; Changkakati, R. Serum Copper as a Marker of Disease Activity in Rheumatoid Arthritis. *J. Clin. Diag. Res.* **2015**, *9*, BC09-11. [CrossRef]
5. Sorenson, J.R. Copper chelates as possible active forms of the anti-arthritic agents. *J. Med. Chem.* **1976**, *19*, 135–148. [CrossRef]
6. Jackson, G.E.; May, P.M.; Williams, D.R. Metal-ligand complexes involved in rheumatoid arthritis. I. Justification for Copper Administration. *J. Inorg. Nucl. Chem.* **1978**, *40*, 1227–1234. [CrossRef]
7. Zvimba, J.N.; Jackson, G.E. Copper chelating anti-inflammatory agents: N1-(2-aminoethyl)-N2-(pyridin-2-ylmethyl)ethane-1,2-diamine and N-(2-(2-aminoethylamino)ethyl)picolinamide: An in vitro and in vivo study. *J. Inorg. Biochem.* **2007**, *101*, 148–158. [CrossRef]
8. Walker, W.R.; Beveridge, S.J.; Whitehouse, M.W. Anti-inflammatory activity of a dermally applied copper salicylate preparation (Alcusal). *Agents Actions* **1980**, *10*, 38–47. [CrossRef]
9. Puranik, R.; Bao, S.; Bonin, A.M.; Kaur, R.; Weder, J.E.; Casbolt, L.; Hambley, T.W.; Lay, P.A.; Barter, P.J.; Rye, K.A. A novel class of copper(II)- and zinc(II)-bound non-steroidal anti-inflammatory drugs that inhibits acute inflammation in vivo. *Cell Biosci.* **2016**, *6*, 9. [CrossRef]
10. Hostynek, J.J.; Maibach, H.I. *Copper and the Skin*; CRC Press: Boca Raton, FL, USA, 2006.
11. Zvimba, J.N.; Jackson, G.E. Solution equilibria of copper(II) complexation with N,N'-(2,2'-azanediylbis(ethane-2,1-diyl))dipicolinamide: A bio-distribution and dermal absorption study. *J. Inorg. Biochem.* **2007**, *101*, 1120–1128. [CrossRef]
12. Odisitse, S.; Jackson, G.E. In vitro and in vivo studies of the dermally absorbed Cu(II) complexes of N_5O_2 donor ligands—Potential anti-inflammatory drugs. *Inorg. Chim. Acta* **2009**, *362*, 125–135. [CrossRef]
13. Odisitse, S.; Jackson, G.E. In vitro and in vivo studies of N,N'-bis[2(2-pyridyl)-methyl]pyridine-2,6-dicarboxamide-copper(II) and rheumatoid arthritis. *Polyhedron* **2008**, *27*, 453–464. [CrossRef]
14. Dermal Exposure Assessment: Principles and Applications. January 1992 United States Environmental Protection Agency, EPA /600/8-91/011B. Interim Report. Available online: https://cfpub.epa.gov/ncea/risk/recordisplay.cfm?deid=12188 (accessed on 17 February 2021).
15. Zhang, K.; Chen, M.; Scriba, G.K.E.; Abraham, M.H.; Fahr, A.; Liu, X. Human Skin Permeation of Neutral Species and Ionic Species: Extended Linear Free-Energy Relationship Analyses. *J. Pharm. Sci.* **2012**, *101*, 2034–2044. [CrossRef]
16. Geinoz, S.; Guy, R.H.; Testa, B.; Carrupt, P.A. Quantitative structure–permeation relationships (QSPeRs) to predict skin permeation: A critical evaluation. *Pharm. Res.* **2004**, *21*, 83–92. [CrossRef]
17. Potts, R.O.; Guy, R.H. Predicting Skin permeability. *Pharm. Res.* **1992**, *9*, 663–669. [CrossRef]
18. Kobayashi, Y.; Komatsu, T.; Sumi, M.; Numajiri, S.; Miyambo, M.; Kobayashi, D.; Sugibayashi, K.; Morimoto, Y. In vitro permeation of several drugs through the human nail plate: Relationship between physicochemical properties and nail permeability of drugs. *Eur. J. Pharm. Sci.* **2004**, *21*, 471–477. [CrossRef]
19. Zhang, K.; Chen, M.; Scriba, K.E.G.; Abraham, M.H.; Fahr, A.; Liu, X. Linear Free Energy Relationship Analysis of Retention Factors in Cerasome Electrokinetic Chromatography Intended for Predicting Drug Skin Permeation. *J. Pharm. Sci.* **2011**, *100*, 3105–3113. [CrossRef]
20. Krulikowska, M.; Arct, J.; Lucova, M.; Cetner, B.; Majewski, S. Artificial membranes as models in penetration Investigations. *Skin Res. Technol.* **2013**, *19*, 139–145. [CrossRef]
21. Jackson, G.E.; Linder, P.W.; Voye, A. A potentiometric and spectroscopic study of copper(ii) diamidodiamino complexes. *J. Chem. Soc. Dalton Trans.* **1996**, *24*, 4605–4612. [CrossRef]
22. Bartzatt, R. Determination of dermal permeability coefficient (Kp) by utilizing multiple descriptors in artificial neural network analysis and multiple regression analysis. *J. Sci. Res. Rep.* **2014**, *3*, 2884–2899. [CrossRef]
23. Korinth, G.; Schaller, K.H.; Drexler, H. Is the permeability coefficient Kp a reliable tool in percutaneous absorption studies? *Arch. Toxicol.* **2005**, *79*, 155–159. [CrossRef] [PubMed]
24. Mazurowska, L.; Nowak-Buciak, K.; Mojski, M. ESI-MS method for in vitro investigation of skin penetration by copper-amino acid complexes: From an emulsion through a model membrane. *Anal. Bioanal. Chem.* **2007**, *388*, 1157–1163. [CrossRef]
25. Mazurowska, L.; Mojski, M. Biological activities of selected peptides: Skin penetration ability of copper complexes with peptide. *J. Cosmet. Sci.* **2008**, *59*, 59–69. [PubMed]
26. Mazurowska, L.; Mojski, M. ESI-MS study of the mechanism of glycyl-l-histidyl-l-lysine-Cu(II) complex transport through model membrane of stratum corneum. *Talanta* **2007**, *72*, 650–654. [CrossRef]
27. Zvimba, J.N.; Jackson, G.E. Thermodynamic and spectroscopic study of the interaction of Cu(II), Ni(II), Zn(II) and Ca(II) ions with 2-amino-N-(2-oxo-2-(2-(pyridin-2-yl)ethyl amino)ethyl)acetamide, a pseudo-mimic of human serum albumin. *Polyhedron* **2007**, *26*, 2395–2404. [CrossRef]
28. Wajda, R. Cerasomes—Liposomes with membranes formed from stratum corneum lipids. *Euro Cosmet.* **2001**, *4*, 1–2.
29. Leo, A.; Hansch, A.C.; Elkins, D. Partition Coefficient and their uses. *Chem. Rev.* **1971**, *71*, 525–616. [CrossRef]

Disclaimer/Publisher's Note: The statements, opinions and data contained in all publications are solely those of the individual author(s) and contributor(s) and not of MDPI and/or the editor(s). MDPI and/or the editor(s) disclaim responsibility for any injury to people or property resulting from any ideas, methods, instructions or products referred to in the content.

Article

Oxidation of Phospholipids by OH Radical Coordinated to Copper Amyloid-β Peptide—A Density Functional Theory Modeling [†]

Alberto Rovetta, Laura Carosella, Federica Arrigoni, Jacopo Vertemara, Luca De Gioia, Giuseppe Zampella and Luca Bertini *

Department of Biotechnologies and Biosciences, University of Milano-Bicocca, Piazza della Scienza 2, 20126 Milan, Italy
* Correspondence: luca.bertini@unimib.it
† This paper is dedicated to Professor Wolfgang Weigand on the occasion of his 65th birthday.

Abstract: Oxidative stress and metal dyshomeostasis are considered crucial factors in the pathogenesis of Alzheimer's disease (AD). Indeed, transition metal ions such as Cu(II) can generate reactive oxygen species (ROS) via O_2 Fenton-like reduction, catalyzed by Cu(II) coordinated to the amyloid-beta (Aβ) peptide. Despite intensive efforts, the mechanisms of ROS-induced molecular damage remain poorly understood. In the present paper, we investigate, on the basis of Density Functional Theory (DFT) computations, a possible mechanism of the OH radical propagation toward membrane phospholipid polar head and fatty acid chains starting from the end-product of the OH radical generation by Cu(II)-Aβ. Using phosphatidylcholine as a model of a single unit inside a membrane, we evaluated the thermochemistry of the OH propagation with the oxidation of a C-H bond and the formation of the radical moiety. The DFT results show that Cu(II)-Aβ-OH can oxidize only sn-2 C-H bonds of the polar head and can easily oxidize the C-H bond adjacent to the carbon–carbon double bond in a mono or bis unsaturated fatty acid chain. These results are discussed on the basis of the recent literature on in vitro Aβ metal-catalyzed oxidation and on the possible implications in the AD oxidative stress mechanism.

Keywords: copper amyloid peptide; Alzheimer's disease; oxidative stress; CuAβ hypothesis; phospholipid membrane; Density Functional Theory; molecular modeling

Citation: Rovetta, A.; Carosella, L.; Arrigoni, F.; Vertemara, J.; De Gioia, L.; Zampella, G.; Bertini, L. Oxidation of Phospholipids by OH Radical Coordinated to Copper Amyloid-β Peptide—A Density Functional Theory Modeling. *Inorganics* **2023**, *11*, 227. https://doi.org/10.3390/inorganics11060227

Academic Editors: Christelle Hureau, Ana Maria Da Costa Ferreira, Gianella Facchin and Vladimir Arion

Received: 2 March 2023
Revised: 26 April 2023
Accepted: 22 May 2023
Published: 25 May 2023

Copyright: © 2023 by the authors. Licensee MDPI, Basel, Switzerland. This article is an open access article distributed under the terms and conditions of the Creative Commons Attribution (CC BY) license (https://creativecommons.org/licenses/by/4.0/).

1. Introduction

Alzheimer's disease (AD) is the most common form of dementia that contributes to 60–70% of cases of dementia in the elderly. Among the numerous data reported in the annual 2022 Alzheimer's disease facts and figures [1] drafted by the U.S. non-profit Alzheimer's Association, there is one that is particularly striking: it is estimated that there are 6.5 million Americans living with AD, which could grow up to 13.8 million in 2060; this is equivalent to the population of the U.S. state of Pennsylvania (Census.gov). This compelling data highlight the need for a therapeutic approach to AD since one is sadly still lacking.

AD is a multifactorial disorder characterized by several features such as deposits of senile plaques, hyperphosphorylated tau protein, and extensive synaptic/neuronal loss. Since its publication in 1992, the amyloid cascade hypothesis [2] became dominant in the description of AD pathogenesis and has driven drug development. It postulates that amyloid-beta (Aβ) peptide, a 39–42-residue peptide that comes from the amyloidogenic proteolytic cleavage of the neuronal APP membrane protein, and triggers neuron impairment and death in a variety of forms. Over the years, intense research activity has proposed several therapeutic strategies whose clinical trials have not been successful, as in the case of one of the anti-beta amyloid monoclonal antibodies recently tested [3]. These facts have further highlighted the need to broaden the perspective of the amyloid hypothesis alone [4–7].

It is well established that, within the multifactorial nature of AD, one of the major histopathological hallmarks is the presence of amyloid senile plaques, which are large extracellular aggregates of the Aβ. Prompted by the stabilization of a misfolded hairpin form of Aβ [8], small Aβ oligomers [9–11] are formed, which subsequently further aggregate to fibrils and finally plaques.

There are two important pieces of experimental evidence in defining what has been recently named the CuAD hypothesis [12], i.e., alterations in copper homeostasis [13–20] and the oxidative stress observed in post-mortem AD brain tissue [21,22]. In particular, copper ions could induce oxidative stress, giving rise to lively research work that goes from mechanistic studies [23–26] to the design of pharmacological strategies [15–17,27].

There is a large body of evidence that copper binds Aβ peptide [13,14,25–28] and that Cu(II)-Aβ species are redox-active [29]. In vitro and in the presence of molecular oxygen and a reducing agent such as ascorbate, Cu(II)-Aβ can be a direct source of reactive oxygen species (ROS) [30,31]. The mechanism is a three-step Fenton-like process in which the redox cycling of Cu(II)-Aβ/Cu(I)-Aβ in aerobic conditions results in the production of superoxide [32], hydrogen peroxide [33], and the highly reactive hydroxyl radical [34]. These ROS can further propagate toward the biological substrates, inducing oxidative stress through lipid peroxidation, nucleic acid [35], and protein oxidation [36] in the AD brain [37–39]. The in silico modeling approach has given an important contribution in revealing the active molecular mechanisms that drive AD [11,40,41], particularly including those related to Cu/Aβ interactions.

The complete mechanism of ROS production was also investigated in our laboratory using quantum chemistry and molecular modeling tools, since most of the species involved are transiently populated and thus difficult to detect at a spectroscopical level. Following the approach of Giacovazzi et al. [42], we investigated, at the Density Functional Theory [43–45] level (DFT), the mechanism of O_2 reduction assisted by the Cu(II)-Aβ(1-2) model system in the presence of ascorbate as a reducing agent.

The scheme of this mechanism is depicted in Figure 1. Briefly, the first step of this process (Figure 1, green) is also the rate-determining one: Cu(II)-Aβ undergoes reduction to Cu(I) by one equivalent of ascorbate and successively binds one O_2 molecule, which is the slow endergonic step (free energy barrier around 24.8 kcal/mol [44]). At this point, superoxide is formed and can eventually leave the Cu(II) coordination sphere (Figure 1, blue). The two next steps (Figure 1, yellow) are both exergonic, with the formation of the hydroperoxide HO_2^- and the hydroxyl anion plus a hydroxyl radical. In this Cu(II)-Aβ-OH form (vide infra), the OH radical is coordinated to copper and can propagate, inducing oxidative stress at the cellular level. We estimated a standard reduction potential of the couple Cu(II)-Aβ-OH/Cu(II)-Aβ-OH$^-$ of 1.4 V, which results in roughly 1.3 V less oxidating than free solvated OH radical (OH/OH$^-$ 2.73 V) [45].

At this point, a question arises spontaneously: is this Cu(II)-Aβ-OH form able to induce oxidative stress by propagating the OH radical toward phospholipid membranes? The purpose of this paper is to attempt to answer this question through DFT calculation by using the same approach recently adopted to investigate amino acid oxidation [46]. In the following, starting from the Cu(II)-Aβ-OH form [45], we investigate its binding and the OH propagation toward a phosphatidylcholine single molecule, focusing on either the propagation toward the dipolar head group or the aliphatic chain of selected fatty acids.

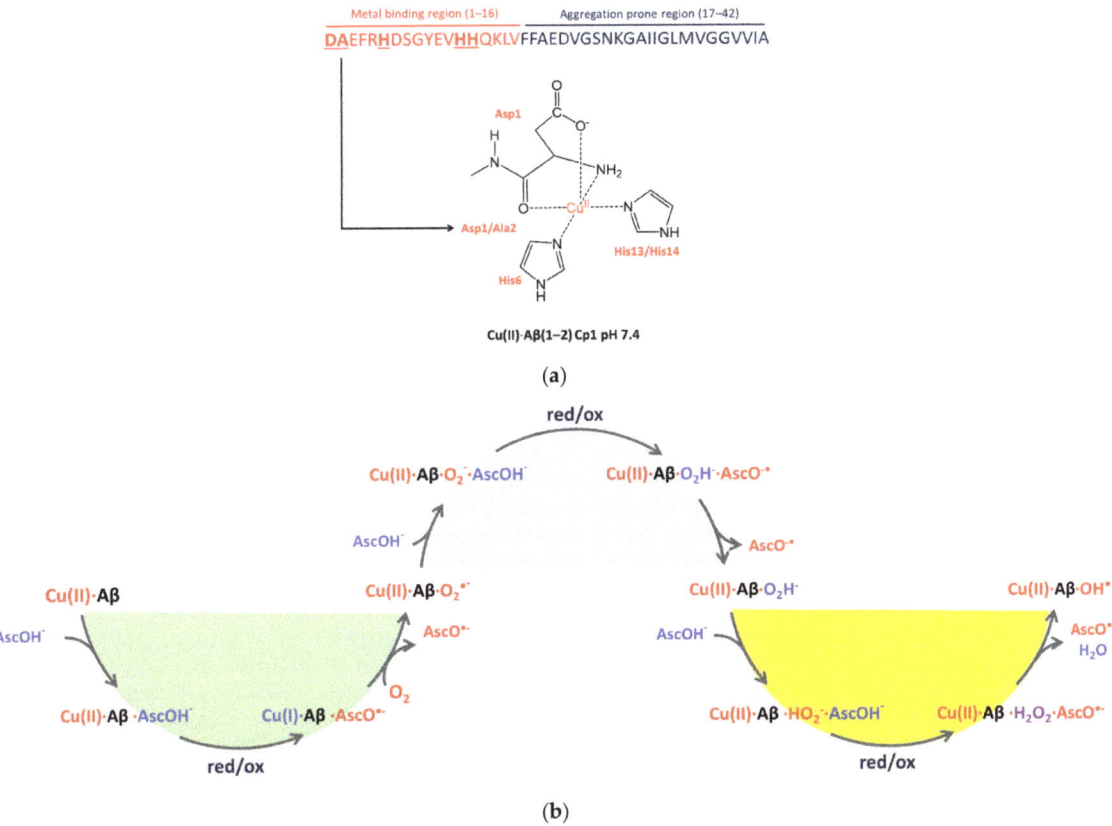

Figure 1. (**a**) Structure of the pH 7.4 Component 1 (Cp1) Cu(II)-Aβ(1-2) investigated in Refs. [43–45]. In the Aβ primary structures, the metal binding region is evidenced (in red), with the three His residues that are involved in the Cu(II) binding. (**b**) Scheme of the three-step Fenton-like O$_2$ reduction mechanism by ascorbate (AscOH$^-$) assisted by Cu(II)-Aβ.

2. Results

2.1. The Electronic Structure of Cu(II)-Aβ-OH

The molecular shape and the electronic structure of the Cu(II)-Aβ-OH investigated in Ref. [45] are peculiar. The Cu(II) coordination includes, in addition to the Asp1 side chain and the amino terminal, two OH ligands; formally, one is a radical and the other is an OH$^-$ anion (Figure 2). In the five-coordinated form, the apical ligand can be an imidazole ring of histidine residues. The electronic structure can be described as follows. The electron vacancy that defines the nature of the OH radical is delocalized over the two oxygen atoms that belong to the OH groups and on the amino terminal with the formation of an amino radical species (see Figure 2A). In addition, the Cu(II) d^9 couples with the unpaired electron on the three centers, which results in a global Cu(II)-Aβ-OH singlet ground state (Figure 2B) according to DFT Natural Bond Orbitals (NBO) atomic and spin populations. This result implies that roughly one-third of the 1e vacancy is delocalized on each of the three ligands. This fact is nicely put in evidence by the frontier MOs reported in Figure 2C, which are those that mix, in a bonding way, the Cu(II) d orbitals with p orbitals on oxygen and nitrogen atoms. For these reasons, the oxidizing power of this species is less than free-solvated OH by 1.3 V [45].

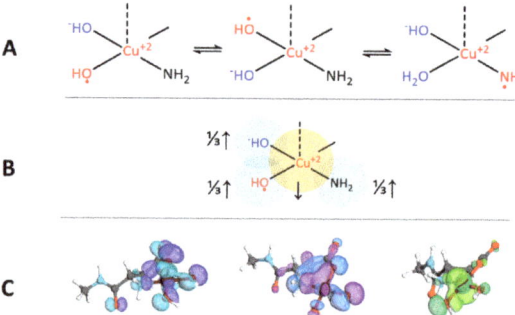

Figure 2. (**A**) Structure of the Cu(II) coordination in Cu(II)-Aβ-OH, evidencing that the negative charge (in blue) and the electron vacancy (in red) can be delocalized on three centers; (**B**) scheme of S = 0 Cu(II)-Aβ-OH electronic structure ground state; (**C**) frontier MOs (cutoff 0.05 a.u.) that describe the delocalization as described in (**A**,**B**).

To further investigate the structure of the Cu(II)-Aβ-OH species, we carried out a DFT conformational sampling on the Cu(II)-Aβ(1-7)(OH)$_2^-$ (Figure 3), where as many as three charged residues are found. Moreover, these computations will allow us to compute the binding energy between Cu(II)-Aβ-OH and phosphatidylcholine moiety. The details of this conformational search are reported in the Supplementary Materials.

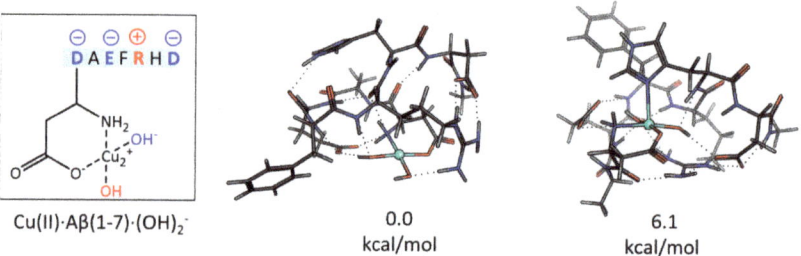

Figure 3. Left: Cu(II)-Aβ(1-7)(OH)$_2^-$ model in which charged residues are evidenced. **Right**: Most stable 4-coordinated and 5-coordinated forms identified upon DFT speciation. The energy difference is computed (in kcal/mol) with respect to the 4-coordinated form which is also the most stable of all those identified. The dotted lines evidence the H-bonding network.

According to this DFT speciation study, (a) the lowest energy form has the two OH groups in the cis position with respect to Cu(II); (b) the lowest energy form is four-coordinated; (c) the most stable five-coordinated form is 6.1 kcal/mol higher in energy with the His6 imidazole ring that occupies the Cu(II) apical position. It is interesting that most of the low-energy forms are characterized by Arg5/Glu3 or Arg5/Asp7 salt bridges.

2.2. Cu(II)-Aβ-OH Propagation Mechanism

In this paper, we investigated the radical propagation of Cu(II)-Aβ-OH toward the C-H bonds that belong to methyl and methylene groups of phospholipid polar head and fatty acid chains. In general, this is a spin-forbidden hydrogen abstraction process, as sketched in Figure 4.

Starting from Cu(II)-Aβ-OH/substrate adduct, the S = 0 singlet potential energy surface (PES) along the C–H bond elongation coordinate, which describes the radical propagation, crosses the S = 1 triplet PES. The molecular structure at this S = 0/S = 1 crossing point can be considered the transition state of the process. By continuing to lengthen the C–H distance, the energy of the system decreases, reaching the S = 1 final product in which one unpaired electron is localized at the aliphatic carbon atom under OH attack and one is at the Cu(II) center.

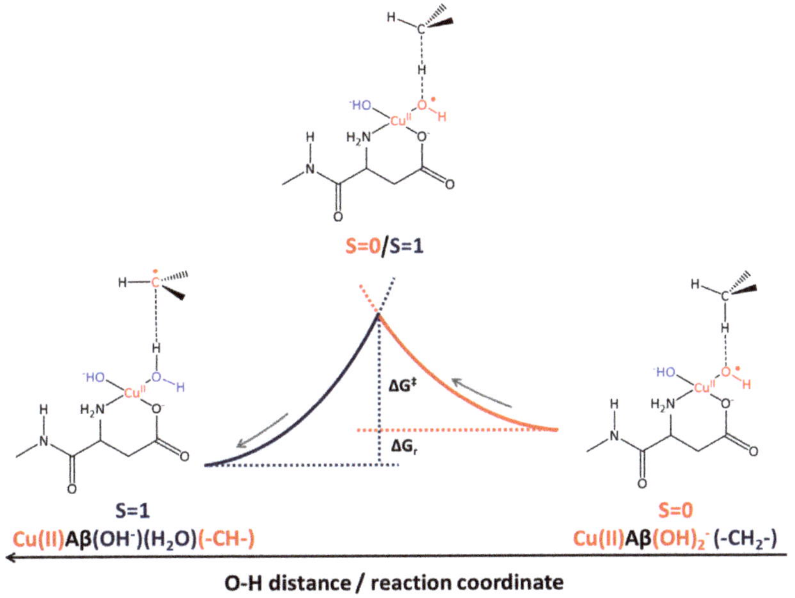

Figure 4. General trend of the singlet- and triplet-state PESs along the O–H reaction coordinate that describes the hydrogen abstraction from a methylene group (that belongs to the polar head or to the aliphatic chain) to the OH of Cu(II)-Aβ-OH coordination. Starting from S = 0 state, upon OH propagation, the electron vacancy is shifted on the carbon radical, resulting in a final product with triplet ground state. The structure at the crossing between S = 0 and S = 1 PES represents the transition state of the process.

2.3. Phosphatidylcholine Moiety, Structure, and Reactivity toward Cu(II)-Aβ-OH

In this section, we explore the reactivity of Cu(II)-Aβ-OH toward a phosphatidylcholine (PC) model (see Figure 5) with two acetyl groups instead of the fatty acid aliphatic chains.

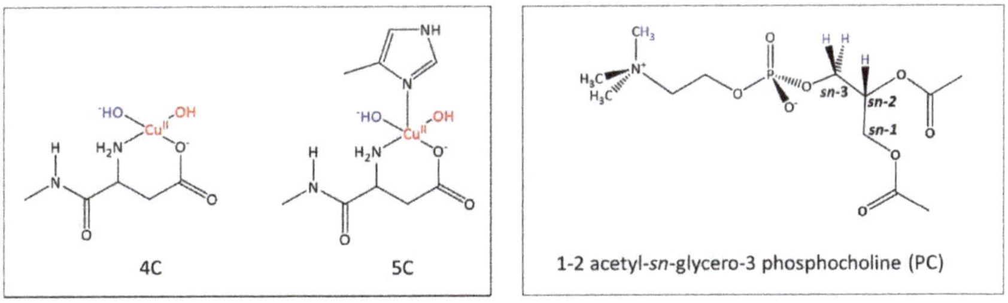

Figure 5. Left: Four- and five-coordinated Cu(II)·Aβ·(OH)$_2$$^{\bullet-}$ models considered in this paper. Model 5C has one imidazole ring bound to Cu(II) in the apical position which is derived from one of the histidine residues of Aβ peptide (very likely His6). **Right**: The phosphatidylcholine (PC) model used to explore the Cu(II)·Aβ·(OH)$_2$$^{\bullet-}$ propagation toward the polar head of membranes. In the PC structure, the possible C-H bonds oxidized by OH radical are evidenced in blue ("*sn*" = stereospecifically numbered): the methyl group of the quaternary ammonium of the choline portion, and the CH$_2$ and CH in the *sn*-2 and *sn*-3 positions of the glycerol portion.

We evaluate the binding between PC and Cu(II)-Aβ-OH (Figure 6A) starting from the Cu(II)-Aβ(1-7)(OH)$_2^-$ model coordination already presented above according to the following equation:

$$\text{Cu(II)·Aβ(1-7)·(OH)}_2^- + \text{PC} \rightleftarrows \text{Cu(II)·Aβ(1-7)·(OH)}_2^-\text{·PC}$$

and considering the total binding energy for 63 different structures (Figure 6B, see Supplementary Materials for details on the conformational search approach).

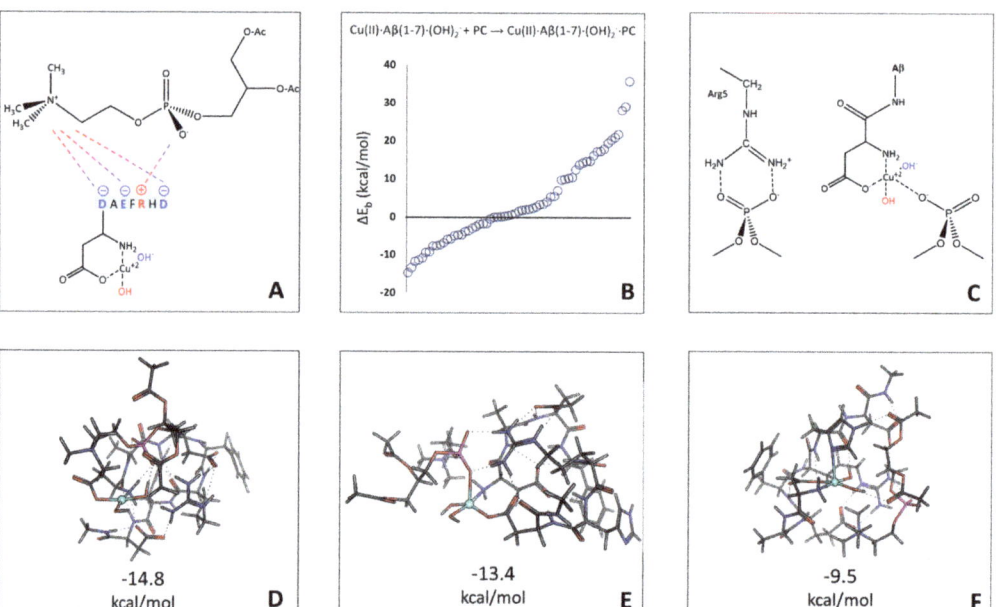

Figure 6. (**A**) Model adopted for the Cu(II)-Aβ(1-7)(OH)$_2^-$ PC interaction; (**B**) total binding energy (in kcal/mol) of the Cu(II)·Aβ(1-7)·(OH)$_2^-$·PC adducts; (**C**) recurring molecular motifs in the adducts; (**D–F**) lowest energy, 4-coordinated, 5-coordinated (apical PO$_4$), and 5-coordinated (apical His6), respectively, along with their adduct binding energy.

The most stable form is four-coordinated (Figure 6D), as in free Cu(II)-Aβ(1-7)(OH)$_2^-$. The phosphate group in this structure forms two H-bond interactions (see Supplementary Materials, Table S2) with one OH of the Cu(II) coordination and one with the Arg5 side chain. The most stable five-coordinated form (+1.4 kcal/mol) has a phosphate group bound to Cu(II) in the apical position (Figure 6C, Cu-OP distance equal to 2.516 Å). Finally, the most stable five-coordinate form with apical His residue is 5.3 kcal/mol higher in energy (Figure 6F).

To investigate the thermochemistry of the OH propagation process, we considered three smaller Cu(II)·Aβ(1-2)·(OH)$_2^{\bullet-}$·PC model adducts. The most stable adduct is the one in which the quaternary ammonium portion forms three hydrogen bonds with Cu(II)·Aβ (one with an OH ligand, one with Asp1 amino terminal, and one with the Asp1-Ala2 amide bond). The free binding energy is −2.9 kcal/mol. The structure with phosphate coordinated to Cu(II) in the apical position is 3.4 kcal/mol higher in energy.

The phosphocholine polar head group can be the target of OH radical propagation [47,48]. In the case of the PC moiety, three sites for OH attack via hydrogen abstraction reaction can be identified, namely the methyl groups at the quaternary ammonium and the *sn*-3 CH$_2$ and *sn*-2 CH glycerol backbone positions (Figure 5). In all possible scenarios, OH radical propagates with the formation of alkyl radicals, which are reactive species susceptible to further radical propagation. According to Ref. [48], one possibility is that a second reduction equivalent reduces the radical carbon with the detachment of a portion of the phosphocholine. This latter step would result in the general destabilization of the phosphocholine unit, increasing the plasticity of the membrane and inducing damage (Figure 7).

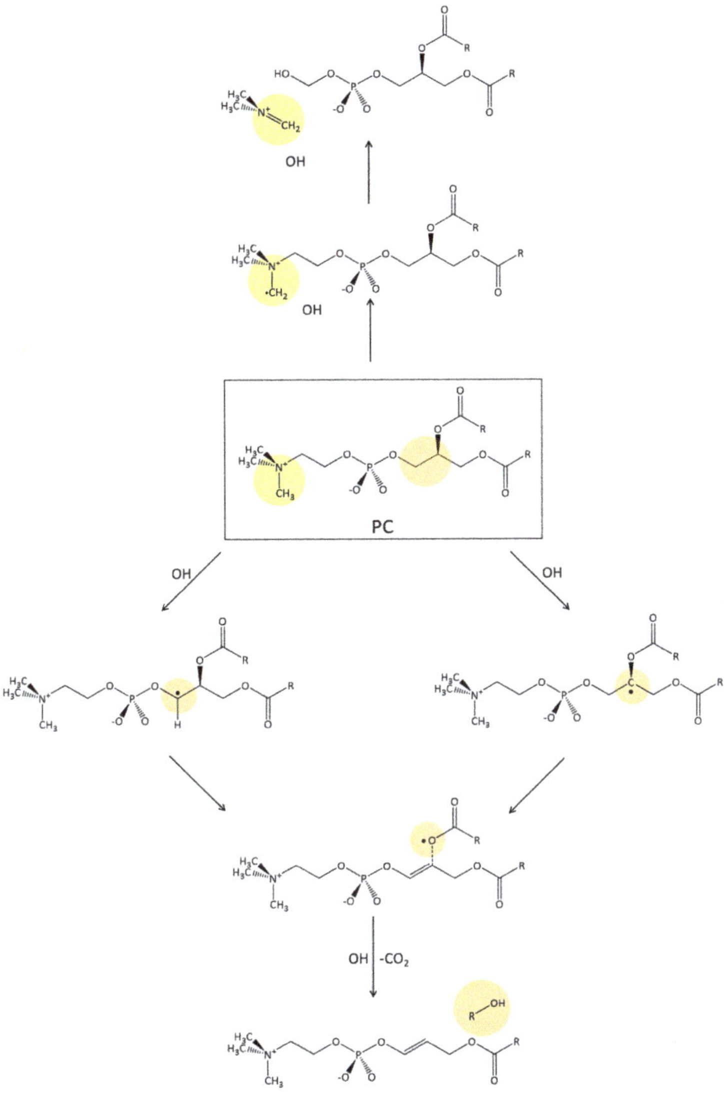

Figure 7. OH propagation reaction mechanisms toward the phosphocholine (PC) head group at the methyl of the quaternary ammonium portion or at the sn-3 CH_2/sn-2 CH group of the glycerol portion (according to Ref. [48]). In all three cases, the first step is the H atom abstraction with the propagation of radical vacancy on the PC unit and the formation of a water molecule. In the case of attack at the quaternary ammonium, a second OH radical induces the detachment of $NC_3H_8^+$ dimethylaminomethylene iminium cation and the formation of a hydroxyl group. In the case of attack at the glycerol unit, the formation of a C=C double bond induces the detachment of a CO_2 molecule; finally, the attack of a second OH brings to the formation of an alcohol.

Binding, activation, and reaction free energies for each adduct are reported in Table 1, while the scans of singlet and triplet PESs along the radical propagation coordinate are reported in Figure 8.

Table 1. Molecular interaction and radical propagation for the Cu(II)·Aβ·(OH)$_2$•$^-$·PC models and fatty acid chains for 4C and 5C. Binding, activation, and reaction free energy differences (ΔG$_b$, ΔG‡, and ΔG$_r$, respectively) are in kcal/mol.

Cu(II) Coordination	Attack Site	ΔG$_b$	ΔG‡	ΔG$_r$
4C	N-CH$_3$	−2.9	26.9	22.8
	CH$_2$	6.8	23.6	5.3
	CH	3.7	15.1	6.9
5C	CH	0	15.9	−1.2
4C	16:0	8.5	16.6	16.2
	18:1	6.3	11.7	−1.4
	18:2	4.5	2.7	−15.4
5C	16:0	11.4	10.6	11.1

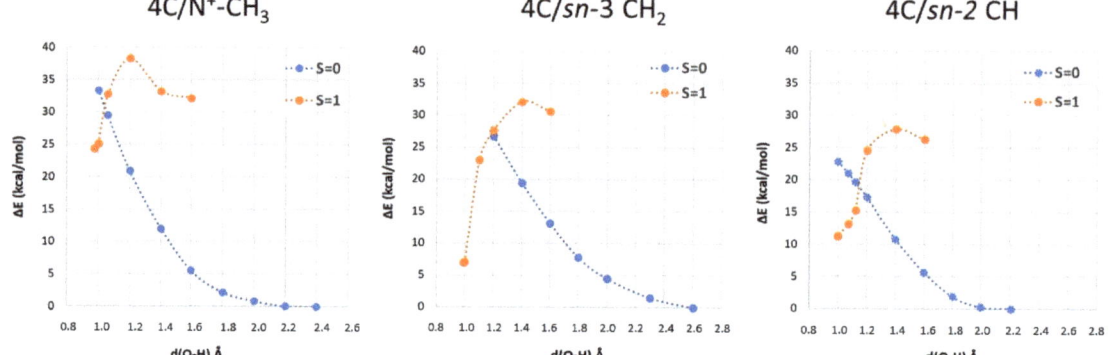

Figure 8. S = 0 and S = 1 PES scanning along the approaching coordinate between the oxygen atom of the OH, coordinated to Cu(II), and the H atom that belongs to the aliphatic carbon atom of the quaternary ammonium group (N$^+$-CH$_3$), and the CH$_2$ sn-3 and the CH sn-2 of the glycerol portion. 4C refers to the 4-coordinated Cu(II)·Aβ(1-2)·(OH)$_2$•$^-$ model. The scan for 5C is reported in the Supplementary Materials. Distances in Å, total energy differences in kcal/mol. For each scan, the corresponding free energies (binding, barrier, and reaction) are reported in Table 1.

Upon hydrogen abstraction, the final product is the Cu(II)·Aβ·(H$_2$O)(OH)$^-$·PC• adduct in which the OH radical is reduced to a water molecule, leaving a radical carbon on the oxidized PC. The Cu atomic charges are slightly reduced upon H abstraction (on average, 1.20 e) with two unpaired electrons localized on the radical carbon atom and on the metal ion (see Figure S5 in Supplementary Materials). The S = 0 broken-symmetry and S = 1 solutions are very close in energy. The propagation toward CH sn-2 implies an activation free energy barrier of 15.1 kcal/mol, which makes this process chemically feasible. Oxidation of CH$_2$ sn-3 and quaternary ammonium groups are much less favorable, with activation barriers of 23.6 and 26.9 kcal/mol, respectively.

In the case of the propagation toward the CH sn-2 group, we investigated the reactivity of the 5C coordination (see Figure S7 in Supplementary Materials) with the aim of elucidating the influence of the fifth His ligand on the reaction free energy, according to the following mechanism:

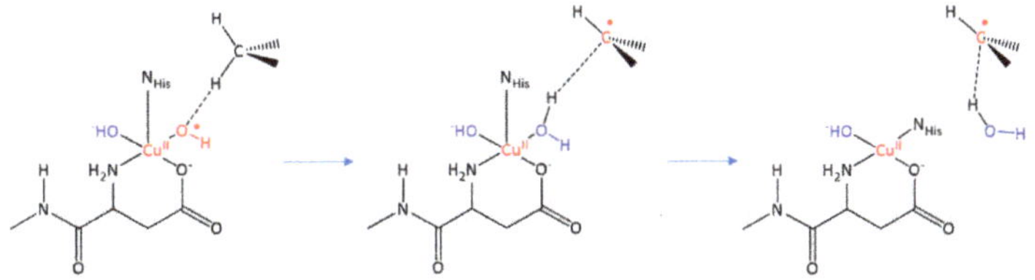

where after OH propagation, the H₂O ligand leaves the Cu(II) coordination sphere with the formation of a four-coordinated form. Here, the activation barrier does not change significantly (+0.8 kcal/mol), but the ΔG_r value decreases by 8.1 kcal/mol.

2.4. Fatty Acid Aliphatic Chains, Structure, and Reactivity toward Cu(II)-Aβ-OH

In this section, we report the results of the reactivity of the Cu(II)·Aβ·(OH)$_2$$^{\bullet-}$ models toward the saturated and polyunsaturated fatty acid chains considering the first step of lipid peroxidation [49], an oxidative chain reaction responsible for membrane damage induced by oxidative stress. The first step of this process is the hydrogen abstraction from a methylene C-H bond adjacent to the carbon–carbon double bond (monoallylic or bis-allylic position), especially in polyunsaturated fatty acids. Upon hydrogen abstraction by OH, a secondary CH$^\bullet$ radical is formed, which then further propagates in the presence of an oxygen molecule, yielding various products, including highly reactive electrophilic aldehydes (such as malondialdehyde and 4-hydroxynonenal) [50,51].

We considered the methyl ester of three fatty acids, reported in Figure 9, chosen on the basis of the lipid raft composition (in mole %) of the human frontal brain cortex in AD patients [52] (16:0 is present at 23.6%, 18:1 (ω-9) at 15.2% and 18:1 (ω-6) at 0.8%).

Figure 9. The structure of the three-fatty-acid methyl ester considered in this paper to model the initial step of the lipid peroxidation process in the presence of Cu(II)·Aβ·(OH)$_2$$^{\bullet-}$. The 16:0 chain is not the target of the lipid peroxidation but it allows a direct comparison of the oxidation power of OH coordinated to Cu(II) vs. free OH species; in 18:1 and 18:2, the attacks to the allylic and bis-allylic CH$_2$ positions are considered, respectively, with the formation of an allyl and a pentadienyl radical species.

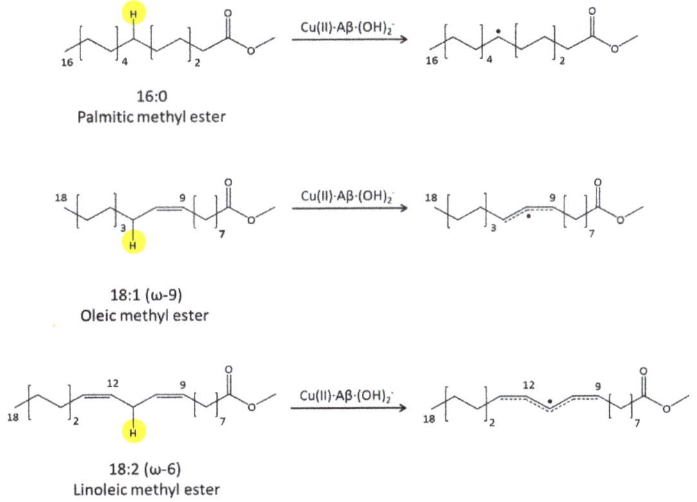

The Cu(II)·Aβ·(OH)$_2$$^{\bullet-}$ fatty acid adducts (see Figure S4 in Supplementary Materials) have been generated by searching the minimum energy structure in which the OH ligand on Cu(II) approached the hydrogen atom of the C-H under attack. All the structures considered are characterized by a hydrogen bond between the ligand coordinated to Cu(II) and the carbonyl of the ester group, but the binding is not favored by 7.6 kcal/mol on average. The singlet and triplet PES scans along the OH propagation coordinates are reported in Figure 10.

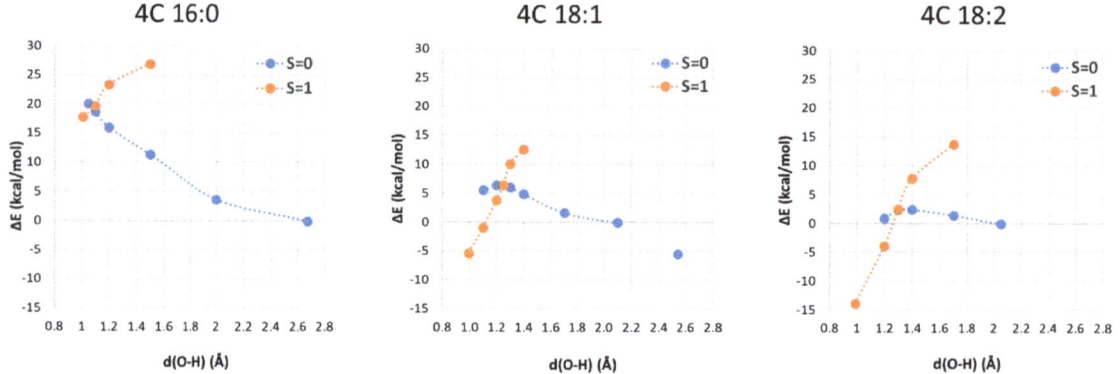

Figure 10. S = 0 and S = 1 PES scanning along the approaching coordinate between the oxygen atom of the OH, coordinated to Cu(II), and the H atom that belong to a CH_2 group of the fatty acid methyl ester aliphatic chain (see Figure 9). 4C refers to the 4-coordinated $Cu(II) \cdot A\beta \cdot (OH)_2^{\bullet-}$ portion. Distances in Å, total energy differences in kcal/mol. For each scan, the corresponding binding barrier and reaction free energies are reported in Table 1.

By increasing the unsaturation number, we observed that the transition state structure becomes closer to the initial reactant (increasing the O···H distance), and consequently, the free energy barrier decreases. For 16:0, we compute a free energy barrier of 16.6 kcal/mol, suggesting that the attack toward a saturated fatty acid chain is still chemically feasible although slow, while for 18:1 and 18:2, the barriers are lower (11.7 and 2.7 kcal/mol, respectively), so these processes are much faster. The case of 18:1 is peculiar. Along the coordinates, we found a second local minimum higher in energy compared to the first one by 5.5 kcal/mol. The final product is the $Cu(II) \cdot A\beta \cdot (H_2O)(OH)^- \cdot (-C\bullet H-)$ adduct in which the OH radical is reduced to a water molecule, leaving a secondary carbon radical on the aliphatic chain.

In the case of 18:1 and 18:2, we observe the formation of allylic and bis-allylic radicals, respectively, which stabilize the final product by resonance. S = 1 and broken symmetry S = 0 states are very close in energy. The copper ion is still in a Cu(II) redox state, with one unpaired electron localized on the radical carbon atom and the other on the copper ion (see Figure S6 in Supplementary Materials).

We finally considered the comparison between 4C and 5C coordination in the case of the 16:0 moiety (see Figure S8 in Supplementary Materials), as in the case of the PC oxidation pathways reported above. Starting from 5C coordination, the H abstraction process is thermodynamically and kinetically more favored ($\Delta G^\ddagger$ and ΔG_r decrease by 6.0 and 5.1 kcal/mol, respectively).

3. Discussion and Conclusions

The DFT modeling proposed led to a picture of the OH reactivity toward phospholipid oxidation that reveals new and interesting aspects.

The approach of PC moiety to Cu(II)-Aβ-OH coordination has been investigated using the $Cu(II) \cdot A\beta(1\text{-}7) \cdot (OH)_2^{\bullet-}$ model, observing that the phosphate group is very likely to form an ion–ion interaction with the positively charged side chain of Arg5 or a Cu-OPO$_3$ bond. These interactions stabilize the adducts between PC and copper OH amyloid coordination, and the Cu-OPO$_3$ one could contribute to the broadening of the [31]P

NMR spectra observed by Gehman et al. [53] in model membranes associated with copper Aβ complexes.

We investigated the OH propagation toward three possible sites of a phosphatidylcholine model (Figure 5) and toward the aliphatic chain of three fatty acids with increasing unsaturation (Figures 7 and 9). Once Cu(II)·Aβ·(OH)$_2$$^{•-}$ has reached the PC, only the hydrogen abstraction at sn-2 CH is viable (ΔG‡ = 15.9 kcal/mol) due to the formation of a more stable tertiary radical—CH$_2$-C•-CH$_2$-, and when 5C coordination is involved, the process is also slightly exergonic.

Next, we considered the hydrogen atom abstraction from one saturated and two unsaturated fatty acid aliphatic chains (Figure 9). We observed that the ΔG‡ decreases almost linearly with the increase in unsaturation, due to the higher stability of the allyl and then pentadienyl radicals. Remarkably, saturated fatty acid chains are not the target of lipid peroxidation, but the computation of ΔG‡ for 16:0 is a precise estimation of the Cu(II)·Aβ·(OH)$_2$$^{•-}$ less oxidizing power (see Figure S9 in Supplementary Materials), a fact that also implies that Cu(II)-Aβ-OH propagation processes are slower than those induced by solvated OH.

Also noteworthy is that if on the one hand, Cu(II)·Aβ is a protective agent against radical species, on the other hand, it is also paradoxically able to increase the lifetime of the OH radicals and thus increase the risk of oxidative damage at the cellular level. Indeed, Cu(II)·Aβ has a sort of redox neuroprotective system since it is less efficient in OH production than the free Cu(II) in solution. At the same time, Cu(II)·Aβ·(OH)$_2$$^{•-}$ is less oxidant than free OH but it is still a sufficiently strong oxidant to induce oxidative stress. In particular, it has a much longer lifetime, allowing it to diffuse into the cell membrane lipid bilayer, where it starts lipid peroxidation. These results can also be seen in the context of the mechanism of Aβ neurotoxicity. It has recently been proposed that the Aβ small oligomers may penetrate [54] and aggregate within the membrane, forming unregulated and heterogeneous ion channels [55–57] which would allow ion dyshomeostasis [58], leading to cellular degeneration. In this context, it is useful to remind the reader that numerous studies [59–61] underline the important role of neuronal Ca^{2+} dyshomeostasis in AD. A recent investigation [62] showed that oxidized membranes are prone to form native and oxidized domains which can induce a poration process [63]. In the hypothesis of an active role of Cu-Aβ-OH in the oxidation of the neuronal membrane, as suggested by our simulations, this would favor the poration and therefore the dyshomeostasis of calcium.

In conclusion, the understanding of the molecular mechanism of Cu-Aβ ROS production and propagation is crucial in the perspective of a better comprehension of AD etiology and hopefully for a therapeutic strategy.

It is well established that Cu(II)-Aβ is redox-active and catalyzes the formation of OH coordinated to the metal ion. OH radicals produced by Cu(II)-Aβ are added to those normally produced at the cellular level [64] but most likely in the extracellular area. Since OH radicals coordinated to Cu(II)-Aβ have an attenuated redox potential, one may wonder if they can attack cell membranes by radical propagation. DFT investigation of such reactivity suggests that Cu(II)-Aβ-OH can attack the sn-2 CH of the phosphocholine polar head and the mono- and bis-unsaturated fatty acid chains on a slower time scale compared to the free solvated OH radical. We can then infer that the Cu(II)-Aβ-OH structure described in this paper could be one of the copper Aβ species responsible for oxidative stress in AD.

4. Computational Details

The DFT computations were performed using the pure GGA BP86 [65,66] DFT functional and RI technique [67], as implemented in TURBOMOLE [68]. Basis sets of triple-ζ plus polarization split valence quality [69] (def2-TZVP) were adopted for all atoms. The solvent effect was accounted for by using the COSMO (Conductor-like Screening Model) approach [70]. Water solvation was considered by setting the dielectric constant equal to 80 in the investigation of the PC reactivity. In the case of fatty acids, we considered a dielectric constant value of 4, thus mimicking the interior of a membrane. More in general, this computational setting provides ground-state geometry parameters in good agreement with experimental X-ray values [71]. Charge distribution was evaluated using natural bond orbital analysis (NBO). Ground-state geometry optimizations were carried out with convergence criteria fixed to 10^{-6} hartree for the energy and 0.001 hartree bohr for the gradient norm vector. For investigation of the fatty acid chain oxidation, D3 Grimme empirical dispersion correction [72] was adopted. The effects of ZPE, thermal, and entropic contributions on the purely electronic total energy values to compute free energies were investigated by means of evaluation of the approximated roto-translational partition function of each molecular species, at T = 298 K and P = 1 bar.

Supplementary Materials: The following supporting information can be downloaded at: https://www.mdpi.com/article/10.3390/inorganics11060227/s1, Table S1: DFT speciation of $Cu(II)(OH)_2^-\cdot A\beta(1-7)$; Table S2–Figure S3: $Cu(II)(OH)_2^-\cdot A\beta(1-7)\cdot PC$; Figure S4: Structures of the $Cu(II)(OH)_2^-\cdot A\beta(1-2)$ fatty acid chain adducts; Figures S5 and S6: NBO spin population of the final products; $Cu(II)(OH)_2^-\cdot A\beta(1-2)\cdot PC$ PES scans for 4C and 5C coordination; Figures S7–S9: Reactivity of free OH radical with 16:0 palmitic acid.

Author Contributions: L.B., F.A., L.D.G., J.V. and G.Z. designed the simulations; A.R. and L.C. performed the simulations; L.B., A.R. and L.C. analyzed the results; L.B. and F.A. wrote the manuscript. All authors have read and agreed to the published version of the manuscript.

Funding: This work was supported by the PRIN Project 2020BKK3W9.

Data Availability Statement: Not applicable.

Conflicts of Interest: The authors declare no conflict of interest.

References

1. Alzheimer's Association. 2022 Alzheimer's Disease Facts and Figures. *Alzheimers. Dement.* **2022**, *18*, 700–789. [CrossRef]
2. Hardy, J.A.; Higgins, G.A. Alzheimer's Disease: The Amyloid Cascade Hypothesis. *Science* **1992**, *256*, 184–185. [CrossRef]
3. Budd Haeberlein, S.; Aisen, P.S.; Barkhof, F.; Chalkias, S.; Chen, T.; Cohen, S.; Dent, G.; Hansson, O.; Harrison, K.; von Hehn, C.; et al. Two Randomized Phase 3 Studies of Aducanumab in Early Alzheimer's Disease. *J. Prev. Alzheimers Dis.* **2022**, *9*, 197–210. [CrossRef]
4. Makin, S. The Amyloid Hypothesis on Trial. *Nature* **2018**, *559*, S4–S7. [CrossRef]
5. Doig, A.J.; del Castillo-Frias, M.P.; Berthoumieu, O.; Tarus, B.; Nasica-Labouze, J.; Sterpone, F.; Nguyen, P.H.; Hooper, N.M.; Faller, P.; Derreumaux, P. Why Is Research on Amyloid Failing to Give New Drugs for Alzheimers Disease? *ACS Chem. Neurosci.* **2017**, *8*, 1435–1437. [CrossRef]
6. Lei, P.; Ayton, S.; Bush, A.I. The Essential Elements of Alzheimer's Disease. *J. Biol. Chem.* **2021**, *296*, 100105. [CrossRef]
7. Ayton, S.; Bush, A.I. β-Amyloid: The Known Unknowns. *Ageing Res. Rev.* **2021**, *65*, 101212. [CrossRef]
8. Abelein, A.; Abrahams, J.P.; Danielsson, J.; Gräslund, A.; Jarvet, J.; Luo, J.; Tiiman, A.; Wärmländer, S.K.T.S. The Hairpin Conformation of the Amyloid β Peptide Is an Important Structural Motif along the Aggregation Pathway. *J. Biol. Inorg. Chem.* **2014**, *19*, 623–634. [CrossRef]
9. Cline, E.N.; Bicca, M.A.; Viola, K.L.; Klein, W.L. The Amyloid-β Oligomer Hypothesis: Beginning of the Third Decade. *J. Alzheimer's Dis.* **2018**, *64*, S567–S610. [CrossRef]
10. Michaels, T.C.T.; Šarić, A.; Curk, S.; Bernfur, K.; Arosio, P.; Meisl, G.; Dear, A.J.; Cohen, S.I.A.; Dobson, C.M.; Vendruscolo, M.; et al. Author Correction: Dynamics of Oligomer Populations Formed during the Aggregation of Alzheimer's Aβ42 Peptide. *Nat. Chem.* **2020**, *12*, 497. [CrossRef]
11. Nguyen, P.H.; Ramamoorthy, A.; Sahoo, B.R.; Zheng, J.; Faller, P.; Straub, J.E.; Dominguez, L.; Shea, J.-E.; Dokholyan, N.V.; De Simone, A.; et al. Amyloid Oligomers: A Joint Experimental/Computational Perspective on Alzheimer's Disease, Parkinson's Disease, Type II Diabetes, and Amyotrophic Lateral Sclerosis. *Chem. Rev.* **2021**, *121*, 2545–2647. [CrossRef]
12. Squitti, R.; Faller, P.; Hureau, C.; Granzotto, A.; White, A.R.; Kepp, K.P. Copper Imbalance in Alzheimer's Disease and Its Link with the Amyloid Hypothesis: Towards a Combined Clinical, Chemical, and Genetic Etiology. *J. Alzheimers. Dis.* **2021**, *83*, 23–41. [CrossRef]
13. Lovell, M.A.; Robertson, J.D.; Teesdale, W.J.; Campbell, J.L.; Markesbery, W.R. Copper, Iron and Zinc in Alzheimer's Disease Senile Plaques. *J. Neurol. Sci.* **1998**, *158*, 47–52. [CrossRef]
14. Atrián-Blasco, E.; Gonzalez, P.; Santoro, A.; Alies, B.; Faller, P.; Hureau, C. Cu and Zn Coordination to Amyloid Peptides: From Fascinating Chemistry to Debated Pathological Relevance. *Coord. Chem. Rev.* **2018**, *375*, 38–55. [CrossRef]
15. Atrián-Blasco, E.; Cerrada, E.; Conte-Daban, A.; Testemale, D.; Faller, P.; Laguna, M.; Hureau, C. Copper(I) Targeting in the Alzheimer's Disease Context: A First Example Using the Biocompatible PTA Ligand. *Metallomics* **2015**, *7*, 1229–1232. [CrossRef]

16. Faller, P.; Hureau, C. A Bioinorganic View of Alzheimer's Disease: When Misplaced Metal Ions (Re)direct the Electrons to the Wrong Target. *Chem. Eur. J.* **2012**, *18*, 15910–15920. [CrossRef]
17. Pardo-Moreno, T.; González-Acedo, A.; Rivas-Domínguez, A.; García-Morales, V.; García-Cozar, F.J.; Ramos-Rodríguez, J.J.; Melguizo-Rodríguez, L. Therapeutic Approach to Alzheimer's Disease: Current Treatments and New Perspectives. *Pharmaceutics* **2022**, *14*, 1117. [CrossRef]
18. Dong, J.; Atwood, C.S.; Anderson, V.E.; Siedlak, S.L.; Smith, M.A.; Perry, G.; Carey, P.R. Metal Binding and Oxidation of Amyloid-Beta within Isolated Senile Plaque Cores: Raman Microscopic Evidence. *Biochemistry* **2003**, *42*, 2768–2773. [CrossRef]
19. Greenough, M.A.; Camakaris, J.; Bush, A.I. Metal Dyshomeostasis and Oxidative Stress in Alzheimer's Disease. *Neurochem. Int.* **2013**, *62*, 540–555. [CrossRef]
20. Kepp, K.P.; Squitti, R. Copper Imbalance in Alzheimer's Disease: Convergence of the Chemistry and the Clinic. *Coord. Chem. Rev.* **2019**, *397*, 168–187. [CrossRef]
21. Bradley, M.A.; Xiong-Fister, S.; Markesbery, W.R.; Lovell, M.A. Elevated 4-Hydroxyhexenal in Alzheimer's Disease (AD) Progression. *Neurobiol. Aging* **2012**, *33*, 1034–1044. [CrossRef]
22. Butterfield, D.A.; Reed, T.; Perluigi, M.; De Marco, C.; Coccia, R.; Cini, C.; Sultana, R. Elevated Protein-Bound Levels of the Lipid Peroxidation Product, 4-Hydroxy-2-Nonenal, in Brain from Persons with Mild Cognitive Impairment. *Neurosci. Lett.* **2006**, *397*, 170–173. [CrossRef]
23. Collin, F. Chemical Basis of Reactive Oxygen Species Reactivity and Involvement in Neurodegenerative Diseases. *Int. J. Mol. Sci.* **2019**, *20*, 2407. [CrossRef]
24. Butterfield, D.A.; Halliwell, B. Oxidative Stress, Dysfunctional Glucose Metabolism and Alzheimer Disease. *Nat. Rev. Neurosci.* **2019**, *20*, 148–160. [CrossRef]
25. Cheignon, C.; Tomas, M.; Bonnefont-Rousselot, D.; Faller, P.; Hureau, C.; Collin, F. Oxidative Stress and the Amyloid Beta Peptide in Alzheimer's Disease. *Redox Biol.* **2018**, *14*, 450–464. [CrossRef]
26. Furlan, S.; Hureau, C.; Faller, P.; La Penna, G. Modeling Copper Binding to the Amyloid-β Peptide at Different pH: Toward a Molecular Mechanism for Cu Reduction. *J. Phys. Chem. B* **2012**, *116*, 11899–11910. [CrossRef]
27. Singh, S.K.; Balendra, V.; Obaid, A.A.; Esposto, J.; Tikhonova, M.A.; Gautam, N.K.; Poeggeler, B. Copper-Mediated β-Amyloid Toxicity and Its Chelation Therapy in Alzheimer's Disease. *Metallomics* **2022**, *14*, mfac018. [CrossRef]
28. Mutter, S.T.; Turner, M.; Deeth, R.J.; Platts, J.A. Metal Binding to Amyloid-β: A Ligand Field Molecular Dynamics Study. *ACS Chem. Neurosci.* **2018**, *9*, 2795–2806. [CrossRef]
29. Rauk, A. The Chemistry of Alzheimer's Disease. *Chem. Soc. Rev.* **2009**, *38*, 2698–2715. [CrossRef]
30. Cheignon, C.; Collin, F.; Faller, P.; Hureau, C. Is Ascorbate Dr Jekyll or Mr Hyde in the Cu(Aβ) Mediated Oxidative Stress Linked to Alzheimer's Disease? *Dalton Trans.* **2016**, *45*, 12627–12631. [CrossRef]
31. Cheignon, C.; Jones, M.; Atrián-Blasco, E.; Kieffer, I.; Faller, P.; Collin, F.; Hureau, C. Identification of Key Structural Features of the Elusive Cu–Aβ Complex That Generates ROS in Alzheimer's Disease. *Chem. Sci.* **2017**, *8*, 5107–5118. [CrossRef]
32. Reybier, K.; Ayala, S.; Alies, B.; Rodrigues, J.V.; Bustos Rodriguez, S.; La Penna, G.; Collin, F.; Gomes, C.M.; Hureau, C.; Faller, P. Free Superoxide Is an Intermediate in the Production of H2O2 by Copper(I)-Aβ Peptide and O$_2$. *Angew. Chem. Int. Ed. Engl.* **2016**, *55*, 1085–1089. [CrossRef]
33. Cassagnes, L.-E.; Hervé, V.; Nepveu, F.; Hureau, C.; Faller, P.; Collin, F. The Catalytically Active Copper-Amyloid-Beta State: Coordination Site Responsible for Reactive Oxygen Species Production. *Angew. Chem. Int. Ed. Engl.* **2013**, *52*, 11110–11113. [CrossRef]
34. La Penna, G.; Hureau, C.; Andreussi, O.; Faller, P. Identifying, By First-Principles Simulations, Cu[Amyloid-β] Species Making Fenton-Type Reactions in Alzheimer's Disease. *J. Phys. Chem. B* **2013**, *117*, 16455–16467. [CrossRef]
35. Bradley-Whitman, M.A.; Timmons, M.D.; Beckett, T.L.; Murphy, M.P.; Lynn, B.C.; Lovell, M.A. Nucleic Acid Oxidation: An Early Feature of Alzheimer's Disease. *J. Neurochem.* **2014**, *128*, 294–304. [CrossRef]
36. Sultana, R.; Allan Butterfield, D. Oxidative Modification of Brain Proteins in Alzheimer's Disease: Perspective on Future Studies Based on Results of Redox Proteomics Studies. *J. Alzheimer's Dis.* **2012**, *33*, S243–S251. [CrossRef]
37. Butterfield, D.A.; Lauderback, C.M. Lipid Peroxidation and Protein Oxidation in Alzheimer's Disease Brain: Potential Causes and Consequences Involving Amyloid β-Peptide-Associated Free Radical Oxidative Stress. *Free. Radic. Biol. Med.* **2002**, *32*, 1050–1060. [CrossRef]
38. Butterfield, D.A.; Allan Butterfield, D.; Reed, T.; Newman, S.F.; Sultana, R. Roles of Amyloid β-Peptide-Associated Oxidative Stress and Brain Protein Modifications in the Pathogenesis of Alzheimer's Disease and Mild Cognitive Impairment. *Free Radic. Biol. Med.* **2007**, *43*, 658–677. [CrossRef]
39. Peña-Bautista, C.; Tirle, T.; López-Nogueroles, M.; Vento, M.; Baquero, M.; Cháfer-Pericás, C. Oxidative Damage of DNA as Early Marker of Alzheimer's Disease. *Int. J. Mol. Sci.* **2019**, *20*, 6136. [CrossRef]
40. Nasica-Labouze, J.; Nguyen, P.H.; Sterpone, F.; Berthoumieu, O.; Buchete, N.-V.; Coté, S.; De Simone, A.; Doig, A.J.; Faller, P.; Garcia, A.; et al. Amyloid β Protein and Alzheimer's Disease: When Computer Simulations Complement Experimental Studies. *Chem. Rev.* **2015**, *115*, 3518–3563. [CrossRef]
41. Strodel, B.; Coskuner-Weber, O. Transition Metal Ion Interactions with Disordered Amyloid-β Peptides in the Pathogenesis of Alzheimer's Disease: Insights from Computational Chemistry Studies. *J. Chem. Inf. Model.* **2019**, *59*, 1782–1805. [CrossRef]

42. Giacovazzi, R.; Ciofini, I.; Rao, L.; Amatore, C.; Adamo, C. Copper–amyloid-β Complex May Catalyze Peroxynitrite Production in Brain: Evidence from Molecular Modeling. *Phys. Chem. Chem. Phys.* **2014**, *16*, 10169–10174. [CrossRef]
43. Prosdocimi, T.; De Gioia, L.; Zampella, G.; Bertini, L. On the Generation of OH(·) Radical Species from H_2O_2 by Cu(I) Amyloid Beta Peptide Model Complexes: A DFT Investigation. *J. Biol. Inorg. Chem.* **2016**, *21*, 197–212. [CrossRef]
44. Arrigoni, F.; Prosdocimi, T.; Mollica, L.; De Gioia, L.; Zampella, G.; Bertini, L. Copper Reduction and Dioxygen Activation in Cu-Amyloid Beta Peptide Complexes: Insight from Molecular Modelling. *Metallomics* **2018**, *10*, 1618–1630. [CrossRef]
45. Arrigoni, F.; Di Carlo, C.; Rovetta, A.; De Gioia, L.; Zampella, G.; Bertini, L. Superoxide Reduction by Cu-Amyloid Beta Peptide Complexes: A Density Functional Theory Study. *Eur. J. Inorg. Chem.* **2022**, *2022*, e202200245. [CrossRef]
46. Arrigoni, F.; Rizza, F.; Tisi, R.; De Gioia, L.; Zampella, G.; Bertini, L. On the Propagation of the OH Radical Produced by Cu-Amyloid Beta Peptide Model Complexes. *Insight Mol. Modelling Met.* **2020**, *12*, 1765–1780.
47. Maciel, E.; da Silva, R.N.; Simões, C.; Domingues, P.; Domingues, M.R.M. Structural Characterization of Oxidized Glycerophosphatidylserine: Evidence of Polar Head Oxidation. *J. Am. Soc. Mass. Spectrom.* **2011**, *22*, 1804–1814. [CrossRef]
48. Yusupov, M.; Wende, K.; Kupsch, S.; Neyts, E.C.; Reuter, S.; Bogaerts, A. Effect of Head Group and Lipid Tail Oxidation in the Cell Membrane Revealed through Integrated Simulations and Experiments. *Sci. Rep.* **2017**, *7*, 5761. [CrossRef]
49. Yin, H.; Xu, L.; Porter, N.A. Free Radical Lipid Peroxidation: Mechanisms and Analysis. *Chem. Rev.* **2011**, *111*, 5944–5972. [CrossRef]
50. Barrera, G.; Pizzimenti, S.; Daga, M.; Dianzani, C.; Arcaro, A.; Cetrangolo, G.P.; Giordano, G.; Cucci, M.A.; Graf, M.; Gentile, F. Lipid Peroxidation-Derived Aldehydes, 4-Hydroxynonenal and Malondialdehyde in Aging-Related Disorders. *Antioxidants* **2018**, *7*, 102. [CrossRef]
51. Solís-Calero, C.; Ortega-Castro, J.; Frau, J.; Muñoz, F. Nonenzymatic Reactions above Phospholipid Surfaces of Biological Membranes: Reactivity of Phospholipids and Their Oxidation Derivatives. *Oxid. Med. Cell. Longev.* **2015**, *2015*, 319505. [CrossRef] [PubMed]
52. Martín, V.; Fabelo, N.; Santpere, G.; Puig, B.; Marín, R.; Ferrer, I.; Díaz, M. Lipid Alterations in Lipid Rafts from Alzheimer's Disease Human Brain Cortex. *J. Alzheimers. Dis.* **2010**, *19*, 489–502. [CrossRef] [PubMed]
53. Gehman, J.D.; O'Brien, C.C.; Shabanpoor, F.; Wade, J.D.; Separovic, F. Metal Effects on the Membrane Interactions of Amyloid-β Peptides. *Eur. Biophys. J.* **2008**, *37*, 333–344. [CrossRef] [PubMed]
54. Lockhart, C.; Klimov, D.K. Alzheimer's Aβ10–40 Peptide Binds and Penetrates DMPC Bilayer: An Isobaric–Isothermal Replica Exchange Molecular Dynamics Study. *J. Phys. Chem. B* **2014**, *118*, 2638–2648. [CrossRef] [PubMed]
55. Sciacca, M.F.M.; Kotler, S.A.; Brender, J.R.; Chen, J.; Lee, D.-K.; Ramamoorthy, A. Two-Step Mechanism of Membrane Disruption by Aβ through Membrane Fragmentation and Pore Formation. *Biophys. J.* **2012**, *103*, 702–710. [CrossRef]
56. Jang, H.; Zheng, J.; Nussinov, R. Models of β-Amyloid Ion Channels in the Membrane Suggest That Channel Formation in the Bilayer Is a Dynamic Process. *Biophys. J.* **2007**, *93*, 1938–1949. [CrossRef]
57. Sepehri, A.; Lazaridis, T. Putative Structures of Membrane-Embedded Amyloid β Oligomers. *ACS Chem. Neurosci.* **2023**, *14*, 99–110. [CrossRef]
58. Supnet, C.; Bezprozvanny, I. The Dysregulation of Intracellular Calcium in Alzheimer Disease. *Cell. Calcium* **2010**, *47*, 183–189. [CrossRef]
59. LaFerla, F.M. Calcium Dyshomeostasis and Intracellular Signalling in Alzheimer's Disease. *Nat. Rev. Neurosci.* **2002**, *3*, 862–872. [CrossRef]
60. Kawahara, M.; Negishi-Kato, M.; Sadakane, Y. Calcium Dyshomeostasis and Neurotoxicity of Alzheimer's β-Amyloid Protein. *Expert. Rev. Neurother.* **2009**, *9*, 681–693. [CrossRef]
61. Chami, M. Calcium Signalling in Alzheimer's Disease: From Pathophysiological Regulation to Therapeutic Approaches. *Cells* **2021**, *10*, 140. [CrossRef] [PubMed]
62. Oliveira, M.C.; Yusupov, M.; Bogaerts, A.; Cordeiro, R.M. Molecular Dynamics Simulations of Mechanical Stress on Oxidized Membranes. *Biophys. Chem.* **2019**, *254*, 106266. [CrossRef] [PubMed]
63. Haluska, C.K.; Baptista, M.S.; Fernandes, A.U.; Schroder, A.P.; Marques, C.M.; Itri, R. Photo-Activated Phase Separation in Giant Vesicles Made from Different Lipid Mixtures. *Biochim. Biophys. Acta* **2012**, *1818*, 666–672. [CrossRef] [PubMed]
64. Ayala, A.; Muñoz, M.F.; Argüelles, S. Lipid Peroxidation: Production, Metabolism, and Signaling Mechanisms of Malondialdehyde and 4-Hydroxy-2-Nonenal. *Oxid. Med. Cell. Longev.* **2014**, *2014*, 360438. [CrossRef] [PubMed]
65. Becke, A.D. Density-Functional Exchange-Energy Approximation with Correct Asymptotic Behavior. *Phys. Rev. A Gen. Phys.* **1988**, *38*, 3098–3100. [CrossRef]
66. Perdew, J.P. Density-Functional Approximation for the Correlation Energy of the Inhomogeneous Electron Gas. *Phys. Rev. B Condens. Matter* **1986**, *33*, 8822–8824. [CrossRef]
67. Eichkorn, K.; Weigend, F.; Treutler, O.; Ahlrichs, R. Auxiliary Basis Sets for Main Row Atoms and Transition Metals and Their Use to Approximate Coulomb Potentials. *Theor. Chem. Acc.* **1997**, *97*, 119–124. [CrossRef]
68. Ahlrichs, R.; Bär, M.; Häser, M.; Horn, H.; Kölmel, C. Electronic Structure Calculations on Workstation Computers: The Program System Turbomole. *Chem. Phys. Lett.* **1989**, *162*, 165–169. [CrossRef]
69. Schäfer, A.; Huber, C.; Ahlrichs, R. Fully Optimized Contracted Gaussian Basis Sets of Triple Zeta Valence Quality for Atoms Li to Kr. *J. Chem. Phys.* **1994**, *100*, 5829–5835. [CrossRef]

70. Bertini, L.; Bruschi, M.; Romaniello, M.; Zampella, G.; Tiberti, M.; Barbieri, V.; Greco, C.; La Mendola, D.; Bonomo, R.P.; Fantucci, P.; et al. Copper Coordination to the Putative Cell Binding Site of Angiogenin: A DFT Investigation. In *Highlights in Theoretical Chemistry*; Springer: Berlin/Heidelberg, Germany, 2012; pp. 255–269.
71. Klamt, A.; Schüürmann, G. COSMO: A New Approach to Dielectric Screening in Solvents with Explicit Expressions for the Screening Energy and Its Gradient. *J. Chem. Soc. Perkin Trans.* **1993**, *2*, 799–805. [CrossRef]
72. Grimme, S. Semiempirical GGA-Type Density Functional Constructed with a Long-Range Dispersion Correction. *J. Comput. Chem.* **2006**, *27*, 1787–1799. [CrossRef] [PubMed]

Disclaimer/Publisher's Note: The statements, opinions and data contained in all publications are solely those of the individual author(s) and contributor(s) and not of MDPI and/or the editor(s). MDPI and/or the editor(s) disclaim responsibility for any injury to people or property resulting from any ideas, methods, instructions or products referred to in the content.

Article

Exploration of Lycorine and Copper(II)'s Association with the N-Terminal Domain of Amyloid β

Arian Kola *, Ginevra Vigni and Daniela Valensin

Department of Biotechnology, Chemistry and Pharmacy, University of Siena, Via Aldo Moro 2, 53100 Siena, Italy; ginevra.vigni2@unisi.it (G.V.); daniela.valensin@unisi.it (D.V.)
* Correspondence: arian.kola@unisi.it; Tel.: +39-0577-232428

Abstract: Lycorine (LYC) is an active alkaloid first isolated from Narcissus pseudonarcissus and found in most Amaryllidaceae plants. It belongs to the same family as galantamine, which is the active component of a drug used for the treatment of Alzheimer's disease. Similarly to galantamine, LYC is able to suppress induced amyloid β (Aβ) toxicity in differentiated SH-SY5Y cell lines and it can weakly interact with the N-terminal region of Aβ via electrostatic interactions. The N-terminal Aβ domain is also involved in Cu(II)/Cu(I) binding and the formed complexes are known to play a key role in ROS production. In this study, the Aβ–LYC interaction in the absence and in the presence of copper ions was investigated by using the N-terminal Aβ peptide encompassing the first 16 residues. NMR analysis showed that Aβ can simultaneously interact with Cu(II) and LYC. The Cu(II) binding mode remains unchanged in the presence of LYC, while LYC association is favored when an Aβ–Cu(II) complex is formed. Moreover, UV-VIS studies revealed the ability of LYC to interfere with the catalytic activities of the Aβ–Cu(II) complexes by reducing the ascorbate consumption monitored at 265 nm.

Keywords: lycorine; copper; amyloid; ternary association; ROS; natural compounds; Alzheimer's disease

1. Introduction

Alzheimer's disease (AD) is a progressive neurodegenerative disorder that is the most common cause of dementia in the world. It is characterized by a gradual decline in cognitive function, including memory, language, problem-solving, and judgment. As the disease progresses, people with AD may have difficulty performing everyday tasks and may become dependent on others for care [1].

Presently, over 55 million individuals globally are affected by dementia, with the majority, exceeding 60%, residing in low- and middle-income nations, as reported by the World Health Organization (WHO). Additionally, each year witnesses the onset of nearly 10 million new cases. Beyond the severity of the inexorable increase in cases of neurodegenerative diseases, it is important to ascertain the incidence of the costs involved. In fact, during the year 2019, dementia incurred a global economic cost of 1.3 trillion US dollars. Around half of this financial burden is associated with informal caregivers (such as family members and close friends) who, on average, devote 5 h per day to caregiving and supervision [2,3].

In addition to the suffering of patients and family members, the socioeconomic impact is devastating. Prevention must therefore be strengthened in order to delay and slow down symptoms. At the same time, it is necessary to invest in drug research even if developing a drug takes 13 years from preclinical studies to FDA approval. The high rate of failure of AD drug development is partly responsible for the high costs of advancing AD drug development [4–6]. It is advisable to increase research funds to counter this inexorable trend and at the same time not to neglect research on natural molecules that can offer a valuable therapeutic contribution [7–10].

Citation: Kola, A.; Vigni, G.; Valensin, D. Exploration of Lycorine and Copper(II)'s Association with the N-Terminal Domain of Amyloid β. *Inorganics* **2023**, *11*, 443. https://doi.org/10.3390/inorganics11110443

Academic Editors: Isabel Correia and Vladimir Arion

Received: 2 October 2023
Revised: 23 October 2023
Accepted: 15 November 2023
Published: 18 November 2023

Copyright: © 2023 by the authors. Licensee MDPI, Basel, Switzerland. This article is an open access article distributed under the terms and conditions of the Creative Commons Attribution (CC BY) license (https://creativecommons.org/licenses/by/4.0/).

The exact cause of AD is unknown, but it is thought to be caused by a combination of genetic and environmental factors. Some of the known risk factors for AD include (1) age: AD is most common in older adults, with the risk of developing the disease increasing with age [11–13]; (2) family history: people with a family history of AD are more likely to develop the disease themselves [14,15]; (3) genetic mutations: some genetic mutations have been linked to an increased risk of developing AD [16,17]; (4) head injuries: a history of head injury may increase the risk of developing AD [18]; and (5) cardiovascular disease: cardiovascular disease risk factors, such as high blood pressure, high cholesterol, and diabetes, have been linked to an increased risk of AD [19].

AD is characterized by two hallmark neuropathological features: amyloid plaques and tau tangles. Amyloid plaques are formed by the buildup of amyloid beta protein outside of the nerve cells. Tau tangles are formed by the buildup of tau protein inside of the nerve cells. Amyloid beta (Aβ) is a peptide that is produced by the normal processing of the amyloid precursor protein (APP), leading to the formation of Aβ42 and Aβ40 peptides. These fragments differ in length, aggregation propensity, and toxicity, the former being more prone to form aggregates [20]. APP is a protein that is found on the surface of the nerve cells. Aβ is normally produced and cleared from the brain, but in people with AD, Aβ accumulates in the brain and forms plaques [21]. Although the exact role of Aβ in AD is not fully understood, there is evidence supporting its central role in the disease process, showing a correlation between the amount of Aβ in the brain and the severity of AD symptoms [22]. The accumulation and formation of beta-amyloid plaques in the brain is also correlated to the oxidative damage caused by Reactive Oxygen Species (ROS). ROS are chemically reactive molecules (superoxide anions, hydroxyl radicals, and hydrogen peroxide) that can cause damage to various cellular components, including DNA, proteins, and lipids. The body has defense mechanisms, such as antioxidants, to neutralize ROS and prevent their harmful effects. However, when ROS levels are chronically elevated or antioxidant defenses are overwhelmed, it can lead to pathological conditions [23–26]. ROS can trigger oxidative stress processes that damage brain cells and induce inflammatory reactions. At the same time, amyloid accumulation can lead to an imbalance in brain metal homeostasis, creating conditions that favor ROS production. This detrimental cycle can amplify cellular damage and the cognitive decline observed in neurodegenerative diseases.

Nowadays, there is no cure for Alzheimer's disease, and the available treatments focus on alleviating symptoms, slowing the progression of the disease, and enhancing the individual's quality of life [27]. The currently approved medications are cholinesterase inhibitors. Acetylcholinesterase is an enzyme that breaks acetylcholine in the synaptic cleft, reducing its availability for nerve communication with negative consequences for memory, learning, and other cognitive functions [28]. Cholinesterase inhibitors, including drugs like donepezil, rivastigmine, and galantamine, work by blocking this enzyme's activity to promote greater availability of the neurotransmitter to the neurons [29,30]. Memantine is the ultimate drug approved by the FDA for moderate to severe AD and works by blocking excessive activity of glutamate, an excitatory neurotransmitter [31]. It helps regulate glutamate levels in the brain, potentially protecting nerve cells from further damage. Memantine is often used in combination with cholinesterase inhibitors for a more comprehensive approach to symptom management.

Among the four approved drugs, galantamine (GAL) is the only one of natural origin; in fact, it is an alkaloid that derives from the family of Amaryllidaceae plants [32]. We recently investigated and compared the behavior of GAL with lycorine (LYC), another alkaloid from the same family plant [33]. The interest in LYC is derived from the interesting features exhibited by this natural alkaloid against different pathologies [34–36]. GAL and LYC were studied by evaluating their neuroprotective effects, antioxidant properties, and beta-amyloid-binding abilities [33]. Using a combined ligand- and peptide-based approach, we analyzed the atomic and molecular interactions of LYC and GAL with the pathogenic Aβ40 peptide, revealing that both alkaloids possess the ability to selectively induce changes in Aβ40 resonances [33]. The protective effect of these two alkaloids on SH-SY5Y differentiated cells previously intoxicated with Aβ42 were also evaluated. Surprisingly, our data indicated that LYC exhibits a greater ability to attenuate Aβ42-induced cytotoxicity in SH-SY5Y cells compared to GAL [33]. In this study, according to the investigation methods, Aβ42 or Aβ40 isoforms were differentially used. Spectroscopic analysis was mainly performed on Aβ40, which is less prone to aggregation compared to Aβ42 and therefore more stable and easy to handle. On the other hand, cellular studies were performed by using Aβ42, exhibiting a greater tendency to form aggregates and being more toxic than Aβ40.

Given the highly promising outcomes exhibited by LYC, particularly its capability to engage with the N-terminal section of Aβ via electrostatic interactions with residues also involved in copper binding [37], we opted to focus our study on a comprehensive exploration of the molecular interactions between LYC and Aβ, both in the presence and absence of copper(II). For this study, we decided to utilize the Aβ16 peptide, a fragment encompassing the N-terminal domain of Aβ, acting as the minimal binding motif for Cu(II) [38]. The interactions of LYC with the apo- and copper(II)-bound forms of Aβ16 were investigated by using NMR and UV-VIS techniques, providing new insights into the chemical and reactivity features of Aβ–Cu(II)–LYC associations.

2. Results

The interaction between Aβ16, LYC, and Cu(II) ions was first evaluated using NMR spectroscopy. Compared to the longest Aβ fragments, Aβ42 and Aβ40, Aβ16 has a good solubility in water at a physiological pH and does not form oligomeric or aggregated species in solutions. The NMR assignment of the Aβ16 signals was obtained via the analysis of ^{1}H-^{1}H TOCSY and NOESY spectra and it is reported in Table S1. From the analysis of the NMR spectra, the lack of the amide signals corresponding to Ala2, His6, Asp7, Ser8, His13, His14, and Gln15 is evident. The absence of NH resonances is generally observed for flexible peptides at a physiological pH due to their lability and exchange with water protons. In this case, the NMR data are consistent with a larger solvent exposure of Ala, Asp, His, Ser, and Gln residues leading to faster amide proton exchange rates known to be dependent on the amide pKa [39]. On the other hand, NMR investigations performed on an acetylated Aβ16 system at a lower temperature (T = 278 K) revealed the presence of all nitrogen backbone main-chain protons, strongly indicating the influence of temperature on amide–water proton exchange [40].

2.1. Study of Aβ16–LYC Interaction

Upon the full NMR assignment of the Aβ16 spectra, the effects of LYC were evaluated by looking at the variations in the chemical shifts and line broadening of both the Aβ16 and LYC signals. As shown in Table 1 and Figure 1, the LYC protons were slightly perturbed in presence of Aβ16 and, as expected, the chemical shift variations were more pronounced at a the larger Aβ16:LYC ratio. Moreover, Table 1 points out that larger effects are exhibited by the protons in the proximity of the nitrogen atom in position 6, in agreement with the data recorded for the system Aβ40L–YC [33].

Table 1. Chemical shifts of LYC protons in absence and in presence of different Aβ concentrations. T = 298 K, pH 7.5, phosphate buffer 30 mM.

Atom Type	ppm Values		
LYC Protons	LYC	LYC (0.4 eqs) + Aβ	LYC (1.0 eqs) + Aβ
H12	7.03	7.03	7.03
H8	6.84	6.83	6.83
H10	6.01	6.00	6.00
H3	5.74	5.73	5.73
H1	4.65	4.65	4.65

Table 1. Cont.

Atom Type	ppm Values		
LYC Protons	LYC	LYC (0.4 eqs) + Aβ	LYC (1.0 eqs) + Aβ
H2	4.34	4.33	4.33
H7″	4.25	4.24	4.24
H7′	4.02	3.99 (−0.03 ppm) [1]	4.00
H5″	3.50	3.49	3.49
H3a1	3.33	3.29 (−0.04 ppm) [1]	3.30
H12b	2.88	2.87	2.87
H4	2.82	2.81	2.81
H5′	2.75	2.74	2.74

[1] Chemical shift variations are calculated by subtracting the chemical shift ppm values of LYC in presence and in absence of Aβ16.

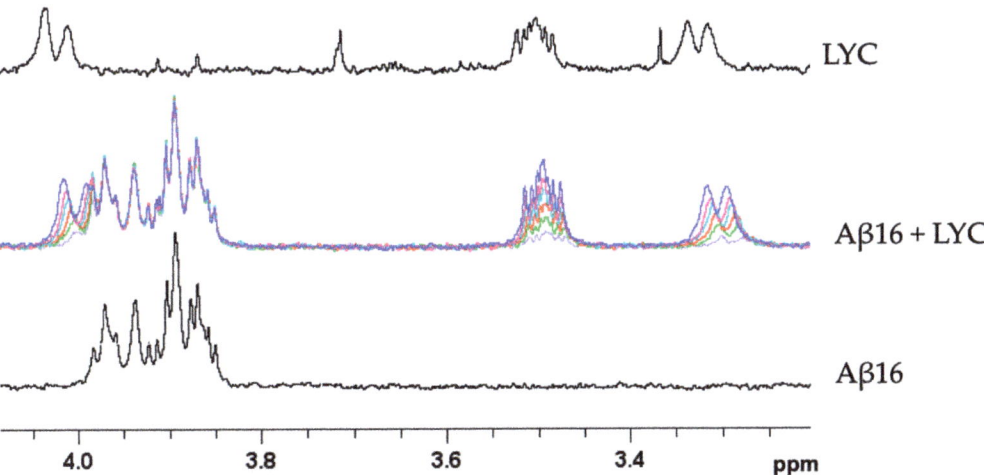

Figure 1. Superimposition of selected regions of ^{1}H-NMR spectra of Aβ16 0.5 mM (lower trace), LYC (upper trace), and Aβ16:LYC solutions (middle traces) at different ratios. Aβ16:LYC ratios are shown as the following: violet 1.0:0.2; green 1.0:0.4; red 1.0:0.6; cyan 1.0:0.8; magenta 1.0:1.0; and blue 1.0:1.2. T = 298 K, pH 7.5, 30 mM phosphate buffer.

Beyond the results obtained on the LYC resonances, the comparison between the NMR spectra of the Aβ16 in the absence and in the presence of LYC indicates His residues as the most affected ones, being weakly downfield-shifted by increasing the LYC concentration up to 1.2 eqs. (Figure 2A). On the other hand, LYC causes the line broadening of selected Aβ16 cross-peaks of the ^{1}H-^{1}H TOCSY (Figure 2B,C). In particular, upon LYC addition, we observed the disappearance of the correlations belonging to Asp1, Glu3, Arg5, Glu11, Val12, and Lys16. The observed variations agree with the effects recorded on the Aβ40–LYC system, strongly indicating that the Aβ–LYC interaction occurs at the N-terminal Aβ region [33].

In order to better evaluate the LYC-induced structural rearrangements, the CD spectra of Aβ16 in the presence and in the absence of LYC were collected. The CD spectra of Aβ16 showed the typical features of a disordered and flexible peptide exhibiting a negative absorption at 198 nm (Figure S1). The addition of 0.5 and 1.0 LYC equivalents lead to subtle changes in the CD spectra. In both cases, we observed a slightly increased absorption at 198 nm. No new absorptions were visible strongly indicating the absence of significant structural rearrangements of Aβ16 (Figure S1).

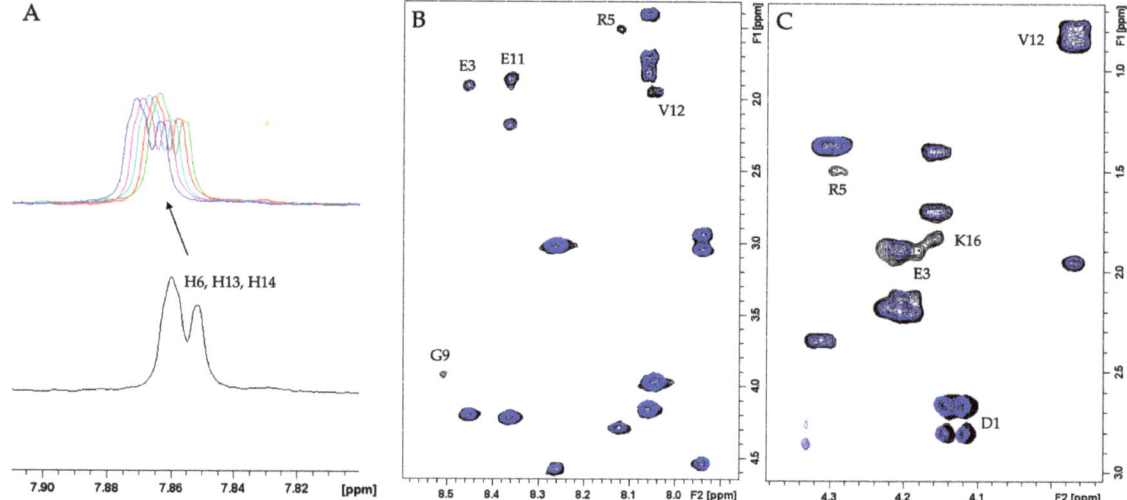

Figure 2. Superimposition of selected regions of NMR spectra of Aβ16 alone and with LYC. (**A**) Aromatic region of ^{1}H NMR spectra of Aβ16 0.5mM in absence (black) and in presence of 0.16 (mauve), 0.32 (lime), 0.48 (red), 0.64 (cyan), 0.80 (magenta) and 0.96 (blue) LYC eqs. (**B,C**) ^{1}H-^{1}H TOCSY NMR spectra of Aβ16 0.5 mM in absence (black traces) and in presence of 1.0 LYC eqs. (blue traces). T = 298 K, pH 7.5, 30 mM phosphate buffer.

2.2. Study of Aβ16–Cu(II) Interaction

Cu(II)/Cu(I) binding to Aβ peptides has been extensively investigated in recent years as nicely described in recent review papers [23,41–43]. The binding domains of both copper oxidation states are located at the N-terminus, and it is well accepted that His acts as a copper-anchoring site. In order to evaluate the ability of LYC to interfere with the Aβ–Cu(II) interaction, ^{1}H-NMR analysis on the Aβ16–Cu(II) system was first performed. In agreement with previous studies, the presence of substoichiometric Cu(II) ions in the Aβ16 solutions caused extensive line broadening on the His residues (Figure S2). In addition to the effects recorded on the His protons, the disappearance of the ^{1}H-^{1}H TOCSY correlations belonging to Asp1, Glu3, Arg5, Val12, Gln15, and Lys16 was observed (Figure 3). All these findings confirmed the involvement of the N-terminal and imidazole nitrogen in the copper coordination sphere, together with the carboxylate oxygens of Asp1 and Glu3.

2.3. Study of the Ternary Association between Aβ16, Cu(II), and LYC

The NMR spectra of the ternary systems were compared with the correspondent NMR spectra recorded only in the presence of Cu(II) or LYC. Both Aβ16 and LYC NMR signals were monitored for insights into the formation of ternary adducts. As shown in Figures S3 and 4A, the copper-induced line broadening was completely conserved in the sample containing LYC as well, strongly indicating that copper coordination is unaltered by the presence of LYC, and pointing out the ability of copper to bind Aβ16 regardless of LYC's presence. In fact, the two ^{1}H-^{1}H TOCSY experiments of Aβ16 recorded using Cu(II) and LYC or using Cu(II) only almost overlapped, except for the LYC signals that were present in the sample containing LYC only (Figure 4A). At the same time, the LYC NMR signals were monitored in the presence and absence of Cu(II) ions. As shown in Figure 4B,C, LYC protons experience a larger up-field shift when Cu(II) is coordinated to Aβ16. These findings suggest that upon Cu(II) coordination, Aβ16 retains its ability to associate with LYC. Moreover, the large shift observed in the LYC protons (Figure 5) suggests that the Aβ16–LYC interaction is more efficient when the peptide is bound to copper ions.

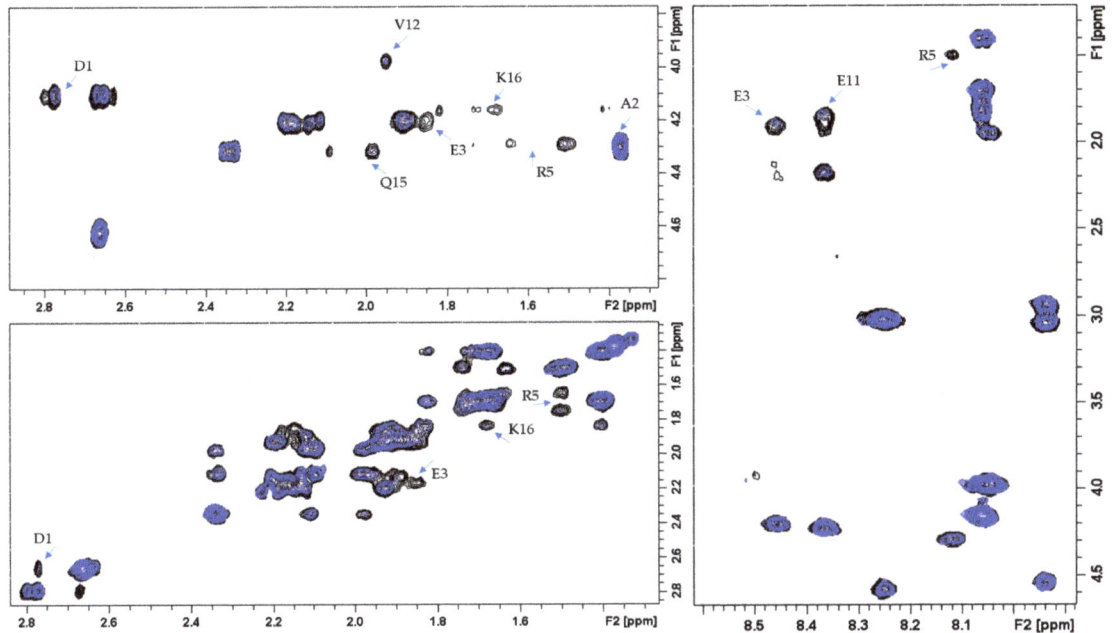

Figure 3. Superimposition of selected regions of ^{1}H-^{1}H TOCSY spectra of Aβ16 0.5 mM in absence (black) and in presence of 0.1 Cu(II) eqs. (blue) T = 298 K, pH 7.5, 30 mM phosphate buffer.

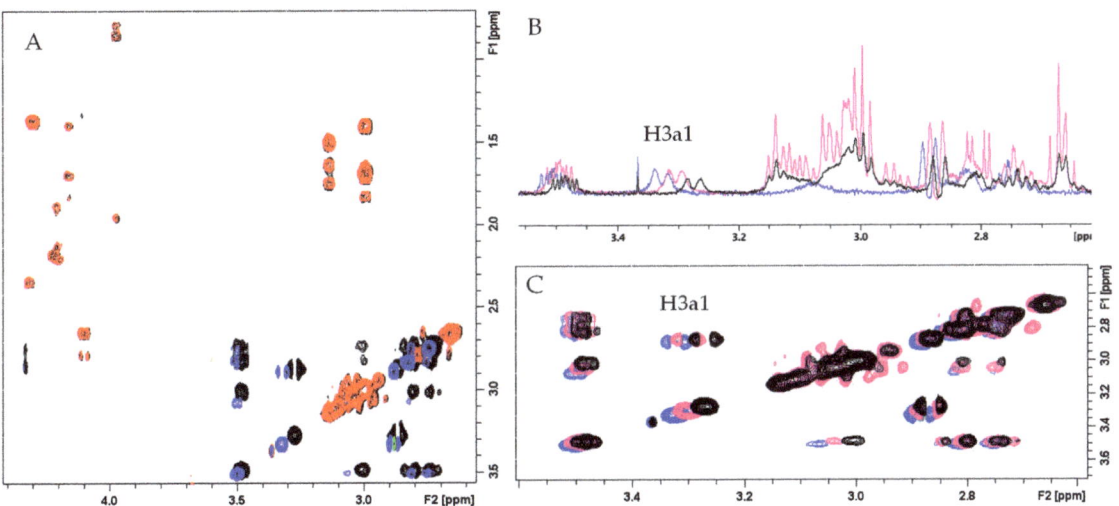

Figure 4. Comparison of NMR spectra of Aβ16, Cu(II), and LYC systems at different concentrations (**A**) ^{1}H-^{1}H TOCSY spectra of Aβ16 0.5 mM alone (black contours), in presence of 0.1 Cu(II) eqs. (red contours), and in presence of 0.1 Cu(II) and 1.0 LYC eqs. (blue contours). (**B**) 1D and (**C**) ^{1}H-^{1}H TOCSY spectra of LYC 0.5 mM alone (blue), in presence of 1.0 Aβ16 eqs. (magenta), and in presence of 1.0 Aβ16 and 0.1 Cu(II) eqs. (black). T = 298 K, pH 7.5, 30 mM phosphate buffer.

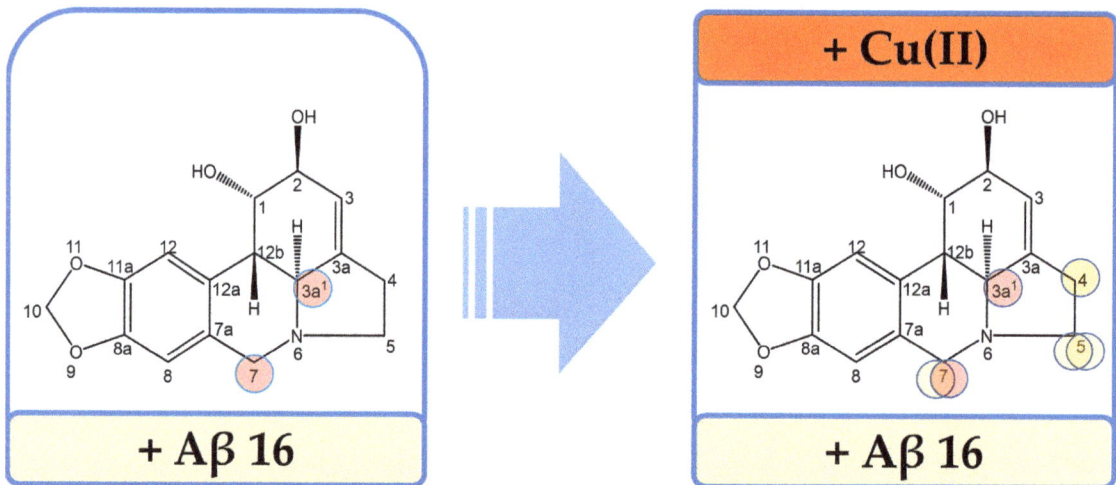

Figure 5. Comparison of the significant chemical shift variations of LYC protons upon Aβ16 and Aβ16/Cu(II) additions. The most shifted protons are shown as colored circles; the larger the variations, the more intense the color.

Although NMR experiments provided evidence of a ternary interaction between Aβ16, Cu(II), and LYC, further analysis was needed for a better understanding of the features associated with these adducts. By considering the ability of Aβ16–Cu(II) complexes to generate ROS, we decided to gain more insights into the impact of the Aβ16–Cu(II)–LYC system by analyzing the effects of LYC on the ascorbate prooxidant activity, both in the presence and absence of Aβ16. Redox active metal ions, like Cu(II), have the capacity to expedite the oxidation process of ascorbate when exposed to oxygen. This acceleration results in the generation of ROS via Fenton-type reactions [44,45]. The consumption of ascorbate can be effectively monitored by measuring its absorption at 265 nm as a function of time. This method provides a characteristic kinetic curve, the slope of which is directly associated with the reaction rate.

Figure 6 shows that LYC delays the consumption/oxidation of ascorbate, strongly indicating a protective role of LYC against ROS species, usually formed by ascorbate in the presence of copper(II) and molecular oxygen [44–46]. Such effects are dependent on the LYC concentration and are much more evident in the system containing Aβ16 and LYC. In particular, the changes observed on the slope of the kinetic curve (Figure 6A) reveal that the Aβ16–Cu(II)–LYC adduct is able to impact the kinetic rate of the ascorbate oxidation. Finally, the effects measured on LYC alone allowed us to independently evaluate LYC's impact on the ascorbate–Cu(II) system. As shown in Figure 6B, the absence of Aβ16 results in a completely different LYC behavior, thus indicating that the observed ROS protection is mainly dependent on the Aβ1–LYC interaction. These findings agree with the NMR observations and indicate that the Aβ–LYC association, albeit weak, is able to interfere with Aβ16's ability to generate ROS.

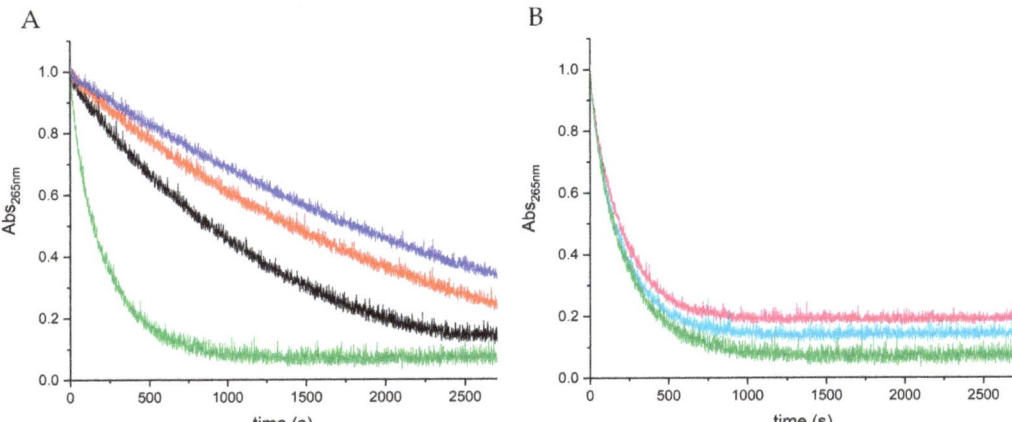

Figure 6. UV–VIS kinetic curves of the systems composed of ascorbate 20 µM and Cu(II) 1 µM in the presence of Aβ16 and LYC (**A**) or LYC only (**B**). The curve corresponding to ascorbate in the presence of copper(II) is shown in green, while the other colors refer to samples in the simultaneous presence of LYC and Aβ16, together or alone: Aβ16 10 µM (black); Aβ16 10 µM + LYC 5 µM (red); Aβ16 10 µM + LYC 10 µM (blue); LYC 5 µM (light blue); LYC 10 µM (magenta). Room temperature, pH 7.5, 1 mM phosphate buffer.

3. Discussion

In this study, the ability of LYC to interact with Aβ16 in the absence and presence of Cu(II) was investigated using NMR spectroscopy. Our findings indicate that LYC weakly associates with Aβ16 as shown by the variations in the NMR parameters of both LYC and Aβ16 (Figures 1 and 2). In fact, the NMR signals of LYC showed slight chemical shift variations together with the His aromatic protons of Aβ16, while other Aβ16 residues, like Asp1, Glu3, Arg5, Glu11, and Val12, exhibited decreased intensity signals upon LYC addition. These data are in good agreement with the recent features shown by an Aβ40–LYC system [33] and indicate that LYC is also able to interact with the monomeric, disordered, and flexible Aβ16 form. The interaction takes place at the N-atom at position 6 of the LYC as shown by the largest effects displayed by the protons located nearby (Figure 5).

The NMR data collected on the Aβ16–LYC and Aβ16–Cu(II) systems indicate that both LYC and Cu(II) share a similar Aβ16-binding domain, mainly encompassing the N-terminal and His residues (Figures 2 and 3). Despite the evidence of a correspondence between the LYC and Cu(II) association sites, the two species experience different binding modes since Cu(II) is able to form very stable coordination complexes while LYC weakly interacts with Aβ16 via electrostatic interaction. In this scenario, the NMR behavior of solutions containing Aβ16, copper, and LYC was investigated with the aim to evaluate the possible existence of a ternary association involving all three analyzed species.

The analysis of the NMR spectra reported in Figures 3 and 4 points out that LYC's association with Aβ16 is also conserved in presence of Cu(II). Moreover, the largest shifts measured in the ternary system containing Aβ16, Cu(II), and LYC (Figures 4 and 5) gave evidence of stronger Aβ16–LYC associations when the peptide was bound to the cupric ion. This phenomenon could be explained by considering the different conformation assumed by the peptide in the apo- or metal-complexed form. In fact, previous CD investigations have shown that upon Cu(II) binding, both Aβ16 and Aβ26 assume a more ordered structure [47], which in turn might favor the interaction with LYC.

The formation of an Aβ16–Cu(II)–LYC adduct was also confirmed by the UV-VIS kinetic curve, indicating that the ternary system is capable of interfering with the prooxidant activity of the ascorbate (Figure 6A). In fact, the protective effects of LYC are tangible only in the presence of Aβ16, probably due to LYC's influence in favoring peptide conformations less suitable for Cu(II)/Cu(I) redox cycling. The importance of backbone structural rearrangements is also supported by measuring the ascorbate oxidation in the presence of the His–LYC system (Figure S4). The choice of using His was made in order to evaluate the effects of LYC in a system able to strongly bind both Cu(II) and Cu(I), such as His, but at the same time not able to interact with LYC or undergo structural changes upon LYC association. The obtained UV-VIS kinetic curves point out that the same LYC amounts used for the Aβ–LYC system lead to completely different results when Aβ is substituted with His. In fact, the presence of LYC in the solution does not yield a slowing down of ascorbate oxidation but it rather induces a mild increase. Moreover, the lack of LYC concentration dependence suggests that the observed changes can be considered negligible.

The role of copper ions in AD is well documented in the literature [48–53]. Altered copper levels have been measured in the serum, cerebrospinal fluid, and post-mortem brains of AD patients [54,55]. Copper is also involved in several AD processes, such as oligomer and fibril Aβ formation [56–58], Aβ proteolysis and clearance [59,60], and oxidative stress [26,61–63]. At the same time, copper binding to Aβ peptides has been extensively investigated in recent years. It is well accepted that copper forms stable metal complexes at the N-terminal region of Aβ in both oxidation states, and the formed metal complexes are able to catalyze ROS production in vitro in the presence of molecular oxygen and ascorbate [26]. ROS production is mediated by the redox cycling between the Cu(II) and Cu(I) oxidation states occurring in the presence of ascorbate. Recently, it has been shown that ROS production is catalyzed by a low-populated copper binding state, different from the Cu(II) and Cu(I) binding modes observed in the "resting state" [64].

The copper-induced line broadening of the NMR signals allowed us to identify and compare the metal coordination sphere in the presence and absence of LYC. Our findings indicate that LYC has no effect on a Cu(II) binding mode of Aβ16, consisting of the three His imidazoles together with the Asp1 and Glu3 carboxylic groups, in agreement with previous results reporting copper(II) coordination to N and O donor atoms from His, N-terminus amine, and Asp and Glu carboxylate groups in a distorted square-pyramidal geometry [61].

In conclusion, our findings strongly suggest LYC's ability to function against oxidative stress via its interaction with Aβ–Cu(II) complexes, which are known to be able to catalyze ROS production. Similarly to LYC, GAL was also found inhibit Aβ-mediated ROS accumulation [65], thus possibly explaining the neuroprotection exhibited by both alkaloids against Aβ toxicity and providing new insights into a deeper understanding of AD progression and the molecular basis of GAL and LYC in neuroprotection.

4. Materials and Methods

4.1. Materials

The $CuSO_4$ solution (4% w/v, prepared from copper(II) sulfate pentahydrate), ascorbic acid (≥99%), lycorine hydrochloride (≥98% TLC), and L-Histidine, phosphate buffer and water for chromatography (LC-MS-grade) were all supplied by Sigma-Aldrich (Schnelldorf, Germany). The Aβ16 peptide was purchased from DBA Italia (Segrate, Italy).

4.2. NMR Experiments

The NMR experiments were performed at 14.1 T using a Bruker Avance III 600 MHz spectrometer and a 5 mm BBI (Broadband Inverse) probe. All the experiments were collected and carried out at controlled temperature T = 298 K ± 0.2 K. The chemical shifts were referenced against external 2-(Trimethylsilyl)-propionic-2,2,3,3-d_4 acid sodium salt (TMSP-d_4). The 1D spectra were recorded by using standard pulse sequences, and were analyzed by using the TopSpin 4.1.4 software. The residual water signal was suppressed using an excitation sculpting pulse program, applying a selective 2 ms long square pulse to water [66]. The TOCSY spectra were obtained using the MLEV-17 pulse sequence with a mixing time of 60 ms. The NOESY spectra were obtained using different mixing times to ascertain the best one. The NMR tubes were prepared by using a stock solution of Aβ16 peptide to achieve a final concentration of 0.5 mM. LYC and Cu(II) stock solutions were used to obtained the desired stoichiometric ratios Aβ16:Cu(II) and Aβ16:LYC in the NMR tubes. All the samples were prepared in phosphate buffer 30 mM at a pH of 7.5 with 10% D_2O.

4.3. UV-VIS Measurements

The absorption spectra and the kinetic curves (45 min, 2700 s) were recorded on a Perkin Elmer Lambda 900 UV/VIS/NIR spectrophotometer. The UV-VIS samples were prepared by using ascorbate, Aβ16, L-His, and Cu(II) stock solutions to generate the final concentrations 20 μM, 10 μM, and 1 μM for ascorbate, Aβ16/L-His, and Cu(II), respectively. The stoichiometric ratios Aβ16/L-His:LYC were 1:0.5 and 1:1 during all the experiments. The samples were prepared in phosphate buffer 1 mM, pH 7.5. In order to avoid any sample contamination interfering with the ascorbate oxidation, all the stock solutions, the buffer, and the UV-VIS samples were prepared by using water for chromatography.

4.4. CD Studies

The Circular Dichroism (CD) spectra were acquired using a Jasco J-815 spectropolarimeter at room temperature. A 1 cm cell path length was used for data between 190 and 260 nm, with a 1 nm sampling interval. Four scans were collected for each sample, with a scan speed of 100 nm min^{-1} and a bandwidth of 1 nm. The baseline spectra were subtracted from each spectrum and data were smoothed with the Savitzky–Golay method [67]. The data were processed using the Origin 5.0 spread sheet/graph package. The Aβ16 samples were prepared to obtain a final concentration 10 μM in the cuvette. LYC addition was performed to obtain the Aβ16:LYC ratios 1:0.5 and 1:1. The samples were prepared in phosphate buffer 1 mM, pH 7.5.

Supplementary Materials: The following supporting information can be downloaded at https://www.mdpi.com/article/10.3390/inorganics11110443/s1: Figure S1: CD spectra of Aβ16 in absence (black lines) and presence of 0.5 (red lines) and 1.0 LYC eqs. (blue lines). Aβ16 concentration 10 μM, phosphate buffer 1 mM, T = 298 K; Figure S2: NMR spectra of Aβ16 0.5 mM in absence (black traces) and presence of 1.2 LYC eqs. (blue traces), 1.0 LYC eqs. (magenta traces), 0.8 LYC eqs. (cyan traces), 0.6 LYC eqs. (red traces), 0.4 LYC eqs. (green traces), and 0.2 LYC eqs. (gray traces). T = 298 K, pH 7.5, 20 mM phosphate buffer; Figure S3: Superimposition of selected regions of ^1H-^1H TOCSY spectra of Aβ16 0.5 mM (black), Aβ16 0.5 mM with 0.1 Cu(II) eqs. in absence (blue) and presence of 1.0 LYC eqs. (magenta). T = 298 K, pH 7.5, 20 mM phosphate buffer; Figure S4: UV-VIS kinetic curves of the systems composed of ascorbate 20 μM and Cu(II) 1 μM in the presence of His and LYC. The green curve corresponds to ascorbate in the presence of only copper(II), while the other colors refer to samples in the simultaneous presence of His and Aβ16, together or alone. Specifically, His 10 μM (blue), His 10 μM + LYC 5 μM (black), His 10 μM + LYC 10 μM (black); Table S1: ^1H chemical shift assignment of Aβ16 0.5 mM, T = 298 K, pH 7.5, 20 mM phosphate buffer.

Author Contributions: Conceptualization, A.K. and D.V.; methodology, A.K. and D.V.; validation, A.K., G.V. and D.V.; formal analysis, A.K., G.V. and D.V.; investigation, A.K., G.V. and D.V.; resources, A.K. and D.V.; data curation, A.K., G.V. and D.V.; writing—original draft preparation, A.K. and D.V.; writing—review and editing, A.K., G.V. and D.V.; visualization, A.K., G.V. and D.V.; supervision, A.K.; project administration, D.V. All authors have read and agreed to the published version of the manuscript.

Funding: This research received no external funding.

Data Availability Statement: Data are contained within the article and supplementary materials.

Acknowledgments: The Consorzio Interuniversitario Risonanze Magnetiche di Metallo Proteine (CIRMMP) is acknowledged for the scholarship support.

Conflicts of Interest: The authors declare no conflict of interest.

References

1. Scheltens, P.; De Strooper, B.; Kivipelto, M.; Holstege, H.; Chételat, G.; Teunissen, C.E.; Cummings, J.; van der Flier, W.M. Alzheimer's Disease. *Lancet* **2021**, *397*, 1577–1590. [CrossRef]
2. Dementia. Available online: https://www.who.int/news-room/fact-sheets/detail/dementia (accessed on 29 September 2023).
3. World Health Organization. *Global Action Plan on the Public Health Response to Dementia 2017–2025*; World Health Organization: Geneva, Switzerland, 2017; ISBN 978-92-4-151348-7.
4. Scott, T.J.; O'Connor, A.C.; Link, A.N.; Beaulieu, T.J. Economic Analysis of Opportunities to Accelerate Alzheimer's Disease Research and Development. *Ann. N. Y. Acad. Sci.* **2014**, *1313*, 17–34. [CrossRef] [PubMed]
5. Cummings, J.L.; Morstorf, T.; Zhong, K. Alzheimer's Disease Drug-Development Pipeline: Few Candidates, Frequent Failures. *Alzheimer's Res. Ther.* **2014**, *6*, 37. [CrossRef] [PubMed]
6. Cummings, J.; Reiber, C.; Kumar, P. The Price of Progress: Funding and Financing Alzheimer's Disease Drug Development. *A&D Transl. Res. Clin. Interv.* **2018**, *4*, 330–343. [CrossRef]
7. International, A.D.; Patterson, C. *World Alzheimer Report 2018: The State of the Art of Dementia Research: New Frontiers*; Alzheimer's Disease International (ADI): London, UK, 2018.
8. Xiao, J.; Tundis, R. Natural Products for Alzheimer's Disease Therapy: Basic and Application. *J. Pharm. Pharmacol.* **2013**, *65*, 1679–1680. [CrossRef] [PubMed]
9. Palmioli, A.; Mazzoni, V.; De Luigi, A.; Bruzzone, C.; Sala, G.; Colombo, L.; Bazzini, C.; Zoia, C.P.; Inserra, M.; Salmona, M.; et al. Alzheimer's Disease Prevention through Natural Compounds: Cell-Free, In Vitro, and In Vivo Dissection of Hop (*Humulus lupulus* L.) Multitarget Activity. *ACS Chem. Neurosci.* **2022**, *13*, 3152–3167. [CrossRef] [PubMed]
10. da Rosa, M.M.; de Amorim, L.C.; Alves, J.V.d.O.; Aguiar, I.F.d.S.; Oliveira, F.G.d.S.; da Silva, M.V.; dos Santos, M.T.C. The Promising Role of Natural Products in Alzheimer's Disease. *Brain Disord.* **2022**, *7*, 100049. [CrossRef]
11. Braak, H.; Thal, D.R.; Ghebremedhin, E.; Del Tredici, K. Stages of the Pathologic Process in Alzheimer Disease: Age Categories from 1 to 100 Years. *J. Neuropathol. Exp. Neurol.* **2011**, *70*, 960–969. [CrossRef]
12. Abate, G.; Vezzoli, M.; Sandri, M.; Rungratanawanich, W.; Memo, M.; Uberti, D. Mitochondria and Cellular Redox State on the Route from Ageing to Alzheimer's Disease. *Mech. Ageing Dev.* **2020**, *192*, 111385. [CrossRef]
13. Azam, S.; Haque, M.E.; Balakrishnan, R.; Kim, I.-S.; Choi, D.-K. The Ageing Brain: Molecular and Cellular Basis of Neurodegeneration. *Front. Cell Dev. Biol.* **2021**, *9*, 683459. [CrossRef]
14. Talboom, J.S.; Håberg, A.; De Both, M.D.; Naymik, M.A.; Schrauwen, I.; Lewis, C.R.; Bertinelli, S.F.; Hammersland, C.; Fritz, M.A.; Myers, A.J.; et al. Family History of Alzheimer's Disease Alters Cognition and Is Modified by Medical and Genetic Factors. *eLife* **2019**, *8*, e46179. [CrossRef]
15. Green, R.C.; Cupples, L.A.; Go, R.; Benke, K.S.; Edeki, T.; Griffith, P.A.; Williams, M.; Hipps, Y.; Graff-Radford, N.; Bachman, D.; et al. Risk of Dementia among White and African American Relatives of Patients with Alzheimer Disease. *JAMA* **2002**, *287*, 329–336. [CrossRef] [PubMed]
16. Silva, M.V.F.; Loures, C.d.M.G.; Alves, L.C.V.; de Souza, L.C.; Borges, K.B.G.; das Graças Carvalho, M. Alzheimer's Disease: Risk Factors and Potentially Protective Measures. *J. Biomed. Sci.* **2019**, *26*, 33. [CrossRef] [PubMed]
17. Bellenguez, C.; Küçükali, F.; Jansen, I.E.; Kleineidam, L.; Moreno-Grau, S.; Amin, N.; Naj, A.C.; Campos-Martin, R.; Grenier-Boley, B.; Andrade, V.; et al. New Insights into the Genetic Etiology of Alzheimer's Disease and Related Dementias. *Nat. Genet.* **2022**, *54*, 412–436. [CrossRef]
18. Schneider, A.L.C.; Selvin, E.; Latour, L.; Turtzo, L.C.; Coresh, J.; Mosley, T.; Ling, G.; Gottesman, R.F. Head Injury and 25-year Risk of Dementia. *Alzheimer's Dement.* **2021**, *17*, 1432–1441. [CrossRef] [PubMed]
19. Abubakar, M.B.; Sanusi, K.O.; Ugusman, A.; Mohamed, W.; Kamal, H.; Ibrahim, N.H.; Khoo, C.S.; Kumar, J. Alzheimer's Disease: An Update and Insights Into Pathophysiology. *Front. Aging Neurosci.* **2022**, *14*, 742408. [CrossRef]
20. Gu, L.; Guo, Z. Alzheimer's Aβ42 and Aβ40 Peptides Form Interlaced Amyloid Fibrils. *J. Neurochem.* **2013**, *126*, 305–311. [CrossRef]
21. Kepp, K.P.; Robakis, N.K.; Høilund-Carlsen, P.F.; Sensi, S.L.; Vissel, B. The Amyloid Cascade Hypothesis: An Updated Critical Review. *Brain* **2023**, *146*, awad159. [CrossRef] [PubMed]
22. Näslund, J.; Haroutunian, V.; Mohs, R.; Davis, K.L.; Davies, P.; Greengard, P.; Buxbaum, J.D. Correlation Between Elevated Levels of Amyloid β-Peptide in the Brain and Cognitive Decline. *JAMA* **2000**, *283*, 1571–1577. [CrossRef]
23. Falcone, E.; Hureau, C. Redox Processes in Cu-Binding Proteins: The "in-between" States in Intrinsically Disordered Peptides. *Chem. Soc. Rev.* **2023**, *52*, 6595–6600. [CrossRef]
24. Wärmländer, S.K.T.S.; Österlund, N.; Wallin, C.; Wu, J.; Luo, J.; Tiiman, A.; Jarvet, J.; Gräslund, A. Metal Binding to the Amyloid-β Peptides in the Presence of Biomembranes: Potential Mechanisms of Cell Toxicity. *J. Biol. Inorg. Chem.* **2019**, *24*, 1189–1196. [CrossRef]
25. Esmieu, C.; Guettas, D.; Conte-Daban, A.; Sabater, L.; Faller, P.; Hureau, C. Copper-Targeting Approaches in Alzheimer's Disease: How To Improve the Fallouts Obtained from in Vitro Studies. *Inorg. Chem.* **2019**, *58*, 13509–13527. [CrossRef]
26. Cheignon, C.; Tomas, M.; Bonnefont-Rousselot, D.; Faller, P.; Hureau, C.; Collin, F. Oxidative Stress and the Amyloid Beta Peptide in Alzheimer's Disease. *Redox Biol.* **2018**, *14*, 450–464. [CrossRef] [PubMed]
27. Buccellato, F.R.; D'Anca, M.; Tartaglia, G.M.; Del Fabbro, M.; Scarpini, E.; Galimberti, D. Treatment of Alzheimer's Disease: Beyond Symptomatic Therapies. *Int. J. Mol. Sci.* **2023**, *24*, 13900. [CrossRef] [PubMed]

28. Hampel, H.; Mesulam, M.-M.; Cuello, A.C.; Farlow, M.R.; Giacobini, E.; Grossberg, G.T.; Khachaturian, A.S.; Vergallo, A.; Cavedo, E.; Snyder, P.J.; et al. The Cholinergic System in the Pathophysiology and Treatment of Alzheimer's Disease. *Brain* **2018**, *141*, 1917–1933. [CrossRef]
29. Zemek, F.; Drtinova, L.; Nepovimova, E.; Sepsova, V.; Korabecny, J.; Klimes, J.; Kuca, K. Outcomes of Alzheimer's Disease Therapy with Acetylcholinesterase Inhibitors and Memantine. *Expert. Opin. Drug Saf.* **2014**, *13*, 759–774.
30. Marucci, G.; Michela Buccioni, M.; Ben, D.D.; Lambertucci, C.; Volpini, R.; Amenta, F. Efficacy of Acetylcholinesterase Inhibitors in Alzheimer's Disease. *Neuropharmacology* **2021**, *190*, 108352. [CrossRef]
31. Guzior, N.; Wieckowska, A.; Panek, D.; Malawska, B. Recent Development of Multifunctional Agents as Potential Drug Candidates for the Treatment of Alzheimer's Disease. *Curr. Med. Chem.* **2015**, *22*, 373–404. [CrossRef]
32. Vrabec, R.; Blunden, G.; Cahlíková, L. Natural Alkaloids as Multi-Target Compounds towards Factors Implicated in Alzheimer's Disease. *Int. J. Mol. Sci.* **2023**, *24*, 4399. [CrossRef] [PubMed]
33. Kola, A.; Lamponi, S.; Currò, F.; Valensin, D. A Comparative Study between Lycorine and Galantamine Abilities to Interact with AMYLOID β and Reduce In Vitro Neurotoxicity. *Int. J. Mol. Sci.* **2023**, *24*, 2500. [CrossRef] [PubMed]
34. Nair, J.J.; van Staden, J. Insight to the Antifungal Properties of Amaryllidaceae Constituents. *Phytomedicine* **2020**, *73*, 152753. [CrossRef]
35. Roy, M.; Liang, L.; Xiao, X.; Feng, P.; Ye, M.; Liu, J. Lycorine: A Prospective Natural Lead for Anticancer Drug Discovery. *Biomed. Pharmacother.* **2018**, *107*, 615–624. [CrossRef] [PubMed]
36. Xiao, H.; Xu, X.; Du, L.; Li, X.; Zhao, H.; Wang, Z.; Zhao, L.; Yang, Z.; Zhang, S.; Yang, Y.; et al. Lycorine and Organ Protection: Review of Its Potential Effects and Molecular Mechanisms. *Phytomedicine* **2022**, *104*, 154266. [CrossRef] [PubMed]
37. Rana, M.; Sharma, A.K. Cu and Zn Interactions with Aβ Peptides: Consequence of Coordination on Aggregation and Formation of Neurotoxic Soluble Aβ Oligomers. *Metallomics* **2019**, *11*, 64–84. [CrossRef]
38. De Gregorio, G.; Biasotto, F.; Hecel, A.; Luczkowski, M.; Kozlowski, H.; Valensin, D. Structural Analysis of Copper(I) Interaction with Amyloid β Peptide. *J. Inorg. Biochem.* **2019**, *195*, 31–38. [CrossRef] [PubMed]
39. Fogolari, F.; Esposito, G.; Viglino, P.; Briggs, J.M.; McCammon, J.A. pKa Shift Effects on Backbone Amide Base-Catalyzed Hydrogen Exchange Rates in Peptides. *J. Am. Chem. Soc.* **1998**, *120*, 3735–3738. [CrossRef]
40. Zirah, S.; Kozin, S.A.; Mazur, A.K.; Blond, A.; Cheminant, M.; Ségalas-Milazzo, I.; Debey, P.; Rebuffat, S. Structural Changes of Region 1-16 of the Alzheimer Disease Amyloid Beta-Peptide upon Zinc Binding and in Vitro Aging. *J. Biol. Chem.* **2006**, *281*, 2151–2161. [CrossRef]
41. Park, S.; Na, C.; Han, J.; Lim, M.H. Methods for Analyzing the Coordination and Aggregation of Metal-Amyloid-β. *Metallomics* **2023**, *15*, mfac102. [CrossRef]
42. Sóvágó, I.; Várnagy, K.; Kállay, C.; Grenács, Á. Interactions of Copper(II) and Zinc(II) Ions with the Peptide Fragments of Proteins Related to Neurodegenerative Disorders: Similarities and Differences. *Curr. Med. Chem.* **2023**, *30*, 4050–4071. [CrossRef]
43. Singh, S.K.; Balendra, V.; Obaid, A.A.; Esposto, J.; Tikhonova, M.A.; Gautam, N.K.; Poeggeler, B. Copper-Mediated β-Amyloid Toxicity and Its Chelation Therapy in Alzheimer's Disease. *Metallomics* **2022**, *14*, mfac018. [CrossRef]
44. Atrián-Blasco, E.; Del Barrio, M.; Faller, P.; Hureau, C. Ascorbate Oxidation by Cu(Amyloid-β) Complexes: Determination of the Intrinsic Rate as a Function of Alterations in the Peptide Sequence Revealing Key Residues for Reactive Oxygen Species Production. *Anal. Chem.* **2018**, *90*, 5909–5915. [CrossRef]
45. Shen, J.; Griffiths, P.T.; Campbell, S.J.; Utinger, B.; Kalberer, M.; Paulson, S.E. Ascorbate Oxidation by Iron, Copper and Reactive Oxygen Species: Review, Model Development, and Derivation of Key Rate Constants. *Sci. Rep.* **2021**, *11*, 7417. [CrossRef] [PubMed]
46. Kola, A.; Vigni, G.; Baratto, M.C.; Valensin, D. A Combined NMR and UV-Vis Approach to Evaluate Radical Scavenging Activity of Rosmarinic Acid and Other Polyphenols. *Molecules* **2023**, *28*, 6629. [CrossRef]
47. Syme, C.D.; Nadal, R.C.; Rigby, S.E.J.; Viles, J.H. Copper Binding to the Amyloid-Beta (Abeta) Peptide Associated with Alzheimer's Disease: Folding, Coordination Geometry, pH Dependence, Stoichiometry, and Affinity of Abeta-(1-28): Insights from a Range of Complementary Spectroscopic Techniques. *J. Biol. Chem.* **2004**, *279*, 18169–18177. [CrossRef] [PubMed]
48. Cicero, C.E.; Mostile, G.; Vasta, R.; Rapisarda, V.; Signorelli, S.S.; Ferrante, M.; Zappia, M.; Nicoletti, A. Metals and Neurodegenerative Diseases. A Systematic Review. *Environ. Res.* **2017**, *159*, 82–94. [CrossRef] [PubMed]
49. Kola, A.; Nencioni, F.; Valensin, D. Bioinorganic Chemistry of Micronutrients Related to Alzheimer's and Parkinson's Diseases. *Molecules* **2023**, *28*, 5467. [CrossRef]
50. Gaggelli, E.; Kozlowski, H.; Valensin, D.; Valensin, G. Copper Homeostasis and Neurodegenerative Disorders (Alzheimer's, Prion, and Parkinson's Diseases and Amyotrophic Lateral Sclerosis). *Chem. Rev.* **2006**, *106*, 1995–2044. [CrossRef]
51. Liu, Y.; Nguyen, M.; Robert, A.; Meunier, B. Metal Ions in Alzheimer's Disease: A Key Role or Not? *Acc. Chem. Res.* **2019**, *52*, 2026–2035. [CrossRef]
52. Kozlowski, H.; Luczkowski, M.; Remelli, M.; Valensin, D. Copper, Zinc and Iron in Neurodegenerative Diseases (Alzheimer's, Parkinson's and Prion Diseases). *Coord. Chem. Rev.* **2012**, *256*, 2129–2141. [CrossRef]
53. Wang, L.; Yin, Y.-L.; Liu, X.-Z.; Shen, P.; Zheng, Y.-G.; Lan, X.-R.; Lu, C.-B.; Wang, J.-Z. Current Understanding of Metal Ions in the Pathogenesis of Alzheimer's Disease. *Transl. Neurodegener.* **2020**, *9*, 10. [CrossRef]
54. Scolari Grotto, F.; Glaser, V. Are High Copper Levels Related to Alzheimer's and Parkinson's Diseases? A Systematic Review and Meta-Analysis of Articles Published between 2011 and 2022. *Biometals* **2023**. [CrossRef] [PubMed]

55. Wang, Z.-X.; Tan, L.; Wang, H.-F.; Ma, J.; Liu, J.; Tan, M.-S.; Sun, J.-H.; Zhu, X.-C.; Jiang, T.; Yu, J.-T. Serum Iron, Zinc, and Copper Levels in Patients with Alzheimer's Disease: A Replication Study and Meta-Analyses. *J. Alzheimers Dis.* **2015**, *47*, 565–581. [CrossRef] [PubMed]
56. Atrián-Blasco, E.; Gonzalez, P.; Santoro, A.; Alies, B.; Faller, P.; Hureau, C. Cu and Zn Coordination to Amyloid Peptides: From Fascinating Chemistry to Debated Pathological Relevance. *Coord. Chem. Rev.* **2018**, *375*, 38–55. [CrossRef]
57. Leal, S.S.; Botelho, H.M.; Gomes, C.M. Metal Ions as Modulators of Protein Conformation and Misfolding in Neurodegeneration. *Coord. Chem. Rev.* **2012**, *256*, 2253–2270. [CrossRef]
58. Cherny, R.A.; Atwood, C.S.; Xilinas, M.E.; Gray, D.N.; Jones, W.D.; McLean, C.A.; Barnham, K.J.; Volitakis, I.; Fraser, F.W.; Kim, Y.-S.; et al. Treatment with a Copper-Zinc Chelator Markedly and Rapidly Inhibits β-Amyloid Accumulation in Alzheimer's Disease Transgenic Mice. *Neuron* **2001**, *30*, 665–676. [CrossRef] [PubMed]
59. Mital, M.; Bal, W.; Frączyk, T.; Drew, S.C. Interplay between Copper, Neprilysin, and N-Truncation of β-Amyloid. *Inorg. Chem.* **2018**, *57*, 6193–6197. [CrossRef]
60. Grasso, G.; Pietropaolo, A.; Spoto, G.; Pappalardo, G.; Tundo, G.R.; Ciaccio, C.; Coletta, M.; Rizzarelli, E. Copper(I) and Copper(II) Inhibit Aβ Peptides Proteolysis by Insulin-Degrading Enzyme Differently: Implications for Metallostasis Alteration in Alzheimer's Disease. *Chemistry* **2011**, *17*, 2752–2762. [CrossRef] [PubMed]
61. Cheignon, C.; Jones, M.; Atrián-Blasco, E.; Kieffer, I.; Faller, P.; Collin, F.; Hureau, C. Identification of Key Structural Features of the Elusive Cu-Aβ Complex That Generates ROS in Alzheimer's Disease. *Chem. Sci.* **2017**, *8*, 5107–5118. [CrossRef]
62. Shen, H.; Dou, Y.; Wang, X.; Wang, X.; Kong, F.; Wang, S. Guluronic Acid Can Inhibit Copper(II) and Amyloid-β Peptide Coordination and Reduce Copper-Related Reactive Oxygen Species Formation Associated with Alzheimer's Disease. *J. Inorg. Biochem.* **2023**, *245*, 112252. [CrossRef]
63. Birla, H.; Minocha, T.; Kumar, G.; Misra, A.; Singh, S.K. Role of Oxidative Stress and Metal Toxicity in the Progression of Alzheimer's Disease. *Curr. Neuropharmacol.* **2020**, *18*, 552–562. [CrossRef]
64. Falcone, E.; Nobili, G.; Okafor, M.; Proux, O.; Rossi, G.; Morante, S.; Faller, P.; Stellato, F. Chasing the Elusive "In-Between" State of the Copper-Amyloid β Complex by X-Ray Absorption through Partial Thermal Relaxation after Photoreduction. *Angew. Chem. Int. Ed. Engl.* **2023**, *62*, e202217791. [CrossRef] [PubMed]
65. Jiang, S.; Zhao, Y.; Zhang, T.; Lan, J.; Yang, J.; Yuan, L.; Zhang, Q.; Pan, K.; Zhang, K. Galantamine Inhibits β-Amyloid-Induced Cytostatic Autophagy in PC12 Cells through Decreasing ROS Production. *Cell Prolif.* **2018**, *51*, e12427. [CrossRef] [PubMed]
66. Hwang, T.-L.; Shaka, A.J. Multiple-Pulse Mixing Sequences That Selectively Enhance Chemical Exchange or Cross-Relaxation Peaks in High-Resolution NMR Spectra. *J. Magn. Reson.* **1998**, *135*, 280–287. [CrossRef]
67. Savitzky, A.; Golay, M.J.E. Smoothing and Differentiation of Data by Simplified Least Squares Procedures. *Anal. Chem.* **1964**, *36*, 1627–1639. [CrossRef]

Disclaimer/Publisher's Note: The statements, opinions and data contained in all publications are solely those of the individual author(s) and contributor(s) and not of MDPI and/or the editor(s). MDPI and/or the editor(s) disclaim responsibility for any injury to people or property resulting from any ideas, methods, instructions or products referred to in the content.

Article

Antiparasitic Activity of Oxindolimine–Metal Complexes against Chagas Disease

Marcelo Cecconi Portes [1,†], Grazielle Alves Ribeiro [2,†], Gustavo Levendoski Sabino [1], Ricardo Alexandre Alves De Couto [1], Leda Quércia Vieira [2], Maria Júlia Manso Alves [3] and Ana Maria Da Costa Ferreira [1,*]

[1] Departamento de Química Fundamental, Instituto de Química, Universidade de São Paulo, São Paulo 05508-000, SP, Brazil; marcelo_cecconi@hotmail.com (M.C.P.); ricardo1@iq.usp.br (R.A.A.D.C.)
[2] Departamento de Bioquímica e Imunologia, Universidade Federal de Minas Gerais, Belo Horizonte 31270-901, MG, Brazil; graziellear@yahoo.com.br (G.A.R.); lqvieira.ufmg@gmail.com (L.Q.V.)
[3] Departamento de Bioquímica, Instituto de Química, Universidade de São Paulo, São Paulo 05508-000, SP, Brazil; mjmalves@iq.usp.br
* Correspondence: amdcferr@iq.usp.br; Tel.: +55-11-3091-9147 or +55-11-2648-1681
† These authors contributed equally to this work.

Abstract: Some copper(II) and zinc(II) complexes with oxindolimine ligands were tested regarding their trypanocidal properties. These complexes have already shown good biological activity in the inhibition of tumor cell proliferation, having DNA and mitochondria as main targets, through an oxidative mechanism, and inducing apoptosis. Herein, we demonstrate that they also have significant activity against the infective trypomastigote forms and the intracellular amastigote forms of *T. cruzi*, modulated by the metal ion as well as by the oxindolimine ligand. Selective indexes (LC_{50}/IC_{50}) determined for both zinc(II) and copper(II) complexes, are higher after 24 or 48 h incubation with trypomastigotes, in comparison to traditional drugs used in clinics, such as benznidazole, and other metal-based compounds previously reported in the literature. Additionally, tests against amastigotes indicated infection index <10% (% of infected macrophages/average number of amastigotes per macrophage), after 24 or 48 h in the presence of zinc(II) (60–80 µM) or analogous copper(II) complexes (10–25 µM). The copper complexes exhibit further oxidative properties, being able to damage DNA, proteins and carbohydrates, in the presence of hydrogen peroxide, with the generation of hydroxyl radicals. This redox reactivity could explain its better performance towards the parasites in relation to the zinc analogs. However, both copper and zinc complexes display good selective indexes, indicating that the influence of the ligand is also crucial, and is probably related to the inhibition of some crucial proteins.

Keywords: oxindolimine ligands; metal complexes; Chagas disease; *T. cruzi*; trypanocidal activity; mechanism of action

1. Introduction

Chagas disease, also known as American trypanosomiasis, is among the potentially fatal neglected tropical diseases. According to World Health Organization (WHO) data [1], it afflicts more than 6 million persons in extremely poor areas worldwide, especially in Latin America, where it is present in 20 countries, with an estimated 70 million people at risk of infection [2]. This disease was named after Carlos Ribeiro Justiniano Chagas, a Brazilian physician and researcher who discovered it in 1909. Further, Chagas disease has been described in non-endemic areas due to migration [3,4], and so it has become a public health issue even in developed nations [5]. Cure is only possible if treatment is administered soon after infection; thus, exposure to the parasite mostly leads to chronic infection. The reasons for this picture involve different factors, such as insufficient knowledge of the disease, a deficient market that makes the development of new drugs or new treatments not financially attractive for the pharmaceutical industry [6], and inadequate or non-existing efficient drugs once infection is established. Some recent reviews report efforts to elucidate the fundamental molecular and cell biology of parasitic trypanosomatids as well as the diseases they can cause [7,8].

The parasite responsible for this disease is the flagellate protozoa *Trypanosoma cruzi*, injected in the blood of human beings by triatominae bugs, in its trypomastigote form [9]. Transported through the blood stream, the parasites can enter host cells, including macrophages, where they are transformed into their amastigote forms. A recent review focused on the *T. cruzi* life cycle [10], providing a comprehensive update on its morphological forms and genetic diversity, aiming at identifying intervention points to cure the disease.

Current clinical treatments of Chagas disease are based on drugs that have been developed many decades ago, such as benznidazole (N-benzyl-2-nitroimidazole-1-acetamide) or nifurtimox (4[(5-nitrofurfurylidene)amino]-3-methylthiomorpholine-1,1-dioxide). Although active in earlier phases of the disease, these drugs show severe toxic side effects, and are inactive in the chronic phase of infection [11]. Therefore, many efforts have been made by several research groups in developing new, more efficient, and less toxic, usually organic compounds [12,13]. A wide range of compounds have been developed, with quite different structural features, and some of them have been clinically tested against trypanosomiases [14]. Among these chemotherapeutic agents there are natural as well as synthetic compounds, including naphthoquinones [15,16], alkaloids, antihistaminics, antibiotics, thiazolidinones [17], aminoquinolines, and thiazoles [18]. More recently, leucinostatins, natural products derived from fungi collected from soil samples, showed potent activity against intracellular and replicative amastigote forms of the parasite, with no host cell toxicity up to 1.5 µM [19]. On the other hand, biological studies on macrophage-derived peroxynitrite ($ONOO^-$ and $ONOOH$) formation, a strong oxidant arising from the reaction of nitric oxide ($^{\cdot}NO$) with superoxide radical ($O_2^{\cdot -}$), revealed that $^{\cdot}NO$ plays a central role in the control of acute infection by *T. cruzi* [20]. When those reactive species are formed simultaneously, the generation of peroxynitrite leads to severe cellular oxidative damage and morphological disturbance in internalized parasites [21].

Among synthetic compounds, different metal complexes with diverse ligands have been reported as showing significant trypanocidal activity. Some ternary nickel(II) complexes with imine and azapurine derivatives [22] showed high antitrypanosomatid activity against the epimastigote, amastigote, and trypomastigote forms of the parasite, after 72 h of culture, with IC_{50} in the 1–90 µM range, lower than those of the reference drug, benznidazole (BZ). A gold(III) complex with tridentate thiosemicarbazone ligands coordinated by an *ONS* donor set, [AuCl(L^{Me})] was found to be more active and more selective than its precursor ligand and the standard drug benznidazole with a selective index SI (trypomastigote/amastigote) higher than 200 [23]. Some of these complexes, however, despite showing a suppressive effect on the parasitemia, were not curative, since there are several reasons contributing to the incurability of the disease.

Therefore, the list of complexes tested in the development of new antiparasitic drugs against trypanosomiasis includes Co(II), or Cu(II) with triazole derivatives [24], Pt(II) or Pd(II) with thiosemicarbazones [25], Ru(II) with lapachol [26,27] or thiosemicarbazones [28], and vanadium with polypyridyl ligands [29]. Those studies are based on the action of such complexes toward different targets, such as cysteine proteases [30], hypoxanthine−guanine phosphoribosyl-transferases (HGPRTs) [31], and DNA [32].

Some recent studies reported the survival of *T. cruzi* exposed to benznidazole (BZ), using genetically modified parasites that overexpress different DNA repair proteins [33]. These investigations indicated that this drug causes double-stranded DNA breaks in the parasite, reinforcing its mechanism of action by reactive oxygen species (ROS) formation, particularly hydroxyl radical [34]. The importance of ROS in *T. cruzi* infections has been emphasized by showing that high levels of ROS are deleterious to the parasite. However, when ROS production was inhibited in the host cell, a significant reduced proliferation of wild-type parasites was also reported [35]. Further, overexpression of mitochondrial DNA repair proteins increases parasite survival upon exposure to benznidazole, indicating that mitochondrial DNA is also a target. More recent studies [36] described an intimate relation between the parasite and the host protein U2AF35 that binds to RNA at the polypyrimidine tract [37], and is essential for initiating the RNA processing, significantly affecting the host cell in their functions.

Since our group has developed some oxindolimine–metal complexes capable of generating ROS and showing significant antitumor properties [38], based on oxidative damage to DNA and mitochondria, besides their inhibition of selected proteins, we decided to test them as potential antiparasitic agents. Previously, some oxindolimine-copper, zinc, and vanadyl complexes were tested against *Schistosoma mansoni* worms. Copper(II) complexes showed 50% inhibitory concentrations of 30 to 45 µM, and demonstrated greater antischistosomal properties than the analogous zinc and vanadyl complexes regarding lethality, reduction in motor activity, and oviposition [39]. Analogous zinc complexes were active after 72 h treatment, and vanadyl complexes were inactive up to 500 µM, even if they quite inhibited oviposition. Results showed that both copper and zinc easily cross the cell membrane and induce severe tegumental damage in schistosomes.

Herein, the reactivity of four such oxindolimine compounds, metalated with copper(II) or zinc(II), against the trypomastigote and amastigote forms of *T. cruzi* is reported.

2. Experimental Section
2.1. Synthesis of the Ligands

Two different oxindolimine ligands were previously prepared by condensation reaction of 2,3-dioxindole with 1,3-diaminopropane (isapn), or 2-(2-aminoethyl) pyridine (isaepy). *Briefly*, 1.47 g isatin (10 mmol) was dissolved in 40 mL ethanol, in a 125 mL flask. To this solution, 420 µL (5 mmol) 1,3-diaminopropane was added, adjusting the final pH to 5 with a few drops of 0.1 mol/L HCl solution. The reaction solution was maintained under stirring for 6 h, until yellow crystals of the isapn ligand were formed. The precipitate was filtered, washed with cold ethanol and ethyl ether, and stored in a desiccator under reduced pressure. Yield: 79%. The other ligand isaepy was analogously prepared; yield 72%. Analytical data: for isapn, yellow powder, 68.66%C, 4.85 %H, 16.86 %N; Calc. for $C_{19}H_{16}N_4O_2$, 67.64%C, 4.72 %H, 16.52 %N; MS (ESI+): m/z = 333.1, [M + 1]+ in CH_3OH, MW = 332.36 g/mol for $C_{19}H_{16}N_4O_2$; for *isaepy*, yellow powder, 71.70%C, 5.21 %H, 16.72 %N; Calc. for $C_{15}H_{13}N_3O$, 71.17%C, 5.32 %H, 16.45 %N.

2.2. Syntheses of the Metal Complexes

The corresponding metal complexes, [Cu(isapn)](ClO$_4$)$_2$ **1**, [Zn(isapn)]ClO$_4$ **2** [Cu(isaepy)H$_2$O]ClO$_4$ **3**, [Zn(isaepy)Cl$_2$] **4**, and [Cu(isaepy)$_2$](ClO$_4$)$_2$ **5** (shown in Figure 1), have been prepared by metalation in situ of these ligands with suitable metal salts, as reported in previous studies [40,41]. According to the pH adjusted at the metalation step, the keto or the enol form was preferentially obtained, although in a solution at pH 7.4 both forms are detected, as indicated by mass spectrometry measurements. The crystals formed were filtered and washed with a few mL ethanol and ethyl ether, and afterwards dried under suction. The corresponding products (yields in the range of 65 to 90%) were stored in a desiccator under reduced pressure. They were identified by UV/Vis, IR spectroscopies, and mass spectrometry, in addition to elemental analyses. The copper(II) complexes were further characterized using EPR, and the analogous zinc(II) using NMR spectroscopy. Analytical data: Complex **1** [Cu(isapn)](ClO$_4$)$_2$, brown crystals, 85% yield, MW 594.80 g/mol. Experim. data: 39.33%C, 2.88%H, 9.69%N, 10.58%Cu; Calc. for $C_{19}H_{16}N_4O_{10}Cl_2Cu$, 38.36%C, 2.71 %H, 9.42%N, 10.72%Cu; MS (ESI+): m/z found: 395.02 (calcd.: 395.07, for $C_{19}H_{16}N_4O_2Cu$). Complex **2** [Zn(isapn)](ClO$_4$), orange solid, 85% yield—Experim. data: 45.71%C, 3.48%H, 11.31%N; Calc. for $C_{19}H_{15}N_4O_6ClZn$, 45.99%C, 3.05%H, 11.29%N; MS (ESI+) at pH 7: m/z = 396.9, in CH_3OH–H_2O, MW = 396.05, fragment monocation [$C_{19}H_{16}N_4O_2$ ^{64}Zn]; 398.9 [MW = 398.05 for the isotopic pattern of the (^{66}Zn) monocation] [38]. Complex **3** [Cu(isaepy)(H$_2$O)]ClO$_4$, Yield 66%, MW 449.31 g/mol. Experim. data: 40.10%C, 3.59%H, 9.35%N; Calc. for $C_{15}H_{16}N_3O_7ClCu$, 40.73%C, 3.41 %H,

9.18%N. MS (ESI+): m/z = 314.1 [MW = 431.29, in CH_3OH/H_2O, fragment $C_{15}H_{12}N_3OCu$]; 316.1 [isotopic pattern ($Cu^{63/65}$) monocation]; 564.1 [keto-form, fragment $C_{30}H_{24}N_6O_2Cu$ (compound **5**, MW = 566.11)]; 566.2 [isotopic pattern ($Cu^{63/65}$) monocation]. Complex **4** [Zn(isaepy)Cl$_2$], orange solid, yield 90%, MW 387.56 g/mol. Experim. 46.49%C, 3.38%H, 10.84%N; Calc. for $C_{15}H_{13}N_3OCl_2Zn$, 46.45%C, 3.41%H, 11.01%N. MS (ESI$^+$): m/z = 350.0 MW = 387.58, in CH_3OH/H_2O, monocation fragment [M+1]$^+$; [Zn(isaepy)(H$_2$O)]$^+$·H$_2$O, [$C_{15}H_{16}N_3O_3^{64}$Zn], and 352.0 [MW = 352.04 g/mol [$C_{15}H_{16}N_3O_3^{66}$Zn]. Complex **5** [Cu(isaepy)$_2$](ClO$_4$)$_2$·2 H$_2$O, brown crystals, yield 86%, MW 765.02 g. Experim. C, 45.69; H, 3.52; N, 10.08%. Calc. for $C_{30}H_{26}N_6O_2Cu(ClO_4)_2$·2H$_2$O: C, 44.98%C; H, 3.76 %H; 10.45 %N. MS (ESI+): m/z = 565.12 [MW = 565.15, in CH$_3$CN, fragment $C_{30}H_{26}N_6O_2Cu$]; 563.12 [isotopic pattern ($Cu^{63/65}$) monocation]; 312.02 [MW = 312.81, in CH$_3$CN, fragment $C_{15}H_{11}N_3OCu$] [42].

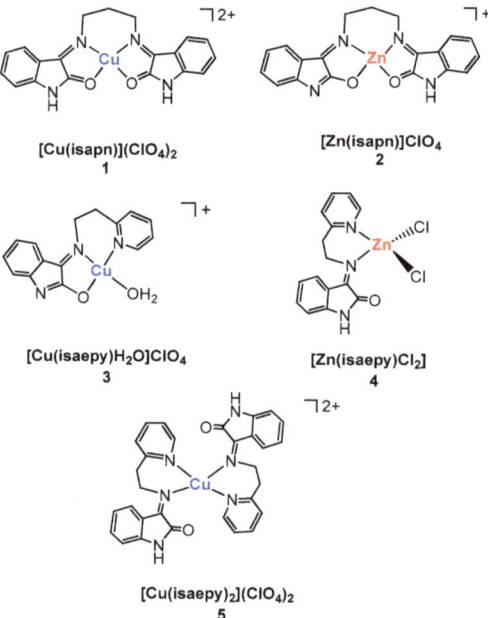

Figure 1. Structures of the oxindolimine–metal complexes, as isolated in solid state.

2.3. Materials and Methods

Most of the reagents used in the syntheses of the metal complexes were purchased from Merck or Sigma-Aldrich Co. Elemental analyses using a 2400 CNH Elemental Analyzer (Perkin-Elmer, Billerica, MA, USA), or metal analyses (ICP-OES, Spectro Arcos, Spectro/AMETEK, Kleve, Germany), and NMR spectra, using a DRX-500 instrument (from Bruker, Karlsruhe, Germany), operating at 500 MHz, were performed at the *Central Analítica* of our Institution (Facility Center, https://www.iq.usp.br/portaliqusp/?q=en/services/analytical-center, accessed on 29 September 2023). IR spectra were recorded in a BOMEM 3.0 (diffuse reflectance) instrument (Quebec, QC, Canada), in the range of 4000–400 cm^{-1}, while UV/Vis spectra were recorded in an UV-1650PC equipment from Shimadzu (Kyoto, Japan). EPR spectra were registered using an EMX spectrometer from Bruker Instruments (Karlsruhe, Germany), operating at X-band (9.5 GHz), 100 kHz modulation frequency, and 20.0 mW power, using standard Wilmad quartz tubes and quartz Dewar (Vineland, NJ, USA). DPPH (α, α′–diphenyl-β-picrylhydrazyl) was used as the frequency calibrant (g = 2.0036), with samples as frozen CH$_3$OH or CH$_3$CH$_2$OH/H$_2$O (4:1) solutions, at 77 K. A modulation amplitude of 15 G, and 3.56 × 10^2 receiver gain were usually employed. Simulation and analyses of spectra were provided by the EasySpin 5.2.35 software package [42], in a MatLab environment.

2.4. Cells and Parasites

The studied *T. cruzi* parasites were from Y strain, classified as TcII among the six discrete typing unit (DTU) groups [43]. Mouse peritoneal macrophages were used in MTT assays. Biological tests of the viability of different forms of the parasites, in the presence of the metal complexes or the free ligands, were carried out by using a Neubauer chamber or by MTT assays, as described next. All experiments were carried out in triplicate. Original LLC-MK2 cell lines (Macaca mulatta, code 0146, purchased from Banco de Células do Rio de Janeiro, RJ, Brazil—https://bcrj.org.br/accessed on 29 September 2023) in DME medium, supplemented by 10% fetal bovine serum at 37 °C, and 5% CO_2, were used as control and as cells to be infected by the parasites. Graphical treatments and statistical analyses were performed with the GraphPad Prism version 5.0 or 8.0.

2.5. Viability Test in Mouse Peritoneal Macrophages (MTT Assay)

C57BL/6 mice were stimulated with 3% thioglycolate, three days before obtaining the peritoneal macrophages. To verify the effect of the compounds on the viability of macrophages, an assay with MTT [3-(4,5-dimethyl-2-thiazol-2-yl)-2,5-diphenyl-2H-tetrazolium bromide] was performed [44,45]. A total of 2×10^5 cells were plated in each well of a 96-well culture plate, and after adherence of the cells, 200 µL of DME (Dulbecco's modified Eagle's) medium, supplemented with 10% fetal bovine serum, was added, containing different concentrations of the compounds. Then, the culture plate was incubated for a period of 24 or 48 h in an oven at 37 °C, with 5% CO_2. Subsequently, 22 µL of MTT (5 mg/mL) was added to each well and the plate was incubated for an additional 4 h. After this incubation period, the supernatant was removed from each well and 80 µL of DMSO was added; 5 min later, the absorbance of each well in the plate was measured in a spectrophotometer at 492 nm. The protocol was approved by the UFMG ethical committee (CEUA 2/2018).

2.6. Effect of Compounds on the Trypomastigote Forms of T. cruzi

Trypomastigotes were obtained by infection of LLC-MK2 cell lines (*Macaca mulatta*) in DME medium, that were also used as control, supplemented with 10% fetal bovine serum at 37 °C, and 5% CO_2, as previously described [46]. Five days after infection, trypomastigotes released into the medium were collected, washed with PBS by centrifugation, pelleted at $10,000\times$ rpm for 10 min, and re-suspended to adequate cell density of the experiment in DME medium supplemented with 10% fetal bovine serum at 37 °C for 24 or 48 h, in the presence of aliquots of different concentrations of each copper or zinc compound, solubilized in aqueous solution containing 1% DMSO. To each well of a 96-well culture plate, 100 µL of culture medium containing 2×10^5 parasites in the trypomastigote form and 100 µL of the complex solutions in different concentrations were added. After an incubation period of 24 or 48 h, the viability of *T. cruzi* trypomastigotes was assessed by verifying the mobility of the parasites using optical microscopy, through counting in a Neubauer chamber.

2.7. Evaluation of the Effect of Compounds on Amastigote Forms of T. cruzi

To perform this test, 5×10^5 peritoneal macrophages were plated on coverslips in 24-well plates and incubated at 37 °C in a 5% CO_2 oven for 2 h. Subsequently, *T. cruzi* trypomastigotes were added in a 5:1 ratio (parasite/macrophage) and the cultures were incubated for another 2 h. After incubation, the wells were washed three times with RPMI medium, and supplemented with 10% fetal bovine serum at 37 °C to remove free parasites. Then, 1 mL of different concentrations of the compounds of copper (10–25 µmol/L) and zinc (60–155 µmol/L) in 1% DMSO aqueous solution was added, and the plate incubated for 24 h or 48 h. We also used benznidazole as a control in our tests. After the incubation period, the coverslips were fixed, using the fast panoptic staining kit, for later verification of the inhibitory activity of the compounds against the parasite. To determine the inhibitory activity, the amastigote-infected macrophages and uninfected macrophages were counted, totaling 300 macrophages per count.

2.8. MTT Assay with Trypomastigote Parasites

T. cruzi trypomastigotes, Y strain, were maintained by infection in LLC-MK2 cells as described above (item 2.6). Approximately 1×10^7 trypomastigotes were added to 16 wells of a 24-well plate, to which the compounds studied were added at concentrations ranging from 10 to 100 µM at first, and 1 to 10 µM depending on the results of the initial screening. Two of the sixteen wells containing trypomastigotes were used as control for the DMSO concentration at 1%, used for the solubilization of the compounds, and two other wells were used as control for the trypomastigotes' viability.

The trypomastigotes were incubated at 37 °C and 5% CO_2. After 24 h incubation, biological activity was measured using a colorimetric MTT assay (3-(4,5-dimethylthiazol-2-yl)-2,5-diphenyltetrazolium bromide; 2.5 mg/mL). Readings were conducted in a Tecan Infinite F200 microplate reader at a wavelength of 565 nm. Assays were performed at least in triplicate. The absorbance values of wells containing only medium and reagents were used as blank for this assay.

3. Results and Discussion

3.1. Characterization and Stability of the Metal Complexes

The oxindolimine complexes investigated in our studies are very stable, and can be isolated in different tautomeric forms, depending on the pH adjusted during the metalation step in their syntheses (see Figure 2). Both species, the keto and the enol forms, co-exist in solution, depending on the pH, as detected by EPR spectra in the case of copper(II) species [40], or NMR spectra for the zinc(II) analogs [41], corroborated by ESI-MS data.

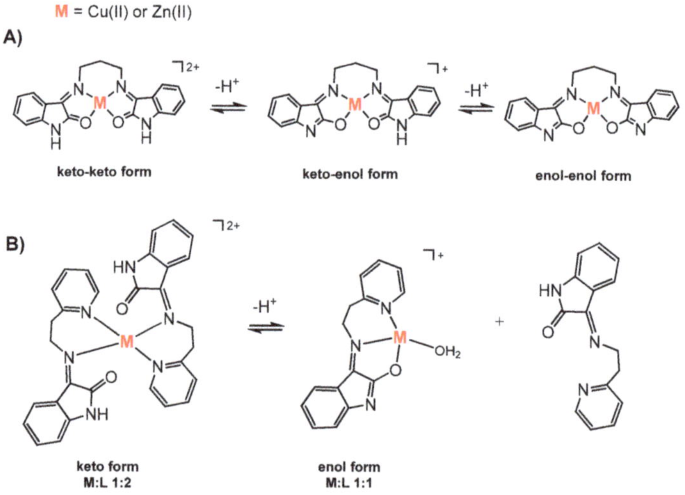

Figure 2. Tautomeric equilibria, depending on the pH, verified with copper(II) or zinc(II) oxindolimine complexes: (**A**) with the *isapn* ligand, and (**B**) with the *isaepy* ligand.

In the case of the copper complexes **1**, **3**, and **5**, the corresponding EPR hyperfine structure parameters (see Table 1) have been already reported [47], showing that the determined values for the $g_{//}/A_{//}$ ratio are much lower for the enol forms (around 120 cm) than for the corresponding keto forms (around 190 cm), indicating a more tetragonal or planar geometry. This $g_{//}/A_{//}$ ratio is frequently used to estimate the tetrahedral distortion around a copper ion in a tetragonal environment [48].

Table 1. Determined values of spectroscopic parameters (g) and hyperfine constants (A) for oxindolimine–copper(II) complexes [#].

Complexes at Different pHs	$g_\perp$	$g_{//}$	$A_{//}$, G	$A_{//}$,* 10^{-4}cm^{-1}	$g_{//}/A_{//}$ cm
[Cu(isaepy)H$_2$O] pH = 3, keto-form	2.086		112	127	191
pH = 7	2.058	2.246	173	181	124
pH = 10, enol-form	2.059	2.252	177	186	121
[Cu(isapn)] pH = 4, keto-keto form	2.101	2.445	115	131	187
pH = 7, keto-enol form	2.112	2.256	186	196	115
pH = 10, enol-enol form	2.092	2.262	184	194	116

* $A_{//}(10^{-4}$ cm$^{-1}) = g_{//} \beta\, A_{//}(G) = 0.46686 \times 10^{-4}\, g_{//} A_{//}(G)$, where $\beta = 1.39969$ MHz/G; [#] results adapted and simplified from ref. [49].

^{1}H NMR spectra of the zinc(II) complex, [Zn(isapn)]ClO$_4$ **2** in MeOH-d4, at different pHs in the range of 5 to 9 also indicated the presence of tautomeric equilibria, due to the deprotonation of the NH group at indole ring, as previously verified [42]. Similar results were observed for [Zn(isaepy)Cl$_2$] complex **4**. In this case, three species were detected by ESI-MS data, as shown in Figure S1, corroborated by NMR spectra in Figure S2, in the Supplementary Material.

A signal corresponding to the free ligand *isaepy* occurred at $m/z = 252.1$ [M + 1]$^+$, [MW = 251.29], attributed to fragment [C$_{15}$H$_{13}$N$_3$O]; another fragment at $m/z = 350.0$ corresponding to the monocation [Zn(isaepy)(H$_2$O)]$^+ \cdot$H$_2$O or [C$_{15}$H$_{16}$N$_3$O$_3$ ^{64}Zn] and 352.0, [MW = 352.04], for [C$_{15}$H$_{16}$N$_3$O$_3$ ^{66}Zn]. Finally, a species Zn:L 1:2 was also verified, since a fragment at $m/z = 564.9$, [MW = 564.12], was assigned to the monocation [Zn(isaepy)$_2$]$^+$ [C$_{30}$H$_{24}$N$_6$O$_2$ ^{64}Zn] and $m/z = 566.9$ [MW = 566.12 for [C$_{30}$H$_{24}$N$_6$O$_2$ ^{66}Zn].

At physiological conditions (pH 7), the keto-enol form seems to be predominant over the corresponding keto-keto or enol-enol forms for both copper(II) and zinc(II) complexes with the ligand *isapn*, as indicated by EPR or NMR data, respectively. For complex [Cu(isaepy)H$_2$O]$^+$ **3**, the enol form, more planar, with *isaepy* acting as a tridentate ligand, is predominant. For complexes [Zn(isaepy)]$^+$ **4**, and [Cu(isaepy)$_2$]$^{2+}$ **5**, although isolated as keto forms in solution at pH 7, the enol form seem to be dominant.

All these metal complexes have already shown high stability in solution [48], with relative stability constants of the same order as those of copper(II) or zinc(II) ions inserted in human serum albumin, for which log K$_{[Cu(has)]}$ = 12.0 for copper [50], or log K$_{[hasHSA)]}$ = 7.2 for zinc, have been reported [51].

3.2. Evaluation of Trypanocidal Activity

Firstly, complexes **1**, **2**, **4**, and **5** had their toxicity verified against trypomastigote forms of *T. cruzi*, as shown in Figure 3. The corresponding IC$_{50}$ results, corresponding to 50% inhibition of the parasites' viability, are displayed in Table 2.

Table 2. IC$_{50}$ values for trypomastigote forms of *T. cruzi* viability (IC$_{50}$ ± SD *) after 24 or 48 h incubation with the oxindolimine–metal complexes, at 37 °C.

IC$_{50}$ µmol/L	[Cu(isapn)] (ClO$_4$)$_2$	[Cu(isaepy)$_2$] (ClO$_4$)$_2$	[Zn(isapn)] ClO$_4$	[Zn(isaepy)Cl$_2$]
24 h	15.5 ± 5.5	10.7 ± 3.8	32.9 ± 14.1	80.2 ± 52.6
48 h	2.7 ± 1.0	3.0 ± 1.0	11.3 ± 3.6	56.2 ± 23.0

* SD = standard deviation.

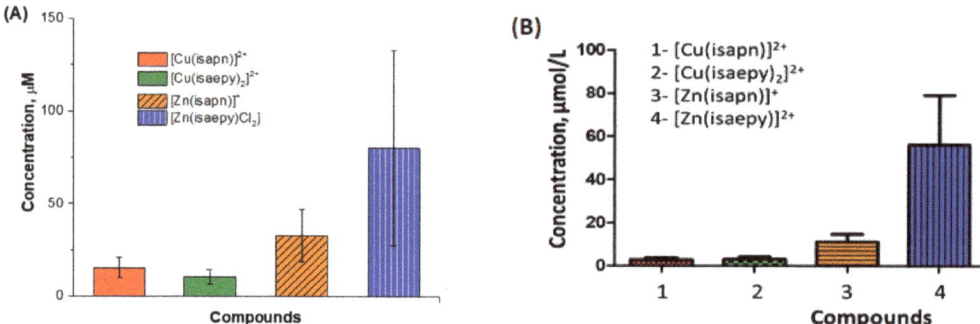

Figure 3. IC$_{50}$ values for trypomastigote forms of *T. cruzi* viability, after (**A**) 24 h and (**B**) 48 h incubation with the oxindolimine–metal complexes.

In parallel experiments, the toxicity of such complexes toward macrophages was also determined, as shown in Figure 4 and Table 3.

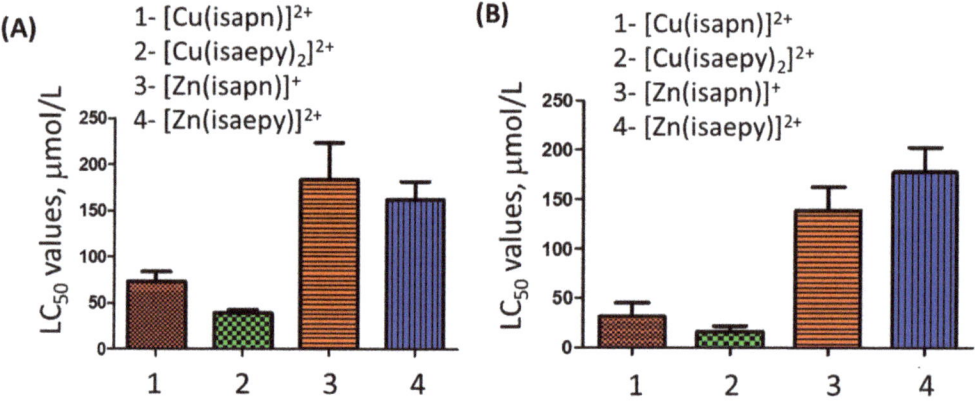

Figure 4. Lethal concentrations to 50% of non-infected macrophages (LC$_{50}$) were compared after (**A**) 24 h and (**B**) 48 h incubation with the metal complexes at 37 °C.

Table 3. Values of lethal concentration to 50% of non-infected macrophages (LC$_{50}$ ± SD *) incubated with these complexes for 24 h or 48 h at 37 °C.

LC$_{50}$ µmol/L	[Cu(isapn)](ClO$_4$)$_2$	[Cu(isaepy)$_2$](ClO$_4$)$_2$	[Zn(isapn)]ClO$_4$	[Zn(isaepy)Cl$_2$]
24 h	73.3 ± 10.4	39.1 ± 3.5	183.8 ± 39.9	162.8 ± 18.8
48 h	31.3 ± 14.0	16.2 ± 5.2	138.9 ± 23.8	177.8 ± 25.0

* SD = standard deviation.

Further, the toxicity of such metal complexes against amastigotes in macrophages of mouse C57BL/6 was also verified. The results are shown in Figure 5 (copper complexes) and Figure 6 (zinc complexes).

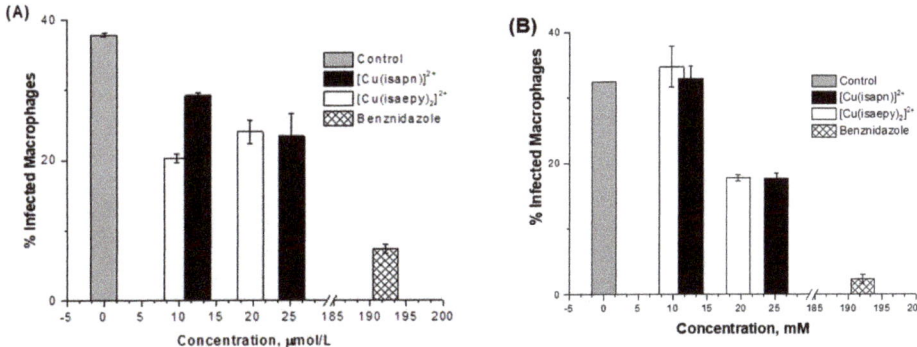

Figure 5. Percentage of infected macrophages after (**A**) 24 h and (**B**) 48 h incubation at 37 °C with copper(II) complexes **1** and **5**, in comparison with benznidazole.

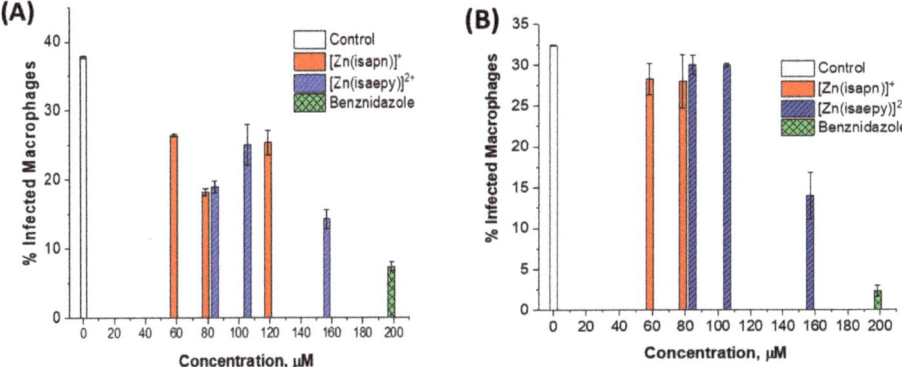

Figure 6. Percentage of infected macrophages after (**A**) 24 h and (**B**) 48 h incubation at 37 °C with zinc(II) complexes **2** and **4**, in comparison with benznidazole.

In Table 4, the results and the corresponding selective indexes, macrophages versus trypomastigotes, verified with these complexes are compared to benznidazole.

Table 4. IC$_{50}$ values (μM) and corresponding selective indexes (SI) verified against infected macrophages and trypomastigotes, in the presence of the studied metal complexes, after an incubation period of 24 or 48 h, at 37 °C, in comparison to benznidazole.

Complex	IC$_{50}$ (μM) after 24 h Incubation			IC$_{50}$ (μM) after 48 h Incubation		
	Macrophages	Trypomastigotes	S.I.	Macrophages	Trypomastigotes	S.I.
[Cu(isapn)] (ClO$_4$)$_2$ **1**	73.3 ± 10.4	15.5 ± 5.5	4.8	31.3 ± 14.0	2.7 ± 1.0	11.6
[Zn(isapn)] ClO$_4$ **2**	183.8 ± 39.9	32.9 ± 14.1	5.6	138.9 ± 23.8	11.3 ± 3.6	12.4
[Cu(isaepy)$_2$] (ClO$_4$)$_2$ **5**	39.1 ± 3.5	10.7 ± 3.8	3.7	16.2 ± 5.2	3.0 ± 1.0	5.4
[Zn(isaepy)Cl$_2$] **4**	162.8 ± 18,8	80.2 ± 52.6	2.0	177.8 ± 25	56.2 ± 23	3.2
Benznidazole [#]		30.3 ± 2.83	2.7			

[#] from ref. [52].

All those metal complexes showed to be efficient toward the trypomastigote forms of parasites, and more active than the free ligands (see Figure 7). Particularly, complexes **1** and **2** were the most active in the series. Copper(II) complexes were more active than the corresponding zinc(II) ones with the same ligand, and the ability of copper compounds to generate ROS is probably responsible for their better performance.

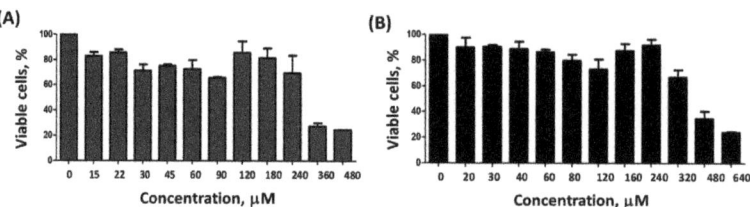

Figure 7. Viability of infected macrophages after 48 h incubation with the free ligands (**A**) isapn, and (**B**) isaepy.

Many studies in the literature, however, reported values against the epimastigote forms of the parasite. In Table 5, some already reported data for other metal complexes toward trypomastigotes are displayed, for comparison. In many of them, the incubation times are longer than 24 or 48 h.

Table 5. IC$_{50}$ values (µM) and corresponding selective indexes reported in the literature for different metal complexes, at different incubation times (h), against the trypomastigote (or epimastigote [#]) forms of the parasite.

Complexes	IC$_{50}$ (µM) Trypomastigotes	Selective Index (S.I.)	Incubation Time
[Cu(dmtp)$_4$(H$_2$O)$_2$] (ClO$_4$)$_2$ dmtp = 5,7-dimethyl-1,2,4-triazolo[1,5-a]pyrimidine	25.4 ± 2.3	16.5	72 h [a]
[Zn(dmtp)$_2$(H$_2$O)$_4$] (ClO$_4$)$_2$	19.2 ± 1.1	3.8	72 h [a]
[Cu(4-MH)(dmb)(ClO$_4$)$_2$]·2H$_2$O 4-MH = 4-methoxybenzhydrazide; dmb = 4-4′-dimethoxy-2-2′-bipyridine	14.0	12.9	72 h [b]
trans-[Ru(tzdt)(PPh$_3$)$_2$(bipy)]PF$_6$ tzdtH = 1,3-thiazolidine-2-thione	0.01	34	24 h [c]
[Au^{III}(Hdamp)(L1)]NO$_3$ (4-NO$_3$) Hdamp = dimethylaminomethylphenyl	16.9	5.1	48 h [d]
[Pt(HL1)(L1)]Cl [#] L1 = thiosemicarbazone derivative of 1-indanone	(8.7)	(8.8)	120 h [e] (Epimastigote form)
[Pd(HL2)(L2)]Cl [#] L2 = thiosemicarbazone derivative of 1-indanone	(2.3)	(9.5)	120 h [e] (Epimastigote form)

[a] ref. [53]; [b] ref. [54]; [c] ref. [55]; [d] ref. [56]; [#] epimastigote form, [e] ref. [25].

These metal complexes were shown to be quite toxic toward both macrophages and trypomastigotes, with good selectivity indexes (S.I.), although after longer times of incubation. Among our oxindolimine complexes, the two derivatives of the *isapn* ligand, complexes **1** and **2**, stand out, because despite presenting less efficiency with higher IC$_{50}$ values, they showed better selectivity indexes after shorter times of incubation.

In further studies, the macrophages viability was also tested in the presence of free ligands isapn and isaepy, as shown in Figure 7.

The cytotoxicity verified in the presence of the free ligands were similar, in the range of 80 to 160 µg/mL, or 320 to 640 µmol/mL for ligand isaepy and 240 to 480 µmol/mL for ligand isapn, respectively, much lower than that of the corresponding zinc(II) or copper(II) complexes. In Figure 8, the corresponding infection indexes verified after 24 h incubation of macrophages with the free ligands are displayed.

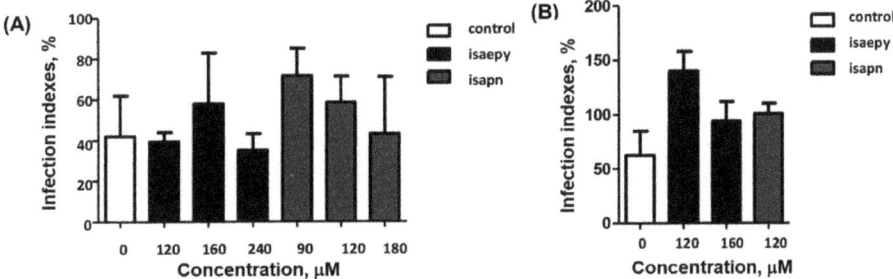

Figure 8. Infection indexes after (**A**) 24 h or (**B**) 48 h incubation of the infected macrophages with the free ligands isapn and isaepy at 37 °C.

These results are shown in Table 6.

Table 6. Comparison of IC_{50} values at 24 h assays, between two different techniques to measure the trypomastigotes' viability toward the studied copper(II) compounds.

Complexes	Trypomastigotes (Neubauer Chamber)	Trypomastigotes (MTT)	Correlation (Neubauer Chamber/MTT)
[Cu(isapn)](ClO$_4$)$_2$	15.5 ± 5.5 µM	6.11 ± 0.44 µM	2.54
[Cu(isaepy)$_2$](ClO$_4$)$_2$	10.7 ± 3.8 µM		
[Cu(isaepy)H$_2$O]ClO$_4$		1.37 ± 0.12 µM	7.81

We can see that the IC_{50} values obtained by the MTT technique were lower than those by reading in the Neubauer chamber, in a ratio that ranged from 2.5× for complex **1** to 7.8× for complex **3** or **5**. This discrepancy can be explained by the difference between the techniques. The optical microscopy with Neubauer chamber counting depends more on the visual acuity of the operator [57], and even if the counting is done meticulously, it is subject to more errors depending on the operator than the MTT method, which carries out the spectrometric reading of the sample operator-independent. Despite this, MTT is not free from errors, requiring extreme care to eliminate influences from the culture medium and possible absorbance of the complexes. The trypomastigotes do not adhere to the well wall, making it impossible to wash them before reading as in the viability test for the macrophage, and requiring the use of controls to reduce the influence of the absorbance of the medium in the measurements.

The oxindolimine–metal complexes also showed toxicity toward the amastigote forms of *T. cruzi*, as shown in Figure 9. The copper complexes were more active (up to 20 µM) than the analogous zinc ones (60 to 120 µM), although less active than benznidazole (160 µM).

Although the generation of ROS can provide a good explanation for the better activity of the copper compounds in comparison to zinc, selected parasite proteins can also be important targets for such metal complexes. Both copper and zinc compounds showed high selective indexes. Kinases have been reported as potential targets for trypanocidal drugs, since there are ~190 protein kinases encoded for *T. cruzi* genomes [58]. Another target ubiquitously studied in the literature is the parasite cruzain protein, essential for the development and survival of the parasite within the host cells [59].

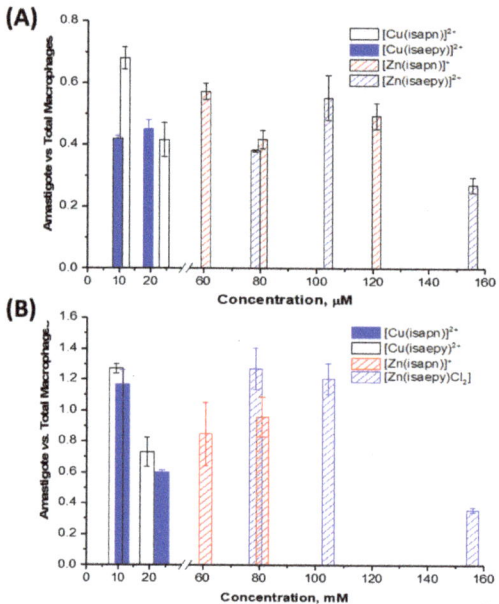

Figure 9. Toxicities of oxindolimine–copper(II) and –zinc(II) complexes toward the amastigote forms of *T. cruzi*, after incubation for (**A**) 24 h and (**B**) 48 h at 37 °C.

4. Conclusions

All metal complexes reported here showed good activity against the trypomastigote and amastigote forms of *T. cruzi*. For both parasite forms found in humans, the determined IC_{50} values in the presence of the copper(II) complexes are lower than for the analogous zinc(II) compounds, attesting the better activity of copper compounds. However, the estimated selective indexes are better for both copper and zinc compounds with the ligand isapn. In comparison to other metal complexes described in the literature (see Tables 4 and 5), their selective indexes were more favorable after only 24 or 48 h incubation. Further, they were more reactive against protozoa *T. cruzi* than toward *Schistosoma* worms, and their modes of action probably differ in both cases.

Those complexes are very stable thermodynamically, with formation constants of the same order as a copper ion inserted in the N-terminal site of human albumin (log $K_{Cu(HSA)}$ = 12.0) or zinc ion inserted in this protein (log $K_{Zn(HSA)}$ = 7.1), as already reported elsewhere [48]. Further, the active tautomeric forms of these complexes at physiological pH 7.4 are probably the enol ones, corresponding to the deprotonation of the NH group at the indole ring, as demonstrated by the EPR hyperfine parameters, in the case of the copper complexes (Figure 2 and Table 1), or by the NMR spectra, in the case of the zinc complexes [35]. More planar or tetragonal species seem to be more active toward the parasites. The compounds with the isapn ligand were more reactive than the analogous ones with the isaepy ligand, in both copper and zinc complexes.

Additionally, the copper(II) complexes were shown in previous studies to have significant oxidant properties, being able to damage DNA and HSA, in the presence of hydrogen peroxide, with the formation of hydroxyl radicals. Therefore, an explanation for its antiparasitic activity could be the induction of oxidative stress, damaging membranes, and vital molecules in the parasite, through ROS formation [38]. However, previous results also revealed that interactions of such oxindolimine complexes with specific proteins are in the same way determinant of their biological activity. The same order of reactivity has been demonstrated for such metal complexes regarding the inhibition of topoisomerase IB protein [41]. Also in these studies, the [Cu(isapn)] complex **1** was a more active inhibitor than

[Cu(isaepy)] complex **3** or **5**, and the copper species with each of these ligands were more efficient than the corresponding zinc species. Nevertheless, further studies are necessary to elucidate the probable modes of action of these oxindolimine–metal complexes, and to provide other new and more efficient compounds as trypanocydal agents, based on our results. Further mechanistic studies are in progress in our laboratory to identify probable parasite targets.

5. Patents

The University of São Paulo has filed patent applications (AUCANI—USP Innovation Agency) related to the antiparasitic activity of the oxindolimine–metal complexes under study in our laboratory (INPI, **BR 10 2013 026558 6**). This patent has not been conceded yet, it is still under evaluation. A related Brazilian patent, regarding anticancer activity of this class of metal complexes, was conceded on 24 March 2020 (**BR 2006 00985-A**).

Supplementary Materials: The following supporting information can be downloaded at: https://www.mdpi.com/article/10.3390/inorganics11110420/s1, Figure S1: Mass spectrogram of complex [Zn(isaepy)Cl$_2$] **4** in methanol:water (9:1) solution; Figure S2. ^{1}H NMR spectra of (A) isaepy free ligand, and (B) complex **4** [Zn(isaepy)Cl$_2$] in D$_2$O.

Author Contributions: A detailed description of the diverse contributions of the co-authors to the published work. G.L.S.: investigation, formal analysis, validation, and writing—original draft preparation. M.C.P.: methodology, investigation, formal analysis, validation, and writing—original draft preparation. G.A.R.: investigation, methodology, and formal analysis. R.A.A.D.C.: methodology, investigation, formal analysis, and visualization. L.Q.V.: conceptualization, investigation, writing—review and editing, and resources. M.J.M.A.: conceptualization, investigation, resources, and writing—review and editing. A.M.D.C.F.: conceptualization, methodology, supervision, writing—review and editing, and funding acquisition. All authors have read and agreed to the published version of the manuscript.

Funding: This research was funded by Brazilian entities: FAPESP—São Paulo State Research Foundation (grants 2011/50318-1, and 21/10572-8); CEPID-Redoxoma Project (FAPESP, grant 2013/07937-8); CAPES—Coordenação de Aperfeiçoamento de Pessoal de Nível Superior (grant 1457853); CNPq—Conselho Nacional de Desenvolvimento Científico e Tecnológico (grants 134508/2011-4; and 312954/2021-2) and FAPEMIG—Fundação de Amparo à Pesquisa do Estado de Minas Gerais (grant RED-00313-16).

Data Availability Statement: Data are available on request.

Acknowledgments: The authors are grateful to the Brazilian entities São Paulo State Research Foundation (FAPESP, grant 2011/50318-1), and CEPID-Redoxoma Project (FAPESP, grant 2013/07937-8) for their financial support. M.C.P. is grateful to Coordenação de Aperfeiçoamento de Pessoal de Nível Superior (CAPES, grant 1457853) for fellowships during his Ph.D. studies. G.L.S. thanks CNPq (grant 134508/2011-4) for his M.Sc. degree fellowships. L.Q.V. is a CNPq fellow, grant number 312954/2021-2, and this work was partially supported by FAPEMIG (grant RED-00313-16). M.J.M.A. is grateful to FAPESP (grant 21/10572-8).

Conflicts of Interest: The authors declare no conflict of interest.

Abbreviations

Cruzain	a recombinant form of protein cruzipain, EC 3.4.22.51
DPPH	α,α'-diphenyl-β-picrylhydrazyl
EPR	electron paramagnetic resonance
HGPRTs	hypoxanthine–guanine phosphoribosyl-transferases
isaepy	(E)-3-((2-(pyridin-2-yl)ethyl)imino)indolin-2-one; oxindolimine ligand obtained from isatin and 2-(2-aminoethyl)pyridine
isapn	(3E,3'E)-3,3'-(propane-1,3-diylbis(azaneylylidene)bis(indolin-2-one); oxindolimine ligand obtained from isatin and 1,3-diaminopropane
U2AF35	host protein that binds to RNA at the polypyrimidine tract

References

1. Available online: https://www.who.int/news-room/fact-sheets/detail/chagas-disease-(american-trypanosomiasis) (accessed on 7 August 2023).
2. Available online: https://dndi.org/diseases/chagas/ (accessed on 7 August 2023).
3. Conners, E.E.; Vinetz, J.M.; Weeks, J.R.; Brouwer, K.C. A global systematic review of Chagas disease prevalence among migrants. *Acta Tropica* **2016**, *156*, 68–78. [CrossRef]
4. Monge-Maillo, B.; Lopez-Velez, R. Challenges in the management of Chagas disease in Latin-American migrants in Europe. *Clin. Microbiol. Infect.* **2017**, *23*, 290–295. [CrossRef]
5. Bern, C.; Messenger, L.A.; Whitman, J.D.; Maguire, J.H. Chagas disease in the United States: A public health approach. *Clin. Microbiol. Rev.* **2020**, *33*, e00023-19. [CrossRef]
6. Trouiller, P.; Olliaro, P.; Torreele, E.; Orbinski, J.; Laing, R.; Ford, N. Drug development for neglected diseases: A deficient market and a public-health policy failure. *Lancet* **2002**, *359*, 2188–2194. [CrossRef]
7. Horn, D. A profile of research on the parasitic trypanosomatids and the diseases they cause. *PLoS Neg. Trop. Dis.* **2022**, *16*, e0010040. [CrossRef]
8. Parthasarathy, A.; Kalesh, K. Defeating the trypanosomatid trio: Proteomics of the protozoan parasites causing neglected tropical diseases. *RSC Med. Chem.* **2020**, *11*, 625–645. [CrossRef]
9. Barrett, M.P.; Burchmore, R.J.S.; Stich, A.; Lazzari, J.O.; Frasch, A.C.; Cazzulo, J.J.; Krishna, S. The trypanosomiases. *Lancet* **2003**, *362*, 1469–1480. [CrossRef]
10. Martín-Escolano, J.; Marín, C.; Rosales, M.J.; Tsaousis, A.D.; Medina-Carmona, E.; Martín-Escolano, R. An Updated View of the *Trypanosoma cruzi* Life Cycle: Intervention Points for an Effective Treatment. *ACS Infect. Dis.* **2022**, *8*, 1107–1115. [CrossRef]
11. Guedes, P.M.M.; Fietto, J.L.R.; Lana, M.; Bahia, M.T. Advances in Chagas Disease Chemotherapy. *Anti.-Infect. Ag. Med. Chem.* **2006**, *5*, 175–186. [CrossRef]
12. Duschak, V.G.; Couto, A.S. An insight on targets and patented drugs for chemotherapy of Chagas disease. *Recent Pat. Anti.-Infect. Drug Discov.* **2007**, *2*, 19–51. [CrossRef]
13. Njoroge, M.; Njuguna, N.M.; Mutai, P.; Ongarora, D.S.B.; Smith, P.W.; Chibale, K. Recent Approaches to Chemical Discovery and Development against Malaria and the Neglected Tropical Diseases Human African Trypanosomiasis and Schistosomiasis. *Chem. Rev.* **2014**, *114*, 11138–11163. [CrossRef] [PubMed]
14. Coura, J.R.; de Castro, S.L. A critical review on Chagas disease chemotherapy. *Mem. Inst. Oswaldo Cruz* **2002**, *97*, 3–24. [CrossRef] [PubMed]
15. Salas, C.O.; Faundez, M.; Morello, A.; Maya, J.D.; Tapia, R.A. Natural and Synthetic Naphthoquinones Active Against Trypanosoma Cruzi: An Initial Step Towards New Drugs for Chagas Disease. *Curr. Med. Chem.* **2011**, *18*, 144–161. [CrossRef] [PubMed]
16. Pinto, A.V.; de Castro, S.L. The Trypanocidal Activity of Naphthoquinones: A Review. *Molecules* **2009**, *14*, 4570–4590. [CrossRef] [PubMed]
17. Moreira, D.R.M.; Lima Leite, A.C.; Cardoso, M.V.O.; Srivastava, R.M.; Hernandes, M.Z.; Rabello, M.M.; da Cruz, L.F.; Ferreira, R.S.; de Simone, C.A.; Meira, C.S.; et al. Structural Design, Synthesis and Structure—Activity Relationships of Thiazolidinones with Enhanced Anti-*Trypanosoma cruzi* Activity. *ChemMedChem* **2014**, *9*, 177–188. [CrossRef]
18. de Moraes Gomes, P.A.T.; de Oliveira Barbosa, M.; Santiago, E.F.; de Oliveira Cardoso, M.V.; Costa, N.T.C.; Hernandes, M.Z.; Rabello, M.M.; da Cruz, L.F.; Fer-reira, R.S.; de Simone, C.A.; et al. New 1,3-thiazole derivatives and their biological and ultrastructural effects on *Trypanosoma cruzi*. *Eur. J. Med. Chem.* **2016**, *121*, 387–398. [CrossRef]
19. Bernatchez, J.A.; Kil, Y.-S.; da Silva, E.B.; Thomas, D.; McCall, L.I.; Wendt, K.L.; Souza, J.M.; Ackermann, J.; McKerrow, J.H.; Cichewicz, R.H.; et al. Identification of Leucinostatins from *Ophiocordyceps* sp. as Antiparasitic Agents against *Trypanosoma cruzi*. *ACS Omega* **2022**, *7*, 7675–7682. [CrossRef]
20. Gazzinelli, R.T.; Oswald, I.P.; Hieny, S.; James, S.L.; Sher, A. The microbicidal activity of interferon-gamma-treated macrophages against *Trypanosoma cruzi* involves an L-arginine-dependent, nitrogen oxide-mediated mechanism inhabitable by interleukin-10 and transforming growth factor-beta. *Eur. J. Immunol.* **1992**, *22*, 2501–2506. [CrossRef]
21. Alvarez, M.N.; Peluffo, G.; Piacenza, L.; Radi, R. Intraphagosomal Peroxynitrite as a Macrophage-derived Cytotoxin against Internalized *Trypanosoma cruzi*. *J. Biol. Chem.* **2011**, *286*, 6627–6640. [CrossRef]
22. Maldonado, C.R.; Marín, C.; Olmo, F.; Huertas, O.; Quirós, M.; Sánchez-Moreno, M.; Rosales, M.J.; Salas, J.M. In Vitro and in Vivo Trypanocidal Evaluation of Nickel Complexes with an Azapurine Derivative against *T. cruzi*. *J. Med. Chem.* **2010**, *53*, 6964–6972. [CrossRef]
23. Rettondin, A.R.; Carneiro, Z.A.; Gonçalves, A.C.R.; Ferreira, V.F.; Oliveira, C.G.; Lima, A.N.; Oliveira, R.J.; Albuquerque, S.; de Deflon, V.M.; Maia, P.I.; et al. Gold(III) complexes with ONS-Tridentate thiosemicarbazones: Toward selective trypanocidal drugs. *Eur. J. Med. Chem.* **2016**, *120*, 217–226. [CrossRef] [PubMed]
24. Juan, M.; Salas, J.M.; Ana, B.; Caballero, A.B.; Esteban-Parra, G.M.; Méndez-Arriaga, J.M. Leishmanicidal and Trypanocidal Activity of Metal Complexes with 1,2,4-Triazolo[1,5-a]pyrimidines: Insights on their Therapeutic Potential against Leishmaniasis and Chagas Disease. *Curr. Med. Chem.* **2017**, *24*, 2796–2806.

25. Santosa, D.; Parajón-Costa, B.; Rossi, M.; Caruso, F.; Benítez, D.; Varela, J.; Cerecetto, H.; González, M.; Gómez, N.; Caputto, M.E.; et al. Activity on *Trypanosoma cruzi*, erythrocytes lysis and biologically relevant physicochemical properties of Pd(II) and Pt(II) complexes of thiosemicarbazones derived from1-indanones. *J. Inorg. Biochem.* **2012**, *117*, 270–276. [CrossRef] [PubMed]
26. Navarro, M.; Gabbiani, C.; Messori, L.D.; Gambino, D. Metal-based drugs for malaria, trypanosomiasis and leishmaniasis: Recent achievements and perspectives. *Drug Discov. Today* **2010**, *15*, 1070–1078. [CrossRef]
27. Barbosa, M.I.F.; Corrêa, R.S.; de Oliveira, K.M.; Rodrigues, C.; Ellena, J.; Nascimento, O.R.; Rocha, V.P.C.; Nonato, F.R.; Macedo, T.S.; Barbosa-Filho, J.M.; et al. Antiparasitic activities of novel ruthenium/lapachol complexes. *J. Inorg. Biochem.* **2014**, *136*, 33–39. [CrossRef]
28. Sarniguet, C.; Toloza, J.; Cipriani, M.; Lapier, M.; Vieites, M.; Toledano-Magaña, Y.; Otero, L. Water-Soluble Ruthenium Complexes Bearing Activity Against Protozoan Parasites. *Biol. Trace Elem. Res.* **2014**, *159*, 379–392. [CrossRef]
29. Benítez, J.; Becco, L.; Correia, I.; Leal, S.M.; Guiset, H.; Costa Pessoa, J.; Lorenzo, J.; Tanco, S.; Escobar, P.; Moreno, V.; et al. Vanadium polypyridyl compounds as potential antiparasitic and antitumoral agents: New achievements. *J. Inorg. Biochem.* **2011**, *105*, 303–312. [CrossRef]
30. Fricker, S.P.; Mosi, R.M.; Cameron, B.R.; Baird, I.; Zhu, Y.; Anastassov, V.; Cox, J.; Doyle, P.S.; Hansell, E.; Lau, G.; et al. Metal compounds for the treatment of parasitic diseases. *J. Inorg. Biochem.* **2008**, *102*, 1839–1845. [CrossRef]
31. Glockzin, K.; Kostomiris, D.; Minnow, Y.V.T.; Suthagar, K.; Clinch, K.; Gai, S.; Buckler, J.N.; Schramm, V.L.; Tyler, P.C.; Meek, T.D. Kinetic Characterization, and Inhibition of *Trypanosoma cruzi* Hypoxanthine−Guanine Phosphoribosyltransferases. *Biochemistry* **2022**, *61*, 2088–2105. [CrossRef]
32. Sánchez-Delgado, R.A.; Anzellotti, A. Metal Complexes as Chemotherapeutic Agents Against Tropical Diseases: Trypanosomiasis, Malaria and Leishmaniasis. *Mini.-Rev. Med. Chem.* **2004**, *4*, 23–30.
33. Rajão, M.A.; Furtado, C.; Alves, C.L.; Passos-Silva, D.G.; de Moura, M.B.; Schamber-Reis, B.L.; Kunrath-Lima, M.; Zuma, A.A.; Vieira-da-Rocha, J.P.; Garcia, J.B.F.; et al. Unveiling Benznidazole's mechanism of action through overexpression of DNA repair proteins in *Trypanosoma cruzi*. *Environ. Mol. Mutagen.* **2014**, *55*, 309–321. [CrossRef] [PubMed]
34. Docampo, R. Sensitivity of parasites to free radical damage by antiparasitic drugs. *Chem.-Biol. Interact.* **1990**, *73*, 1–27. [CrossRef] [PubMed]
35. Goes, G.R.; Rocha, P.S.; Diniz, A.R.S.; Aguiar, P.H.N.; Machado, C.R.; Vieira, L.Q. *Trypanosoma cruzi* Needs a Signal Provided by Reactive Oxygen Species to Infect Macrophages. *PLoS Neg. Trop. Dis.* **2016**, *10*, e0004555. [CrossRef] [PubMed]
36. Gachet-Castro, C.; Freitas-Castro, F.; Gonzáles-Córdova, R.A.; da Fonseca, C.I.K.; Gomes, M.D.; Ishikawa-Ankerhold, H.C.; Baqui, M.M.A. Modulation of the Host Nuclear Compartment by *T. cruzi* Uncovers Effects on Host Transcription and Splicing Machinery. *Front. Cell. Infect. Microbiol.* **2021**, *11*, 718028. [CrossRef]
37. Tronchre, H.; Wang, J.; Fu, X.-D. A protein related to splicing factor U2AF^{35} that interacts with U2AF^{65} and SR proteins in splicing of pre-mRNA. *Nature* **1997**, *388*, 397–400. [CrossRef]
38. de Paiva, R.E.F.; Vieira, E.G.; da Silva, D.R.; Wegermann, C.A.; da Costa Ferreira, A.M. Anticancer compounds based on isatin-derivatives: Strategies to ameliorate selectivity and efficiency. *Front. Mol. Biosci.* **2021**, *7*, 627272.
39. de Moraes, J.; Dario, S.; Couto, R.A.A.; Pinto, P.L.S.; Da Costa Ferreira, A.M. Antischistosomal Activity of Oxindolimine-Metal Complexes. *Antimicrob. Agents Chemother.* **2015**, *59*, 6648–6652. [CrossRef]
40. da Silveira, V.C.; Luz, J.S.; Oliveira, C.C.; Graziani, I.; Ciriolo, M.R.; Da Costa Ferreira, A.M. Double-strand DNA cleavage induced by oxindole-Schiff base copper(II) complexes with potential antitumor activity. *J. Inorg. Biochem.* **2008**, *102*, 1090–1103. [CrossRef]
41. Katkar, P.; Coletta, A.; Castelli, S.; Sabino, G.L.; Couto, R.A.A.; Da Costa Ferreira, A.M.; Desideri, A. Effect of oxindolimine copper(II) and zinc(II) complexes on human topoisomerase I activity. *Metallomics* **2014**, *6*, 117–125. [CrossRef]
42. Stoll, S.; Schweiger, A. EasySpin, a comprehensive software package for spectral simulation and analysis in EPR. *J. Magn. Reson.* **2006**, *178*, 42–55. [CrossRef]
43. Ribeiro, A.R.; Lima, L.; de Almeida, L.A.; Monteiro, J.; Moreno, C.J.G.; Nascimento, J.D.; de Araújo, R.F.; Mello, F.; Martins, L.P.A.; Graminha, M.A.S.; et al. Biological and Molecular Characterization of *Trypanosoma cruzi* Strains from Four States of Brazil. *Am. J. Trop. Med. Hyg.* **2018**, *98*, 453–463. [CrossRef] [PubMed]
44. Mosmann, T. Rapid colorimetric assay for cellular growth and survival: Application to proliferation and cytotoxicity assays. *J. Immunol. Meth.* **1983**, *65*, 55–63. [CrossRef] [PubMed]
45. van Meerloo, J.; Kaspers, G.J.L.; Cloos, J. Cell Sensitivity Assays: The MTT Assay. *Meth. Mol. Biol.* **2011**, *731*, 237–245.
46. Andrews, N.W.; Colli, W. Adhesion and interiorization of *Trypanosoma cruzi* in mammalian cells. *J. Protozool.* **1982**, *29*, 264–269. [CrossRef] [PubMed]
47. Miguel, R.B.; Petersen, P.A.D.; Gonzales-Zubiate, F.A.; Oliveira, C.C.; Kumar, N.; do Nascimento, R.R.; Petrilli, H.M.; Da Costa Ferreira, A.M. Inhibition of cyclin-dependent kinase CDK1 by oxindolimine ligands and corresponding copper and zinc complexes. *J. Biol. Inorg. Chem.* **2015**, *20*, 1205–1217. [CrossRef]
48. Sakaguchi, U.; Addison, A.W. Spectroscopic and Redox Studies of Some Copper(II) Complexes with Biomimetic Donor Atoms: Implications for Protein Copper Centres. *J. Chem. Soc. Dalton Trans.* **1979**, *1979*, 600–608. [CrossRef]
49. Da Costa Ferreira, A.M.; Petersen, P.A.D.; Petrilli, H.M.; Ciriolo, M.R. Molecular basis for anticancer and antiparasite activities of copper-based drugs. In *Redox-Active Therapeutics*; Batinic-Haberle, I., Reboucas, J.S., Spasojevic, I., Eds.; Springer: Berlin/Heidelberg, Germany, 2016; Chapter. 12; pp. 287–309.

50. Rozga, M.; Sokołowska, M.; Protas, A.M.; Bal, W. Human serum albumin coordinates Cu(II) at its N-terminal binding site with 1 pM affinity. *J. Biol. Inorg. Chem.* **2007**, *12*, 913–918. [CrossRef]
51. Blindauer, C.A.; Harvey, I.; Bunyan, K.E.; Stewart, A.J.; Sleep, D.; Harrison, D.J.; Berezenko, S.; Sadler, P.J. Structure, properties, and engineering of the major zinc binding site on human albumin. *J. Biol. Chem.* **2009**, *284*, 23116–23124. [CrossRef]
52. Ciccarelli, A.B.; Frank, F.M.; Puente, V.; Malchiodi, E.L.; Batle, A.; Lombardo, M.E. Antiparasitic effect of vitamin B12 on *T. cruzi*. *Antimicrob. Ag. Chemother.* **2012**, *56*, 5315–5320. [CrossRef]
53. Caballero, A.B.; Rodriguez-Diéguez, A.; Quirós, M.; Salas, J.M.; Huertas, O.; Ramírez-Macías, I.; Olmo, F.; Marín, C.; Chaves-Lemaur, G.; Gutierrez-Sánchez, R.; et al. Triazolopyrimidine compounds containing first-row transition metals and their activity against the neglected infectious Chagas disease and Leishmaniasis. *Eur. J. Med. Chem.* **2014**, *85*, 526–534. [CrossRef]
54. Paixão, D.A.; Lopes, C.D.; Carneiro, Z.A.; Sousa, L.M.; de Oliveira, L.P.; Lopes, N.P.; Pivatto, M.; Chaves, J.D.S.; de Almeida, M.V.; Ellena, J.; et al. In vitro anti-Trypanosoma cruzi activity of ternary copper(II) complexes and in vivo evaluation of the most promising complex. *Biomed. Pharmacother.* **2019**, *109*, 157–166. [CrossRef]
55. Corrêa, R.S.; da Silva, M.M.; Graminha, A.E.; Meira, C.S.; dos Santos, J.A.F.; Moreira, D.R.M.; Soares, M.B.P.; von Poelhsitz, G.; Castellano, E.E.; Bloch, C., Jr.; et al. Ruthenium(II) complexes of 1,3-thiazolidine-2-thione: Cytotoxicity against tumor cells and anti-*Trypanosoma cruzi* activity enhanced upon combination with benznidazole. *J. Inorg. Biochem.* **2016**, *156*, 153–163. [CrossRef] [PubMed]
56. Maia, P.I.D.S.; Carneiro, Z.A.; Lopes, C.D.; Oliveira, C.G.; Silva, J.S.; de Albuquerque, S.; Hagenbach, A.; Gust, R.; Deflon, V.M.; Abram, U. Organometallic gold(III) complexes with hybrid SNS-donating thiosemicarbazone ligands: Cytotoxicity and anti-Trypanosoma cruzi activity. *Dalton Trans.* **2017**, *46*, 2559–2571. [CrossRef] [PubMed]
57. Muelas-Serrano, S.; Nogal-Ruiz, J.J.; Gómez-Barrio, A. Setting of a colorimetric method to determine the viability of *Trypanosoma cruzi* epimastigotes. *Parasitol. Res.* **2000**, *86*, 999–1002. [CrossRef] [PubMed]
58. Merritt, C.; Silva, L.E.; Tanner, A.L.; Stuart, K.; Pollastri, M.P. Kinases as Druggable Targets in Trypanosomatid Protozoan Parasites. *Chem. Rev.* **2014**, *114*, 11280–11304. [CrossRef] [PubMed]
59. da Silva, E.B.; Dall, E.; Briza, P.; Brandstetter, H.; Ferreira, R.S. Cruzain structures: Apocruzain and cruzain bound to S-ethylthiomethanesulfonate and implications for drug design. *Acta Cryst.* **2019**, *F75*, 419–427.

Disclaimer/Publisher's Note: The statements, opinions and data contained in all publications are solely those of the individual author(s) and contributor(s) and not of MDPI and/or the editor(s). MDPI and/or the editor(s) disclaim responsibility for any injury to people or property resulting from any ideas, methods, instructions or products referred to in the content.

Article

Pentadentate and Hexadentate Pyridinophane Ligands Support Reversible Cu(II)/Cu(I) Redox Couples

Glenn Blade [1], Andrew J. Wessel [2], Karna Terpstra [1] and Liviu M. Mirica [1,*]

[1] Department of Chemistry, University of Illinois Urbana-Champaign, 600 S. Matthews Ave, Urbana, IL 61801, USA
[2] Department of Chemistry, Washington University, One Brookings Drive, St. Louis, MO 63130, USA
* Correspondence: mirica@illinois.edu

Abstract: Two new ligands were synthesized with the goal of copper stabilization, N,N′-(2-methylpyridine)-2,11-diaza[3,3](2,6)pyridinophane (^{Pic}N4) and N-(methyl),N′-(2-methylpyridine)-2,11-diaza[3,3](2,6)pyridinophane (^{PicMe}N4), by selective functionalization of ^{H}N4 and ^{TsH}N4. These two ligands, when reacted with various copper salts, generated both Cu(II) and Cu(I) complexes. These ligands and Cu complexes were characterized by various methods, such as NMR, UV-Vis, MS, and EA. Each compound was also examined electrochemically, and each revealed reversible Cu(II)/Cu(I) redox couples. Additionally, stability constants were determined via spectrophotometric titrations, and radiolabeling and cytotoxicity experiments were performed to assess the chelators relevance to their potential use in vivo as ^{64}Cu PET imaging agents.

Keywords: bioinorganic chemistry; pyridinophane ligands; copper(II) complexes; copper(I) complexes; cyclic voltammetry; radiolabeling; ^{64}Cu PET imaging agents; reversibility

Citation: Blade, G.; Wessel, A.J.; Terpstra, K.; Mirica, L.M. Pentadentate and Hexadentate Pyridinophane Ligands Support Reversible Cu(II)/Cu(I) Redox Couples. *Inorganics* **2023**, *11*, 446. https://doi.org/10.3390/inorganics11110446

Academic Editors: Christelle Hureau, Ana Maria Da Costa Ferreira and Gianella Facchin

Received: 5 October 2023
Revised: 31 October 2023
Accepted: 17 November 2023
Published: 20 November 2023

Copyright: © 2023 by the authors. Licensee MDPI, Basel, Switzerland. This article is an open access article distributed under the terms and conditions of the Creative Commons Attribution (CC BY) license (https:// creativecommons.org/licenses/by/ 4.0/).

1. Introduction

Mononuclear copper complexes have been extensively utilized throughout various areas of inorganic chemistry: synthesis of structural or functional biomimetic inorganic complexes of Cu-containing enzymes [1–4], cation detection or sequestration [5,6], development of metal-based therapeutic or diagnostic compounds [7–10], and many others. In particular, the development of ^{64}Cu-based positron emission tomography (PET) agents has garnered significant attention in recent years as an alternative to shorter-lived radionuclides ^{11}C and ^{18}F, which are commonly used in PET imaging [11–13]. However, these ^{64}Cu PET agents still face challenges presented by the possibility of in vivo decomplexation. Ideally, a suitable chelator should demonstrate high thermodynamic stability and kinetic inertness in order to avoid this problem, which has been the focus of many studies in recent years [14]. However, a problem faced by some of the most common ^{64}Cu chelators is the issue of reduction-induced demetallation. Given the reducing environment of cells and the presence of in vivo bioreductants, the ideal chelator should be able to avoid this issue by remaining stable even upon reduction to Cu^I [15]. As such, ligands with flexible donor arms that have the ability to stabilize both Cu^{II} and Cu^I are good candidates for chelating ^{64}Cu [16]. Several studies have been published in recent years focused on the coordination chemistry of ^{64}Cu complexes with macrocyclic ligands substituted with pendant arms like 2-pyridylmethyl, picolinate, thiazolyl, and others [17–21].

In that vein, two new pyridinophane ligand systems, inspired by previous macrocyclic polydentate ligands, have been synthesized [22–24]. By substituting non-interacting groups (i.e. Me or tBu) for groups that can interact with the metal center, greater binding modes than usual for ^{R}N4 ligands can be achieved with altered characteristics of the resultant complexes. Interacting groups like 2-methylpyridyl, picolyl—"Pic", could bind directly with the metal center while being easily synthetically attached to the N4 backbone.

When both alkyl groups are chosen to be the picolyl fragment, the resultant hexadentate ligand N,N'-(2-methylpyridine)-2,11-diaza[3,3](2,6)pyridinophane, ^{Pic}N4, offers the possibility of a distorted octahedral environment around the metal center while simultaneously shielding from inner sphere interactions. An asymmetric version could also be synthesized using previously reported methods to make N-(methyl),N'-(2-methylpyridine)-2,11-diaza[3,3](2,6)pyridinophane, ^{PicMe}N4, which could act as a pentadentate ligand and leave one coordination site available for an exogenous ligand. This flexible pentadentate ligand could more easily adapt to geometries other than a distorted octahedral arrangement. When this ligand was bound to copper, both ^{Pic}N4 and ^{PicMe}N4 were able to stabilize both Cu^{II} and Cu^{I} oxidation states, with each complex being crystallographically characterized. Each of these four complexes was spectroscopically scrutinized by various techniques, including NMR, EPR, ESI-MS, and UV-Vis. Cyclic voltammetry experiments were able to show that the conversion between Cu^{II} and Cu^{I} was remarkably reversible for both systems, as a consequence of the flexible nature of the picolyl arms being able to come off the Cu center. The ^{Pic}N4$Cu^{II/I}$ couple was also low at $E_{1/2} = -1.1$ V vs. Fc/Fc$^+$. Calculation of Cu^{II} stability constants using spectrophotometric titrations also revealed the moderate ability of the complexes to stabilize both Cu^{II} and Cu^{I} complexes. Finally, preliminary radiolabeling studies showed that both ^{Pic}N4 and ^{PicMe}N4 can quickly and efficiently be radiolabeled with ^{64}Cu, making these ligands potentially relevant chelators for use in ^{64}Cu PET imaging studies.

2. Results and Discussion

2.1. Synthesis

The ligand synthesis of ^{Pic}N4 was a two-step development. The first attempt at the synthesis of ^{Pic}N4 involved an S_N2-based mechanism utilizing 2-chloromethyl pyridine under basic conditions at a roughly 80% yield [25]. Multiple bases were tested for this synthesis; between sodium carbonate, potassium carbonate, and Hünig's base (diisopropylethylamine), Hünig's base gave the highest yield of 81%, while the carbonates gave smaller yields of around 40%. Additional synthetic attempts using reductive amination have given inconsistent results and a maximum yield of 46%. Full synthetic details and product descriptions can be found in the Supporting Information.

Ligand synthesis for ^{PicMe}N4 was achieved by two different methods, as depicted in Scheme 1. In the first pathway, the direct functionalization of ^{MeH}N4 by placing the 2-methylpyridyl on the secondary amine was performed to make the product. This pathway requires making ^{MeH}N4, a product synthesized by previously discussed methods [23]. The second pathway utilized ^{TsH}N4 for functionalization to yield ^{PicTs}N4. The tosyl deprotection reaction using concentrated sulfuric acid did not degrade the ligand significantly, providing a good yield of ^{PicH}N4. The penultimate product, ^{PicH}N4, was then methylated to yield the final product, ^{PicMe}N4. Since both pathways showed that the products ^{TsMe}N4 and ^{TsPic}N4 could survive the harsh sulfuric acid conditions of the detosylation reaction, the second pathway was chosen. Functionalization of the secondary amines occurred by one of two methods: reductive amination or S_N2. Both methods achieved high yields (75% and 81%, respectively), but the S_N2 was much more consistent and less reliant on the purity of reagents.

The syntheses of the **1·(OTf)$_2$** and **2·OTf** complexes was achieved by mixing the appropriate triflate salt with the ligand in MeCN (Scheme 2). Cu^{II}(OTf)$_2$ and ^{Pic}N4 were mixed overnight and either crashed out of solution by trituration with diethyl ether or recrystallized via diethyl ether diffusion, with an 88% yield of **1·(OTf)$_2$**. While most studies in this paper utilize the triflate complex, other salts like Cu^{II}(ClO$_4$)$_2$ or Cu^{II}(PF$_6$)$_2$ were also employed with similar yields. Similarly, [(MeCN)$_4$Cu^I]OTf and ^{Pic}N4 were mixed in MeCN for one hour and recrystallized via ether diffusion for a 55% yield of **2·OTf**.

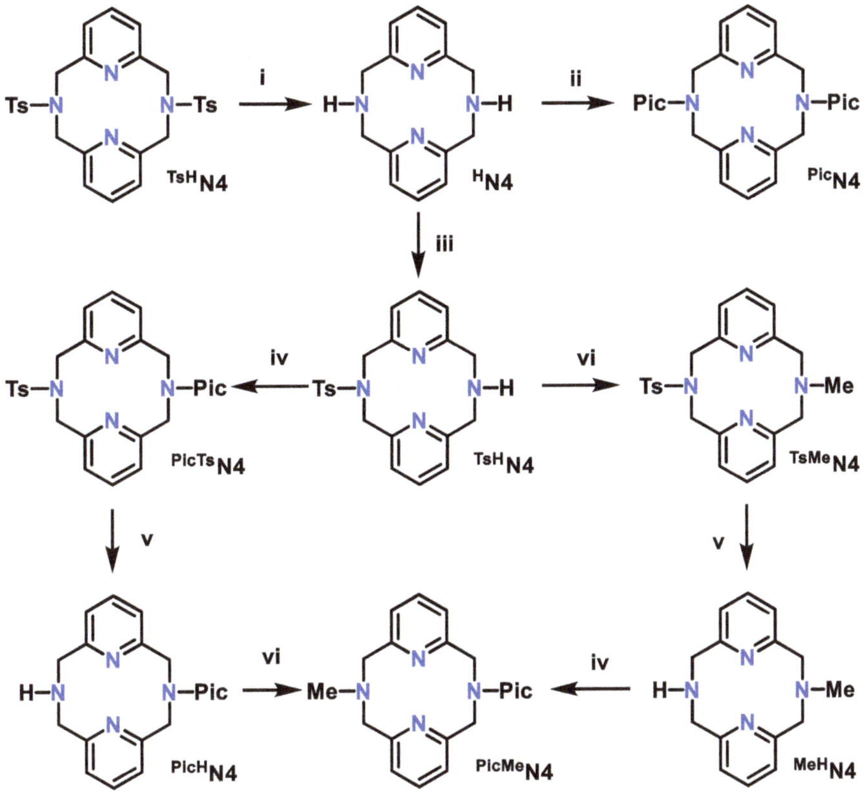

Scheme 1. Synthesis of ᴾⁱᶜN4 and ᴾⁱᶜᴹᵉN4. (i) 90% H_2SO_4, reflux 2.5 h; 88% (ii) 2-(methylchloro)pyridine HCl, iPr$_2$EtN, MeCN, 48 h; 81% (iii) TsCl, DCM, 0 C, 3 h; 44% (iv) 2-(methylchloro)pyridine HCl, iPr$_2$EtN, MeCN, 48 h; 82% (v) 90% H_2SO_4, reflux, overnight; 89% (vi) formic acid, formaldehyde, reflux, overnight; 82%.

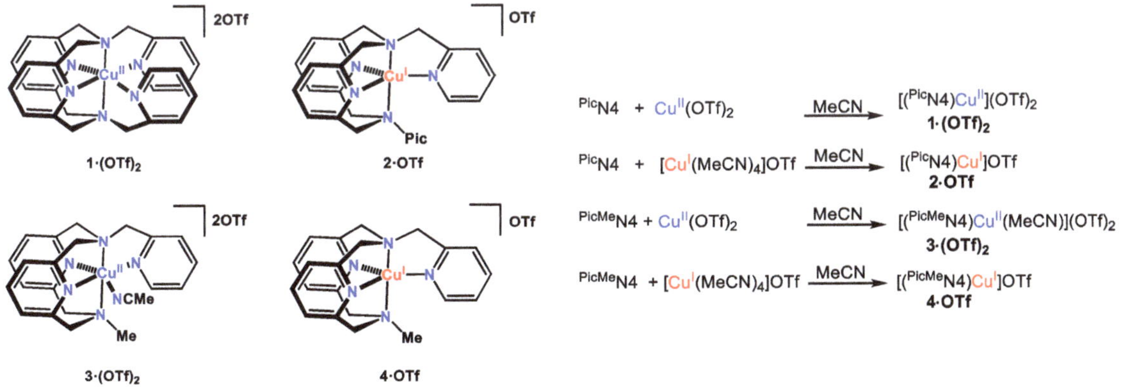

$$^{Pic}N4 + Cu^{II}(OTf)_2 \xrightarrow{MeCN} [(^{Pic}N4)Cu^{II}](OTf)_2$$
$$\mathbf{1 \cdot (OTf)_2}$$

$$^{Pic}N4 + [Cu^{I}(MeCN)_4]OTf \xrightarrow{MeCN} [(^{Pic}N4)Cu^{I}]OTf$$
$$\mathbf{2 \cdot OTf}$$

$$^{PicMe}N4 + Cu^{II}(OTf)_2 \xrightarrow{MeCN} [(^{PicMe}N4)Cu^{II}(MeCN)](OTf)_2$$
$$\mathbf{3 \cdot (OTf)_2}$$

$$^{PicMe}N4 + [Cu^{I}(MeCN)_4]OTf \xrightarrow{MeCN} [(^{PicMe}N4)Cu^{I}]OTf$$
$$\mathbf{4 \cdot OTf}$$

Scheme 2. Preparation of Copper Complexes.

The synthesis of complexes **3·(OTf)₂** and **4·OTf** were similarly completed: the relevant copper triflate salt and the ligand were mixed in MeCN (Scheme 2). While crystals for the Cu^{II} salt were not easily obtained, a green solid was crashed out from MeCN with toluene and rinsing with pentane (55% isolated yield). When attempting to get crystals of **3·(OTf)₂**, the use of sodium tetraphenylborate, NaBPh₄, generated orange crystals of **4·OTf** in low yield. A more acceptable approach to preparing **4·OTf** was ^{PicMe}N4 and [(MeCN)₄Cu^I]OTf mixed in MeCN for one hour in the dark and recrystallized via diethyl ether diffusion at −35 °C (77% yield).

2.2. Structural Characterization of Metal Complexes

X-ray diffraction-quality crystals of the copper complexes were obtained by diethyl ether diffusion into MeCN solutions at room temperature or −35 °C (Figure 1 and Table 1). Full crystallographic details are provided in the Supplemental Information. The crystal structure of **1²⁺** had a similar distorted octahedral environment to the one observed for the analogous ^{tBu}N4 complex [22,26,27]. The inclusion of the two pyridine moieties on the metal center preclude the need of additional exogenous ligands, such as solvent or triflates directly bound to the metal center. The Cu^{II} center exhibits a Jahn-Teller like distortion, with the four pyridine nitrogens having relatively short bond lengths to copper (2.00–2.06 Å), while the amine nitrogens form much longer bonds (2.28, 2.35 Å).

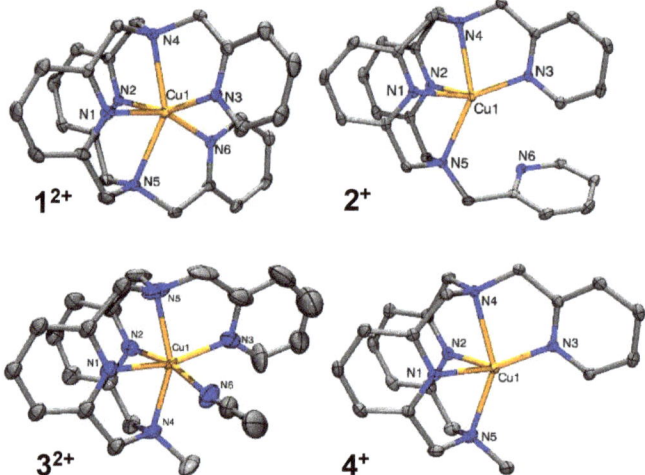

Figure 1. ORTEP plots (50% probability ellipsoids) of cations **1²⁺**, **2⁺**, **3²⁺**, and **4⁺**. Counterions and H atoms are omitted for clarity. The crystallographic datasets for **1·(OTf)₂**, **2·OTf**, **3·(OTf)₂**, and **3·OTf** have been deposited at CCDC under the record numbers 2049802, 2049803, 2049804, and 2049805.

Table 1. Selected bond distances (Å) and angles (°) of cations **1–4**.

	1²⁺	**2⁺**	**3²⁺**	**4⁺**
Cu-N1	2.056(4)	2.1341(1)	1.944(9)	2.1286(1)
Cu-N2	2.028(4)	2.0817(1)	2.173(7)	2.0768(2)
Cu-N3	2.003(4)	1.9640(1)	1.967(8)	1.9461(1)
Cu-N4	2.276(4)	2.3983(1)	2.258(7)	2.3957(2)
Cu-N5	2.348(4)	2.3456(1)	2.165(8)	2.262(2)
Cu-N6	2.017(4)	3.323	2.219(9)	---
N2-Cu-N1	84.43	81.36	82.9	83.07
N4-Cu-N5	148.38	146.78	152.1	147.85
φ(°) ^{a}	86.60, 84.19	86.91, 87.93	88.07, 89.28	88.19, 87.00
θ(°) ^{b}	36.19	28.92	20.39	31.80

^{a} φ (°) designates the angles between the average plane of two pyridine rings and a mean equatorial plane; ^{b} θ (°) designates the angle between the equatorial plane made between atoms N1, N2, and Cu and the plane made between atoms N3, Cu, and N6 for Cu^{II} complexes and N3, Cu, and para-carbon on picolyl arm for Cu^I.

Upon reduction to 2^+, the coordination environment changes to a pentadentate distorted square pyramid geometry with a structural parameter $\tau_5 = 0.13$.[28] As expected for Cu^I structures, the $Cu-N_{eq}$ bond lengths averaged a shorter value of 2.00 Å, while the $Cu-N_{ax}$ bond lengths were much longer at 2.34 Å. The non-coordinating picolyl nitrogen was sufficiently far away to not interact with the Cu center at 3.32 Å [4,22].

Supposing that crystals of 4^+ was more stable than 3^{2+}, the ligand was mixed with Cu^IOTf in MeCN and recrystallized with ether diffusion at room temperature to yield large orange crystals. The cation of 4^+ adopts a distorted square pyramid pentadentate geometry with a structural parameter $\tau_5 = 0.12$, similar to 2^+ [28]. It has a similar delineation of Cu-N bonds around the Cu: $Cu-N_{eq}$ bond lengths averaged 2.00 Å, while $Cu-N_{ax}$ bond lengths averaged 2.33 Å. Unlike in **2**, the methylamine on the ^{PicMe}N4 backbone was less sterically restricted than the picolyl functionalized amines.

In an attempt to obtain crystals of $3·(PF_6)_2$, the blue solid was dissolved in MeCN with two equivalents of NaBPh$_4$ and subjected to diethyl ether diffusion at room temperature. When isolated, there were primarily orange crystals of $4·PF_6$ present with a blue solution. Crystals of $3·(BPh_4)_2$ were eventually recovered under these conditions as a mixture of Cu^{II} and Cu^I crystals. Although it was unknown exactly how the Cu^{II} complex was reduced in the solution, it was suspected that the NaBPh$_4$ and the BPh$_3$ impurity promoted the reduction of the Cu^{II} center. Regardless, 3^{2+} displays a Jahn-Teller like distorted octahedral geometry with one exogenous MeCN bound to the Cu^{II} center. Interestingly, the Cu^{II} species seemed to exhibit Jahn-Teller like compression along the N1-Cu-N3 axis where the Cu-N bond lengths are around 1.95 Å, while the other four Cu-N bonds are longer at an average of 2.20 Å.

2.3. Complex Characterization

To better understand the solution state characteristics of these complexes, several techniques were utilized, including electron paramagnetic resonance (EPR), NMR, and electrochemical studies. The paramagnetic d^9 Cu^{II} complexes 1^{2+} and 3^{2+} were characterized by EPR in a fashion similar to previous Cu^{II} species, and the EPR spectra are shown in Figure 2 and the EPR parameters are summarized in Table 2. Analysis by Evan's method measured in CD$_3$CN yielded the expected values for these d^9 Cu^{II} centers: 1.80 μ_B and 1.71 μ_B for 1^{2+} and 3^{2+}, respectively [29]. The EPR spectrum for $1·(OTf)_2$ exhibited values of $g_x = 2.070$, $g_y = 2.055$, and $g_z = 2.259$ ($A_z = 144.5$ G) and $g_x = 2.067$, $g_y = 2.056$, and $g_z = 2.264$ ($A_z = 152.5$ G) for $3·(OTf)_2$, which is consistent with what is expected of a distorted octahedral Cu^{II} center, and in line with the solid state structural data [22,29,30].

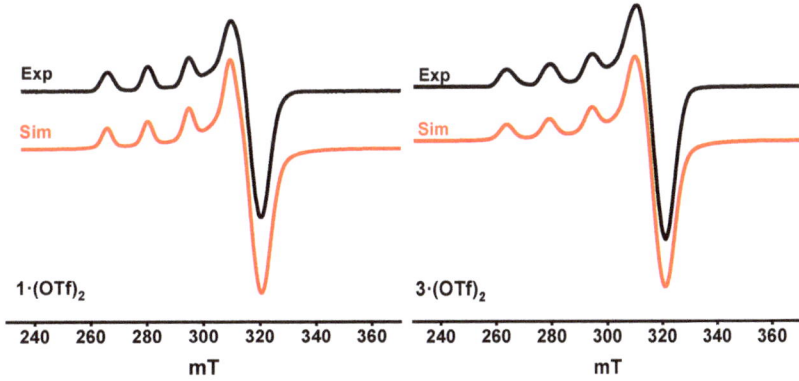

Figure 2. EPR spectrum (black) and simulation (red) of $1·(OTf)_2$ (**left**) and $3·(OTf)_2$ (**right**) in MeCN:PrCN (1:3) at 77K.

Table 2. Selected EPR Data for Paramagnetic Complexes.

	g_x	g_y	g_z	A_z (G)
1·(OTf)$_2$	2.070	2.055	2.259	145
3·(OTf)$_2$	2.067	2.056	2.264	152

While the paramagnetic ^{1}H NMR spectrum of 1^{2+} did not afford much information (Figure S14), the spectra of the d^{10} Cu^I species could generally be assigned with the help of a gCOSY 2D spectrum (Figures S9–S13). The assignment of the 2·OTf spectrum gathered in CD$_3$CN was assigned as so: the four most downfield aromatic peaks corresponded to the different pyridine hydrogens on the picolyl arm, while the two upfield sets of aromatic multiplets corresponded to the pyridine hydrogens on the N4 backbone [31]. The methylene region contained five total peaks: a singlet (4.62 ppm) matched to the two picolyl methylene hydrogens and a pair of doublets matched to the N4 methylene hydrogens. Further assignment of the N4 methylene hydrogens could not be easily discerned due to the symmetry and structure of the molecule. The integration and the 2D NMR corroborated this assignment (Figure S11). Since the methylene on the picolyl arm appears as a singlet, this implies there was rapid exchange between the bound arm and the unbound arm which was faster than the NMR time-scale.

The assignment of the 4·OTf spectrum followed in a similar way. The three most downfield aromatic peaks matched the four pyridine hydrogens on the picolyl arm, while the two down-field aromatic multiplets corresponded with the para- and meta-hydrogens on the N4 pyridine backbone. The two singlet peaks in the aliphatic region corresponded to the two methylamines: the methylene moiety on the picolyl arm (4.469 ppm) and the methyl group (3.313 ppm). In a similar fashion to the ^{Pic}N4 complex, the methylene on the picolyl arm was not fixed in spaced which allowed resolution into a singlet. The remaining four sets of doublets ($J_{avg} \approx 15$ Hz, geminal) corresponded to the methylene protons fixed in place on the N4 backbone. Based on the gCOSY crossover peaks (Figure S13), the doublet pairs 4.248 and 3.668 ppm correspond to interaction protons, while 4.177 and 4.040 ppm are also coupled.

Cyclic voltammetry (CV) for 1·(OTf)$_2$ featured a couple at -0.752 V vs. Fc$^{0/+}$ (Figure 3), corresponding to the Cu$^{II/I}$ couple with a quasi-reversible nature ($\Delta E_p = 97$ mV) as well as an irreversible oxidation at $+1.047$ V (Figure S15). In order to confirm the reversibility of the Cu$^{II/I}$ couple, 2·OTf was also scrutinized to yield a similar couple at -0.716 V vs. Fc$^{0/+}$ ($\Delta E_p = 176$ mV). A similar analysis for 3·(OTf)$_2$ found a quasi-reversible Cu$^{II/I}$ couple at -0.468 V vs. Fc$^{0/+}$ ($\Delta E_p = 105$ mV) along with an irreversible Cu^{III} oxidation at 1.552 V (Figure S16). Confirming the reversibility of this quasi-reversible Cu$^{II/I}$ couple, a CV of 4·OTf showed the couple at -0.441 V vs. Fc$^{0/+}$ ($\Delta E_p = 96$ mV).

Notably, all four copper complexes exhibit larger ΔE_p values than the values expected for fully reversible redox processes (Table 3). However, the measured ΔE_p values for the Fc$^{0/+}$ couple in both sets of experimental conditions, 129 mV and 176 mV, are also larger than standard values, indicating that the large peak-to-peak separation may not necessarily imply redox irreversibility of the copper complexes. Furthermore, the discrepancies between the ΔE_p values of the complexes can be explained by the different sets of experimental conditions for each: Cu^{II} complexes are air stable and can be analyzed on the bench top, while Cu^I complexes are very air sensitive and required rigorous anaerobic conditions of a glovebox.

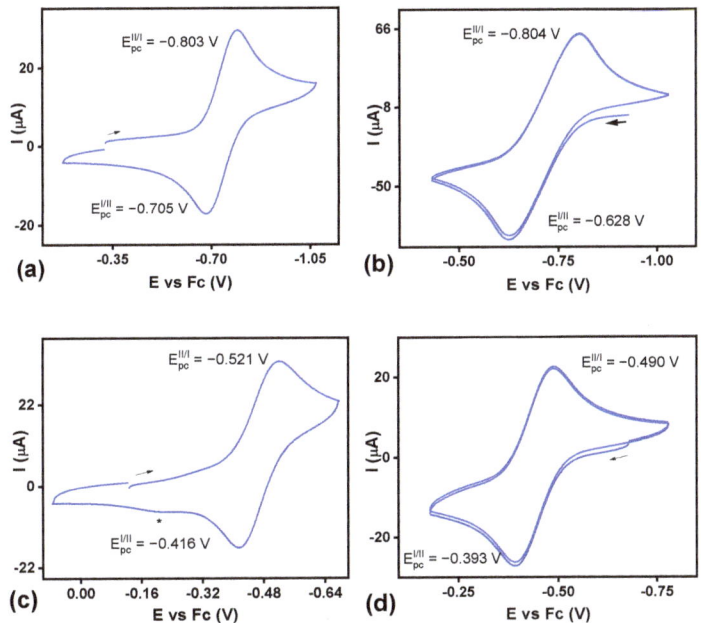

Figure 3. Cyclic voltammetry of the copper complexes **1·(Otf)₂** (a), **2·Otf** (b), **3·(Otf)₂** (c), and **4·Otf** (d) (0.1 M Bu₄NClO₄/CH₃CN; arrow indicates the initial scan direction). The asterisk (*) corresponds to a trace amount of $^{PicMe}N4Cu^{II}(H_2O)$ complex.

Table 3. Selected Physical Parameters of Complexes 1–4.

1^{2+}	2^{+}	3^{2+}	4^{+}
E, V (ΔE_p, mV) ^{a,b,c}			
$E_{1/2} = -0.752$ (97) $E_{ox} = 1.047$	$E_{1/2} = -0.716$ (176)	$E_{1/2} = -0.468$ (105) $E_{ox} = 1.552$	$E_{1/2} = -0.441$ (96)
UV-Vis, λ_{max}, nm (ε, M^{-1} cm^{-1}), MeCN			
257 (22,775), 340 (446), 717 (146)	250 (8395), 362 (2438), 444 (893)	258 (12,251), 322 (671), 687 (91)	246 (11,731), 332 (3331), 370 (3854), 435 (1611)
μ_{eff} (μ_B) at 293 K, Evans' Method, CD₃CN			
1.80	N/A	1.71	N/A

^{a} Redox Potentials (vs. Fc/Fc+), 0.1 M tBAP/MeCN, 0.01 M Ag/AgNO₃ or Ag wire reference, Δep is the separation between anodic and cathodic waves in mV, measured at 100 mV/s. ^{b} **1²⁺** & **3²⁺** had 3-segment sweep. ^{c} **2⁺** & **4⁺** had 5-segment sweep. N/A: not applicable.

Additionally, both complexes were subjected to conditions with increasing concentrations of water in MeCN. The Cu$^{II/I}$ couple for ^{Pic}N4 was only shifted slightly to −0.800 V vs. Fc$^{0/+}$ even in a 70% water to MeCN solution (with 0.2 M TBAP). The reversibility of the couple remained stable throughout the course of the experiment (Figure S17). The Cu$^{II/I}$ couple of ^{PicMe}N4 shifted less drastically to −0.600V vs. Fc$^{0/+}$ after adding up to 70% water to an MeCN solution (with 0.2 M TBAP). The system was overall reversible but at higher concentrations of water, and additional oxidation and reduction peaks appeared probably due to water binding to the metal center over MeCN (Figure S18).

2.4. Ligand Acidity Constants and Complex Stability Constants

To determine the acidity constants (pKa) of ^{Pic}N4 and ^{PicMe}N4, UV-Vis spectrophotometric titrations were performed and the changes in the spectra were monitored. To a solution of either ^{Pic}N4 or ^{PicMe}N4 in 0.1 M KCl, aliquots of 0.15 M KOH were added and the UV-Vis spectra were recorded at each pH. For the ^{Pic}N4 ligand, the increase of the solution's pH results in the steady decrease of the π—π^* transition band at 264 nm, until around pH 7, at which point the absorbance begins to increase (Figure 4). The data was then simulated in the HypSpec 2014 program (Protonic Software, UK) [32], which afforded the species distribution plot (Figure 4) and three pKa values: 8.94, 5.32, and 3.60. These values are tentatively assigned to the tertiary amine nitrogen, pyridine on the N4 backbone, and picolyl nitrogen, respectively. Despite containing six potential sites for protonation, only three pKa values were determined. This is likely due to the increased electrostatic repulsion that occurs upon sequential protonation steps, making it difficult to observe higher charged species in the pH range of the titration [33].

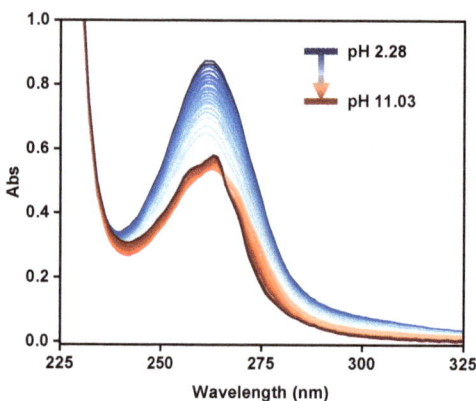 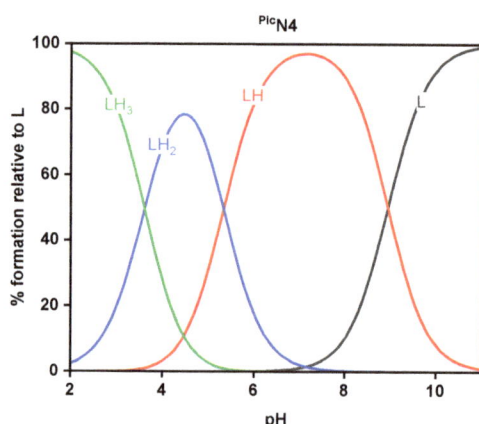

Figure 4. Variable pH (2.28–11.03) UV-Vis spectra of ^{Pic}N4 in 0.1 M KCl at 25 °C (**left**) and its species distribution plot (**right**). [^{Pic}N4]$_{tot}$ = 60 μM.

For the ^{PicMe}N4 ligand, a similar decrease in the peak at 261 nm occurred upon the increase in pH, until the lowest absorbance was observed at approximately pH 4 (Figure S25). Thereafter, the absorbance was observed to increase until a plateau at pH 7. Analysis using HypSpec provided four pKa values, the highest of which, 11.13, is assigned to the deprotonation of the methyl amine nitrogen (Table 4). The next highest value, 9.16, is assigned to the deprotonation of the tertiary amine amended with the 2-methylpyridine arm, which has previously been shown to lower the basicity of amine nitrogens attached to it [34–36]. This assignment also aligns with the highest pKa observed in ^{Pic}N4, which contains two similar amine sites.

Table 4. Acidity constants (pKa) of ligands.

	^{Pic}N4	^{PicMe}N4
$[H_4L]^{4+} = [H_3L]^{3+} + H^+$	-	2.47(9)
$[H_3L]^{3+} = [H_2L]^{2+} + H^+$	3.60(3)	5.46(9)
$[H_2L]^{2+} = [HL]^+ + H^+$	5.32(0)	9.16(9)
$[HL] = [L] + H^+$	8.94(6)	11.13(8)

To obtain the Cu^{II} stability constants for the complexes, similar spectrophotometric pH titrations were performed for a 1:1 mixture of Cu^{2+} and ligand in 0.1M KCl (Figure 5). Analysis of the spectral changes occurring in the UV for each complex gave a series of stability constants, as summarized in Table 5. The log($K_{Cu(II)L}$) values reveal that ^{Pic}N4 is able to form slightly more stable copper complexes than ^{PicMe}N4. In the case of ^{PicMe}N4, a value corresponding to the deprotonation of water was also obtained (7.75), but this was not observed for ^{Pic}N4. This could likely be attributed to the open coordination site available for ^{PicMe}N4, as evidenced in the crystal structure, which would allow for the binding and subsequent deprotonation of a water molecule.

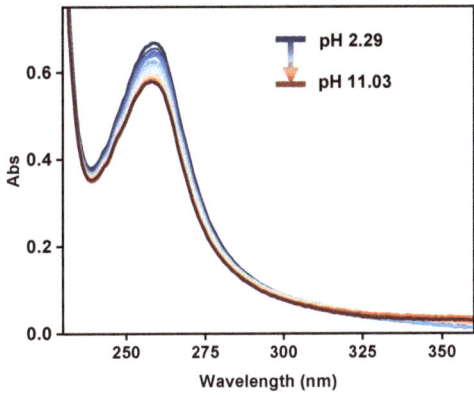 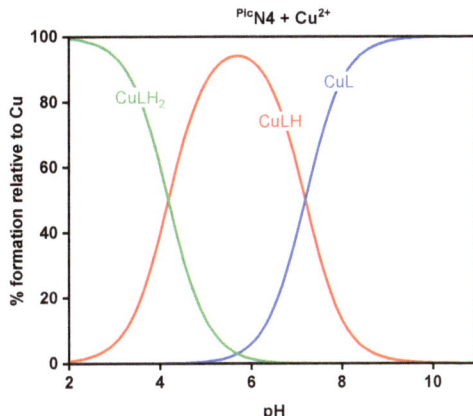

Figure 5. Variable pH (2.29–11.03) UV-Vis spectra of the ^{Pic}N4 + Cu^{2+} system in 0.1 M KCl at 25 °C (**left**) and its species distribution plot (**right**). $[Cu^{2+}]_{tot} = [^{Pic}N4]_{tot} = 50$ µM.

Table 5. Stability constants (logK values) and calculated pM values for Cu and Zn complexes. Errors reported for the last digit.

	^{Pic}N4 + Cu^{2+}	^{PicMe}N4 + Cu^{2+}	^{Pic}N4 + Zn^{2+}	^{PicMe}N4 + Zn^{2+}
$M^{2+} + H_2L^+ = [MH_2L]^{4+}$	4.13(3)	-	-	-
$M^{2+} + HL^+ = [MHL]^{3+}$	7.40(1)	4.54(1)	2.67(2)	9.28(7)
$M^{2+} + L = [ML]^{2+}$	**17.96(3)**	**17.07(1)**	**11.45(4)**	**10.41(7)**
$[ML(H_2O)]^{2+} = [ML(OH)]^+ + H^+$	-	7.75(2)	-	-
pM^{2+} (pH 7.4) [a]	16.81	12.90	8.87	8.02
log($K_{Cu(II)L}$)	17.96	17.07	-	-
log($K_{Cu(I)L}$)	7.05	9.46	-	-

[a] Values calculated as $-\log[M]_{free}$, where $[M^{2+}] = 10^{-6}$ M, $[L] = 10^{-5}$ M.

The determination of the Cu^I stability constants for each complex relied on the Cu^{II} stability constant and the $E_{1/2}$ values determined in the aqueous CV experiments. CVs of the complexes in aqueous conditions with 0.1 M NaOAc as a supporting electrolyte revealed $E_{1/2}$ values of −0.415 V and −0.220 V vs. Ag/AgCl for **1**$^{2+}$ and **3**$^{2+}$, respectively (Figures S19 and S20). It is worth noting that the aqueous CV data of **1**$^{2+}$ showed an additional reversible redox couple at −0.245 V vs. Ag/AgCl, which could be attributed to an alternative coordination mode, perhaps from the binding of an acetate ion present in solution. Another possible explanation for the observation of two species in solution could be two protonation states, as the species distribution for this complex shows nearly equal amounts of CuL and CuLH species at pH 7. Nevertheless, for the purpose of determining Cu^I stability constants, the more negative reduction potential was used in the Nernst equation. Stability constants for Cu^I could then be obtained by applying a Nernstian relationship using the reduction potentials, which results in log($K_{Cu(I)L}$) values of 7.05 and 9.46 for ^{Pic}N4 and ^{PicMe}N4, respectively.

Another important consideration for ^{64}Cu chelators is the possibility of transmetalation with biogenic metals in vivo. Therefore, the stability constants of the ligands ^{Pic}N4 and $^{Pic Me}$N4 towards Zn^{2+} were also determined. Both ligands have significantly lower Zn stability constants as compared to those for Cu, and the affinity towards Zn is also markedly lower at biological pH. This data indicates that the copper complexes of ^{Pic}N4 and ^{PicMe}N4 are unlikely to undergo transmetalation with zinc, a promising trait for potential ^{64}Cu chelators.

When comparing the $\log(K_{Cu(II)L})$ to other commonly used ^{64}Cu chelators, it is observed that ^{Pic}N4 and ^{PicMe}N4 have moderately lower stability constants than the other chelators (Table 6). Many of these chelators have N,O-based donor sets, but notably also have irreversible reduction potentials (e.g. DOTA, TETA). The chelators described in this work have the added benefit of reversible $Cu^{II/I}$ redox couples, allowing for the ability to form stable Cu^{I} complexes.

Table 6. Comparison of $\log(K_{Cu(II)L})$ values of commonly used ^{64}Cu chelators.

Chelator	Log($K_{Cu(II)L}$)	Ref.
^{Pic}N4	17.96	This work
^{PicMe}N4	17.07	This work
YW-15-Me	14.7	[7]
DO4S	19.6	[16]
PCTA	19.1	[37,38]
EDTA	19.2	[37,39,40]
TETA	21.1	[39–42]
DOTA	22.2	[39–41]
cyclen	24.6	[39,40]

2.5. Radiolabeling Studies

The radiolabeling capabilities of ^{Pic}N4 and ^{PicMe}N4 were also evaluated. Using a stock solution of ^{64}CuCl$_2$ diluted in ammonium acetate buffer (pH 5.5), mixtures of the ligands and ^{64}CuCl$_2$ were incubated at 45 °C for 30 min. These relatively mild conditions are comparable to those used for the common ^{64}Cu chelators like NOTA and DOTA [38]. The radiolabeled compounds were then analyzed by radio-HPLC using water (0.1% TFA) and acetonitrile (0.1% TFA) as the mobile phase with a gradient of 0–100% acetonitrile over 15 min (Figure 6). A control of only ^{64}CuCl$_2$ in ammonium acetate was also analyzed to compare the retention times. Both ligands showed complete conversion to ^{64}Cu complexes, with no remaining free ^{64}Cu being observed in either radio-HPLC trace. While ^{PicMe}N4 shows one peak in the chromatogram, ^{Pic}N4 shows two peaks close to one another. One possible explanation for this observation is an alternative coordination environment around the copper center, such as a different counterion (e.g. chloride, acetate) bound to the metal center. This observation is consistent with the aqueous CV studies that show ^{Pic}N4 also having two species present in solution (Figure S19).

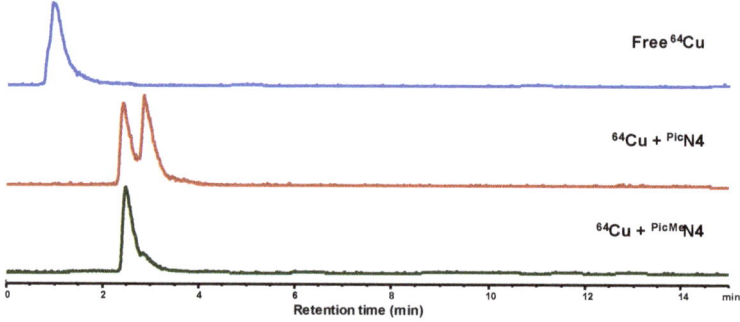

Figure 6. Radio-HPLC chromatograms for the ^{64}Cu labeled complexes of ^{Pic}N4 and ^{PicMe}N4.

After confirming that the ligands were able to be radiolabeled with ^{64}Cu, the lipophilicity of the complexes was determined by measuring the octanol/PBS partition coefficient (logD$_{oct}$, Table 7). Both dicationic complexes are particularly hydrophilic, with ^{64}Cu-^{PicMe}N4 having a more negative partition coefficient than ^{64}Cu-^{Pic}N4.

Table 7. Molecular weight and measured LogD values of the ^{64}Cu complexes.

^{64}Cu Complex	MW (g/mol)	log D$_{oct}$
^{64}Cu-^{Pic}N4	556.98	−1.564 ± 0.26
^{64}Cu-^{PicMe}N4	479.90	−2.171 ± 0.09

2.6. Cytotoxicity Studies

In order to test the plausibility of in vivo applications for these two chelators, cytotoxicity studies were performed using an Alamar blue assay on mouse neuroblastoma Neuro2a (N2a) cells. Cells were treated with each compound of their Cu^{II} complexes and cell viability was evaluated after a 48 h incubation period. The percentage of cell viability, as summarized in Figure 7, revealed that both ligands ^{Pic}N4 and ^{PicMe}N4 are toxic at higher concentrations, but ^{PicMe}N4 is significantly less toxic than ^{Pic}N4. Notably, the addition of Cu greatly reduces the toxicity of these ligands, with both complexes showing extremely high cell viability across the board, even at concentrations of 20 µM.

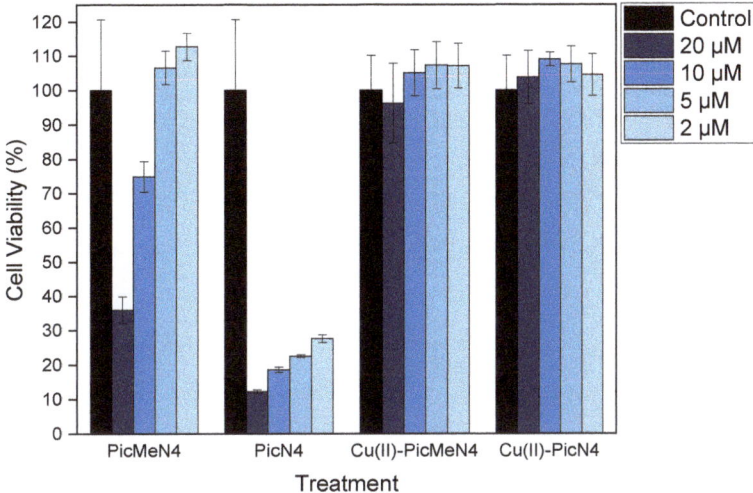

Figure 7. Cell viability (% of control) of Neuro2A cells upon incubation with ^{PicMe}N4, ^{Pic}N4, and their Cu^{II} complexes at 2, 5, 10, and 20 µM concentrations.

3. Conclusions

Inspired by previous pyridinophane ligands, herein we report two new ligand systems, ^{Pic}N4 and ^{PicMe}N4. The 2-methylpyridyl arms of these ligands bind to the Cu center in place of exogenous ligands and allow for a polydentate binding mode greater than ^{tBu}N4. The hexadentate ^{Pic}N4 ligand offers the metal center a fully bound, distorted octahedral geometry, which can shield the metal center from side reactions. The asymmetric ^{PicMe}N4 ligand offers five coordinating atom donors with the option of binding one exogenous ligand. This flexible pentadentate ligand can adopt geometries other than distorted octahedral and could be used to probe electrocatalytic transformations.

When bound to copper, ^{Pic}N4 and ^{PicMe}N4 both stabilized Cu^{II} and Cu^{I} centers, which were characterized by various spectroscopic means. Crystal structures were also obtained for all four compounds, showing a preferred geometry of distorted octahedral for the Cu^{II} complexes and a distorted square pyramidal geometry for the Cu^{I} complexes. Electrochemically, $^{Pic}N4$ exhibits a reversible $Cu^{II/I}$ couple at a low potential of -0.1 V vs. SHE. Conversely, the $^{PicMe}N4$ $Cu^{II/I}$ couple was also reversible, but the ability to bind an exogenous ligand caused other redox features to appear. Both Cu^{I} and Cu^{II} stability constants for each ligand were also determined, as were their affinities for Cu^{II} at biologically relevant pHs.

Supplementary Materials: The following supporting information can be downloaded at https://www.mdpi.com/article/10.3390/inorganics11110446/s1, Figures S1–S33, Tables S1–S11, References [43–49] are cited in the Supplementary Materials.

Author Contributions: Conceptualization, L.M.M.; methodology, G.B., A.J.W. and K.T.; validation, G.B., A.J.W. and K.T.; formal analysis, G.B., A.J.W. and K.T.; investigation, G.B., A.J.W. and K.T.; writing—original draft preparation, A.J.W. and G.B.; writing—review and editing, G.B., A.J.W., K.T. and L.M.M.; visualization, G.B., A.J.W. and K.T.; supervision, L.M.M.; project administration, L.M.M.; funding acquisition, L.M.M. All authors have read and agreed to the published version of the manuscript.

Funding: This research was funded by the US National Institutes of Health (R01GM114588 and RF1AG083937 to L.M.M.).

Data Availability Statement: All research data can be found in the Supplementary Materials or can be requested from the corresponding author.

Acknowledgments: We would like to thank the National Institutes of Health (R01GM114588 and RF1AG083937 to L.M.M.) for financial support for this project. We would also like to thank Nigam P. Rath (Univ. of Missouri—St. Louis, USA) for assistance with X-ray structure analysis.

Conflicts of Interest: The authors declare no competing financial interest.

References

1. Krylova, K.; Kulatilleke, C.P.; Heeg, M.J.; Salhi, C.A.; Ochrymowycz, L.A.; Rorabacher, D.B. A Structural Strategy for Generating Rapid Electron-Transfer Kinetics in Copper(II/I) Systems. *Inorg. Chem.* **1999**, *38*, 4322. [CrossRef]
2. Rorabacher, D.B. Electron Transfer by Copper Centers. *Chem. Rev.* **2004**, *104*, 651. [CrossRef]
3. Himes, R.A.; Karlin, K.D. Copper–dioxygen complex mediated C–H bond oxygenation: Relevance for particulate methane monooxygenase (pMMO). *Curr. Opin. Chem. Biol.* **2009**, *13*, 119. [CrossRef] [PubMed]
4. Mirica, L.M.; Ottenwaelder, X.; Stack, T.D.P. Structure and Spectroscopy of Copper-Dioxygen Complexes. *Chem. Rev.* **2004**, *104*, 1013. [CrossRef]
5. You, Y.; Han, Y.; Lee, Y.-M.; Park, S.Y.; Nam, W.; Lippard, S.J. Phosphorescent Sensor for Robust Quantification of Copper(II) Ion. *J. Am. Chem. Soc.* **2011**, *133*, 11488. [CrossRef]
6. Jung, H.S.; Kwon, P.S.; Lee, J.W.; Kim, J.I.; Hong, C.S.; Kim, J.W.; Yan, S.; Lee, J.Y.; Lee, J.H.; Joo, T.; et al. Coumarin-Derived Cu2+-Selective Fluorescence Sensor: Synthesis, Mechanisms, and Applications in Living Cells. *J. Am. Chem. Soc.* **2009**, *131*, 2008. [CrossRef] [PubMed]
7. Terpstra, K.; Wang, Y.; Huynh, T.T.; Bandara, N.; Cho, H.-J.; Rogers, B.E.; Mirica, L.M. Divalent 2-(4-Hydroxyphenyl)benzothiazole Bifunctional Chelators for 64Cu PET Imaging in Alzheimer's Disease. *Inorg. Chem.* **2022**, *61*, 20326–20336. [CrossRef]
8. Cho, H.J.; Huynh, T.T.; Rogers, B.E.; Mirica, L.M. Design of a multivalent bifunctional chelator for diagnostic (64)Cu PET imaging in Alzheimer's disease. *Proc. Natl. Acad. Sci. USA* **2020**, *117*, 30928. [CrossRef] [PubMed]
9. Guillou, A.; Lima, L.M.P.; Esteban-Gómez, D.; Le Poul, N.; Bartholomä, M.D.; Platas-Iglesias, C.; Delgado, R.; Patinec, V.; Tripier, R. Methylthiazolyl Tacn Ligands for Copper Complexation and Their Bifunctional Chelating Agent Derivatives for Bioconjugation and Copper-64 Radiolabeling: An Example with Bombesin. *Inorg. Chem.* **2019**, *58*, 2669. [CrossRef]
10. Pena-Bonhome, C.; Fiaccabrino, D.; Rama, T.; Fernández-Pavón, D.; Southcott, L.; Zhang, Z.; Lin, K.-S.; de Blas, A.; Patrick, B.O.; Schaffer, P.; et al. Toward 68Ga and 64Cu Positron Emission Tomography Probes: Is H2dedpa-N,N'-pram the Missing Link for dedpa Conjugation? *Inorg. Chem.* **2023**, *ASAP*. [CrossRef]
11. Cai, Z.; Anderson, C.J. Chelators for copper radionuclides in positron emission tomography radiopharmaceuticals. *J. Label. Compd. Radiopharm.* **2014**, *57*, 224. [CrossRef]
12. Bandara, N.; Sharma, A.K.; Krieger, S.; Schultz, J.W.; Han, B.H.; Rogers, B.E.; Mirica, L.M. Evaluation of ^{64}Cu-based Radiopharmaceuticals That Target Aβ Peptide Aggregates as Diagnostic Tools for Alzheimer's Disease. *J. Am. Chem. Soc.* **2017**, *139*, 12550. [CrossRef] [PubMed]
13. Morfin, J.-F.; Lacerda, S.; Geraldes, C.F.G.C.; Tóth, É. Metal complexes for the visualisation of amyloid peptides. *Sens. Diagn.* **2022**, *1*, 627. [CrossRef]

14. Uzal-Varela, R.; Patinec, V.; Tripier, R.; Valencia, L.; Maneiro, M.; Canle, M.; Platas-Iglesias, C.; Esteban-Gómez, D.; Iglesias, E. On the dissociation pathways of copper complexes relevant as PET imaging agents. *J. Inorg. Biochem.* **2022**, *236*, 111951. [CrossRef]
15. Tosato, M.; Franchi, S.; Isse, A.A.; Del Vecchio, A.; Zanoni, G.; Alker, A.; Asti, M.; Gyr, T.; Di Marco, V.; Mäcke, H. Is Smaller Better? Cu2+/Cu+ Coordination Chemistry and Copper-64 Radiochemical Investigation of a 1,4,7-Triazacyclononane-Based Sulfur-Rich Chelator. *Inorg. Chem.* **2023**; *ASAP*. [CrossRef]
16. Tosato, M.; Dalla Tiezza, M.; May, N.V.; Isse, A.A.; Nardella, S.; Orian, L.; Verona, M.; Vaccarin, C.; Alker, A.; Mäcke, H.; et al. Copper Coordination Chemistry of Sulfur Pendant Cyclen Derivatives: An Attempt to Hinder the Reductive-Induced Demetalation in 64/67Cu Radiopharmaceuticals. *Inorg. Chem.* **2021**, *60*, 11530. [CrossRef]
17. Matz, D.L.; Jones, D.G.; Roewe, K.D.; Gorbet, M.J.; Zhang, Z.; Chen, Z.Q.; Prior, T.J.; Archibald, S.J.; Yin, G.C.; Hubin, T.J. Synthesis, structural studies, kinetic stability, and oxidation catalysis of the late first row transition metal complexes of 4,10-dimethyl-1,4,7,10-tetraazabicyclo[6.5.2]pentadecane. *Dalton Trans.* **2015**, *44*, 12210. [CrossRef] [PubMed]
18. Knighton, R.C.; Troadec, T.; Mazan, V.; Le Saëc, P.; Marionneau-Lambot, S.; Le Bihan, T.; Saffon-Merceron, N.; Le Bris, N.; Chérel, M.; Faivre-Chauvet, A.; et al. Cyclam-Based Chelators Bearing Phosphonated Pyridine Pendants for 64Cu-PET Imaging: Synthesis, Physicochemical Studies, Radiolabeling, and Bioimaging. *Inorg. Chem.* **2021**, *60*, 2634. [CrossRef]
19. Lima, L.M.P.; Halime, Z.; Marion, R.; Camus, N.; Delgado, R.; Platas-Iglesias, C.; Tripier, R. Monopicolinate Cross-Bridged Cyclam Combining Very Fast Complexation with Very High Stability and Inertness of Its Copper(II) Complex. *Inorg. Chem.* **2014**, *53*, 5269. [CrossRef]
20. Hierlmeier, I.; Guillou, A.; Earley, D.F.; Linden, A.; Holland, J.P.; Bartholomä, M.D. HNODThia: A Promising Chelator for the Development of 64Cu Radiopharmaceuticals. *Inorg. Chem.* **2023**; *ASAP*. [CrossRef]
21. Brudenell, S.J.; Spiccia, L.; Tiekink, E.R.T. Binuclear Copper(II) Complexes of Bis(pentadentate) Ligands Derived from Alkyl-Bridged Bis(1,4,7-triazacyclonane) Macrocycles. *Inorg. Chem.* **1996**, *35*, 1974. [CrossRef]
22. Khusnutdinova, J.R.; Luo, J.; Rath, N.P.; Mirica, L.M. Late First-Row Transition Metal Complexes of a Tetradentate Pyridinophane Ligand: Electronic Properties and Reactivity Implications. *Inorg. Chem.* **2013**, *52*, 3920. [CrossRef] [PubMed]
23. Wessel, A.J.; Schultz, J.W.; Tang, F.; Duan, H.; Mirica, L.M. Improved synthesis of symmetrically & asymmetrically N-substituted pyridinophane derivatives. *Org. Biomol. Chem.* **2017**, *15*, 9923.
24. Huang, Y.; Huynh, T.T.; Sun, L.; Hu, C.-H.; Wang, Y.-C.; Rogers, B.E.; Mirica, L.M. Neutral Ligands as Potential [64]Cu Chelators for Positron Emission Tomography Imaging Applications in Alzheimer's Disease. *Inorg. Chem.* **2022**, *61*, 4778. [CrossRef] [PubMed]
25. Halfen, J.A.; Tolman, W.B.; Weighardt, K. C2-Symmetric 1,4-Diisopropyl-7-R-1,4,7-Triazacyclononanes. In *Inorganic Syntheses*; John Wiley & Sons, Inc.: Hoboken, NJ, USA, 2007.
26. Li, Y.; Lu, X.-M.; Sheng, X.; Lu, G.-Y.; Shao, Y.; Xu, Q. DNA cleavage promoted by Cu2+ complex of cyclen containing pyridine subunit. *J. Inclusion Phenom. Macrocyclic Chem.* **2007**, *59*, 91. [CrossRef]
27. Che, C.M.; Li, Z.Y.; Wong, K.Y.; Poon, C.K.; Mak, T.C.W.; Peng, S.M. A simple synthetic route to N,N'-dialkyl-2,11-diaza[3.3](2,6)pyridinophanes. Crystal structures of N,N'-di-tert-butyl-2,11-diaza[3.3](2,6)pyridinophane and its copper(II) complex. *Polyhedron* **1994**, *13*, 771. [CrossRef]
28. Addison, A.W.; Rao, T.N.; Reedijk, J.; van Rijn, J.; Verschoor, G.C. Synthesis, structure, and spectroscopic properties of copper(II) compounds containing nitrogen-sulphur donor ligands; the crystal and molecular structure of aqua[1,7-bis(N-methylbenzimidazol-2[prime or minute]-yl)-2,6-dithiaheptane]copper(II) perchlorate. *J. Chem. Soc. Dalton Trans.* **1984**, *7*, 1349. [CrossRef]
29. Cotton, F.A.; Wilkinson, G. *Advanced Inorganic Chemistry*, 5th ed.; Wiley–Interscience: New York, NY, USA, 1988.
30. Pratt, R.C.; Mirica, L.M.; Stack, T.D.P. Snapshots of a metamorphosing Cu(II) ground state in a galactose oxidase-inspired complex. *Inorg. Chem.* **2004**, *43*, 8030. [CrossRef]
31. Bottino, F.; Di Grazia, M.; Finocchiaro, P.; Fronczek, F.R.; Mamo, A.; Pappalardo, S. Reaction of Tosylamide Monosodium Salt with Bis(halomethyl) Compounds: An Easy Entry to Symmetrical N-tosylazamacrocycles. *J. Org. Chem.* **1988**, *53*, 3521. [CrossRef]
32. Alderighi, L.; Gans, P.; Ienco, A.; Peters, D.; Sabatini, A.; Vacca, A. Hyperquad simulation and speciation (HySS): A utility program for the investigation of equilibria involving soluble and partially soluble species. *Coord. Chem. Rev.* **1999**, *184*, 311. [CrossRef]
33. Green, K.N.; Pota, K.; Tircsó, G.; Gogolák, R.A.; Kinsinger, O.; Davda, C.; Blain, K.; Brewer, S.M.; Gonzalez, P.; Johnston, H.M.; et al. Dialing in on pharmacological features for a therapeutic antioxidant small molecule. *Dalton Trans.* **2019**, *48*, 12430. [CrossRef]
34. Marlin, A.; Koller, A.; Madarasi, E.; Cordier, M.; Esteban-Gómez, D.; Platas-Iglesias, C.; Tircsó, G.; Boros, E.; Patinec, V.; Tripier, R. H3nota Derivatives Possessing Picolyl and Picolinate Pendants for Ga[3+] Coordination and 67Ga[3+] Radiolabeling. *Inorg. Chem.* **2023**; *ASAP*. [CrossRef] [PubMed]
35. Nonat, A.M.; Gateau, C.; Fries, P.H.; Helm, L.; Mazzanti, M. New Bisaqua Picolinate-Based Gadolinium Complexes as MRI Contrast Agents with Substantial High-Field Relaxivities. *Eur. J. Inorg. Chem.* **2012**, *2012*, 2049. [CrossRef]
36. Nonat, A.; Gateau, C.; Fries, P.H.; Mazzanti, M. Lanthanide Complexes of a Picolinate Ligand Derived from 1,4,7-Triazacyclononane with Potential Application in Magnetic Resonance Imaging and Time-Resolved Luminescence Imaging. *Chem. Eur. J.* **2006**, *12*, 7133. [CrossRef]
37. Brasse, D.; Nonat, A. Radiometals: Towards a new success story in nuclear imaging? *Dalton Trans.* **2015**, *44*, 4845. [CrossRef]
38. Price, E.W.; Orvig, C. Matching chelators to radiometals for radiopharmaceuticals. *Chem. Soc. Rev.* **2014**, *43*, 260. [CrossRef]

39. Wadas, T.J.; Wong, E.H.; Weisman, G.R.; Anderson, C.J. Coordinating Radiometals of Copper, Gallium, Indium, Yttrium, and Zirconium for PET and SPECT Imaging of Disease. *Chem. Rev.* **2010**, *110*, 2858. [CrossRef] [PubMed]
40. Woodin, K.S.; Heroux, K.J.; Boswell, C.A.; Wong, E.H.; Weisman, G.R.; Niu, W.; Tomellini, S.A.; Anderson, C.J.; Zakharov, L.N.; Rheingold, A.L. Kinetic Inertness and Electrochemical Behavior of Copper(II) Tetraazamacrocyclic Complexes: Possible Implications for in Vivo Stability. *Eur. J. Inorg. Chem.* **2005**, *2005*, 4829. [CrossRef]
41. Anderegg, G.; Arnaud-Neu, F.; Delgado, R.; Felcman, J.; Popov, K. Critical evaluation of stability constants of metal complexes of complexones for biomedical and environmental applications* (IUPAC Technical Report). *Pure Appl. Chem.* **2005**, *77*, 1445. [CrossRef]
42. Martell, A.E.; Motekaitis, R.J.; Clarke, E.T.; Delgado, R.; Sun, Y.; Ma, R. Stability constants of metal complexes of macrocyclic ligands with pendant donor groups. *Supramol. Chem.* **1996**, *6*, 353. [CrossRef]
43. Irangu, J.; Ferguson, M.J.; Jordan, R.B. Reaction of copper(II) with ferrocene and 1,1'-dimethylferrocene in aqueous acetonitrile: The copper(II/I) self-exchange rate. *Inorg. Chem.* **2005**, *44*, 1619. [CrossRef]
44. Evans, D.F. Determination of the Paramagnetic Susceptibility of Substances in Solution By NMR. *J. Chem. Soc.* **1959**, 2003. [CrossRef]
45. De Buysser, K.; Herman, G.G.; Bruneel, E.; Hoste, S.; Van Driessche, I. Determination of the Number of Unpaired Electrons in Metal-Complexes. A Comparison Between the Evans' Method and Susceptometer Results. *Chem. Phys.* **2005**, *315*, 286. [CrossRef]
46. Bain, G.A.; Berry, J.F. Diamagnetic Corrections and Pascal's Constants. *J. Chem. Educ.* **2008**, *85*, 532. [CrossRef]
47. Krause, L.; Herbst-Irmer, R.; Sheldrick, G.M.; Stalke, D. Comparison of silver and molybdenum microfocus X-ray sources for single-crystal structure determination. *J. Appl. Crystallogr.* **2015**, *48*, 3. [CrossRef] [PubMed]
48. Sheldrick, G. SHELXT—Integrated space-group and crystal-structure determination. *Acta Cryst. Sect. A* **2015**, *71*, 3. [CrossRef]
49. Bagchi, P.; Morgan, M.T.; Bacsa, J.; Fahrni, C.J. Robust Affinity Standards for Cu(I) Biochemistry. *J. Am. Chem. Soc.* **2013**, *135*, 18549. [CrossRef]

Disclaimer/Publisher's Note: The statements, opinions and data contained in all publications are solely those of the individual author(s) and contributor(s) and not of MDPI and/or the editor(s). MDPI and/or the editor(s) disclaim responsibility for any injury to people or property resulting from any ideas, methods, instructions or products referred to in the content.

Article

Effect of Metal Environment and Immobilization on the Catalytic Activity of a Cu Superoxide Dismutase Mimic

Micaela Richezzi [1], Joaquín Ferreyra [1], Sharon Signorella [1], Claudia Palopoli [1], Gustavo Terrestre [1], Nora Pellegri [2], Christelle Hureau [3] and Sandra R. Signorella [1,*]

[1] IQUIR (Instituto de Química Rosario), Consejo Nacional de Investigaciones Científicas y Técnicas (CONICET), Facultad de Ciencias Bioquímicas y Farmacéuticas, Universidad Nacional de Rosario, Suipacha 531, Rosario 2000, Argentina; richezzi@iquir-conicet.gov.ar (M.R.); ferreyra@iquir-conicet.gov.ar (J.F.); ssignorella@iquir-conicet.gov.ar (S.S.); palopoli@iquir-conicet.gov.ar (C.P.); terrestre@iquir-conicet.gov.ar (G.T.)

[2] IFIR (Instituto de Física Rosario), Consejo Nacional de Investigaciones Científicas y Técnicas (CONICET), Facultad de Ciencias Exactas, Ingeniería y Agrimensura, Universidad Nacional de Rosario, 27 de Febrero 210 bis, Rosario 2000, Argentina; pellegri@ifir-conicet.gov.ar

[3] LCC (Laboratoire de Chimie de Coordination) CNRS, Université de Toulouse, 205 Route de Narbonne, 31077 Toulouse, France; christelle.hureau@lcc-toulouse.fr

* Correspondence: signorella@iquir-conicet.gov.ar

Abstract: The Cu(II)/Cu(I) conversion involves variation in the coordination number and geometry around the metal center. Therefore, the flexibility/rigidity of the ligand plays a critical role in the design of copper superoxide dismutase (SOD) mimics. A 1,3-Bis[(pyridin-2-ylmethyl)(propargyl)amino]propane (pypapn), a flexible ligand with an N_4-donor set, was used to prepare [Cu(pypapn)(ClO$_4$)$_2$], a trans-Cu(II) complex whose structure was determined by the X-ray diffraction. In DMF or water, perchlorate anions are exchanged with solvent molecules, affording [Cu(pypan)(solv)$_2$]$^{2+}$ that catalyzes O$_2$$^{\bullet-}$ dismutation with a second-order rate constant $k_{\mathrm{McF}} = 1.26 \times 10^7$ M^{-1} s^{-1}, at pH 7.8. This high activity results from a combination of ligand flexibility, total charge, and labile binding sites, which places [Cu(pypapn)(solv)$_2$]$^{2+}$ above other mononuclear Cu(II) complexes with more favorable redox potentials. The covalent anchoring of the alkyne group of the complex to azide functionalized mesoporous silica through "click" chemistry resulted in the retention of the SOD activity and improved stability. A dicationic Cu(II)-N_4-Schiff base complex encapsulated in mesoporous silica was also tested as an SOD mimic, displaying higher activity than the free complex, although lower than [Cu(pypapn)(solv)$_2$]$^{2+}$. The robustness of covalently attached or encapsulated doubly charged Cu(II) complexes in a mesoporous matrix appears as a suitable approach for the design of copper-based hybrid catalysts for O$_2$$^{\bullet-}$ dismutation under physiological conditions.

Keywords: Cu-based SOD mimic; structure; mesoporous silica; click chemistry; SOD activity

1. Introduction

Superoxide radical (O$_2$$^{\bullet-}$) is part of aerobic life. It is formed during the O$_2$ metabolism under normal physiological conditions and is the major initial form of other reactive oxygen species (ROS) associated with oxidative stress, such as H$_2$O$_2$ and HO$^{\bullet}$ radicals [1]. Superoxide dismutase (SOD) enzymes are efficient endogenous defenses that catalyze O$_2$$^{\bullet-}$ dismutation, regulating its concentration and keeping it at a tolerable level [2]. The imbalanced production of O$_2$$^{\bullet-}$ and other ROS associated with a number of neurodegenerative diseases [3] has stimulated the research of synthetic catalytic antioxidants to assist the endogenous counterparts in the suppression of ROS [4]. Among SOD enzymes, copper–zinc superoxide dismutase enzyme (CuZnSOD) catalyzes the proton-dependent dismutation of O$_2$$^{\bullet-}$ at a bimetallic site, where the copper ion is the redox partner of O$_2$$^{\bullet-}$, and Zn(II) plays a structural role in the enzyme stability [2,5]. In the active site of the oxidized form of the enzyme, the Cu(II) ion is bound to the N_4-donor set of four histidine residues and one water molecule, adopting a distorted square–pyramidal geometry [6]. The Cu(II) ion is bridged to the Zn(II) ion through the imidazolate ring from one of the histidine residues, and the Zn(II) coordination sphere is completed by two histidine and one aspartate residues disposed in a distorted tetrahedral geometry (Figure 1) [5,6].

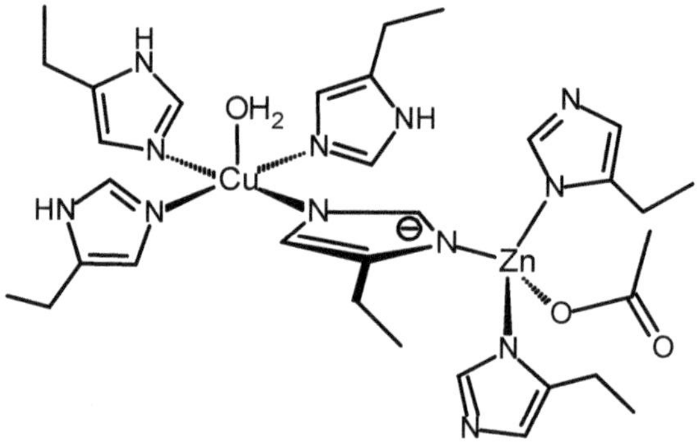

Figure 1. Active site of CuZnSOD.

In the search for efficient SOD mimics, a number of Cu(II) complexes with non-heme ligand scaffolds bearing an N_4-donor set have been synthesized and their SOD activity evaluated [7–17]. During the redox cycle, the switch between Cu(II) and Cu(I) oxidation states involves changes in coordination number and geometry around the metal ion, i.e., from tetragonal to tetrahedral [18,19]. Therefore, more flexible ligands can be thought to be better candidates to speed up electron transfer with a low reorganization energy barrier [20,21]. To serve as a therapeutic antioxidant, besides being active, the complex must be stable at physiological pH. However, in a solution, SOD mimics may undergo hydrolysis, metal dissociation, or oligomerization processes during the redox reaction [8,22,23]. Immobilization of a catalyst in a mesoporous solid has proven to be a good strategy for its protection and site isolation, improving its stability while preserving the properties of the homogeneous system [24–28]. Among the solid supports, mesoporous silica particles possess a large contact surface and pore volume, which allows for high catalyst loading [29,30], chemical and mechanical stability [31,32], biocompatibility [33], and controllable geometric parameters that enable a suitable design of different types and sizes of pores [29]. An effective approach to reducing the complex leaching from the silica consists of the covalent attachment of the catalyst to mesoporous silica employing "click" chemistry [34–37]. In this work, the SOD activity of the Cu(II) complex formed with an N_4-tetradentate "clickable" ligand, 1,3-bis[(pyridin-2-ylmethyl)(propargyl)amino]propane (pypapn, Scheme 1) [38], was evaluated in homogeneous phase and covalently grafted to azido functionalized mesoporous silicas, aimed at assessing the role played by this ligand on the catalytic activity and the effect of the covalent anchoring on the catalyst stability. Additionally, the SOD activity of the Cu(II) complex of N,N'-bis(2-pyridylmethylen)propane-1,3-diamine (py_2pn, Scheme 1), an N_4-Schiff base ligand, was evaluated after insertion in mesoporous silica by ionic exchange. Results obtained via employing these two immobilization approaches are compared with the intention to ascertain the effect of ionic vs. covalent binding of the catalyst on the support of the turnover numbers and catalyst recovery.

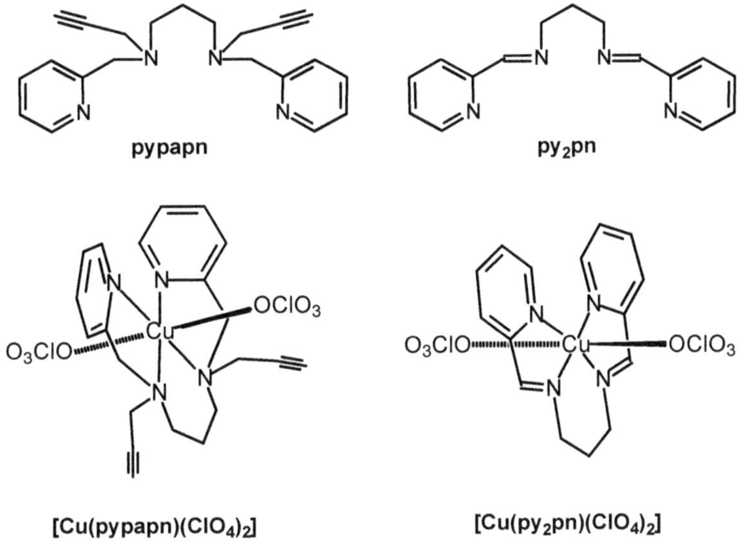

Scheme 1. Ligands and complexes used in this work.

2. Results

2.1. Characterization of the Complex

The reaction of equimolar amounts of pypapn and Cu(OAc)$_2$ in methanol at room temperature afforded a greenish–blue solution. Even when color changes were observed immediately after mixing, it was necessary to add sodium perchlorate and then hexane for the [Cu(pypapn)(ClO$_4$)$_2$] complex to separate from the solution as a violet–blue polycrystalline solid. Single crystals of [Cu(pypapn)(ClO$_4$)$_2$] could be obtained by slow diffusion of the reaction mixture into toluene at 4 °C. X-ray diffraction analysis revealed that the complex crystallizes in the orthorhombic F d d 2 space group with a lattice comprising discrete [Cu(pypapn)(ClO$_4$)$_2$] molecules with Cu(II) bound to the tetradentate ligand and two perchlorate anions, as illustrated in Figure 2.

Bond distances and angles summarized in the caption of Figure 2 indicate that the coordination environment of the Cu(II) ion can be described as an elongated octahedron in which the equatorial plane is defined by the N_4-donor set of the pypapn ligand while the apical positions are occupied by two oxygen atoms belonging to two perchlorate anions. The Cu-O bond distances (2.620 Å) are significantly longer than the equatorial Cu-N ones (Cu-N$_{py}$ 1.987 and Cu-N$_{am}$ 2.057 Å), indicating that the perchlorate anions are weakly bound to the metal ion. The values of the dihedral angles around the Cu(II) center are between 82.28 and 99.53°, evidencing the deviation from an ideal octahedral geometry. In the crystal lattice of [Cu(pypapn)(ClO$_4$)$_2$], the molecules are arranged in layers oriented parallel to the bc plane (Figure S1) and interconnected through H-bond contacts between the perchlorate O-atoms and the alkyne H-atom of a neighbor molecule, with H(C9)···O2 = 2.370 Å and H(C9)···O4 = 3.187 Å distances. The long axes of the complex molecules in each layer are aligned on the crystallographic a-axis intercalated between molecules of adjacent layers.

The FT-IR spectrum of the complex (Figure 3a) displays intense absorption bands at 3262 and 2117 cm^{-1}, assigned to the ≡C-H and C≡C stretching vibrations of the terminal alkyne, and at 1118, 1067, and 622 cm^{-1}, corresponding to the perchlorate anions. Additionally, the shift of the strong in-plane C=N and C=C stretching vibrations of the pyridine ring from 1588 and 1437 cm^{-1} in the ligand to 1611 and 1445 cm^{-1} in the complex is a clear indication of the metal coordination to the N_{py} atom [39], in agreement with the crystal structure of the complex.

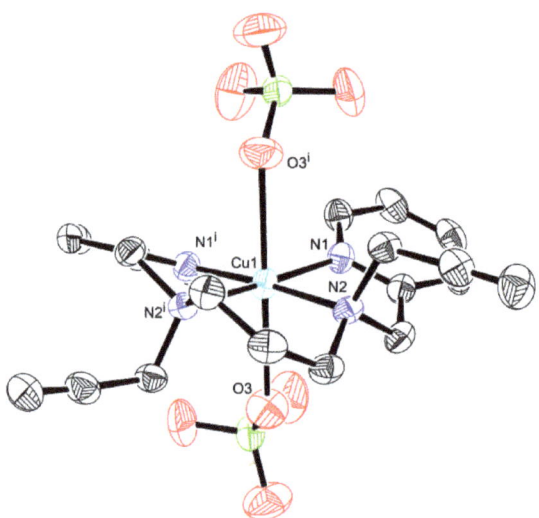

Figure 2. Molecular structure of complex [Cu(pypapn)(ClO$_4$)$_2$] at the 50% probability level with atom numbering. Selected bond lengths (Å) and angles (°): Cu1-N1 1.987(3); Cu1-N2 2.052(3); Cu1-O3 2.620(3); N1-Cu1-N1^i 99.53(19); N1-Cu1-N2^i 175.97(14); N1-Cu1-N2 82.28(10); N2-Cu1-N2^i 96.13(18); N1-Cu1-O3 91.53(13); N2-Cu1-O3 87.68(11); N1^i-Cu1-O3 88.67(12); N2^i-Cu1-O3 92.11(12); O3-Cu1-O3^i 179.69 (18). Standard deviations in parentheses. Symmetry transformations used to generate equivalent i atoms: −x + 1; −y + 1; z.

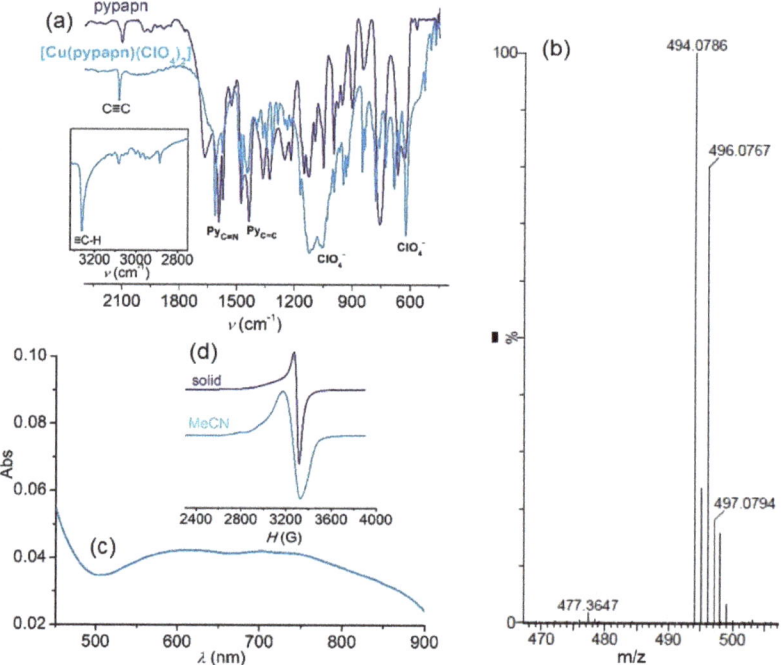

Figure 3. (**a**) FT-IR spectra of pypapn and [Cu(pypapn)(ClO$_4$)$_2$]. (**b**) HRMS of [Cu(pypapn)(ClO$_4$)]$^+$ in MeCN. (**c**) Electronic spectrum of 5×10^{-4} M [Cu(pypapn)(ClO$_4$)$_2$] in MeCN. (**d**) EPR spectra of powdered and frozen MeCN solution of [Cu(pypapn)(ClO$_4$)$_2$]; T = 120 K; ν = 9.51 GHz.

The chemical composition of the complex in solution was confirmed by the high-resolution mass spectrum (HRMS) of the complex in acetonitrile (Figure 3b) that shows the peak at m/z = 494.0786, corresponding to [Cu(pypapn)(ClO$_4$)]$^+$. The full ESI-mass spectrum (Figure S2) also shows peaks at m/z = 197.5634 (12%) and 395.1297 (11%) assigned to [Cu(pypapn)]$^{2+}$ and [Cu(pypapn)]$^+$, respectively, besides the peak of [Cu(pypapn)(ClO$_4$)]$^+$ (10%). These species are present in relatively low proportion compared to the peak at m/z = 440.1276 (100%), corresponding to the [Cu(pypapn)(CHO$_2$)]$^+$ monocation generated by perchlorate exchange with the formate anion present in the spectrometer. It is worth noting that the isotopic distribution of all the peaks is consistent with the simulated spectra. As expected, in the negative mode, the HRMS exhibits the peaks corresponding to the ClO$_4^-$ anion (not shown).

The electronic spectrum of [Cu(pypapn)(ClO$_4$)$_2$] in acetonitrile displays a low-intensity broad absorption band envelope centered at 653 nm (ε = 76 M^{-1} cm^{-1}), characteristic of the d–d transitions of Cu(II) in a tetragonal field. A detailed examination of this region of the spectrum shows that this band splits into two bands of similar intensity (Figure 3c), with energies of 14,184 cm^{-1} and 16,529 cm^{-1}, which can be assigned to d$_{z2}$ → d$_{x2-y2}$ and d$_{xz/yz}$ → d$_{x2-y2}$ transitions, respectively, consistent with an axially elongated octahedral geometry [18]. Other intense absorptions in the UV region (not shown) correspond to ligand-to-metal charge transfers (LMCT) overlapped with intraligand π-π* and n-π* transitions.

The X-band electron paramagnetic resonance (EPR) spectrum of powdered [Cu(pypapn)(ClO$_4$)$_2$] shows an axial signal typical of Cu(II) sites in tetragonal geometry with $g_\parallel$ = 2.20 and $g_\perp$ = 2.06. The spectrum registered on a frozen acetonitrile solution of [Cu(pypapn)(ClO$_4$)$_2$] (Figure 3d) is consistent with a d$_{x^2-y^2}$ ground state in a slightly distorted N$_4$-square–planar geometry, with spectral parameters $g_\parallel$ = 2.25 and $g_\perp$ = 2.08. In this solvent, the spectrum is broad [40], and only two of the four parallel hyperfine features are observed, from which the hyperfine coupling to the Cu nuclear spin $A_\parallel$ = 181 × 10^{-4} cm^{-1} was estimated. Moreover, the empirical distortion index $f(g_\parallel/A_\parallel)$ = 124 cm [41] indicates that in solution, the complex exhibits slight tetrahedral distortion from tetragonal geometry, in agreement with UV-vis results. In this poorly coordinating solvent (DN^{MeCN} = 14), the perchlorate anions remain bound to Cu(II), affording a neutral molecular complex. This was confirmed by the conductivity of the complex measured in acetonitrile, which shows non-electrolytic behavior (same conductivity as for the neat solvent). By contrast, in DMF, a more coordinating solvent (DN^{DMF} = 27), the molar conductivity of the complex is 132 Ω^{-1} cm^2 mol^{-1}, a value expected for a 1:2 electrolyte in this solvent [42]. Therefore, perchlorate is substituted by DMF to afford [Cu(pypapn)(DMF)$_2$]$^{2+}$. The coordination of DMF is also supported by the bathochromic shift of the d–d transition in the visible spectrum, with a broad band centered at ~750 nm as the axial positions are occupied by DMF, lowering the energy of the transition along the z-axis (Figure S3).

The ^{1}H NMR paramagnetic spectrum of [Cu(pypapn)(ClO$_4$)$_2$] in CD$_3$CN (Figure 4a) shows that the pyridine protons undergo broadening as well as a differential isotropic shift depending on the distance between each proton and the metal center. The broad signal at the low field can be assigned to overlapped H$_\alpha$/H$_\alpha$' at 11.0 ppm, with a peak width of 104 Hz. Meanwhile, the β/β' and γ pyridine protons can be observed at 8.61, 7.95, and 7.48 ppm, displaying signals with a peak width of 20 Hz, 24 Hz, and 32 Hz, respectively. This spectral pattern suggests that the two pyridine rings are symmetrically related around the copper center, which is in agreement with the elongated tetragonal geometry proposed from the EPR spectrum of the complex in this solvent. Moreover, signals belonging to the methylene protons broaden and split in the 2.5–4.0 ppm spectral region, with peak widths in the range from 12 to 32 Hz, depending on their relative location. Integration affords an 8:6 ratio for the pyridine and methylene protons adjacent to the donor N atoms, so it is possible that resonances originating from methylene protons closest to the Cu(II) ion are broadened and shifted downfield. The broad resonance at ~13 ppm can account for this (Inset in Figure 4a).

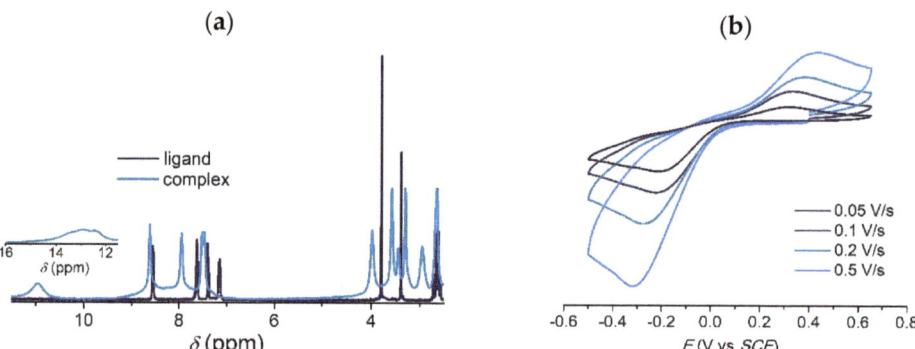

Figure 4. (a) ^{1}H NMR spectrum of [Cu(pypapn)(ClO$_4$)$_2$] in CD$_3$CN. (b) Cyclic voltammogram of 1 mM [Cu(pypapn)(ClO$_4$)$_2$] in acetonitrile, 0.1 M Bu$_4$NPF$_6$, and glassy C/Pt/SCE, at different scan rates.

The donor sites, the geometry, and the flexibility of the ligand are known to modulate the redox behavior of the copper center and, consequently, its reactivity toward superoxide. The redox potential of [Cu(pypapn)(ClO$_4$)$_2$] was determined by cyclic voltammetry. The voltammograms of the complex in acetonitrile at various scan rates (Figure 4b) show the growth of the intensity of the cathodic and anodic peaks as the rate increases, along with a shift of both peaks to more negative and more positive potentials, respectively. The ΔE value (0.55 V at v = 0.1 V/s), its dependence on v, and the I_{pa}/I_{pc} ratio of 0.35 indicate that the reduction in the complex is irreversible and takes place at E_{pc} = −0.22 V vs. SCE (v = 0.1 V/s), while the oxidation of the reduced complex occurs at E_{pa} = 0.33 V vs. SCE. The irreversibility of the process can be attributed to geometry changes experienced by the complex during reduction. Most probably, the reduction in the tetragonal Cu(II) complex is accompanied by perchlorate dissociation to yield Cu(I) in a flattened tetrahedral geometry, and the observed oxidation peak originates from this four-coordinate complex. It must be noted that the absence of a re-dissolution peak upon anodic polarization scans indicates that no copper is released from the Cu(I) complex generated at the electrode. For comparison, [Cu(py$_2$pn)]$^{2+}$, formed with the Schiff base N,N'-bis(2-pyridylmethylen)propane-1,3-diamine, is reduced at −0.044 V vs. SCE [7]. Therefore, the more flexible N_4-diamine/dipyridine ligand stabilizes Cu(II) toward reduction better than the N_4-diimine/dipyridine one, reflecting the higher electron–donor ability of pypapn.

2.2. Synthesis and Characterization of Modified Mesoporous Silicas

2.2.1. Synthesis of Cu-pypntriazole@SBA-15 and Cu-pypntriazole@OP-MS

The [Cu(pypapn)(solv)$_2$]$^{2+}$ complex was covalently anchored to mesoporous silica matrix by reaction of the alkynyl groups with azide functionalized mesoporous silica employing "click" chemistry. Two different approaches were used to prepare the azide-modified mesoporous silica: post-synthetic functionalization of SBA-15 silica; and a co-condensation method [28]. The first method consists of grafting (3-bromopropyl)trichlorosilane to SBA-15 mesoporous silica to yield Brpn@SBA-15, followed by the bromine substitution by azide to afford N$_3$pn@SBA-15 (Scheme 2, route (a)). The other procedure involves a "one-pot" co-condensation reaction using tetraethyl orthosilicate (TEOS) and 3-azidopropyltriethoxysilane (AzPTES) in the presence of the triblock copolymer Pluronic P-123 as surfactant template (Scheme 2, route (b)). In the one-pot methodology, the azidopropyl chains are placed between the copolymer chains favoring the binding of the organic groups to the inner pore walls of the silica particles [43,44]. The material obtained by this procedure exhibits textural properties similar to SBA-15 (see below). Subsequently, the pypapn ligand was covalently linked to the two azide functionalized silicas by the formation of the 1,2,3-triazole ring, using (Ph$_3$P)$_3$CuBr as a catalyst, to afford pypntriazole@SBA-15 and pypntriazole@OP-MS. In each case, the reaction was stopped after the disappearance of the azide band in the ATR-FTIR spectra of the solid samples. The Cu(II) complex was formed in situ by the addition of a solution of Cu(ClO$_4$)$_2$ to a suspension of pypntriazole@SBA-15 or pypntriazole@OP-MS in methanol, yielding Cu-pypntriazole@SBA-15 and Cu-pypntriazole@OP-MS, with similar Cu(II) loading, as described in the experimental part. In both hybrid materials, the ligand is attached to the channel walls at a single binding site (Scheme 2), as evidenced by metal and nitrogen analyses.

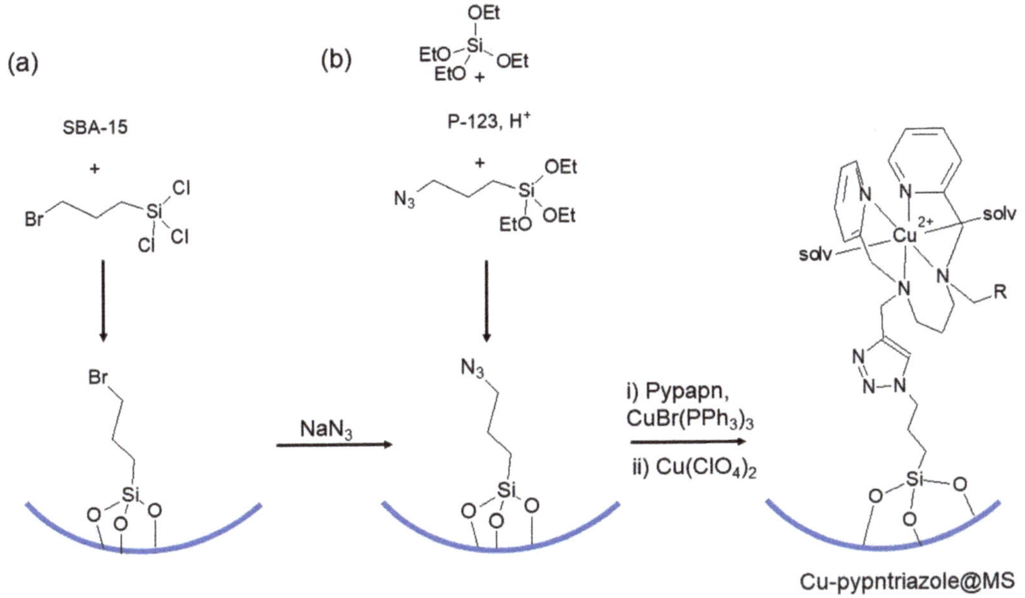

Scheme 2. Synthetic routes toward the hybrid catalyst. MS = SBA-15 or OP-MS.

For comparative purposes, the complex [Cu(py$_2$pn)]$^{2+}$ formed with the N_4-Schiff-base was introduced into the mesoporous matrix of SBA-15 silica by ionic exchange, affording the Cu-py$_2$pn@SBA-15 hybrid material where the divalent complex cation was retained inside the silica channels by electrostatic interaction with the silanolate groups of the pores' surfaces. The FT-IR spectrum of the material confirms that the complex is essentially located inside the pores since it shows only the strong bands of the Si-O-Si, Si-OH$_2$, and Si-OH groups but no or negligible bands belonging to the functional groups of the Schiff-base. The lower proportion of the encapsulated complex in SBA-15 agrees with the smaller decrease in the pore volume and surface area (see below) determined from the adsorption N$_2$ isotherms of this complex compared to Cu-pypntriazole@SBA-15.

2.2.2. Textural Properties and Morphology of the Mesoporous Materials

SBA-15, prepared following a reported methodology [11], shows an ordered mesostructure with a high specific surface area and pore width of 4.9 nm (Table 1) suitable to host the catalysts, which are around 1.0 nm wide (calculated from the crystal structures of [Cu(pypapn)(ClO$_4$)$_2$] and [Cu(py$_2$pn)(ClO$_4$)$_2$] [7]).

Table 1. Textural characterization of mesoporous materials.

	S_{BET} $(m^2\ g^{-1})$	$V_{\mu P}$ $(cm^3\ g^{-1})$	V_{MP} $(cm^3\ g^{-1})$	V_{TP} $(cm^3\ g^{-1})$	w_P (nm)	mmol Complex/ 100 g Material
SBA-15	641	0.03	0.64	0.79	4.9	-
Cu-pypntriazole@SBA-15	310	0.00	0.34	0.41	3.7	24
Cu-py$_2$pn@SBA-15	501	0.00	0.57	0.71	4.4	9.8
N$_3$pn@OP-MS	362	0.00	0.41	0.47	4.8	-
Cu-pypntriazole@OP-MS	433	0.00	0.40	0.46	4.0	18

$V_{TP} = V_{mP} + V_{primary\ MP} + V_{secondary\ MP}$. MP = mesopore. μP = micropore. w_p = pore diameter.

All the materials obtained by catalyst covalent grafting or encapsulation display type IV isotherms with a sharp jump and an H1 hysteresis loop at relative pressure $p/p° = 0.55$–0.75 (Figure 5). This behavior is typical of ordered mesoporous materials, denoting that the hybrids possess mesostructure similar to SBA-15 and uniform distribution of pore sizes, although smaller pore volume and surface area (Table 1) as a consequence of the presence of the organic groups or catalyst inside the channels.

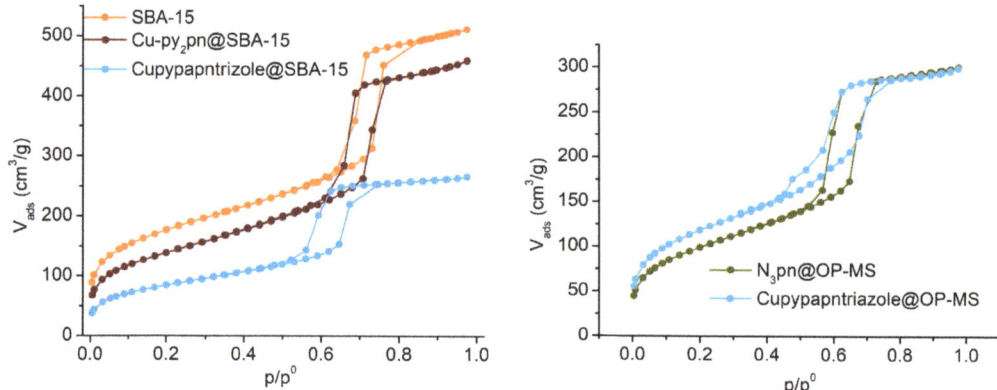

Figure 5. Adsorption–desorption N$_2$ isotherms of mesoporous materials.

The morphology and size of the mesoporous silica particles were analyzed by scanning electron microscopy (SEM, Figure S4). The SBA-15 particles are oblong, with an average size of ~1–1.5 μm long, lined up forming chains of 30–50 μm long. The hybrid materials obtained by either covalent attachment or encapsulation of the catalyst in SBA-15 maintain the morphology of the starting silica particles, although they align to form shorter chains of 15–20 μm long. Fiber-like mesoporous N$_3$pn@OP-MS is made up of a bundle of wires forming long strands of 50–100 μm long, and after the "click" reaction, exhibits a similar morphology. Upon Cu binding, the hybrid catalyst shows rod-shaped particles of 2–3 μm long, forming wheatlike aggregates of 10–15 μm long.

Transmission electron microscopy (TEM) images of the three hybrid materials (Figure 6) corroborate the regular array of cylindrical channels, as well as the hexagonal pore arrangement, confirming a highly ordered mesostructure. The images show that the channels are oriented along the long axis of the particles and that the pore network reaches the particle surface, serving as a substrate entrance channel to interact with the catalyst. From statistical analysis of TEM images of Cu-pypntriazole@SBA-15, pore diameter and wall thickness were calculated using a protocol that takes into account integration over grey scales in carefully selected zones where the electron beam was perpendicular to the channels. The average pore diameter and wall thickness calculated in this way are 3.5 ± 0.6 nm and 4.7 ± 0.5 nm, respectively, in agreement with the pore size calculated from the adsorption isotherm. The histograms are shown in Figure S5.

These thick walls reflect the hydrothermal stability of the mesostructured silica particles without enlargement of the pore size and confer robustness to the catalytic material.

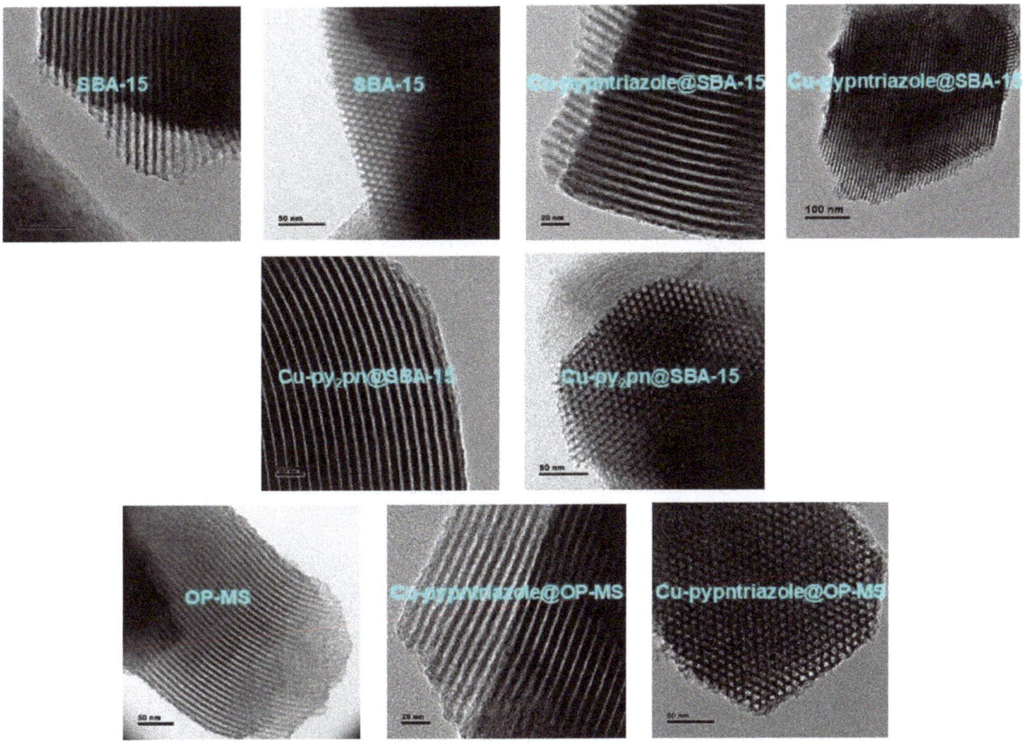

Figure 6. TEM images of hybrid materials showing the parallel channels and their hexagonal arrangement.

The ordered mesostructure of SBA-15 and Cu-pypntriazole@SBA-15 was also verified by low-angle X-ray diffraction (XRD, Figure S6). The XRD pattern indicates that SBA-15 exhibits a long-range mesoscopic ordering of hexagonal p6mm phases, as evidenced by the intense (100) peak and the weaker (110) and (200) reflections [45]. The observed pattern is not modified after anchoring the complex, confirming that the integrity of the porous structure is maintained. The peak at $2\theta = 0.93$ is not shifted after covalent attachment, but it shows a 25% decrease, as well as the higher order reflections, possibly due to a decrease in periodicity [46].

2.3. SOD Activity Studies

The SOD activity of the complex and the hybrid materials was measured in a phosphate buffer of pH 7.8. As described before, when [Cu(pypapn)(ClO$_4$)$_2$] dissolves in MeCN, perchlorate anions remain coordinated to Cu(II), but when it is dissolved in a stronger donor solvent like DMF or water, ligand exchange occurs, and perchlorate is replaced by solvent molecules affording the dicationic [Cu(pypapn)(solv)$_2$]$^{2+}$ complex. For simplicity, hereafter, the solvated complex will be referred to as [Cu(pypapn)]$^{2+}$. Given the low solubility of the complex in aqueous phosphate buffer, it was first dissolved in DMF, and an aliquot of the concentrated solution was diluted with phosphate buffer of pH 7.8 to a final volume of 3.2 mL. In this way, complete dissolution of the complex was achieved, and the stability was verified by monitoring its electronic spectrum for 2 h. Spectra taken at different times after preparing the solution exhibited analogous absorption features and band intensities (Figure 7a). The stability of the hybrid materials was also checked after incubation in a phosphate buffer of pH 7.8 for 2 h, followed by centrifugation and record of the UV-vis spectra of the supernatant. These measurements denoted the absence of complex released into the solution. The SOD activity of [Cu(pypapn)]$^{2+}$ and the hybrid materials was determined by the Beauchamps and Fridovich indirect assay using the NBT reagent [47] at pH 7.8. This assay is based on kinetic competition between NBT and the catalyst for photogenerated superoxide. Therefore, the SOD activity of the catalyst is measured by its ability to inhibit the formation of formazan, the product of the reaction of NBT with superoxide, observed at 560 nm. [Cu(pypapn)]$^{2+}$ displays activity, as shown in Figure 7b. The IC_{50} value, the concentration of catalyst that inhibits by 50% the NBT reduction by $O_2^{\bullet -}$, was determined from the plot of % inhibition vs. [catalyst] and used to calculate the McCord–Fridovich rate constant, k_{McF}, which is independent of the detector [48].

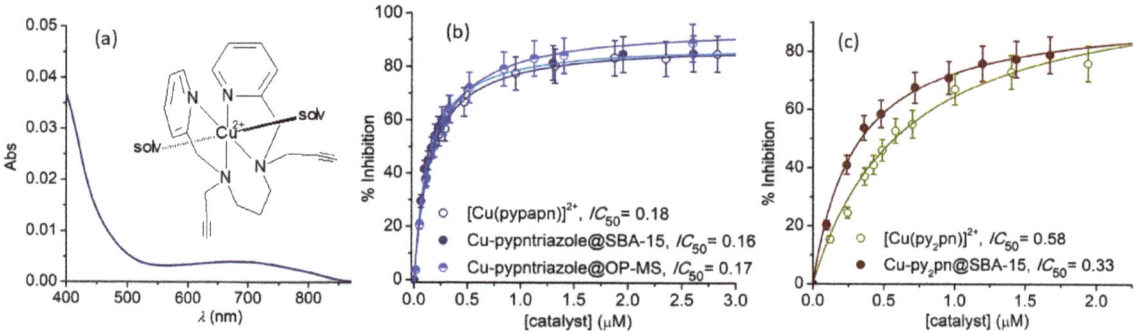

Figure 7. (**a**) UV-vis spectrum of 5×10^{-5} M [Cu(pypapn)]$^{2+}$ in buffer phosphate of pH = 7.8. SOD activity of (**b**) free and covalently attached [Cu(pypapn)]$^{2+}$ and (**c**) free and encapsulated [Cu(py$_2$pn)]$^{2+}$.

Table 2 lists the values of k_{McF} and $E_{1/2}$ (or E_p) redox potential for [Cu(pypapn)]$^{2+}$ and other reported Cu(II) complexes with open-chain ligands (see Chart S1 for their structures) [9–13,49,50]. If the catalyzed $O_2^{\bullet -}$ dismutation occurs through an outer sphere mechanism, a complex will be thermodynamically competent as a catalyst when its redox potential lies between the $E(O_2^{\bullet -}/H_2O_2) = 0.642$ V and $E(O_2/O_2^{\bullet -}) = -0.404$ V vs. SCE redox couples, at pH 7 [51]. On this basis, the best catalysts should be those with a metal-centered redox couple close to 0.12 V vs. SCE (pH 7), the midpoint between the oxidation and reduction in $O_2^{\bullet -}$ radical. However, in Table 2, it can be observed that [Cu(MPBMPA)Cl]$^+$ (E_{pc} = −0.47 V) is twice as active as [Cu(pypapn)]$^{2+}$ (E_{pc} = −0.22 V), even when its redox potential is outside the range for $O_2^{\bullet -}$ dismutation (entries 2,3, Table 2). Furthermore, [Cu(pypapn)]$^{2+}$ exhibits an SOD activity better than [Cu(py$_2$pn)]$^{2+}$ ($E_{1/2}$ = −0.044 V) and similar to [Cu(PBMPA)Cl] ($E_{1/2}$ = 0.213 V), although its redox potential is further away from 0.12 V (entries 3,4,7, Table 2). This behavior can be explained if the reaction takes place through an inner sphere mechanism where the redox potential is not the only relevant factor but also the flexibility of the ligand, steric hindrance, labile coordination sites, and the total charge of the complex. During catalysis, the coordination sphere of Cu in [Cu(pypapn)]$^{2+}$ must change between the tetragonal geometry of Cu(II) to the tetrahedral or tricoordinate geometry preferred by Cu(I). Although both [Cu(pypapn)]$^{2+}$ and [Cu(py$_2$pn)]$^{2+}$ have a similar geometrical arrangement of the N_4-tetradentate ligand around the Cu(II) ion, the diamine pypapn ligand is more flexible than the diimine py$_2$pn and adapts better to the geometrical reorganization required to accommodate the two metal oxidation states during catalysis. This is particularly evident for [Cu(PuPy)]$^{2+}$ (entry 1, Table 2) with a longer and more flexible tetramethylene chain between the imine groups that doubles the SOD activity of [Cu(pypapn)]$^{2+}$. The relevance of steric factors is evident in the four-time decrease in the SOD activity going from [Cu(PuPy)]$^{2+}$ to [Cu(Pu-6-MePy)(H$_2$O)]$^{2+}$, where the o-methyl substituted pyridine derivative hinders the access of $O_2^{\bullet -}$ to the metal center (entries 1,6, Table 2). The number of labile positions available for binding the substrate via ligand exchange is a factor made evident when comparing [Cu(MPBMPA)Cl$_2$] and [Cu(PBMPA)Cl], the first being twice as active as the second (entries 2,4, Table 2). Another example is [CuZn(dien)$_2$(μ-Im)](ClO$_4$)$_3$, which during catalysis, dissociates, converting into [Cu(dien)]$^{2+}$, which can bind O$_2^{\bullet-}$ and promote its dismutation despite its very unfavorable redox potential (entry 5, Table 2). Although with similar ligand flexibility, the SOD activity of [Cu(PBMPA)Cl] is three times higher than that of the related complex [Cu(PClNOL)Cl]$^+$ (entries 4,8, Table 2), probably as a result of the combination of a more favorable redox potential of [Cu(PBMPA)Cl] and a stronger labilizing effect of the carboxylate compared to the alcohol group of [Cu(PClNOL)Cl]$^+$, to facilitate ligand exchange. The more rigid and planar Cu(II) complexes derived from salen and salpn ligands are the least active (entries 9–11, Table 2), even slower than Cu(ClO$_4$)$_2$ (k_{McF} = 2.7 × 10^6 M^{-1} s^{-1}) [52]; hence, reinforcing the ligand flexibility is a key factor in the O$_2^{\bullet-}$ dismutation catalyzed by the Cu(II) complexes.

Table 2. SOD activity of [Cu(pypapn)]$^{2+}$ and other functional SOD models.

Entry	Catalyst	Ligand Donor Sites	k_{McF} (10^6 M^{-1} s^{-1})	$E_{1/2}$ (V vs. SCE)	Ref.
1	[Cu(PuPy)]$^{2+}$	N$_4$	23.6	-	[9]
2	[Cu(MPBMPA)Cl$_2$]	N$_3$	21.2	−0.471	[10]
3	[Cu(pypapn)]$^{2+}$	N$_4$	12.6	−0.22 (E_{pc})	This work
4	[Cu(PBMPA)Cl]	N$_3$O	12.5	0.213	[10]
5	[CuZn(dien)$_2$(μ-Im)(ClO$_4$)$_2$]$^+$	N$_3$N$_{Im}$	6.46	−0.89 (E_{pc})	[11]
6	[Cu(Pu-6-MePy)(H$_2$O)]$^{2+}$	N$_4$	6.3	-	[12]
7	[Cu(py$_2$pn)]$^{2+}$	N$_4$	4.05	−0.044	[7]
8	[Cu(PClNOL)Cl]$^+$	N$_3$O	3.3	−0.416	[13]
9	[Cu(5-EtO-salpn)ZnCl$_2$]	N$_2$O$_2$	2.1	-	[49]
10	[Cu(4-OMe-salchda)ZnCl$_2$]	N$_2$O$_2$	0.87	-	[50]
11	[salpnCuZnCl$_2$]	N$_2$O$_2$	0.85	−0.689	[11]
12	CuZnSOD	N$_4$	2000	0.156	[2]
	Immobilized Catalyst		k_{McF} (10^6 M^{-1} s^{-1})		
13	Cu-pypntriazole@SBA-15		14.2		
14	Cu-pypntriazole@OP-MS		13.3		
15	Cu-py$_2$pn@SBA-15		6.9		

PuPy = N,N'-bis(2-pyridylmethylen)-1,4-butanediamine; HPBMPA, N-propanoate-N,N-bis-(2-pyridylmethyl)amine; dien = diethylenetriamine; HIm = imidazole; Pu-6-MePy = N,N'-bis(2-(6-methylpyridyl)methylen)-1,4-butanediamine; HPClNOL, 1-[bis(pyridin-2-ylmethyl)amino]-3-chloropropan-2-ol; salpn, 1,3-bis(salicylidenamino)propane; 4-OMe-salchda = N,N'-bis(4-methoxysalicylidene)cyclohexane-1,2-diamine.

Even the more flexible and active complexes listed in Table 2 are two orders of magnitude less active than the CuZnSOD enzyme (entry 12, Table 2), where the protein matrix constraints the metal geometry halfway between that of each oxidation state and the small structural reorganization is required for switching between the oxidized and reduced forms of the enzyme results in fast electron transfer.

Aimed at examining the effect of immobilization on the SOD activity and stability of the catalyst, the reactivity of the three hybrid materials toward O$_2^{\bullet-}$ was evaluated (Figure 7b,c). Cu-pypntriazole@SBA-15 and Cu-pypntriazole@OP-MS, with the catalyst covalently attached to the surface of the silica channels, retain the activity of the free complex (entries 13,14, Table 2). This result suggests that the ligand binding through the triazole does not restrain the ligand flexibility or affect the geometry of the metal center. Low-temperature X-band EPR spectra registered after the reaction of these hybrids with an excess of KO$_2$ in DMSO (shown for Cu-pypntriazole@SBA-15 in Figure 8a) confirmed that the immobilized complex keeps the geometrical arrangement of the ligand around the Cu(II) ion. The spectral parameters for Cu-pypntriazole@SBA-15 after reaction with KO$_2$ are $g_\parallel$ = 2.25, $g_\perp$ = 2.06, $A_\parallel$ = 186 × 10^{-4} cm^{-1} and $f(g_\parallel/A_\parallel)$ = 121 cm, denoting that Cu(II) is in a slightly distorted tetragonal geometry, similar to that observed for the complex in a homogeneous phase (Figure 8a). Interestingly, encapsulation of [Cu(py$_2$pn)]$^{2+}$ inside the SBA-15 silica matrix almost doubles the SOD activity of the free complex (entry 15, Table 2), probably because the silanolate–copper interaction renders the copper center more electrophilic, favoring electron transfer from O$_2^{\bullet-}$ to Cu(II). Also, in this case, the low-temperature X-band EPR spectrum registered after the reaction of Cu-py$_2$pn@SBA-15 with an excess of KO$_2$ in DMSO (Figure 8b) accounts for the retention of the complex structure in the silica pores. The calculated spectral parameters of Cu-py$_2$pn@SBA-15 are $g_\parallel$ = 2.23, $g_\perp$ = 2.05, $A_\parallel$ = 182 × 10^{-4} cm^{-1}, and $f(g_\parallel/A_\parallel)$ = 123 cm, analogous to those of the complex in frozen DMSO solution [7]. It must be noted that the absence of the superoxide signal ($g_\parallel$ = 2.1021, $g_\perp$ = 2.003) [53] in the EPR spectra recorded after the reaction of the hybrids with 10-times excess of KO$_2$ in DMSO indicates that these materials act as catalysts for O$_2^{\bullet-}$ dismutation. Aimed at verifying if the SOD activity is retained after several cycles, the NBT conversion was measured with and without the hybrid material after consecutive illuminations of the reaction mixture. In each new illumination, the NBT concentration was kept constant by the addition of the required quantity of NBT to restore the starting concentration. Cu-pypntriazole@SBA-15, Cu-pypntriazole@OP-MS, and Cu-py$_2$pn@SBA-15 retain the activity after several illumination cycles, indicating that, in all cases, immobilization of the catalyst in the mesoporous matrix isolates and protects the complex to react with O$_2^{\bullet-}$, extending the catalyst life and reusability.

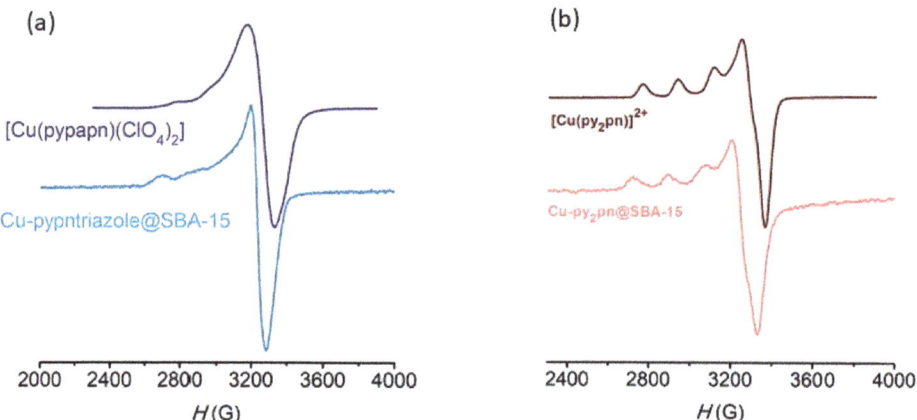

Figure 8. X-band EPR spectra of (**a**) [Cu(pypapn)(ClO$_4$)$_2$] in frozen MeCN solution (ν = 9.5 GHz) and solid Cu-pypntriazole@SBA-15 (ν = 9.31 GHz). (**b**) [Cu(py$_2$pn)]$^{2+}$ in frozen DMSO solution (ν = 9.5 GHz) and solid Cu-py$_2$pn@SBA-15 (ν = 9.31 GHz). T = 120 K.

3. Materials and Methods

In this study, all the used reagents and solvents were commercial products of the highest available purity and, when necessary, were further purified via conventional methods.

3.1. Synthesis of Ligands, Complexes, and Hybrid Materials

The synthesis of the ligands N,N'-bis(2-pyridylmethylen)propane-1,3-diamine (py$_2$pn) [54] and 1,3-bis[(2-pyridilmethyl)(propargyl)amino]propane (pypapn) [38] and the complex [Cu(py$_2$pn)(ClO$_4$)$_2$] [7] were described in previous papers.

3.1.1. Synthesis of [Cu(pypapn)(ClO$_4$)$_2$]

Cu(OAc)$_2$·H$_2$O (30 mg, 0.15 mmol) was dissolved in methanol (5 mL) and added to a solution of pypapn (50 mg, 0.15 mmol) in methanol (2 mL). After stirring the reaction mixture for 1 h at room temperature, NaClO$_4$.H$_2$O (42 mg, 0.30 mmol) in methanol (1 mL) was added. Then, the formed solid was filtered off; hexane was added to the filtrate, and the mixture was left at 4 °C overnight. The solid was filtered, washed with cold methanol and hexane, and dried under a vacuum. Yield: 36 mg (0.06 mmol, 40%). Anal. calcd. for C$_{21}$Cl$_2$CuH$_{24}$N$_4$O$_8$: C 42.4; H 4.03; N 9.40%. Found: C 42.8; H 4.21; N 9.34%. UV-vis, λmax nm (ε M^{-1} cm^{-1}) in acetonitrile: 258 (7600); 280 (sh); 384 (sh); 653 (76). Significant IR bands (KBr, v, cm^{-1}): 3262; 2117; 1611; 1574; 1445; 1118; 622. HRMS (acetonitrile): m/z = 494.0786 [Cu(pypapn)(ClO$_4$)]$^+$; 395.1297 [Cu(pypapn)]$^+$. Violet crystals suitable for X-ray diffraction were obtained after 3 days by slow diffusion of the reaction of mother liquor into toluene at 4 °C.

Caution! *The perchlorate salts used in this study are potentially explosive and should be handled with care.*

3.1.2. Synthesis of Azidopropyl Functionalized Silicas N$_3$pn@SBA-15 and N$_3$pn@OP-MS

A high surface area SBA-15 mesoporous silica was prepared by hydrothermal synthesis, using tetraethoxysilane (TEOS) as Si source and the triblock copolymer Pluronic 123 as template in acid medium, as already reported [11]. The azido-functionalized N$_3$pn@SBA-15 silica was prepared to employ a post-synthetic approach [28] by treating a suspension of 1 g of SBA-15 silica in 30 mL of toluene with 136 mL of 3-(bromopropyl)trichlorosilane, added dropwise, and left stirring for 2 h at 80 °C. The solid was separated by filtration and washed with toluene and ether; then, Soxhlet was extracted with dichloromethane and dried under vacuum at 60 °C. Afterward, the solid was mixed with a saturated solution of NaN$_3$ in dimethylformamide (10.0 mL) and stirred for 2 days. After filtration, the solid was washed with water, acetone, and ethanol and dried at 60 °C to yield 0.95 g of N$_3$pn@SBA-15. Residual mass (%) at 800 °C: 82.5%. IR (KBr): v_{as}(N$_3$) = 2114 cm^{-1}.

The one-pot azido-functionalized mesoporous silica N$_3$pn@OP-MS was prepared by co-condensation of TEOS (2.2 mL) and 3-azidopropyltriethoxysilane (AzPTES, 275 mg) with Pluronic P-123 (1.0 g in 40 mL of aqueous 1.6 M HCl) as surfactant template, following a previously reported methodology [28]. After the template removal by Soxhlet extraction in ethanol, the material was dried at 60 °C to yield 1.1 g of N$_3$pn@OP-MS. Residual mass (%) at 800 °C: 85.3%. IR (KBr): v_{as}(N$_3$) = 2114 cm^{-1}.

3.1.3. Synthesis of Cu-Pypntriazole@SBA-15 and Cu-Pypntriazole@OP-MS

Azide-functionalized mesoporous silicas (350 mg) were suspended in 72 mL of a 75:25 methanol:acetonitrile mixture and stirred for ten minutes. A solution of pypapn in the same solvent mixture (30 mg, 0.09 mmol, 9 mL) was added, and the reaction mixture was stirred for 1 h. Afterward, CuBr(PPh$_3$)$_3$ (84 mg, 0.09 mol) was added and left with stirring at 60 °C. The completion of the reaction was determined by the disappearance of the azide stretching band at 2114 cm^{-1} in the IR spectrum. The solid, pypntriazole@SBA-15 or pypntriazole@OP-MS, was filtered, washed by Soxhlet extraction with methanol, and dried at 60 °C. Then, 250 mg of pypntriazole@SBA-15 was suspended in 25 mL methanol, and a solution of Cu(ClO$_4$)$_2$·6H$_2$O (300 mg, 0.8 mmol) in 5 mL of methanol was added dropwise. The mixture was stirred for one week at room temperature. The solid was filtered, washed with methanol and dichloromethane, and dried at 60 °C to yield 220 mg of Cu-pypntriazole@SBA-15. Anal. (wt.%): N 2.35; Cu 1.5. Catalyst content: 24 mmol/100g. Significant IR bands (KBr, v cm^{-1}): 1640 (d, H-O-H); 1080 (v_{as}, Si-O); 795 (v_s, Si-O); 463 (δ, Si-O-Si). Following the same procedure, 250 mg of pypntriazole@OP-MS yielded 230 mg of Cu-pypntriazole@OP-MS. Anal. (wt.%): N 1.8; Cu 1.15. Catalyst content: 18 mmol/100g. Significant IR bands (KBr, v cm^{-1}): 1640 (δ, H-O-H); 1080 (v_{as}, Si-O); 795 (v_s, Si-O); 463 (δ, Si-O-Si).

3.1.4. Synthesis of Encapsulated Catalyst Cu-py$_2$pn@SBA-15

Complex [Cu(py$_2$pn)(ClO$_4$)$_2$] was inserted in mesoporous silica SBA-15 by the addition of a solution of the complex (100 mg, 0.19 mmol) in methanol (20 mL, 35 °C) to the silica (168 mg). The mixture was stirred for 24 h and filtered. The obtained material was suspended in 5 mL of methanol and left stirring for 24 h. The solid was filtered, washed with methanol, and dried at 60 °C. Yield: 140 mg. Weight loss between 200 and 500 °C: 3.9%. Anal. (wt.%): N 0.55, Cu 0.6. Catalyst content: 9.8 mmol/100g.

3.2. Analytical and Physical Measurements

An inductively coupled plasma mass spectrometer (ICP-MS) PerkinElmer NexION 350× was used to measure the metal content. CHN analyses were performed on a PerkinElmer 2400 series II Analyzer. Infrared spectra were recorded in the 4000–400 cm^{-1} range on a PerkinElmer Spectrum One FTIR spectrophotometer provided with a DTGS detector, resolution = 4 cm^{-1}, and 10 accumulations. FT-IR spectra were registered from KBr sample pellets or ATR-FT-IR. Electronic spectra were recorded on a Jasco V-550 spectrophotometer. Electron Paramagnetic Resonance (EPR) spectra were obtained at 115 K on an Elexsys E 500 Bruker spectrometer, operating at a microwave frequency of ~9.5 GHz, and on a Bruker EMX-Plus spectrometer with a microwave frequency of ~9.3 GHz. Electrospray ionization (ESI) mass spectra were obtained with a Thermo Scientific LCQ Fleet. The solutions for electrospray were prepared from solutions of complex diluted with acetonitrile to a final ~10^{-5} M concentration. ^{1}H NMR spectra were recorded in CD$_3$CN on a Bruker AC 400 NMR spectrometer at ambient probe temperature (ca. 25 °C). Chemical shifts (in ppm) are referenced to tetramethylsilane, and paramagnetic NMR spectra were acquired, employing a superWEFT sequence, with an acquisition time of 270 ms. Conductivity measurements were performed on 1.0 mM solutions of the complexes in MeCN or DMF using a Horiba F-54 BW conductivity meter. The electrochemical experiments were performed with a Princeton Applied Research potentiostat, VERSASTAT II model, with the 270/250 Research Electrochemistry Software. Studies were carried out under Ar in MeCN solution using 0.1 M Bu$_4$NBF$_4$ as a supporting electrolyte and ≈ 10^{-3} M of the complex. The working electrode was a glassy carbon disk, and the reference electrode was a calomel electrode isolated in a fritted bridge with a Pt wire as the auxiliary electrode. Under these conditions, E(ferrocene/ferrocenium) = 388 mV in MeCN at room temperature. The size and morphology of the solid materials were analyzed using an AMR 1000 Leitz scanning electron microscope (SEM) operated at variable accelerating voltages and with EDX detector NORAN System SIX NSS-200. Samples for SEM observation were prepared by dispersing a small amount of powder of dry silica and hybrid samples on double-sided conductive adhesive tabs on top of the SEM sample holders. Then, the samples were covered by a thin layer of gold deposited by sputtering to avoid charge accumulation on the surfaces. The selected accelerating voltage used in the showed images were 20 kV at high vacuum condition. Transmission electron microscopy (TEM) analysis was performed with a TEM/STEM JEM 2100 Plus with the operational voltage of 200 kV (variable), with a LaB6 filament. The samples were prepared by dropping a suspension of material in ethanol over a Formvar/Carbon square mesh Cu, 400 Mesh grids, and let dry. TEM images were processed using the public domain ImageJ program. N$_2$ adsorption–desorption isotherms were obtained at 77 K on a Micrometric ASAP 2020 V4.02 (V4.02 G) apparatus. The samples were degassed at 10^{-3} Torr and 200 °C for 6 h prior to the adsorption experiment. Surface area (S$_{BET}$) was calculated using the Brunauer–Emmett–Teller (BET) [55] equation over the pressure range (p/p°) of 0.05–0.20. The volume of micropores and mesopores (V$_{\mu P}$ and V$_{MP}$) was determined by the alpha-plot method using the standard Licrospher isotherm. The total pore volume (V$_{TP}$) was determined with the Gurvich rule [55] at 0.98 p/p°. The pore size distributions were calculated using the Villarroel–Bezerra–Sapag (VBS) model [56] on the desorption branch of the N$_2$ isotherms.

3.3. Crystal Data Collection and Refinement

Crystallographic data for compound [Cu(pypan)(ClO$_4$)$_2$] were collected at 298(2) K on a Bruker D8 QUEST ECO Photon II CPAD Diffractometer, using graphite monochromated Mo-Kα radiation (λ = 0.71073 Å). Data collection was carried out using the Bruker APEX4 package [57], and cell refinement and data reduction were achieved with the program SAINT V8.40B [58]. The structure was solved by direct methods with SHELXT V 2018/2 [59] and refined by full-matrix least-squares on F^2 data with SHELXL-2019/1 [60]. Molecular graphics were performed with ORTEP-3 [61], with 50% probability displacement ellipsoids. The packing diagrams were generated with SHELXL-2019/1. CCDC-2297367 contains the supplementary crystallographic data for this paper. These data can be obtained free of charge via http://www.ccdc.cam.ac.uk/conts/retrieving.html, accessed on 6 October 2023 (or from the CCDC, 12 Union Road, Cambridge CB2 1EZ, UK; Fax: +44 1223 336033; E-mail: deposit@ccdc.cam.ac.uk).

3.4. Indirect SOD Assay

An indirect assay based on the inhibition of the photoreduction of nitro blue tetrazolium (NBT) was used to test the SOD activity of the free and immobilized complexes [47]. The reaction mixture (3.2 mL) containing riboflavin (3.35 µM), methionine (9.52 mM), NBT (38.2 µM), and different amounts of free or immobilized complex was prepared in phosphate buffer of pH 7.8. Riboflavin was added last, and the mixture was illuminated for 15 min with a 16 W led lamp placed at 30 cm, at 25 °C. The reduction of NBT was measured at 560 nm. Control reactions were performed to verify that the complexes did not react with NBT or riboflavin directly. Inhibition percentage (IC) was calculated according to

$$IC = \frac{\left[(\Delta Abs/t)_{without\ catalyst} - (\Delta Abs/t)_{with\ catalyst}\right] \times 100}{(\Delta Abs/t)_{without\ catalyst}} \qquad (1)$$

The IC$_{50}$ values were determined from plots of % inhibition vs. complex concentration and used to calculate the McCord–Fridovich second-order rate constant (k_{McF}) [48]. At 50% inhibition, the rates of the reactions of O$_2^{\bullet-}$ with NBT and the mimic are identical; therefore, k_{McF} was calculated according to the equation k_{McF} [complex] = k_{NBT} [NBT], with k_{NBT} (pH = 7.8) = 5.94 × 10^4 M^{-1} s^{-1}.

3.5. Preparation of Potassium Superoxide Solutions

The stock solution of KO$_2$ in anhydrous dimethylsulfoxide (DMSO) employed in EPR measurements was prepared by suspending 9.3 mg of KO$_2$ in 5 mL of DMSO, followed by sonication during 15 min and centrifugation at 6000 rpm for 25 min. The concentration of KO$_2$ in the supernatant was calculated using ε = 2686 M^{-1} cm^{-1} in deoxygenated DMSO [62] and confirmed by the assay of the horseradish peroxidase. The saturated KO$_2$ solution in DMSO (0.75 mL) was added to a suspension of 10 mg of the hybrid material in 1.25 mL of DMSO, and the mixture was left stirring for 10 min. The solid was separated by centrifugation, washed with methanol, and dried at 60 °C.

4. Conclusions

[Cu(pypapn)]$^{2+}$ is among the most active Cu(II) complexes formed with open-chain ligands for catalyzing O$_2^{\bullet-}$ dismutation. In this complex, the metal ion is in a tetragonal N$_{2(amine)}$N$_{2(py)}$O$_{2(solvent)}$ environment, with the ligand disposed in the equatorial plane and two labile *trans*-positions for reaction with the substrate. The ligand flexibility seems to play a decisive role in the SOD activity, more than the redox potential, as [Cu(pypapn)]$^{2+}$ exhibits higher activity than [Cu(py$_2$pn)]$^{2+}$, with redox potential closer to the optimum value for O$_2^{\bullet-}$ dismutation but less conformational flexibility in the chelate rings, while displaying lower activity than complexes with a longer, and more flexible, central chain between the N-donor sites. Covalently linked [Cu(pypapn)]$^{2+}$ holds the metal ion geometry inside the pores of the silica matrix and retains the SOD activity exhibited in the homogeneous phase. The silica matrix preserves its ordered mesostructure after functionalization and in the same way as the protein framework, isolates and protects the catalyst from hydrolysis, extending its lifetime. In the hybrid material obtained by encapsulation of [Cu(py$_2$pn)]$^{2+}$, the strong electrostatic interactions between the dicationic catalyst and the surface groups on the pores proved decisive for the full retention of the complex within the silica matrix and activation of the metal center to react with $O_2^{\bullet-}$. In view of the robustness and stability of Cu-pypntriazole@SBA-15, Cu-pypntriazole@OP-MS, and Cu-py$_2$pn@SBA-15, covalent anchoring and electrostatically-driven encapsulation of doubly charged copper complexes, appear suitable strategies for the design of copper-based hybrid catalysts for $O_2^{\bullet-}$ dismutation under physiological conditions.

Supplementary Materials: The following supporting information can be downloaded at https://www.mdpi.com/article/10.3390/inorganics11110425/s1, Table S1: Crystal data and structure refinement for [Cu(pypapn)(ClO$_4$)$_2$]; Figure S1: Crystal packing diagram for [Cu(pypapn)(ClO$_4$)$_2$]; Figure S2: HRMS of [Cy(pypapn)(ClO$_4$)$_2$] in acetonitrile; Figure S3: Electronic spectrum of [Cu(pypapn)(ClO$_4$)$_2$] in DMF; Figure S4: SEM images of the mesoporous silica and the hybrid materials; Figure S5: Histograms of the channel diameter and wall thickness of Cu-pypntriazole@SBA-15; Figure S6: Low angle X-Ray diffractograms of SBA-15 and Cu-pypntriazole@SBA-15; Chart S1: Structures of complexes listed in Table 2.

Author Contributions: Conceptualization, S.R.S. and C.P.; methodology, resources, funding acquisition, S.R.S., C.P., N.P. and C.H.; investigation, M.R., J.F., S.S., G.T., C.P. and N.P.; formal analysis, M.R., S.S., J.F., G.T., S.R.S. and N.P.; writing—original draft preparation, S.R.S.; writing—review and editing, S.R.S., C.H., N.P. and C.P.; supervision, S.R.S. and C.P. All authors have read and agreed to the published version of the manuscript.

Funding: This research was funded by National University of Rosario (PID 80020220700136UR and PID 8002019040023UR), Consejo Nacional de Investigaciones Científicas y Técnicas (CONICET, PIP-0852 and PUE-0068), Centre National de la Recherche Scientifique (CNRS, PICS-07121), and Agencia Nacional de Promoción Científica y Tecnológica (ANPCyT, PICT-2019-03276).

Data Availability Statement: The data presented in this study are available on request from the corresponding author.

Conflicts of Interest: The authors declare no conflict of interest. The funders had no role in the design of this study.

References

1. Yang, B.; Chen, Y.; Shi, J. Reactive Oxygen Species (ROS)-Based Nanomedicine. *Chem. Rev.* **2019**, *119*, 4881–4985. [CrossRef] [PubMed]
2. Abreu, I.A.; Cabelli, D.E. Superoxide dismutases—A review of the metal-associated mechanistic variations. *Biochim. Biophys. Acta* **2010**, *1804*, 263–274. [CrossRef]
3. Batinić-Haberle, I.; Reboucas, J.S.; Spasojević, I. Superoxide dismutase mimics: Chemistry, pharmacology, and therapeutic potential. *Antioxid. Redox. Signal.* **2010**, *13*, 877–918. [CrossRef]
4. Riley, D.P. Functional Mimics of Superoxide Dismutase Enzymes as Therapeutic Agents. *Chem. Rev.* **1999**, *99*, 2573–2587. [CrossRef]
5. Sheng, Y.; Abreu, I.A.; Cabelli, D.E.; Maroney, M.J.; Miller, A.-F.; Teixeira, M.; Valentine, J.S. Superoxide Dismutases and Superoxide Reductases. *Chem. Rev.* **2014**, *114*, 3854–3918. [CrossRef]
6. Tainer, J.A.; Getzoff, E.D.; Beem, K.M.; Richardson, J.S.; Richardson, D.C. Determination and analysis of the 2 Å structure of copper, zinc superoxide dismutase. *J. Mol. Biol.* **1982**, *160*, 181–217. [CrossRef]
7. Richezzi, M.; Ferreyra, J.; Puzzolo, J.; Milesi, L.; Palopoli, C.M.; Moreno, D.M.; Hureau, C.; Signorella, S.R. Versatile Activity of a Copper(II) Complex Bearing a N$_4$-Tetradentate Schiff Base Ligand with Reduced Oxygen Species. *Eur. J. Inorg. Chem.* **2022**, *2022*, e202101042. [CrossRef]
8. Lange, J.; Elias, H.; Paulus, H.; Müller, J.; Weser, U. Copper(II) and Copper(I) complexes with an open-chain N$_4$ Schiff base ligand modeling CuZn superoxide dismutase: Structural and spectroscopic characterization and kinetics of electron transfer. *Inorg. Chem.* **2000**, *39*, 3342–3349. [CrossRef] [PubMed]
9. Müller, J.; Felix, K.; Maichle, C.; Lengfelder, E.; Strähle, J.; Weser, U. Phenyl-Substituted Copper Di-Schiff Base, a Potent CuZn Superoxide Dismutase Mimic Surviving Competitive Biochelation. *Inorg. Chim. Acta* **1995**, *233*, 11–19. [CrossRef]

10. Pap, J.S.; Kripli, B.; Bors, I.; Bogáth, D.; Giorgi, M.; Kaizer, J.; Speier, G. Transition Metal Complexes Bearing Flexible N_3 or N_3O Donor Ligands: Reactivity toward Superoxide Radical Anion and Hydrogen Peroxide. *J. Inorg. Biochem.* **2012**, *117*, 60–70. [CrossRef]
11. Patriarca, M.; Daier, V.; Camí, G.; Pellegri, N.; Rivière, E.; Hureau, C.; Signorella, S. Biomimetic Cu, Zn and Cu_2 Complexes Inserted in Mesoporous Silica as Catalysts for Superoxide Dismutation. *Microporous Mesoporous Mater.* **2019**, *279*, 133–141. [CrossRef]
12. Müller, J.; Schübl, D.; Maichle-Mössmer, C.; Strähle, J.; Weser, U. Structure—Function Correlation of Cu (II)-and Cu (I)-Di-Schiff-Base Complexes during the Catalysis of Superoxide Dismutation. *J. Inorg. Biochem.* **1999**, *75*, 63–69. [CrossRef]
13. Ribeiro, T.P.; Fernandes, C.; Melo, K.V.; Ferreira, S.S.; Lessa, J.A.; Franco, R.W.A.; Schenk, G.; Pereira, M.D.; Horn, A., Jr. Iron, Copper, and Manganese Complexes with in Vitro Superoxide Dismutase and/or Catalase Activities That Keep Saccharomyces Cerevisiae Cells Alive under Severe Oxidative Stress. *Free Radic. Biol. Med.* **2015**, *80*, 67–76. [CrossRef]
14. Mekhail, M.A.; Smith, K.J.; Freire, D.M.; Pota, K.; Nguyen, N.; Burnett, M.E.; Green, K.N. Increased Efficiency of a Functional SOD Mimic Achieved with Pyridine Modification on a Pyclen-Based Copper(II) Complex. *Inorg. Chem.* **2023**, *62*, 5415–5425. [CrossRef]
15. Green, K.N.; Pota, K.; Tircso, G.; Gogolak, R.A.; Kinsinger, O.; Davda, C.; Blain, K.; Brewer, S.M.; Gonzalez, P.; Johnston, H.M.; et al. Dialing in on pharmacological features for a therapeutic antioxidant small molecule. *Dalton Trans.* **2019**, *48*, 12430–12439. [CrossRef]
16. Policar, C.; Bouvet, J.; Bertrand, H.C.; Delsuc, N. SOD mimics: From the tool box of the chemists to cellular studies. *Curr. Opin. Chem. Biol.* **2022**, *67*, 102109. [CrossRef]
17. Martinez-Camarena, Á.; Sanchez-Murcia, P.A.; Blasco, S.; Gonzalez, L.; Garcia-España, E. Unveiling the reaction mechanism of novel copper N-alkylated tetra-azacyclophanes with outstanding superoxide dismutase activity. *Chem. Commun.* **2020**, *56*, 7511–7514. [CrossRef]
18. Vaughn, B.A.; Brown, A.M.; Ahn, S.H.; Robinson, J.R.; Borosm, E. Is Less More? Influence of the Coordination Geometry of Copper(II)Picolinate Chelate Complexes on Metabolic Stability. *Inorg. Chem.* **2020**, *59*, 16095–16108. [CrossRef]
19. Smits, N.W.G.; van Dijk, B.; de Bruin, I.; Groeneveld, S.L.T.; Siegler, M.A.; Hetterscheid, D.G.H. Influence of Ligand Denticity and Flexibility on the Molecular Copper Mediated Oxygen Reduction Reaction. *Inorg. Chem.* **2020**, *59*, 16398–16409. [CrossRef]
20. Stanek, J.; Hoffmann, A.; Herres-Pawlis, S. Renaissance of the entatic state principle. *Coord. Chem. Rev.* **2018**, *365*, 103–121. [CrossRef]
21. Falcone, E.; Hureau, C. Redox processes in Cu-binding proteins: The "in-between" states in intrinsically disordered peptides. *Chem. Soc. Rev.* **2023**, *52*, 6595–6600. [CrossRef]
22. Uzal-Varela, R.; Patinec, V.; Tripier, R.; Valencia, L.; Maneiro, M.; Canle, M.; Platas-Iglesias, C.; Esteban-Gómez, D.; Iglesias, E. On the dissociation pathways of copper complexes relevant as PET imaging agents. *J. Inorg. Biochem.* **2022**, *236*, 111951. [CrossRef]
23. Mohammadnezhad, G.; Amirian, A.M.; Plass, H.G.W.; Sandleben, A.; Schäfer, S.; Klein, A. Redox Instability of Copper(II) Complexes of a Triazine-Based PNP Pincer. *Eur. J. Inorg. Chem.* **2021**, *2021*, 1140–1151. [CrossRef]
24. Mureseanu, M.; Filip, M.; Bleotu, I.; Spinu, C.I.; Marin, A.H.; Matei, I.; Parvulescu, V. Cu(II) and Mn(II) Anchored on Functionalized Mesoporous Silica with Schiff Bases: Effects of Supports and Metal–Ligand Interactions on Catalytic Activity. *Nanomaterials* **2023**, *13*, 1884. [CrossRef]
25. Isa, E.D.M.; Ahmad, H.; Rahman, M.B.A.; Gill, M.R. Progress in Mesoporous Silica Nanoparticles as Drug Delivery Agents for Cancer Treatment. *Pharmaceutics* **2021**, *13*, 152.
26. Cadavid-Vargas, J.F.; Arnal, P.M.; Sepúlveda, R.D.M.; Rizzo, A.; Soria, D.B.; Di Virgilio, A.L. Copper complex with sulfamethazine and 2,2′-bipyridine supported on mesoporous silica microspheres improves its antitumor action toward human osteosarcoma cells: Cyto- and genotoxic effects. *Biometals* **2019**, *32*, 21–32. [CrossRef]
27. Donato, L.; Atoini, Y.; Prasetyanto, E.A.; Chen, P.; Rosticher, C.; Bizarri, C.; Rissansen, K.; De Cola, L. Selective encapsulation and enhancement of the emission properties of a luminescent Cu(I) complex in mesoporous silica. *Helvetica Chim. Acta* **2018**, *101*, e1700273. [CrossRef]
28. Richezzi, M.; Palopoli, C.; Pellegri, N.; Hureau, C.; Signorella, S.R. Synthesis, characterization and superoxide dismutase activity of a biomimetic Mn(III) complex covalently anchored to mesoporous silica. *J. Inorg. Biochem.* **2022**, *237*, 112026. [CrossRef] [PubMed]
29. Pajchel, L.; Kolodziejski, W. Synthesis and characterization of MCM-48/hydroxyapatite composites for drug delivery: Ibuprofen incorporation, location and release studies. *Mater. Sci. Eng. C* **2018**, *91*, 734–742. [CrossRef] [PubMed]
30. Lu, Z.; Wang, J.; Qu, L.; Kan, G.; Zhang, T.; Shen, J.; Li, Y.; Yang, J.; Niu, Y.; Xiao, Z.; et al. Reactive mesoporous silica nanoparticles loaded with limonene for improving physical and mental health of mice at simulated microgravity condition. *Bioact. Mater.* **2020**, *5*, 1127–1137. [CrossRef]
31. Wang, L.S.; Wu, L.C.; Lu, S.Y.; Chang, L.L.; Teng, I.T.; Yang, C.M.; Ho, J.A. Biofunctionalized Phospholipid-Capped Mesoporous Silica Nanoshuttles for Targeted Drug Delivery: Improved Water Suspensibility and Decreased Nonspecific Protein Binding. *ACS Nano* **2010**, *4*, 4371–4379. [CrossRef]
32. Martinez-Carmona, M.; Lozano, D.; Colilla, M.; Vallet-Regí, M. Lectin-conjugated pH-responsive mesoporous silica nanoparticles for targeted bone cancer treatment. *Acta Biomater.* **2018**, *65*, 393–404. [CrossRef]
33. Ha, S.W.; Viggeswarapu, M.; Habib, M.M.; Beck, G.R., Jr. Bioactive effects of silica nanoparticles on bone cells are size, surface, and composition dependent. *Acta Biomater.* **2018**, *82*, 184–196. [CrossRef]

34. Zhao, N.; Yan, L.; Zhao, X.; Chen, X.; Li, A.; Zheng, D.; Zhou, X.; Dai, X.; Xu, F.J. Versatile Types of Organic/Inorganic Nanohybrids: From Strategic Design to Biomedical Applications. *Chem. Rev.* **2019**, *119*, 1666–1762. [CrossRef]
35. Freire, C.; Pereira, C.; Rebelo, S. Green oxidation catalysis with metal complexes: From bulk to nano recyclable hybrid catalysts. *Catalysis* **2012**, *24*, 116–203.
36. Rana, B.S.; Jain, S.L.; Singh, B.; Bhaumik, A.; Sain, B.; Sinha, A.K. Click on silica: Systematic immobilization of Co(II)Schiff bases to the mesoporous silicavia click reaction and their catalytic activity for aerobic oxidation of alcohols. *Dalton Trans.* **2010**, *39*, 7760–7767. [CrossRef]
37. Bagherzadeh, M.; Hosseini, M.; Mortazavi-Manesh, A. Manganese(III) porphyrin anchored onto magnetic nanoparticles via "Click" reaction: An efficient and reusable catalyst for the heterogeneous oxidation of alkenes and sulfides. *Inorg. Chem. Commun.* **2019**, *107*, 107495. [CrossRef]
38. Richezzi, M.; Signorella, S.; Palopoli, C.; Pellegri, N.; Hureau, C.; Signorella, S.R. The Critical Role of Ligand Flexibility on the Activity of Free and Immobilized Mn Superoxide Dismutase Mimics. *Inorganics* **2023**, *11*, 359. [CrossRef]
39. Pat McCurdie, M.; Belfiore, L.A. Spectroscopic analysis of transition-metal coordination complexes based on poly(4-vinylpyridine) and dichlorotricarbonylruthenium(II). *Polymer* **1999**, *40*, 2889–2902. [CrossRef]
40. Subramanian, P.S.; Suresh, E.; Dastidar, P.; Waghmode, S.; Srinivas, D. Conformational Isomerism and Weak Molecular and Magnetic Interactions in Ternary Copper(II) Complexes of [Cu(AA)L']ClO$_4$·nH$_2$O, Where AA = L-Phenylalanine and L-Histidine, L' = 1,10-Phenanthroline and 2,2-Bipyridine, and n = 1 or 1.5: Synthesis, Single-Crystal X-ray Structures, and Magnetic Resonance Investigations. *Inorg. Chem.* **2001**, *40*, 4291–4301.
41. Muthuramalingam, S.; Anandababu, K.; Velusamy, M.; Mayilmurugan, R. Benzene Hydroxylation by Bioinspired Copper(II) Complexes: Coordination Geometry versus Reactivity. *Inorg. Chem.* **2020**, *59*, 5918–5928. [CrossRef]
42. Geary, W.J. The use of conductivity measurements in organic solvents for the characterization of coordination compounds. *Coord. Chem. Rev.* **1971**, *7*, 81–122. [CrossRef]
43. Malvi, B.; Sarkar, B.R.; Pati, D.; Mathew, R.; Ajithkumarb, T.G.; Sen Gupta, S. "Clickable" SBA-15 mesoporous materials: Synthesis, characterization and their reaction with alkynes. *J. Mater. Chem.* **2009**, *19*, 1409–1416. [CrossRef]
44. Mercier, L.; Pinnavaia, T.J. Direct Synthesis of Hybrid Organic-Inorganic Nanoporous Silica by a Neutral Amine Assembly Route: Structure-Function Control by Stoichiometric Incorporation of Organosiloxane Molecules. *Chem. Mater.* **2000**, *12*, 188–196. [CrossRef]
45. Zhao, D.; Feng, J.; Huo, Q.; Melosh, N.; Frederichson, G.H.; Chmelka, B.F.; Stucky, G.D. Triblock Copolymer Syntheses of Mesoporous Silica with Periodic 50 to 300 Angstrom Pores. *Science* **1998**, *279*, 548–552. [CrossRef]
46. Taghavimoghaddam, J.; Knowles, G.P.; Chaffee, A.L. SBA-15 supported cobalt oxide species: Synthesis, morphology and catalytic oxidation of cyclohexanol using TBHP. *J. Mol. Catal. A Chem.* **2013**, *379*, 277–286. [CrossRef]
47. Beauchamps, C.; Fridovich, I. Superoxide dismutase: Improved assays and an assay applicable to acrylamide gels. *Anal. Biochem.* **1971**, *44*, 276–287. [CrossRef]
48. Liao, Z.-R.; Zheng, X.-F.; Luo, B.-S.; Shen, L.-R.; Li, D.-F.; Liu, H.-L.; Zhao, W. SOD-like activities of manganese-containing complexes with N,N,N,N-tetrakis(2-benzimidazolyl methyl)-1,2-ethanediamine (EDTB). *Polyhedron* **2001**, *20*, 2813–2821. [CrossRef]
49. Wang, C.; Li, S.; Shang, D.-J.; Wang, X.-L.; You, Z.-L.; Li, H.-B. Antihyperglycemic and neuroprotective effects of one novel Cu–Zn SOD mimetic. *Bioorg. Med. Chem. Lett.* **2011**, *21*, 4320–4324. [CrossRef]
50. You, Z.-L.; Ni, L.-L.; Hou, P.; Zhang, J.-C.; Wang, C. Synthesis, Crystal Structures, and Superoxide Dismutase Activity of Two Isostructural Copper(II)—Zinc(II) Complexes Derived from N,N'-Bis (4-Methoxysalicylidene) Cyclohexane-1, 2-Diamine. *J. Coord. Chem.* **2010**, *63*, 515–523. [CrossRef]
51. Ivanović-Burmazović, I.; Filipović, M.R. Chapter 3—Reactivity of manganese superoxide dismutase mimics toward superoxide and nitric oxide: Selectivity versus cross-reactivity. *Adv. Inorg. Chem.* **2012**, *64*, 53–95.
52. Diószegi, R.; Bonczidai-Kelemen, D.; Bényei, A.C.; May, N.V.; Fábián, I.; Lihi, N. Copper(II) Complexes of Pyridine-2,6-dicarboxamide Ligands with High SOD Activity. *Inorg. Chem.* **2022**, *61*, 2319–2332. [CrossRef]
53. Bagchi, R.N.; Bond, A.M.; Scholz, F.; Stösser, R. Characterization of the ESR spectrum of the superoxide anion in the liquid phase. *J. Am. Chem. Soc.* **1989**, *111*, 8270–8271. [CrossRef]
54. Ebralidze, I.I.; Leitus, G.; Shimon, L.J.W.; Wang, Y.; Shaik, S.; Neumann, R. Structural variability in manganese(II) complexes of N,N'-bis(2-pyridinylmethylene) ethane (and propane) diamine ligands. *Inorg. Chim. Acta* **2009**, *362*, 4713–4720. [CrossRef]
55. Thommes, M.; Kaneko, K.; Neimark, A.V.; Olivier, J.P.; Rodriguez-Reinoso, F.; Rouquerol, J.; Sing, K.S.W. Physisorption of gases, with special reference to the evaluation of surface area and pore size distribution (IUPAC Technical Report). *Pure Appl. Chem.* **2015**, *87*, 1051–1069. [CrossRef]
56. Villarroel Rocha, J.; Barrera, D.; Sapag, K. Improvement in the Pore Size Distribution for Ordered Mesoporous Materials with Cylindrical and Spherical Pores Using the Kelvin Equation. *Top. Catal.* **2011**, *54*, 121–134. [CrossRef]
57. *Bruker*, APEX4 v2022.10-1; Bruker AXS Inc.: Madison, WI, USA, 2022.
58. *Bruker*, SAINT V8.40B; Bruker AXS Inc.: Madison, WI, USA, 2019.
59. Sheldrick, G.M. SHELXT-Integrated space-group and crystal-structure determination. *Acta Cryst.* **2015**, *A71*, 3–8. [CrossRef]
60. Sheldrick, G.M. Crystal structure refinement with SHELXL. *Acta Cryst.* **2015**, *C71*, 3–8.

61. Farrugia, L.J. ORTEP3 for Windows. *J. Appl. Crystallogr.* **1997**, *30*, 565. [CrossRef]
62. Hyland, K.; Auclair, C. The formation of superoxide radical anions by a reaction between O_2, OH^- and dimethyl sulfoxide. *Biochem. Biophys. Res. Commun.* **1981**, *102*, 531–537. [CrossRef]

Disclaimer/Publisher's Note: The statements, opinions and data contained in all publications are solely those of the individual author(s) and contributor(s) and not of MDPI and/or the editor(s). MDPI and/or the editor(s) disclaim responsibility for any injury to people or property resulting from any ideas, methods, instructions or products referred to in the content.

Review

Probing the Bioinorganic Chemistry of Cu(I) with ^{111}Ag Perturbed Angular Correlation (PAC) Spectroscopy

Victoria Karner [1], Attila Jancso [2] and Lars Hemmingsen [3,*]

[1] TRIUMF, 4004 Wesbrook Mall, Vancouver, BC V6T 2A3, Canada; vkarner@triumf.ca
[2] Department of Molecular and Analytical Chemistry, University of Szeged, Dóm tér 7-8, H-6720 Szeged, Hungary; jancso@chem.u-szeged.hu
[3] Department of Chemistry, University of Copenhagen, Universitetsparken 5, DK-2100 Copenhagen, Denmark
* Correspondence: lhe@chem.ku.dk

Abstract: The two most common oxidation states of copper in biochemistry are Cu(II) and Cu(I), and while Cu(II) lends itself to spectroscopic interrogation, Cu(I) is silent in most techniques. Ag(I) and Cu(I) are both closed-shell d^{10} monovalent ions, and to some extent share ligand and coordination geometry preferences. Therefore, Ag(I) may be applied to explore Cu(I) binding sites in biomolecules. Here, we review applications of ^{111}Ag perturbed angular correlation (PAC) of γ-ray spectroscopy aimed to elucidate the chemistry of Cu(I) in biological systems. Examples span from small blue copper proteins such as plastocyanin and azurin (electron transport) over hemocyanin (oxygen transport) to CueR and BxmR (metal-ion-sensing proteins). Finally, possible future applications are discussed. ^{111}Ag is a radionuclide which undergoes β-decay to ^{111}Cd, and it is a γ-γ cascade of the ^{111}Cd daughter nucleus, which is used in PAC measurements. ^{111}Ag PAC spectroscopy may provide information on the coordination environment of Ag(I) and on the structural relaxation occurring upon the essentially instantaneous change from Ag(I) to Cd(II).

Keywords: copper proteins; copper biochemistry; Cu(I); spectroscopy; metal site structure and rigidity; Ag(I) binding to proteins

Citation: Karner, V.; Jancso, A.; Hemmingsen, L. Probing the Bioinorganic Chemistry of Cu(I) with ^{111}Ag Perturbed Angular Correlation (PAC) Spectroscopy. *Inorganics* **2023**, *11*, 375. https://doi.org/10.3390/inorganics11100375

Academic Editors: Christelle Hureau, Ana Maria Da Costa Ferreira and Gianella Facchin

Received: 28 August 2023
Revised: 18 September 2023
Accepted: 19 September 2023
Published: 23 September 2023

Copyright: © 2023 by the authors. Licensee MDPI, Basel, Switzerland. This article is an open access article distributed under the terms and conditions of the Creative Commons Attribution (CC BY) license (https://creativecommons.org/licenses/by/4.0/).

1. Introduction

Copper ions take part in fundamental biochemical processes such as electron transfer, oxygen transport, enzyme catalyzed redox reactions, and systems to control copper itself in the cell (transcriptional regulators, chaperones, and transmembrane transporters) [1–4]. The most common oxidation states are Cu(I) and Cu(II), and while Cu(II) is observable by several experimental methods, Cu(I) is a closed-shell (d^{10}) ion and therefore silent in most spectroscopic techniques, except nuclear and X-ray based approaches such as EXAFS, XANES, NMR/NQR, and potentially β-NMR [5,6]. Similarly, other closed-shell ions such as Mg(II), Ca(II), and Zn(II) pose a challenge in terms of the spectroscopic characterization of their coordination environments. Substitution of spectroscopically silent native metal ions by active probes such as Co(II), Mn(II), and ^{113}Cd(II) (for NMR) has found widespread use in the characterization of the metal site structure and function of metalloproteins [7,8]. Similarly, Ag(I) has been used in a number of studies to explore copper biochemistry [9–17], despite the fact that Ag(I) in itself is also silent in most spectroscopic analyses.

In perturbed angular correlation (PAC) of γ-ray spectroscopy, ^{111m}Cd has been employed as a means to probe the function of several Zn-dependent enzymes [18], and in the context of this minireview, the less commonly used ^{111}Ag is of particular interest to explore the biochemistry of Cu(I) and potentially in radiopharmaceutical applications [19,20]. As a group 11 element, Ag is a heavier congener of Cu, and not surprisingly Ag(I) and Cu(I) exhibit common properties such as coordination with soft ligands, most notably thiolates in proteins, and geometries of metal sites, although the ionic radius of Ag(I) is larger than that of Cu(I). ^{111}Ag PAC spectroscopy has been applied in a limited number of (bio)inorganic chemistry studies so far, including small inorganic compounds [21–23], Ag(I) chelators with potential radiopharmaceutical applications [19,20], and metalloproteins. One of the key advantages of PAC spectroscopy is that it relies on a radioactive probe; and so akin to radiotracer techniques, only very small amounts of the probe are required to record a spectrum, typically on the order of picomoles (or 10^{11} probe ions or atoms) [18]. Moreover, PAC spectroscopy may be applied to any physical state (solid, liquid, or gas) [18,24], and as most biological material is transparent to γ-rays, even in vivo PAC experiments are possible [25]. ^{111}Ag may be produced at radioactive ion beam facilities such as ISOLDE/CERN or by neutron irradiation of isotopically enriched ^{110}Pd at facilities such as ILL, Grenoble, France. The relatively long half-life ($T_{1/2}$ = 7.45 days) allows for shipping from a production facility to the home lab, and as such, makes ^{111}Ag an attractive PAC probe.

With this work, we review applications of ^{111}Ag PAC spectroscopy within Cu(I) bioinorganic chemistry, which so far only encompasses electron transfer proteins (plastocyanin and azurin) [26–29], an oxygen transport protein (hemocyanin) [30], and metal-ion-sensing proteins functioning as transcriptional regulators (CueR and BxmR) [31–33].

2. PAC Theory

In this section, we provide a brief introduction to PAC spectroscopy with an emphasis on ^{111}Ag PAC; for a more detailed analysis of the technique, we refer the reader to [18].

In PAC spectroscopy, the hyperfine interactions between the nuclear magnetic and/or quadrupole moments of the probe nucleus and the local magnetic fields and/or electric field gradients (EFGs) are measured through the perturbed angular correlation of a γ-γ cascade in the nuclear decay. In this review, we will focus solely on interactions between the nuclear quadrupole moment of the PAC nucleus and the EFG of its surroundings, i.e., the nuclear quadrupole interaction (NQI). An EFG is a signature of the local electronic and molecular structure surrounding the PAC probe site, and if the EFG is time-dependent, dynamics may also be explored typically on the ps–ns time scale. The NQI gives rise to hyperfine splitting of the nuclear energy levels of the probe nucleus, and this energy splitting is measured via PAC spectroscopy for the intermediate level of the γ-γ cascade. For a nucleus with spin $I = 5/2$, such as the relevant nuclear level of many PAC probes including ^{111}Ag, the intermediate energy level is split into three sublevels with transition frequencies of ω_1, ω_2, and ω_3 between these sublevels. Note that this invokes the rule that $\omega_3 = \omega_1 + \omega_2$. These frequencies serve as a fingerprint for the local structure at the probe site.

For randomly oriented molecules and a time-independent NQI, the measured data reflect the so-called perturbation function, which, for $I = 5/2$, is given by:

$$G_2(t) = a_0 + a_1 \cos(\omega_1 t) + a_2 \cos(\omega_2 t) + a_3 \cos(\omega_3 t) \tag{1}$$

where a_i and ω_i depend on only two parameters, the NQI strength typically reported as ν_Q or ω_0, which is proportional to $|V_{zz}|$, and the axial asymmetry parameter $\eta = (V_{yy}-V_{xx})/V_{zz}$, where V_{xx}, V_{yy}, and V_{zz} are the diagonal elements of the EFG tensor in the principal axis system, ordered such that $|V_{zz}| \geq |V_{yy}| \geq |V_{xx}|$.

Figure 1 shows the decay schemes of ^{111}Ag and ^{111m}Cd. Note that ^{111}Ag does not directly undergo the necessary γ-γ cascade for PAC spectroscopy, instead it decays by β^- emission to ^{111}Cd, and with a ~7% chance, the "correct" excited state of ^{111}Cd is populated (342 keV) which then may decay by the successive emission of two γ-rays necessary for PAC. The intermediate level probed in this cascade has a half-life of 85 ns, and it is the NQI that the Cd nucleus experiences in this state which is measured in ^{111}Ag PAC experiments.

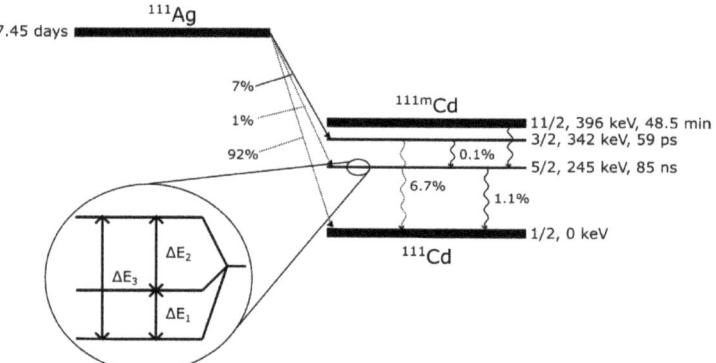

Figure 1. Nuclear decay of ^{111}Ag. The probabilities of the various decay routes are given per ^{111}Ag decay. The decay of ^{111m}Cd (another commonly used PAC isotope) is also indicated to demonstrate that the hyperfine splitting of the same intermediate level ($I = 5/2$, 245 keV) is probed in γ-γ PAC spectroscopy in the two radionuclides, although the first γ differs for the γ-γ cascade in ^{111}Ag PAC and ^{111m}Cd PAC.

If the EFG is time-dependent on the ps–ns time scale, the perturbation function, Equation (1), is affected. If the motion is slow ($1/\tau_c \ll \omega_0$), the perturbation function is exponentially dampened by $\exp(-t/\tau_c)$, where τ_c is the characteristic time of the (stochastic) dynamics, for example, the rotational correlation time of a molecule undergoing rotational diffusion. If the motion is fast ($1/\tau_c \gg \omega_0$), the perturbation function becomes $G_2(t) = \exp(-2.8\omega_0^2 \tau_c (1 + \eta^2/3)t)$, i.e., a purely exponentially decaying function. If the dynamics originate from chemical exchange between two (or more) species, three different scenarios are possible, in analogy to NMR spectroscopy: slow exchange where both species are observed, intermediate exchange where line broadening is pronounced, and fast exchange where a weighted average signal is observed.

3. Examples of Applications of ^{111}Ag PAC Spectroscopy Elucidating Cu(I) Bioinorganic Chemistry

In this section, we aim to provide an overview of the type of questions which can be addressed with ^{111}Ag PAC spectroscopy. We begin with a series of literature examples, and we end with suggestions for future applications of the technique.

3.1. Metal Site Structure in Small Blue Copper Proteins—Electron Transport and Transfer

Small blue copper proteins take part in the transport of electrons. The prototypical example is plastocyanin, transporting electrons in photosynthesis from cytochrome b$_6$/f, diffusing as Cu(I)-plastocyanin to associate with and transfer the electron to photosystem I. The metal site is composed of two histidine residues and one cysteine residue in a distorted trigonal planar structure, with an additional methionine residue at an axial position and with an unusual distance from the metal ion to the thioether sulfur of around 3 Å, see Figure 2. The metal site is assumed to be rigid, and crystallographic data indicate that there is very little difference in the Cu(I) and Cu(II) coordination geometry [34], presumably facilitating rapid electron transfer. In a broader perspective, these so-called type I copper sites are an important class of Cu-binding sites which are present in many copper-containing proteins, most notably in blue multicopper oxidases.

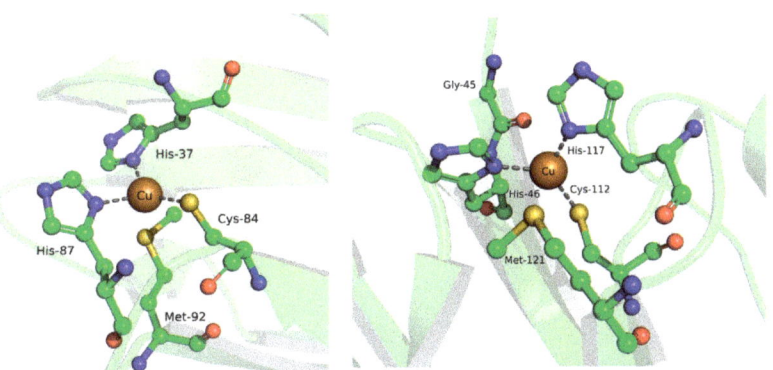

Figure 2. Metal sites of plastocyanin (left) and azurin (right). A copper ion is bound in a distorted trigonal planar structure with two His residues, one Cys residue, and an axial Met residue at an uncommon distance of ca. 3 Å from the metal ion (pdb codes: 1PLC [35] and 1AZC [36]). Figure produced with PyMOL [37].

^{111}Ag and ^{111m}Cd PAC was employed to determine if there is a difference between the metal site structure of the monovalent Ag(I) and the divalent Cd(II)-substituted azurin, or if the protein fully dictates the metal site structure [26]. The PAC spectra are presented in Figure 3, demonstrating that there is very little difference between the recorded NQIs for ^{111}Ag(I) azurin and ^{111m}Cd(II) wild-type (WT) azurin (lower panel). The small (but statistically significant) difference may be accounted for by changes in the ligand–metal–ligand angles of a few degrees [26]. Interestingly, the (small) difference between the NQIs from ^{111}Ag and ^{111m}Cd PAC is most pronounced in the ^{111}Ag PAC data for the first ca. 60 ns after the decay from ^{111}Ag to ^{111}Cd, while at longer times (60–120 ns) after the decay, the NQI approaches the value recorded by ^{111m}Cd PAC. This indicates that although the structures of Ag(I) and Cd(II) are highly similar, (minor) structural relaxation occurs on the nanosecond time scale upon the decay of ^{111}Ag to ^{111}Cd. This model was corroborated by PAC experiments at two different temperatures (1 °C and 25 °C), demonstrating that the rate of relaxation increased by around a factor of 2 when increasing the temperature from 1 to 25 °C. Thus, ^{111}Ag PAC spectroscopy allowed for an estimate of the rate of the (small) structural reorganization occurring when the metal ion changed from monovalent Ag(I) to divalent Cd(II), in analogy to a pump–probe experiment, where the "pump" is the nuclear decay of ^{111}Ag, and the probing is performed by following the time-dependent change in the metal site NQI.

Mutation of the methionine near the metal site to leucine, forming M121L azurin, is a way to probe if M121 significantly affects the metal site structure. Interestingly, ^{111}Ag PAC spectra of the M121L mutant show very little change compared to the WT protein, see Figure 3, top panel, while the ^{111m}Cd PAC spectra exhibit one NQI similar to the WT signal, but another NQI (with significant line broadening, reflecting structural variability) also appears. Thus, it seems that the introduction of both a mutation (M121L) and a non-native metal ion with different combination of electronic structure (d^{10}) and charge (Cd(II)) from Cu(I) and Cu(II) leaves the metal site less well defined. Similarly, ^{111}Ag PAC spectra of the M121H mutant of azurin [27] gave rise to significantly more diverse coordination geometries than observed for the WT protein. As a function of pH from 4.0 to 7.7, four different NQIs were observed and interpreted as [AgHisCys(H$_2$O)$_2$] at low pH, and three different [AgHis$_2$Cys(H$_2$O)] structures in various proportions at the higher pH values, where the changes are controlled by the protonation states of the histidine residues. Thus, M121H mutation significantly alters the metal site properties.

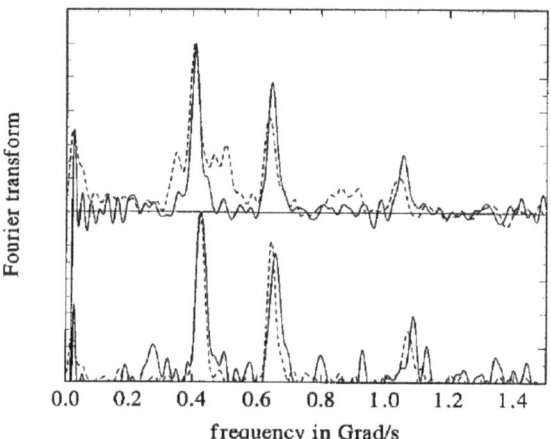

Figure 3. ^{111}Ag and ^{111m}Cd PAC spectra of azurin and M121L azurin. Lower panel: ^{111}Ag PAC spectrum (solid line) and ^{111m}Cd PAC spectrum (dotted line) for wild-type azurin. Upper panel: ^{111}Ag PAC spectrum (solid line) and ^{111m}Cd PAC spectrum (dotted line) for M121L azurin. The Fourier transform for ^{111}Ag PAC is multiplied by -1 for easy comparison with the Fourier transform for ^{111m}Cd PAC. Reprinted with permission from Ref. [26]. Copyright American Chemical Society 1997.

3.2. Plastocynin–Photosystem I Association—Protein–Protein Interactions in Electron Transport

Reduced photosystem I is the physiological partner to which Cu(I)-plastocyanin transfers an electron. In analogy to the study of azurin, vide supra, ^{111}Ag and ^{111m}Cd PAC spectra were recorded for Ag(I)- and Cd(II)-substituted plastocyanin, and the measured NQIs were very similar, indicating that Ag(I) and Cd(II) occupy very similar metal site structures [28] with distorted trigonal planar coordination by two histidine residues and one cysteine residue. In a pH series from 5.0 to 8.5, the low pH spectra indicated significant changes in the first coordination sphere, presumably reflecting dissociation of one of the His residues and coordination by the axial methionine thioether [29], in agreement with data obtained by X-ray diffraction at low pH [38]. The main point of the current subsection is, however, that it is possible to monitor the binding of Ag(I)-plastocyanin and Cd(II)-plastocyanin to reduced photosystem I, see Figure 4 [28]. In the absence of photosystem I (top panels of Figure 4), both the ^{111}Ag(I)-plastocyanin and the ^{111m}Cd(II)-plastocyanin data display heavy damping of the oscillatory signal; this is caused by rotational diffusion of the protein, which is relatively rapid under the experimental conditions. However, upon addition of photosystem I, the oscillations of the observed perturbation function for ^{111}Ag-plastocyanin are recovered, indicating that the Brownian tumbling of the protein is much slower, i.e., that ^{111}Ag-plastocyanin is bound to photosystem I. Contrary to this, no indication of binding of ^{111m}Cd-plastocyanin to photosystem I is observed, in accordance with the function of plastocyanin. Moreover, the binding of ^{111}Ag-plastocyanin to photosystem I eliminates the relaxation of the ^{111}Ag metal site structure to that observed for ^{111m}Cd, implying that the metal site structure and association with photosystem are mutually affecting each other.

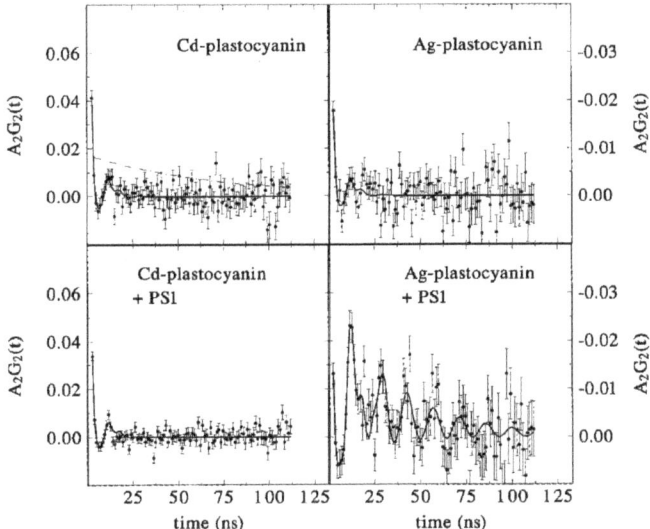

Figure 4. ^{111}Ag- and ^{111m}Cd PAC time traces recorded for plastocyanin in the absence and presence of photosystem I. Left panels: ^{111m}Cd PAC data for plastocyanin in the absence (top) and presence (bottom) of photosystem I. Right panels: ^{111}Ag PAC data for plastocyanin in the absence (top) and presence (bottom) of photosystem I. Data points indicated with error bars, and fit (solid lines) using Equation (1) with exponential damping due to rotational diffusion. The data for ^{111}Ag PAC are multiplied by −1 for easy comparison with the ^{111m}Cd PAC data. Reprinted with permission from ref. [28]. Copyright American Chemical Society 1999.

3.3. Metal Site Structure in Hemocyanin—Oxygen Transport

Hemocyanins (Hcs) transport oxygen in the hemolymph of some mollusks and arthropods. The oxygen binding site is composed of two copper ions coordinated by three histidine residues each, see Figure 5. In the absence of oxygen, both metal ions are reduced, while upon oxygen binding, they are oxidized to Cu(II) and accordingly oxygen is reduced to the peroxide ion.

The binuclear Cu site in deoxy-Hc from the arthropod *Carcinus aestuarii* was characterized by ^{111}Ag PAC spectroscopy by Holm et al. [30] through a series of experiments with varying Ag(I)-to-protein ratios. Note that the amount of radioactive ^{111}Ag is very small, so the Ag(I) concentration is controlled by the addition of non-radioactive Ag(I), typically as a nitrate or perchlorate salt. With 0.1 eq. Ag(I) with respect to 0.5 mM Hc, a PAC signal with $\omega_0 = 0.183(1)$ rad/ns and $\eta = 0.18(3)$ was observed. Increasing the Ag(I) concentration to 2.0 eq. Ag(I) gave rise to a different signal with a larger line width (i.e., increased width in the distribution of NQIs) and an increase in the asymmetry parameter to $\eta = 0.26(4)$, if fitting the data with only one NQI. Although these changes are small, they are statistically significant, and in particular, the increased line width is qualitatively observable in the data. Fitting the 2.0 eq. Ag(I) data with two NQIs instead gives roughly equal amplitudes for the two signals, as expected for full occupation of the two metal sites in Hc. The PAC parameters differ (for both sites) from those observed at 0.1 eq. Ag(I). Thus, it was concluded that the protein displays two slightly different metal ion binding sites, and that the NQI recorded for site 1 depends on whether site 2 is occupied or not, indicating that they are close in space, in agreement with the structure, see Figure 5. An additional experiment was carried out with 1.0 eq. Ag(I), and the spectrum differed from that recorded with 0.1 eq. Ag(I), indicating that both sites are occupied in a significant fraction of the proteins in the presence of 1.0 eq. Ag(I), and consequently that the binding of the two metal ions is likely to be cooperative.

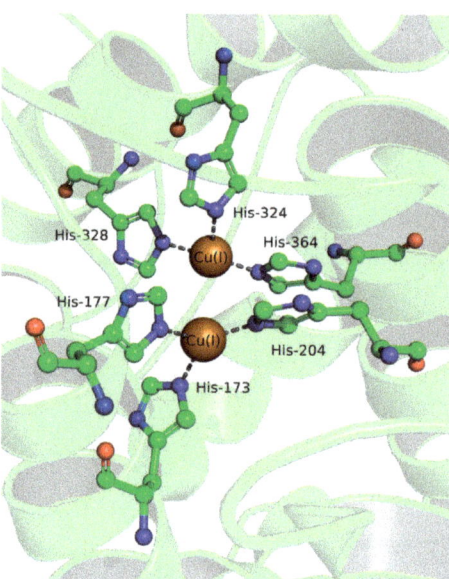

Figure 5. Metal site of hemocyanin. In the resting state of the protein (deoxy-Hc), two Cu(I) are present at the dioxygen binding site, each bound by three histidine residues (pdb code: 1LLA [39]). Figure produced with PyMOL [37].

The structural interpretation of the recorded PAC signals was carried out by Holm et al. [30] using the semiempirical angular overlap model (AOM) [40]. It was concluded that purely three-coordinated Ag(I) does not agree well with the ^{111}Ag PAC spectroscopic data, and that it is likely that a fourth ligand, for example, a water molecule, may be present both with one and with two Ag(I) bound to Hc. In the work by Holm et al. [30], it was assumed that the local Ag(I) metal site structure persisted throughout the PAC measurements, i.e., that little or no structural relaxation occurred, despite the change of element and oxidation state accompanying the nuclear decay of ^{111}Ag(I) to ^{111}Cd(II). Later experiments on the CueR transcriptional regulator [32], vide infra, have demonstrated that this assumption may not always hold. Therefore, it is conceivable that the conclusion of the analysis of the Hc ^{111}Ag PAC spectroscopic data should be re-evaluated, and augmented with an additional possibility: (A) the presence of a water molecule bridging the two Ag(I) ions or (B) a water molecule (or another ligand) might be recruited by Cd(II) rapidly after ^{111}Ag decay. The ^{111}Ag PAC data reflect the structure(s) present from about 10 to 200 ns after the nuclear decay, which could be long enough for a water molecule to migrate to the metal site. The time scale is controlled by the half-life, $T_{1/2}$ = 84 ns, of the intermediate nuclear level of the γ-γ cascade of ^{111}Cd (because $T_{1/2}$ of the initial state of the γ-γ cascade is much shorter, see Figure 1). Note that the other Ag(I) of the binuclear site will (in almost all cases) be non-radioactive, because only a very small amount of ^{111}Ag is present with respect to the total amount of added non-radioactive Ag(I). In summary, the experimental data can be interpreted in two different ways: either Ag(I) in Hc from Carcinus aestuarii is four-coordinated with a water molecule as the fourth ligand, or Ag(I) is three-coordinated by the three histidine residues, and the fourth ligand is recruited after the decay to Cd(II). To discriminate between these two options, it might be useful to carry out experiments at low temperature, where the chance of trapping Ag(I) in the native coordination environment is higher, vide infra.

3.4. Metal Site Structure in Cu(I)-Sensing Proteins—Transcriptional Regulation

3.4.1. BxmR

BxmR is a transcriptional regulator protein of the ArsR family with a complex metal ion response profile. Cu(I), Ag(I), and Cd(II) all bind to the so-called α3N site in which four cysteine residues are available for metal ion coordination.

^{111}Ag PAC spectroscopic data display two NQIs, reflecting two different coordination geometries [31]. ^{111m}Cd PAC spectroscopy, which probes the metal site structure of Cd(II) via the NQIs of the same intermediate nuclear state that is probed in ^{111}Ag PAC, gives spectra that differ from those of ^{111}Ag PAC. This indicates that the NQIs observed by ^{111}Ag PAC are specific for Ag(I) coordination. One of the two NQIs observed by ^{111}Ag PAC spectroscopy has NQI1: $\omega_0 = 0.23$ rad/ns and $\eta = 0.33$, while the other is more difficult to fit, and can either have NQI2: $\omega_0 = 0.42$ rad/ns and $\eta = 0.05$ or NQI2′: $\omega_0 = 0.24$ rad/ns and $\eta = 1$. In the same paper, a comprehensive analysis of the Cu(I) binding (including XANES, EXAFS, UV-Vis absorption, and luminescence) indicated that a Cu(I)$_2$S$_4$ cluster is formed, in which each Cu(I) is tri-coordinated. NQI2 agrees well with a trigonal planar structure for Ag(I). Based on the additional work compiled on ^{111}Ag PAC since this work was published, and on analysis using AOM [40], we tentatively suggest that NQI1 might reflect a Ag(I)$_2$S$_4$ cluster in which the bridging thiolates have weaker bonding to Ag(I) (and Cd(II)) than the cysteines used to parameterize the AOM. Both structural interpretations are in accordance with the formation of a Ag(I)$_2$S$_4$ cluster, although other structures may also underlie the observed NQIs.

3.4.2. CueR

CueR is a transcriptional regulator protein of the MerR family with specificity for monovalent ions of the coinage metals [41]. Cu(I) (and Ag(I) and Au(I)) is found in a near linear structure coordinated by two cysteinates, see Figure 6. Ser77 from the other monomer of the protein dimer is positioned close to the metal site.

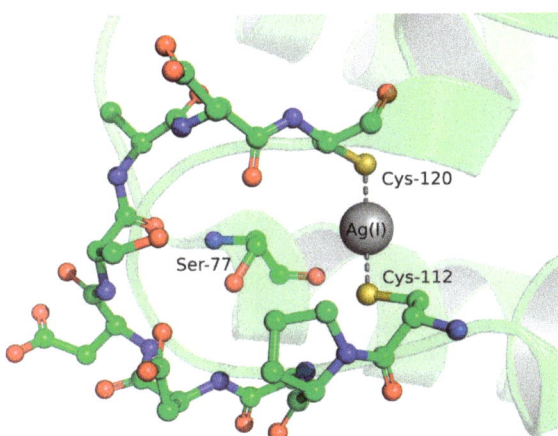

Figure 6. Metal site of CueR. Ag(I) is bound in an almost linear structure by two cysteinates, and the backbone carbonyl oxygen of Ser77 from the other monomer of the homodimeric protein is located near the metal ion (pdb code: 4WLW [42]). Figure produced with PyMOL [37].

^{111}Ag PAC was applied to characterize the CueR metal binding site [32,33], see Figure 7, and the decay of ^{111}Ag to ^{111}Cd(II) provided a means to elucidate the effect of instantaneously changing from Ag(I), which activates the protein function, to Cd(II), which does not. Contrary to the rigid small blue copper proteins, vide supra, significant structural changes are induced at the CueR metal site by the Ag(I) to Cd(II) transition. An experiment conducted at −196 °C displayed two NQIs, reflecting the presence of two different metal site structures. One of the signals is in good agreement with an almost linear structure with two coordinating thiolates, as expected, possibly with a weak ligand in the equatorial plane, such as carbonyl oxygen, suggested to be from Ser77 (high frequency signal, top panel, Figure 7). The other signal indicated a higher coordination number, thus implying that even at this low temperature, Cd(II) does in some cases recruit more ligands to satisfy this metal ion's preference for coordination numbers higher than two. In a subsequent quantum mechanical molecular dynamics (QM/MD) simulation [43], it was demonstrated that Cd(II) may indeed recruit additional ligands, and probably form a four-coordinate site involving, for example, backbone carbonyl oxygens, in addition to maintaining coordination with the two Cys residues. Interestingly, a ^{111}Ag PAC experiment at 1 °C exclusively displayed this second signal, demonstrating that the metal site structure relaxes within ca. 10 ns to a structure accommodating a higher coordination number than two. Thus, the change from Ag(I) to Cd(II) gives rise to rapid structural change, presumably inactivating the protein's function.

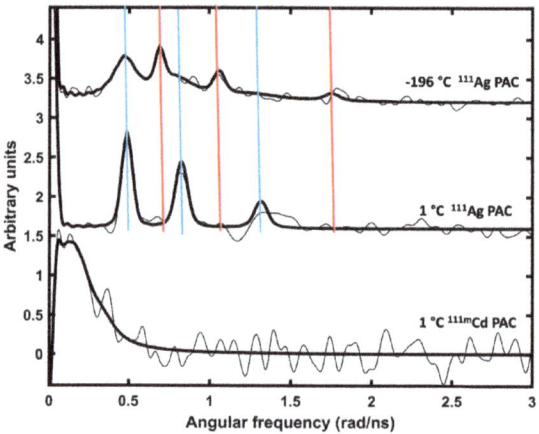

Figure 7. ^{111}Ag and ^{111m}Cd PAC spectra recorded for CueR. Two NQIs are present in the top panel, one (NQI1, red) reflecting the native metal site structure (possibly with an additional weak equatorial ligand) and the other (NQI2, blue) reflecting a higher coordination number. NQI2 is exclusively present in the experiment at 1 °C (middle panel), demonstrating that the daughter nucleus, Cd(II), remodels the metal site within ca. 10 ns of ^{111}Ag decay. The Fourier transform for ^{111}Ag PAC is multiplied by −1 for easy comparison with the Fourier transform for ^{111m}Cd PAC. Reprinted with permission from ref. [32]. Copyright 2020 the authors, published by Wiley-VCH Verlag GmbH & Co.

A ^{111m}Cd PAC experiment, see the lower panel of Figure 7, gave a very low-frequency NQI, presumably originating from coordination by four cysteinates, demonstrating that at thermodynamic equilibrium, Cd(II) most likely recruits two additional Cys ligands, which are available in the C-terminal CCHHRAG fragment.

3.5. Potential Future Applications

^{111}Ag PAC spectroscopy has so far not been applied extensively in bioinorganic chemistry, nor in any other field. Here, we briefly present some examples for which ^{111}Ag could potentially provide useful and unique information.

3.5.1. Cu(I) Binding Sites in Redox Active Proteins

Characterization of the metal site structure of Cu(I) in redox active, copper-dependent proteins is an obvious target for the application of ^{111}Ag PAC. The simplest case would be enzymes with a single copper site, for example, lytic polysaccharide monooxygenases (LPMOs), amine oxidase, heme-copper oxidase, nitrite reductase, and superoxide dismutase (SOD) if the Ag(I)Zn(II)-SOD species can be prepared. Moreover, mixed species of multicopper proteins with Cu(I/II) and Ag(I) might be characterized, assuming that well-defined species may be prepared with certain sites occupied by copper ion(s) and others by silver ion(s) [11]. In particular, this would allow for elucidation of Cu(I) binding site structures of intermediates in reaction cycles where Ag(I) prevents the oxidation at the site it occupies. Although this may be highly challenging and perhaps impossible for multi-copper sites (type II and type III copper sites), Ag(I) substitution in mononuclear copper sites (type I copper sites) of multicopper oxidases might be achieved.

3.5.2. Cu(I) Binding Sites in Cu(I) Transporting ATPases

Cu(I) transport out of cells is accomplished by transmembrane ATP-dependent transporters [44]. The details of Cu(I) coordination at a number of transiently occupied sites within these large proteins might be elucidated by ^{111}Ag PAC spectroscopy. The opportunity to explore metal ion transporters pertains not only to ^{111}Ag(I), but also to other PAC isotopes, such as ^{111m}Cd(II), which might be applied to explore, e.g., Zn(II) transport. Such studies have not been conducted yet.

3.5.3. Methionine Containing Cu(I) Binding Sites

Cu(I) is a relatively soft metal ion, and often displays binding to soft ligands such as sulfur. In the cytosol, cysteine thiolates are commonly found at Cu(I) binding sites of both metallochaperones and metalloregulatory proteins, vide supra. However, under more oxidizing conditions, e.g., extracellularly or in the periplasm of prokaryotic organisms, methionine takes part in the coordination of Cu(I), and even so-called methionine-only binding sites have been identified [11,12,45–56]. It is conceivable that ^{111}Ag PAC spectroscopy can provide further elucidation of the metal site structure and flexibility at such metal sites.

3.5.4. Low Temperature Experiments

The decay of ^{111}Ag(I) to ^{111}Cd(II) is accompanied by an almost instantaneous change of element and oxidation state, and additionally the Cd daughter nucleus receives a significant amount of recoil kinetic energy (up to around 500 kJ/mol) in a direction depending on the direction of emission of the β^- and the antineutrino. It has been demonstrated by QM/MD simulations [43] that the recoil energy is dissipated to the surrounding protein within picoseconds. The nuclear decay gives a unique opportunity to explore these relaxation processes, and thus probe the protein metal site flexibility/rigidity. However, if the aim is to characterize the Cu(I) metal site structure, the nuclear decay is an undesired complication which can, to some extent, be alleviated by conducting the experiments at low sample temperatures. This is illustrated by the ^{111}Ag PAC data recorded for the Cu(I)-sensing CueR protein, vide supra. Thus, there may be a significant advantage to running experiments at low sample temperatures to elucidate Cu(I) metal site structures, and it seems likely that at even lower temperatures, e.g., using a He cryostat, may further enhance the trapping of the native metal site structure.

Funding: LH thanks the Danish Agency for Higher Education and Science for support via the NICE grant, reference number 177363.

Conflicts of Interest: The authors declare no conflict of interest.

References

1. Giedroc, D.P.; Arunkumar, A.I. Metal Sensor Proteins: Nature's Metalloregulated Allosteric Switches. *Dalton Trans.* **2007**, 3107–3120. [CrossRef]
2. Osman, D.; Cavet, J.S. Chapter 8—Copper Homeostasis in Bacteria. In *Advances in Applied Microbiology*; Laskin, A.I., Sariaslani, S., Gadd, G.M., Eds.; Academic Press: Cambridge, MA, USA, 2008; Volume 65, pp. 217–247; ISBN 0065-2164.
3. Lutsenko, S. Human Copper Homeostasis: A Network of Interconnected Pathways. *Curr. Opin. Chem. Biol.* **2010**, *14*, 211–217. [CrossRef] [PubMed]

4. Solomon, E.I.; Heppner, D.E.; Johnston, E.M.; Ginsbach, J.W.; Cirera, J.; Qayyum, M.; Kieber-Emmons, M.; Kjaergaard, C.H.; Hadt, R.G.; Tian, L. Copper Active Sites in Biology. *Chem. Rev.* **2014**, *114*, 3659–3853. [CrossRef] [PubMed]
5. Scott, R.A.; Lukehart, C.M. *Applications of Physical Methods to Inorganic and Bioinorganic Chemistry*; John Wiley & Sons Ltd.: Hoboken, NJ, USA, 2007.
6. Jancso, A.; Correia, J.G.; Gottberg, A.; Schell, J.; Stachura, M.; Szunyogh, D.; Pallada, S.; Lupascu, D.C.; Kowalska, M.; Hemmingsen, L. TDPAC and β-NMR Applications in Chemistry and Biochemistry. *J. Phys. G Nucl. Part. Phys.* **2017**, *44*, 064003. [CrossRef]
7. Summers, M.F. 113Cd NMR Spectroscopy of Coordination Compounds and Proteins. *Coord. Chem. Rev.* **1988**, *86*, 43–134. [CrossRef]
8. Bertini, I.; Luchinat, C. The Reaction Pathways of Zinc Enzymes and Related Biological Catalysts. In *Bioinorganic Chemistry*; University Science Books: Mill Valley, CA, USA, 1994; p. 37.
9. Hay, M.T.; Milberg, R.M.; Lu, Y. Preparation and Characterization of Mercury and Silver Derivatives of an Engineered Purple Copper Center in Azurin. *J. Am. Chem. Soc.* **1996**, *118*, 11976–11977. [CrossRef]
10. Santagostini, L.; Gullotti, M.; Hazzard, J.T.; Maritano, S.; Tollin, G.; Marchesini, A. Inhibition of Intramolecular Electron Transfer in Ascorbate Oxidase by Ag^+: Redox State Dependent Binding. *J. Inorg. Biochem.* **2005**, *99*, 600–605. [CrossRef]
11. Djoko, K.Y.; Chong, L.X.; Wedd, A.G.; Xiao, Z. Reaction Mechanisms of the Multicopper Oxidase CueO from *Escherichia Coli* Support Its Functional Role as a Cuprous Oxidase. *J. Am. Chem. Soc.* **2010**, *132*, 2005–2015. [CrossRef]
12. Singh, S.K.; Roberts, S.A.; McDevitt, S.F.; Weichsel, A.; Wildner, G.F.; Grass, G.B.; Rensing, C.; Montfort, W.R. Crystal Structures of Multicopper Oxidase CueO Bound to Copper(I) and Silver(I): Functional Role of a Methionine-Rich Sequence. *J. Biol. Chem.* **2011**, *286*, 37849–37857. [CrossRef]
13. Wilcoxen, J.; Snider, S.; Hille, R. Substitution of Silver for Copper in the Binuclear Mo/Cu Center of Carbon Monoxide Dehydrogenase from *Oligotropha carboxidovorans*. *J. Am. Chem. Soc.* **2011**, *133*, 12934–12936. [CrossRef]
14. Chauhan, S.; Kline, C.D.; Mayfield, M.; Blackburn, N.J. Binding of Copper and Silver to Single-Site Variants of Peptidylglycine Monooxygenase Reveals the Structure and Chemistry of the Individual Metal Centers. *Biochemistry* **2014**, *53*, 1069–1080. [CrossRef] [PubMed]
15. Puchkova, L.V.; Broggini, M.; Polishchuk, E.V.; Ilyechova, E.Y.; Polishchuk, R.S. Silver Ions as a Tool for Understanding Different Aspects of Copper Metabolism. *Nutrients* **2019**, *11*, 1364. [CrossRef] [PubMed]
16. Nardella, M.I.; Fortino, M.; Barbanente, A.; Natile, G.; Pietropaolo, A.; Arnesano, F. Multinuclear Metal-Binding Ability of the N-Terminal Region of Human Copper Transporter Ctr1: Dependence Upon pH and Metal Oxidation State. *Front. Mol. Biosc.* **2022**, *9*, 897621. [CrossRef] [PubMed]
17. Kircheva, N.; Angelova, S.; Dobrev, S.; Petkova, V.; Nikolova, V.; Dudev, T. Cu^+/Ag^+ Competition in Type I Copper Proteins (T1Cu). *Biomolecules* **2023**, *13*, 681. [CrossRef] [PubMed]
18. Hemmingsen, L.; Sas, K.N.; Danielsen, E. Biological Applications of Perturbed Angular Correlations of γ-Ray Spectroscopy. *Chem. Rev.* **2004**, *104*, 4027–4062. [CrossRef] [PubMed]
19. Tröger, W. Nuclear Probes in Life Sciences. *Hyperfine Interact.* **1999**, *120*, 117–128. [CrossRef]
20. Tosato, M.; Asti, M.; Di Marco, V.; Jensen, M.L.; Schell, J.; Dang, T.T.; Köster, U.; Jensen, M.; Hemmingsen, L. Towards in Vivo Applications of ^{111}Ag Perturbed Angular Correlation of γ-Rays (PAC) Spectroscopy. *Appl. Radiat. Isot.* **2022**, *190*, 110508. [CrossRef]
21. Haas, H.; Shirley, D.A. Nuclear Quadrupole Interaction Studies by Perturbed Angular Correlations. *J. Chem. Phys.* **1973**, *58*, 3339–3355. [CrossRef]
22. Lerf, A.; Butz, T. Nuclear Quadrupole Interactions in Compounds Studied by Time Differential Perturbed Angular Correlations/Distributions. *Hyperfine Interact.* **1987**, *36*, 275–370. [CrossRef]
23. Hansen, B.; Bukrinsky, J.T.; Hemmingsen, L.; Bjerrum, M.J.; Singh, K.; Bauer, R. Effects of the Nuclear Transformation ^{111}Ag(I) to ^{111}Cd(II) in a Single Crystal of Ag[^{111}Ag](Imidazole)$_2$NO$_3$. *Phys. Rev. B* **1999**, *59*, 14182–14190. [CrossRef]
24. Haas, H.; Röder, J.; Correia, J.G.; Schell, J.; Fenta, A.S.; Vianden, R.; Larsen, E.M.H.; Aggelund, P.A.; Fromsejer, R.; Hemmingsen, L.B.S.; et al. Free Molecule Studies by Perturbed γ-γ Angular Correlation: A New Path to Accurate Nuclear Quadrupole Moments. *Phys. Rev. Lett.* **2021**, *126*, 103001. [CrossRef] [PubMed]
25. Mauk, M.R.; Gamble, R.C.; Baldeschwieler, J.D. Vesicle Targeting: Timed Release and Specificity for Leukocytes in Mice by Subcutaneous Injection. *Science* **1980**, *207*, 309–311. [CrossRef] [PubMed]
26. Bauer, R.; Danielsen, E.; Hemmingsen, L.; Bjerrum, M.J.; Hansson, O.; Singh, K. Interplay between Oxidation State and Coordination Geometry of Metal Ions in Azurin. *J. Am. Chem. Soc.* **1997**, *119*, 157–162. [CrossRef]
27. Danielsen, E.; Kroes, S.J.; Canters, G.W.; Bauer, R.; Hemmingsen, L.; Singh, K.; Messerschmidt, A. Coordination Geometries for Monovalent and Divalent Metal Ions in [His121] Azurin. *Eur. J. Biochem.* **1997**, *250*, 249–259. [CrossRef] [PubMed]
28. Danielsen, E.; Scheller, H.V.; Bauer, R.; Hemmingsen, L.; Bjerrum, M.J.; Hansson, O. Plastocyanin Binding to Photosystem I as a Function of the Charge State of the Metal Ion: Effect of Metal Site Conformation. *Biochemistry* **1999**, *38*, 11531–11540. [CrossRef] [PubMed]
29. Sas, K.N.; Haldrup, A.; Hemmingsen, L.; Danielsen, E.; Øgendal, L.H. pH-Dependent Structural Change of Reduced Spinach Plastocyanin Studied by Perturbed Angular Correlation of γ-Rays and Dynamic Light Scattering. *J. Biol. Inorg. Chem.* **2006**, *11*, 409–418. [CrossRef] [PubMed]

30. Holm, J.K.; Hemmingsen, L.; Bubacco, L.; Salvato, B.; Bauer, R. Interaction and Coordination Geometries for Ag(I) in the Two Metal Sites of Hemocyanin. *Eur. J. Biochem.* **2000**, *267*, 1754–1760. [CrossRef]
31. Liu, T.; Chen, X.; Ma, Z.; Shokes, J.; Hemmingsen, L.; Scott, R.A.; Giedroc, D.P. A Cu^I-Sensing ArsR Family Metal Sensor Protein with a Relaxed Metal Selectivity Profile. *Biochemistry* **2008**, *47*, 10564–10575. [CrossRef]
32. Balogh, R.K.; Gyurcsik, B.; Jensen, M.; Thulstrup, P.W.; Köster, U.; Christensen, N.J.; Mørch, F.J.; Jensen, M.L.; Jancsó, A.; Hemmingsen, L. Flexibility of the CueR Metal Site Probed by Instantaneous Change of Element and Oxidation State from Ag^I to Cd^{II}. *Chem. Eur. J.* **2020**, *26*, 7451–7457. [CrossRef]
33. Balogh, R.K.; Gyurcsik, B.; Jensen, M.; Thulstrup, P.W.; Köster, U.; Christensen, N.J.; Jensen, M.L.; Hunyadi-Gulyás, E.; Hemmingsen, L.; Jancsó, A. Tying Up a Loose End: On the Role of the C-Terminal CCHHRAG Fragment of the Metalloregulator CueR. *ChemBioChem* **2022**, *23*, e202200290. [CrossRef]
34. Shepard, W.E.B.; Anderson, B.F.; Lewandoski, D.A.; Norris, G.E.; Baker, E.N. Copper Coordination Geometry in Azurin Undergoes Minimal Change on Reduction of Copper(II) to Copper(I). *J. Am. Chem. Soc.* **1990**, *112*, 7817–7819. [CrossRef]
35. Guss, J.M.; Bartunik, H.D.; Freeman, H.C. Accuracy and Precision in Protein Structure Analysis: Restrained Least-Squares Refinement of the Structure of Poplar Plastocyanin at 1.33 Å Resolution. *Acta Crystallogr. Sect. B Struct. Sci.* **1992**, *48*, 790–811. [CrossRef]
36. Shepard, W.E.B.; Kingston, R.L.; Anderson, B.F.; Baker, E.N. Structure of Apo-Azurin from Alcaligenes Denitrificans at 1.8 Å Resolution. *Acta Crystallogr. Sect. D Biol. Crystallogr.* **1993**, *49*, 331–343. [CrossRef] [PubMed]
37. LLC The PyMOL Molecular Graphics System, Version 2.5.5; Schrödinger: New York, NY, USA, 2015. Available online: https://pymol.org/2/support.html? (accessed on 25 August 2023).
38. Guss, J.M.; Harrowell, P.R.; Murata, M.; Norris, V.A.; Freeman, H.C. Crystal Structure Analyses of Reduced (Cu^I) Poplar Plastocyanin at Six pH Values. *J. Mol. Biol.* **1986**, *192*, 361–387. [CrossRef] [PubMed]
39. Hazes, B.; Kalk, K.H.; Hol, W.G.J.; Magnus, K.A.; Bonaventura, C.; Bonaventura, J.; Dauter, Z. Crystal Structure of Deoxygenated Limulus Polyphemus Subunit II Hemocyanin at 2.18 Å Resolution: Clues for a Mechanism for Allosteric Regulation. *Protein Sci.* **1993**, *2*, 597–619. [CrossRef] [PubMed]
40. Bauer, R.; Jensen, S.J.; Schmidt-Nielsen, B. The Angular Overlap Model Applied to the Calculation of Nuclear Quadrupole Interactions. *Hyperfine Interact.* **1988**, *39*, 203–234. [CrossRef]
41. Changela, A.; Chen, K.; Xue, Y.; Holschen, J.; Outten, C.E.; O'Halloran, T.V.; Mondragón, A. Molecular Basis of Metal-Ion Selectivity and Zeptomolar Sensitivity by CueR. *Science* **2003**, *301*, 1383–1387. [CrossRef]
42. Philips, S.J.; Canalizo-Hernandez, M.; Yildirim, I.; Schatz, G.C.; Mondragón, A.; O'Halloran, T.V. Allosteric Transcriptional Regulation via Changes in the Overall Topology of the Core Promoter. *Science* **2015**, *349*, 877–881. [CrossRef]
43. Fromsejer, R.; Mikkelsen, K.V.; Hemmingsen, L. Dynamics of Nuclear Recoil: QM-BOMD Simulations of Model Systems Following β-Decay. *Phys. Chem. Chem. Phys.* **2021**, *23*, 25689–25698. [CrossRef]
44. Gourdon, P.; Liu, X.-Y.; Skjørringe, T.; Morth, J.P.; Møller, L.B.; Pedersen, B.P.; Nissen, P. Crystal Structure of a Copper-Transporting PIB-Type ATPase. *Nature* **2011**, *475*, 59–64. [CrossRef]
45. Peariso, K.; Huffman, D.L.; Penner-Hahn, J.E.; O'Halloran, T.V. The PcoC Copper Resistance Protein Coordinates Cu(I) via Novel S-Methionine Interactions. *J. Am. Chem. Soc.* **2003**, *125*, 342–343. [CrossRef]
46. Wernimont, A.K.; Huffman, D.L.; Finney, L.A.; Demeler, B.; O'Halloran, T.V.; Rosenzweig, A.C. Crystal Structure and Dimerization Equilibria of PcoC, a Methionine-Rich Copper Resistance Protein from Escherichia coli. *J. Biol. Inorg. Chem.* **2003**, *8*, 185–194. [CrossRef] [PubMed]
47. Arnesano, F.; Banci, L.; Bertini, I.; Mangani, S.; Thompsett, A.R. A Redox Switch in CopC: An Intriguing Copper Trafficking Protein That Binds Copper(I) and Copper(II) at Different Sites. *Proc. Natl. Acad. Sci. USA* **2003**, *100*, 3814–3819. [CrossRef] [PubMed]
48. Banci, L.; Bertini, I.; Ciofi-Baffoni, S.; Katsari, E.; Katsaros, N.; Kubicek, K.; Mangani, S. A Copper(I) Protein Possibly Involved in the Assembly of CuA Center of Bacterial Cytochrome c Oxidase. *Proc. Natl. Acad. Sci. USA* **2005**, *102*, 3994–3999. [CrossRef] [PubMed]
49. Jiang, J.; Nadas, I.A.; Kim, M.A.; Franz, K.J. A Mets Motif Peptide Found in Copper Transport Proteins Selectively Binds Cu(I) with Methionine-Only Coordination. *Inorg. Chem.* **2005**, *44*, 9787–9794. [CrossRef]
50. Zhang, L.; Koay, M.; Maher, M.J.; Xiao, Z.; Wedd, A.G. Intermolecular Transfer of Copper Ions from the CopC Protein of Pseudomonas Syringae. Crystal Structures of Fully Loaded Cu^{I}Cu^{II} Forms. *J. Am. Chem. Soc.* **2006**, *128*, 5834–5850. [CrossRef]
51. Bagai, I.; Liu, W.; Rensing, C.; Blackburn, N.J.; McEvoy, M.M. Substrate-Linked Conformational Change in the Periplasmic Component of a Cu(I)/Ag(I) Efflux System. *J. Biol. Chem.* **2007**, *282*, 35695–35702. [CrossRef]
52. Xue, Y.; Davis, A.V.; Balakrishnan, G.; Stasser, J.P.; Staehlin, B.M.; Focia, P.; Spiro, T.G.; Penner-Hahn, J.E.; O'Halloran, T.V. Cu(I) Recognition via Cation-Pi and Methionine Interactions in CusF. *Nat. Chem. Biol.* **2008**, *4*, 107–109. [CrossRef]
53. Davis, A.V.; O'Halloran, T.V. A Place for Thioether Chemistry in Cellular Copper Ion Recognition and Trafficking. *Nat. Chem. Biol.* **2008**, *4*, 148–151. [CrossRef]
54. Rubino, J.T.; Riggs-Gelasco, P.; Franz, K.J. Methionine Motifs of Copper Transport Proteins Provide General and Flexible Thioether-Only Binding Sites for Cu(I) and Ag(I). *J. Biol. Inorg. Chem.* **2010**, *15*, 1033–1049. [CrossRef]

55. Miranda-Blancas, R.; Avelar, M.; Rodriguez-Arteaga, A.; Sinicropi, A.; Rudiño-Piñera, E. The β-Hairpin from the *Thermus thermophilus* HB27 Laccase Works as a PH-Dependent Switch to Regulate Laccase Activity. *J. Struct. Biol.* **2021**, *213*, 107740. [CrossRef] [PubMed]
56. Roulling, F.; Godin, A.; Feller, G. Function and Versatile Location of Met-Rich Inserts in Blue Oxidases Involved in Bacterial Copper Resistance. *Biochimie* **2022**, *194*, 118–126. [CrossRef] [PubMed]

Disclaimer/Publisher's Note: The statements, opinions and data contained in all publications are solely those of the individual author(s) and contributor(s) and not of MDPI and/or the editor(s). MDPI and/or the editor(s) disclaim responsibility for any injury to people or property resulting from any ideas, methods, instructions or products referred to in the content.

Article

Symmetrical and Unsymmetrical Dicopper Complexes Based on Bis-Oxazoline Units: Synthesis, Spectroscopic Properties and Reactivity

James A. Isaac [1], Gisèle Gellon [1], Florian Molton [1], Christian Philouze [1], Nicolas Le Poul [2], Catherine Belle [1,*] and Aurore Thibon-Pourret [1,*]

[1] Université Grenoble-Alpes, CNRS, Department of Molecular Chemistry (DCM, UMR 5250), 38058 Grenoble, CEDEX 9, France; james.alfisaac@gmail.com (J.A.I.); florian.molton@univ-grenoble-alpes.fr (F.M.); christian.philouze@univ-grenoble-alpes.fr (C.P.)

[2] Université de Bretagne Occidentale, CNRS, Laboratoire de Chimie, Electrochimie Moléculaires et Chimie Analytique (CEMCA, UMR 6521), 29238 Brest, CEDEX 3, France; lepoul@univ-brest.fr

* Correspondence: catherine.belle@univ-grenoble-alpes.fr (C.B.); aurore.thibon@univ-grenoble-alpes.fr (A.T.-P.)

Abstract: Copper–oxygen adducts are known for being key active species for the oxidation of C–H bonds in copper enzymes and their synthetic models. In this work, the synthesis and spectroscopic characterizations of such intermediates using dinucleating ligands based on a 1,8 naphthyridine spacer with oxazolines or mixed pyridine-oxazoline coordination moieties as binding pockets for copper ions have been explored. On the one hand, the reaction of dicopper(I) complexes with O_2 at low temperature led to the formation of a $\mu-\eta^2:\eta^2$ Cu_2:O_2 peroxido species according to UV-Vis spectroscopy monitoring. The reaction of these species with 2,4-di-tert-butyl-phenolate resulted in the formation of the C–C coupling product, but no insertion of oxygen occurred. On the other hand, the synthesis of dinuclear Cu(II) bis-μ-hydroxido complexes based on pyridine–oxazoline and oxazoline ligands were carried out to further generate $Cu^{II}Cu^{III}$ oxygen species. For both complexes, a reversible monoelectronic oxidation was detected via cyclic voltammetry at $E_{1/2}$ = 1.27 and 1.09 V vs. Fc^+/Fc, respectively. Electron paramagnetic resonance spectroscopy (EPR) and UV-Vis spectroelectrochemical methods indicated the formation of a mixed-valent $Cu^{II}Cu^{III}$ species. Although no reactivity towards exogenous substrates (toluene) could be observed, the $Cu^{II}Cu^{III}$ complexes were shown to be able to perform hydroxylation on the methyl group of the oxazoline moieties. The present study therefore indicates that the electrochemically generated $Cu^{II}Cu^{III}$ species described herein are capable of intramolecular aliphatic oxidation of C–H bonds.

Keywords: copper-bioinspired chemistry; mixed-valent $Cu^{II}Cu^{III}$ species; dioxygen activation

Citation: Isaac, J.A.; Gellon, G.; Molton, F.; Philouze, C.; Le Poul, N.; Belle, C.; Thibon-Pourret, A. Symmetrical and Unsymmetrical Dicopper Complexes Based on Bis-Oxazoline Units: Synthesis, Spectroscopic Properties and Reactivity. *Inorganics* **2023**, *11*, 332. https://doi.org/10.3390/inorganics11080332

Academic Editor: Gianella Facchin

Received: 6 July 2023
Revised: 28 July 2023
Accepted: 8 August 2023
Published: 11 August 2023

Copyright: © 2023 by the authors. Licensee MDPI, Basel, Switzerland. This article is an open access article distributed under the terms and conditions of the Creative Commons Attribution (CC BY) license (https://creativecommons.org/licenses/by/4.0/).

1. Introduction

An important part of the research in bio-inorganic chemistry is currently focused on the development of model complexes that mimic the catalytic center of enzymes. In the field of copper-metalloenzymes, low-temperature oxygenation of synthetic Cu^I or Cu^I_2 compounds has provided a variety of structurally and spectroscopically characterized Cu_2O_2 species including high valent species [1] (Scheme 1). The combination with their respective reactivity profiles has provided useful information on dicopper-metalloenzyme structure and function [2,3]. In particular, Cu_2-containing enzymes activate O_2 to generate Cu_2O_2 species capable of oxidizing various substrates varying from catechol to methane [4]. To prepare model complexes, the nature of the supporting ligands employed, the presence of substituents to control the bulkiness and the electronic properties (donor/acceptor groups) and flexibility of the ligand control the final structure of the Cu_2O_2 species. The majority of dinucleating ligands with tridentate arms described in the literature have been used to generate bis-μ-oxido dicopper(III) complex or equilibrium mixtures of ($\mu-\eta^2:\eta^2$

peroxido dicopper(II) and bis-μ-oxido species [5–10] (Scheme 1). By contrast, dinucleating ligands with tetradentate arms have predominantly led to cis-μ-1,2-peroxido dicopper(II) complexes [11] or trans-μ-1,2-peroxido-dicopper(II) complexes [12–16].

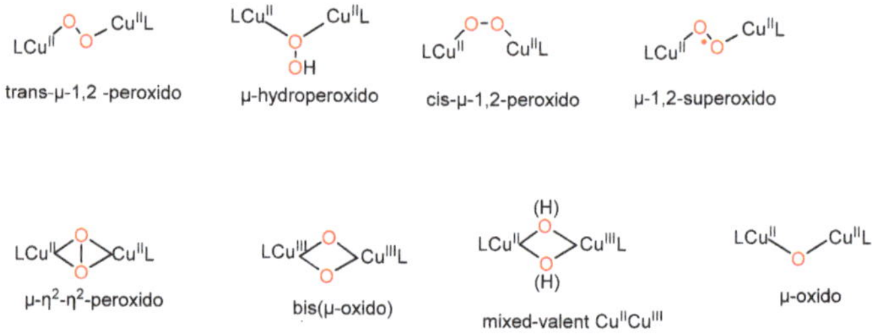

Scheme 1. Representative Cu_2/O_2 adducts characterized from model complexes.

In this context, we reported the synthesis of the symmetrical **Cu_2Py_4** complex (**Py_4** ligand is presented on Scheme 2) containing a naphthyridine spacer (known to be redox-inert, and to promote two well-defined coordination sites with a short metal-metal distance) associated with bis-pyridyl arms [17]. With the **Cu_2Py_4** bis(μ-hydroxido) complex, we showed that a mixed valent $Cu^{II}Cu^{III}$ intermediate could be generated and activate strong sp^3 C–H bonds [18]. In addition, from the corresponding dinuclear **$Cu^I{}_2Py_4$** species, no O_2 adduct was detected [19].

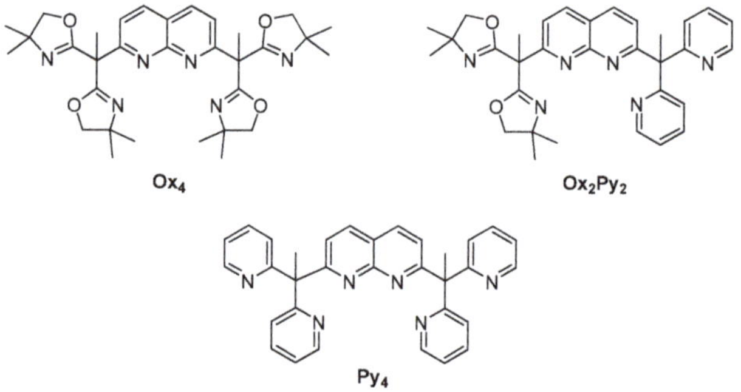

Scheme 2. Ligands used in this work.

With this background, in this study, we describe two novel naphthyridine-bridged ligands (Scheme 2) using bis-oxazoline moieties as binding site. Bis-oxazoline (BOX) ligands have found numerous applications, in particular in the field of asymmetric synthesis [20]. In particular, the steric bulk of the oxazoline group provides hindrance around the copper centers, stabilizing potential intermediates formed. This concept was applied by Meyer et al. in the copper–O_2 activation field showing that such an entity was able to stabilize a μ-η^2:η^2-peroxido-dicopper(II) complex [21].

Here, we have employed **Ox_4** and **Ox_2Py_2** ligands for the preparation of dinuclear copper(I) and (II) complexes. Aiming at generating mixed valent $Cu^{II}Cu^{III}$–oxygen species, we have followed two strategies. In a first approach (Strategy 1, Scheme 3), $Cu^I{}_2$ complexes have been reacted with O_2 in order to generate stable (μ-η^2:η^2) peroxido $Cu^{II}{}_2$ complexes that can be further reduced/protonated. The second approach (Strategy 2) is based on the electrochemical/chemical mono-oxidation of stable $Cu^{II}{}_2$-hydroxido or $Cu^{II}{}_2$-bis(μ-hydroxido) complexes, as previously developed by our group with the **Py_4** ligand [17].

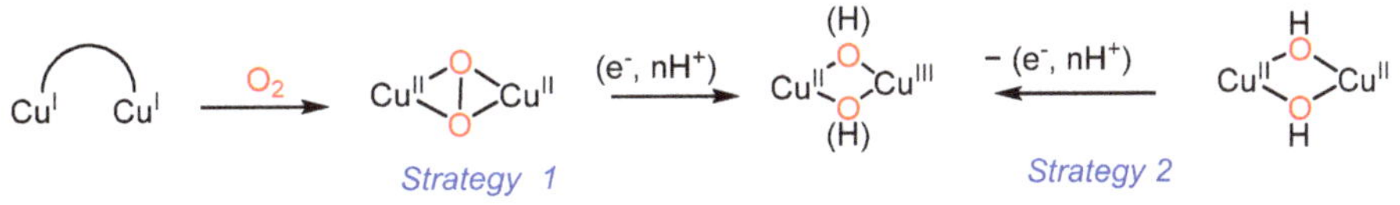

Scheme 3. Possible strategies for formation of $Cu^{II}Cu^{III}$ oxygen–mixed-valent species.

2. Results and Discussion

2.1. Ligands Synthesis

As shown in Scheme 2, the new ligands **Ox$_4$** and **Ox$_2$Py$_2$** display a common 1,8-naphthyridine spacer, the bis-oxazoline (BOX) or pyridine moieties (Py = 2,2′-dipyridylethane) serving as binding pockets (Scheme 2). The bis-oxazoline [22] and the pyridine (Py) [23] entities were synthesized according to procedures described in the literature. **Ox$_4$** was synthesized via lithiation of bis-oxazoline and subsequent reaction with the 2,7-dichloro-1,8-naphthyridine building block (synthesized according to literature procedures, in three steps) [24,25] with 20% yield. The ligand **Ox$_2$Py$_2$** was obtained in the same manner after two successive lithiations with 60% yield (Scheme 4). Corresponding ^{1}H (Figure S1) and ^{13}C (Figure S2) NMR are depicted in the Supplementary Materials along with ESI-MS spectra (Figure S3).

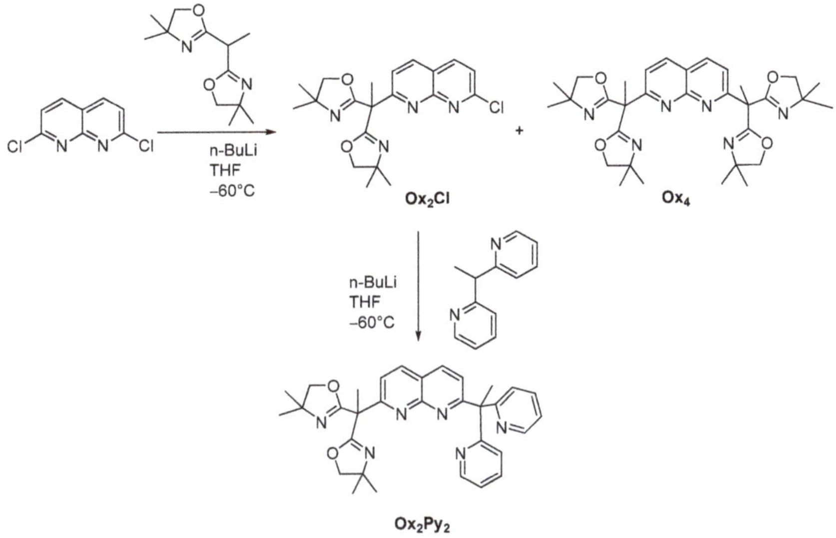

Scheme 4. Synthetic pathway for the ligands **Ox$_4$** and **Ox$_2$Py$_2$**.

2.2. Dicopper(I) Complexes

2.2.1. Synthesis and Characterizations

Complex **Cu$^I_2$Py$_4$** was synthesized according to the procedure described by Tilley et al. [26]. The two air-sensitive Cu(I) complexes based on **Ox$_4$** and **Ox$_2$Py$_2$** ligands (named **Cu$^I_2$Ox$_4$** and **Cu$^I_2$Ox$_2$Py$_2$**, respectively) were prepared using a similar procedure: under argon, one equivalent of ligand (**Ox$_4$** or **Ox$_2$Py$_2$**) was reacted with 2.1 equivalents of [Cu^I(CH$_3$CN)$_4$]OTf in tetrahydrofuran (THF). In both cases, the reaction produced an orange precipitate with 70% or 92% yield, respectively. Unfortunately, no suitable crystal for X-ray diffraction analysis could be obtained. ^{1}H-NMR characterization of the two complexes was carried out in CD$_3$CN (Figure S4). Peaks centered around 4.0 ppm and 1.35 ppm can be attributed to the CH$_2$ and CH$_3$ groups of the oxazoline entities. Two peaks (doublets) were detected around 4 ppm, which did not appear on the ^{1}H-NMR spectrum of the ligand itself. This result is a good indicator for copper(I) coordination to the ligand, as the two diastereotopic protons of the ligand are no longer in an equivalent environment. Two singlets at around 1.35 ppm matching to the CH_3 groups of the oxazoline moieties support this result.

2.2.2. Reactivity

In order to generate Cu_2/O_2 adducts, dioxygen was bubbled through solutions of **$Cu^I_2Ox_4$** and **$Cu^I_2Ox_2Py_2$** complexes in acetone at low temperature (T = 193 K). The reaction was monitored via UV-Vis spectroscopy (Figures 1 and S5). For both complexes, the spectrum of the initial dicopper(I) solution under N_2 was taken as the baseline for more accurate determination of the wavelength of the arising new absorption bands.

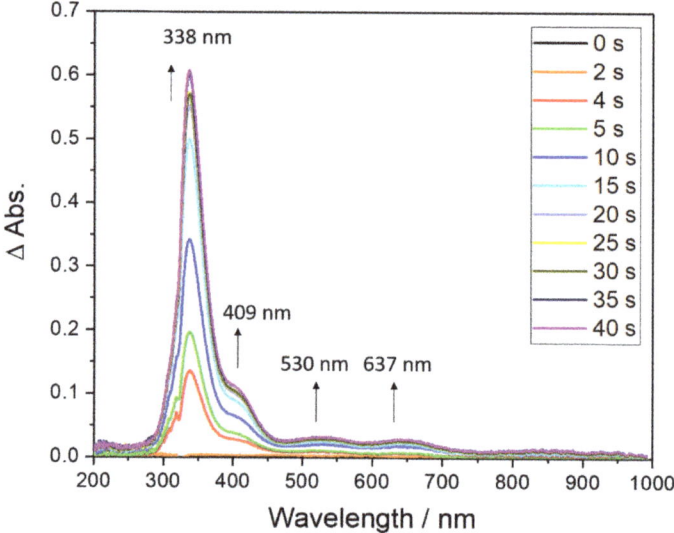

Figure 1. UV-vis monitoring of the addition of O_2 on **$Cu^I_2Ox_4$**; parameters: optical path = 1 cm; solvent: acetone, concentration 0.094 mM, T = 193 K. The baseline was taken on the dicopper(I) complex solution before O_2 addition for better monitoring of the peroxide absorption bands (justifying the use of relative absorbance (ΔAbs.) instead of absorbance (Abs.) on the graph).

For complex **$Cu^I_2Ox_4$**, four new bands with wavelengths ranging between 330 and 700 nm appeared rapidly upon addition of dioxygen. In particular, an intense absorption band at λ_{max} = 338 nm ($\varepsilon \approx 6400$ $M^{-1} \cdot cm^{-1}$) together with a less intense one at λ_{max} = 409 nm ($\varepsilon \approx 1000$ $M^{-1} \cdot cm^{-1}$) were observed. In addition, two low-intensity bands at 530 nm ($\varepsilon \approx 350$ $M^{-1} \cdot cm^{-1}$) and 637 nm ($\varepsilon \approx 312$ $M^{-1} \cdot cm^{-1}$) were detected. These absorptions correspond to low energy d–d transition, as reported in a parent dicopper complex [27]. The main absorption band at 338 nm is in the typical range of $\mu-\eta^2:\eta^2$-peroxido-Cu^{II}_2 oxygen species and attributed to ligand-to-metal charge transfer (LMCT) from the peroxido to the copper centers [3]. A similar behavior was obtained with **$Cu^I_2Ox_2Py_2$**. Indeed, a new absorption band was observed at λ_{max} = 367 nm upon addition of dioxygen ($\varepsilon \approx 2400$ $M^{-1} \cdot cm^{-1}$) [3,28] (Figure S5). As for **$Cu^I_2Ox_4$**, this wavelength value suggests the formation of side-on peroxido dicopper(II) species. However, both **Ox_4** and **Ox_2Py_2**-based peroxides were shown to be poorly stable, hence excluding any further reduction/protonation for generating mixed-valent species such as shown in Scheme 3.

The reactivity of these $\mu-\eta^2:\eta^2$-peroxido-Cu^{II}_2 oxygen adducts towards exogenous substrates was then examined in acetone. Aromatic hydroxylation reactivity is well-known for dicopper(II)-$\mu-\eta^2:\eta^2$-peroxido species, whereas dicopper(III)-bis-μ-oxido species are better for electrophilic reactivity mechanisms such as hydrogen atom abstraction [28–31]. For both species generated here, no reactivity towards PPh$_3$ or benzaldehyde was observed. In contrast, with sodium 2,4-di-butylphenolate as substrate, both dicopper(II)-μ-η^2:η^2-peroxido gave rise to the carbon–carbon coupling product detected via GCMS (Figure S6) over arene oxygenation. These results are in line with those of Meyer et al. [21] using a bis-oxazoline derivatives. As previously suggested, the presence of the dimethyl bulky groups on oxazoline units impacts the reactivity by limiting the access of substrates to the Cu$_2$O$_2$ core.

2.3. Dicopper(II) Complexes

2.3.1. Synthesis and Characterizations

The next step consisted in exploring the redox properties of the dicopper(II) bis(μ-OH) **Ox$_4$** and **Ox$_2$Py$_2$** complexes. For this purpose, the complexation reactions of the two ligands were carried out in a classical manner. One equivalent of the ligand was dissolved in THF, and 2.1 equivalents of triethylamine and water (10 equivalents) were added followed by 2.1 equivalents of Cu(OTf)$_2$. Upon diffusion of di-isopropyl ether into a concentrated acetonitrile solution of complexes, single crystals for X-ray diffraction analysis were obtained with 71% and 75% yield, respectively. Details for X-ray analysis are in Table 1. Complex **1** (**Cu$^{II}_2$Ox$_4$**) crystallizes as a dinuclear complex with the Cu atoms bridged by two hydroxido groups (Figure 2a) with Cu–O–Cu angles close to 90.4°. The Cu1 and Cu2 copper atoms are set at a short distance from each other (2.7537(5) Å). The geometries of both copper atoms are described as a square-based pyramid (τ = 0.07 for Cu1 and 0.08 for Cu2) [32] with N1 and N2 of the naphthyridine spacer occupying the axial position. Although the equatorial distances are in the range of the values obtained for parent/relevant dinuclear complexes (1.9–2.1 Å) [17,33–35], the nitrogen of the naphthyridine is located at 2.43–2.46 Å from the copper(II) centers. These significantly longer distances could indicate a weaker donating group and a possible unbinding of these atoms in solution.

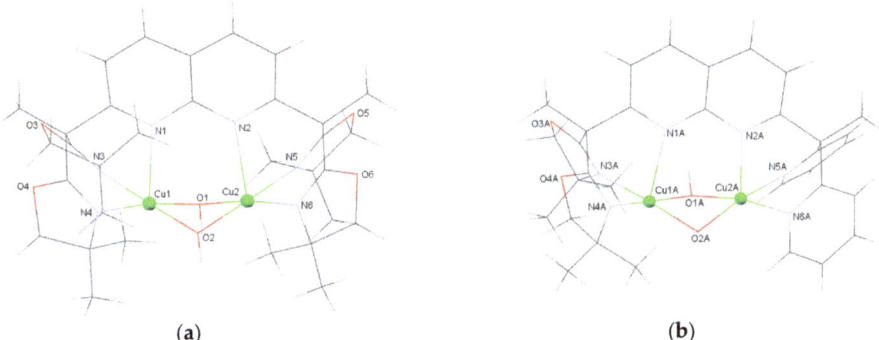

(a)　　　　　　　　　　　　　　(b)

Figure 2. Representation of cationic units of (**a**) [(Cu$_2$(Ox$_4$))(μ-OH)$_2$](CF$_3$SO$_3$)$_2$ (**1**) and (**b**) [(Cu$_2$(Ox$_2$Py$_2$)(μ-OH)$_2$](CF$_3$SO$_3$)$_2$ (**2A**). Selected bond lengths (Å) and angles (°) are reported in ESI (Tables S1 and S2). Others view and complete representation including counterions and solvent are in ESI (Figure S7a,b).

The similarly complexation of ligand **Ox$_2$Py$_2$** afforded compound **2** (**Cu$^{II}_2$Ox$_2$Py$_2$**) as a crystalline material. The unit cell encloses two crystallographically independent [C$_{32}$H$_{44}$Cu$_2$N$_6$O$_6$](CF$_3$SO$_3$)$_2$ entities and one CH$_3$CN as solvate. Each cationic unit displays two triflate counterions. Unit **2A** is shown in Figure 2b (other views on Figure S7b) and unit **2B** is shown in Figure S7c (Supporting Information); each consists of two copper atoms bridged by two hydroxido groups. Selected bond lengths and angles are reported in Tables S1 and S2 in the Supplementary Materials and display no significant difference between **2A** and **2B** dimers. In **2A**, the Cu1A-Cu2A distance is short (2.80) Å but rather long compared to $Cu^{II}_2Py_4$ (complex **3**) [17] and to $Cu^{II}_2Ox_4$, with a distance of 2.75 Å. Each copper atom has a square-based pyramid geometry (τ = 0.01 for Cu1A and 0.02 for Cu2A) [32] with the equatorial positions occupied by the hydroxido groups and the nitrogen atoms of either the pyridines or the BOX entities and the axial positions occupied by the N atoms of the naphthyridine spacer. Bond distances between the Cu atoms and equatorial positions are of 1.9–2.0 Å. The axial bonds are longer at 2.47 Å for Cu1A-N1 (the side of the BOX units) and 2.26 Å for Cu2A-N2 (the side of the pyridine units).

UV-Vis characterization of the dicopper(II) complexes **1** and **2** was carried out in acetonitrile. The complexes exhibited intense electronic transitions in the UV region at 260, 305, 310 and 317 nm ($\varepsilon \sim$ 10 000 M^{-1} cm^{-1}). These bands are similar to those of **Ox₄** and **Ox₂Py₂** free ligands and have been assigned as π–π^* transitions. Large bands at 580 nm and 570 nm, respectively, with weak values of molar absorptivity ($\varepsilon \sim$ 100 M^{-1} cm^{-1}) were observed for each complex and correspond to d–d transitions. These two complexes were also analyzed via ESI-MS in acetonitrile (Figure S8). Peaks were detected at 883 and 843 m/z corresponding to [M-OTf]$^+$ for [(Cu$_2$(**Ox₄**))(μ-OH)$_2$](CF$_3$SO$_3$)$_2$ and [(Cu$_2$(**Ox₂Py₂**))(μ-OH)$_2$](CF$_3$SO$_3$)$_2$, respectively, where M corresponds to the complex. Theoretical isotopic profiles matched experimental ones indicating that the complexes remain dinuclear in solution.

2.3.2. Electrochemical Oxidation of Complexes **1** and **2**

We further investigated the oxidation process for both dicopper(II) complexes **1** and **2**. Attempts to generate a mixed-valent $Cu^{II}Cu^{III}$ species were carried out via cyclic voltammetry (CV) in a CH$_3$CN solution with tetra-n-butyl ammonium perchlorate (TBAP) as the supporting electrolyte.

As shown in Figure 3, the two complexes displayed a reversible oxidation system at $E_{1/2}$ = 1.27 V and 1.09 V vs. Fc$^+$/Fc, respectively, for **1** and **2** at v = 100 mV·s^{-1} rendering them all out of reach of commonly used chemical oxidants [17]. Noteworthy, complex **1** showed a complete loss of reversibility at a lower scan rate (v = 20 mV·s^{-1}) (Figure 3a), contrary to complex **2**, for which reversibility was still observed at v = 5 mV·s^{-1} (Figure 3b). Plots of the normalized anodic peak current $Iv^{-1/2}$ against the scan rate (v) (Figure S9) showed a particular increase of $Iv^{-1/2}$ at low scan rates for **1**. This result is indicative of an ECE mechanism (E = electrochemical and C = chemical) where the oxidation of the complex is followed by a chemical reaction on the experimental time scale, producing new species, which, in turn, can be oxidized. The oxidation of both complexes therefore involves, at room temperature, the formation of a transient species that has a half-life of several seconds. For the oxidation process of each complex, the number of electrons was determined by using the Randles–Sevcik equation [36,37] for scan rate values for which the system remained reversible. I_P was plotted against $v^{1/2}$ for complexes **1** and **2** (Figure S10), yielding values of n = 1.3 and 1.2, respectively, thus indicating a one-electron transfer for both complexes. This behavior is reminiscent of that previously obtained with complex **3** ($Cu^{II}_2Py_4$), which displayed a reversible system at 1.26 V vs. Fc$^+$/Fc in the same conditions [17].

In order to avoid decomposition at room temperature because of the occurrence of an ECE process, the mono-electronic oxidized species were generated at −40 °C via bulk electrolysis of **1** and **2** (0.7 mM) in a 0.1 M NBu$_4$ClO$_4$/acetonitrile solution to the potentials of 1.41 V vs. Fc$^+$/Fc for **1** and 1.26 V vs. Fc$^+$/Fc for **2**. Bulk oxidation was accompanied by a color change from pale blue/colorless to yellow for complex **2** and colorless to brown for complex **1**. During the course of the low-temperature electrolysis, the samples were taken out and immediately frozen in liquid nitrogen for EPR analysis. For control experiments, to get the EPR spectrum of the final product, the oxidized species was also warmed to room temperature and then frozen in liquid nitrogen. Before electrolysis, the two complexes were EPR-silent (CH$_3$CN, 100 K). The EPR spectrum of complex **2** after oxidation via electrolysis is displayed in Figure 4, along with the spectrum of the product warmed to room temperature. Spectra were recorded at 15 K, but the same features were observed from 15 to 100 K. The oxidized complex displayed four clear lines (Figure 4) with coupling constant $A_\parallel$ of 173 G, the expected EPR spectrum for a mononuclear Cu^{II}

complex in an axial geometry (an unpaired electron coupling to the nuclear spin of one copper with $I = 3/2$). This is consistent with a valence-localized $Cu^{II}Cu^{III}$ species (the Cu^{III} being EPR silent), defined in the Robin Day classification system [38] as a class I (classes II and III represents slight and significant delocalization, respectively). The spectrum of the complex **1** after oxidation was less resolved but four lines were still observable ($A_{\parallel} = 177$ G) (Figure S11). The loss of resolution of the signal from this oxidized complex could be due to more instability of the one-oxidized species that can also be stated by its lower reversibility of the CV (Figure 3a) at room temperature compared to complex **2** (Figure 3b). For the mono-oxidized species **1**, **2** and **3**, in all three cases, the EPR parameters of the spectra (see Figures 4 and S11) obtained after simulation show a Cu^{II} in an axial geometry [17]. It is therefore obvious that mono-oxidation of the complexes $Cu^{II}{}_2Ox_2Py_2$ and $Cu^{II}{}_2Ox_4$ leads to the formation of mixed-valent $Cu^{II}Cu^{III}$ species. After the oxidized samples had been warmed to room temperature, the recorded spectra of all complexes at 15K displayed broad signals, indicating a mixture of several Cu^{II} complexes in solution.

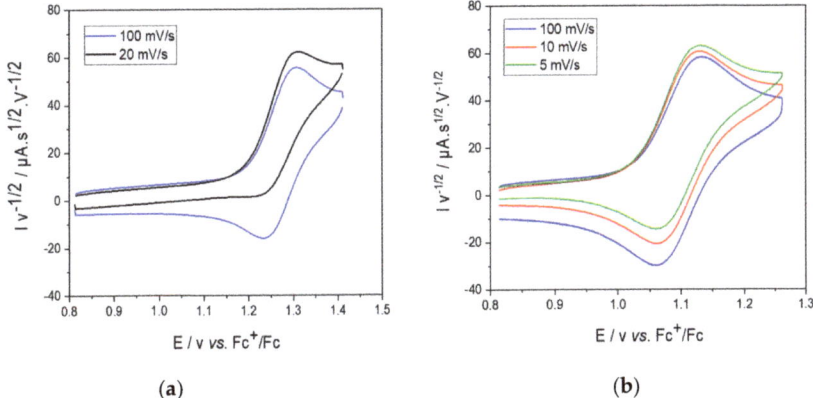

Figure 3. Scan–rate–normalized CV for (**a**) complex **1** and (**b**) complex **2** (for each at 1 mM) in acetonitrile with 0.1 M NBu$_4$ClO$_4$; room temperature; scan rates 100 mV·s^{-1} (blue), 20 mV·s^{-1} (black), 10 mV·s^{-1} (red) and 5 mV·s^{-1} (green).

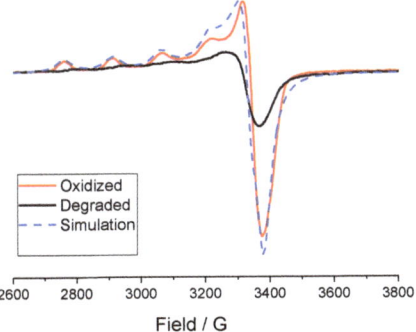

Figure 4. EPR spectra of complex **2** after oxidation after bulk electrolysis at −40 °C (red), and after warming to room temperature (black) in frozen solution (0.1 M NBu$_4$ClO$_4$ in acetonitrile) of 0.7 mM of **2** recorded at 15 K; frequency = 9.419 GHz. Simulation was carried out with the EasySpin program [39] assuming the following parameter: $g_{\perp} = 2.065$, $g_{//} = 2.307$, $A_{\perp} = 0$ G, $A_{//} = 173$ G.

UV-Vis-NIR time-resolved spectroelectrochemistry experiments were carried out at room temperature to further characterize the unstable mono-oxidized species. Upon mono-oxidation of $\mathbf{Cu^{II}_2Ox_4}$ (**1**), new absorption bands were detected at λ_{max} = 380 nm, 480 nm (both $\varepsilon \approx$ 735 M$^{-1}\cdot$cm^{-1}) and 630 nm (weak) (Figure S12). In the NIR region, a very low-intensity and broad band centered at 1680 nm was observed (Figure S12). For this complex, the redox process was found to be irreversible at the defined scan rate (30 mV·s^{-1}), suggesting a fast evolution of the generated mixed-valent species.

For $\mathbf{Cu^{II}Ox_2Py_2}$ (**2**), spectroelectrochemical measurements displayed slightly different features than for **1**. New absorption bands were detected at λ_{max} = 360 nm (3420 M$^{-1}\cdot$cm^{-1}) and 424 nm (sh, 2140 M$^{-1}\cdot$cm^{-1}) (Figures 5 and S13). In the near infrared region, a broad and weak band appeared at λ_{max} = 1185 nm (120 M$^{-1}\cdot$cm^{-1}) (Figure S13). The process was found to be fully reversible as shown by CV and from spectroscopic data since these absorption bands disappeared upon back reduction. Aiming at better analyzing the oxidized product from **2**, we determined the bandwidth at half-height ($\Delta\tilde{\nu}_{1/2}$ = 2460 cm^{-1}) by fitting the NIR experimental curve (Figure S14). From this value and by assuming a Cu–Cu distance of 2.80 Å from the X-ray structure of **2**, we determined the electronic coupling matrix element H_{ab} (determined from the Mulliken-Hush expression [40,41]) as well as the ratio between the experimental and theoretical values of $\Delta\tilde{\nu}_{1/2}$ (Γ parameter). The calculations yielded H_{ab} = 373 cm^{-1} and Γ = 0.45, as typically found for a class II system in the Robin–Day classification, i.e., low delocalization of the charge, here at room temperature. Noteworthy, this result is close to that obtained with $\mathbf{Cu^{II}Py_4}$ (**3**) (H_{ab} = 322 cm^{-1} and Γ = 0.39) [17] and demonstrates the strong similarities between $\mathbf{Cu^{II}Ox_2Py_2}$ and $\mathbf{Cu^{II}Py_4}$ complexes. It likely suggests that the oxidation occurs on the copper/pyridine moieties for complex **2**.

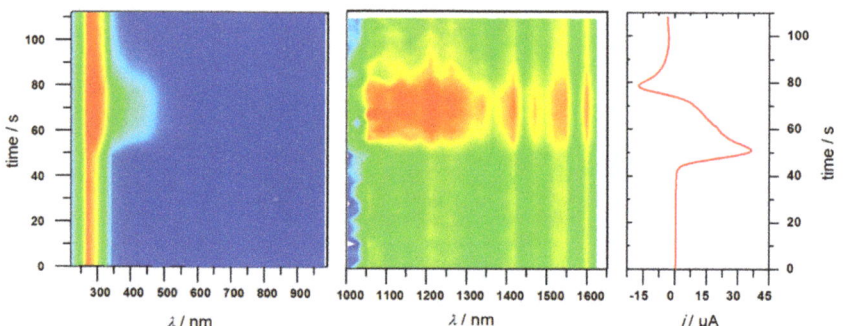

Figure 5. UV-Vis–NIR monitoring of the oxidation of complex **2** via spectroelectrochemistry showing the reversibility of the redox process. Right: current intensity variation with time taken from the CV at v = 30 mV·s^{-1}. Conditions: 7 mM in 0.1 M NBu$_4$ClO$_4$ in acetonitrile, optical path = 0.2 mm, working electrode: Pt, room temperature.

In regard to the observed reactivity toward toluene (bond dissociation energy (BDE) = 89.8 kcal mol^{-1}) [42] of Cu^{II}Cu^{III} species generated electrochemically from $\mathbf{Cu^{II}_2Py_4}$, the reactivity of the mono oxidized species from **1** and **2** were probed via CV with (over 100 equiv.) and without toluene. For both complexes, the CV remained the same when toluene was added, suggesting no reactivity between the Cu^{II}Cu^{III} species and toluene at this timescale. Few Cu^{II}Cu^{III} species are reported in the literature [17,33–35] and besides our previous studies [18], only one result demonstrated a reactivity on dihydroanthracene [33], a rather weak C-H bond.

To test a possible reason of this lack of observed reactivity, the ligands were analyzed after the monoelectronic-oxidation of the complexes (0.7 mM) through exhaustive electrolysis in NBu$_4$ClO$_4$/CH$_3$CN at −40 °C at 1.26 V vs. Fc$^+$/Fc for complex $\mathbf{Cu^{II}_2Ox_2Py_2}$ and at 1.41 V vs. Fc$^+$/Fc for complex $\mathbf{Cu^{II}_2Ox_4}$ under air. After electrolysis, the solutions were warmed to room temperature.

For identification of the oxidation products, the demetallation residue was analyzed via ESI mass spectrometry (Figures S15 and S16). In the case of **Cu$^{II}_2$Ox$_4$**, the protonated ligand [LH + H]$^+$ (where LH represent the ligand) was observed at 575 m/z, but peaks at 591 m/z, 613 m/z and 629 m/z indicated the ligand hydroxylation as [LOH + H]$^+$, [LOH + Na]$^+$ and [LOH + K]$^+$, respectively, corresponding to the addition of an O atom. This behavior is consistent with change of the C(CH$_3$)$_2$ into C(CH$_3$)(CH$_2$OH) as depicted in Figure S17. The peak at 630 m/z was tentatively attributed to [LCHOHCN + H]$^+$, which could be formed by a radical coupling of the oxazoline unit with the CH$_3$CN solvent. Similar products were observed from the residue of **Cu$^{II}_2$Ox$_2$Py$_2$**: peaks at 535 m/z and 557 m/z from the ligand [LH + H]$^+$, [LH + Na]$^+$, respectively, and peaks at 551 m/z and 573 m/z from the hydroxylated product [LOH + H]$^+$ and [LOH + Na]$^+$. Peaks at 590 m/z and 612 m/z could tentatively be assigned to products [LCHOHCN + H]$^+$ and [LCHOHCN + Na]$^+$, respectively. All products are consistent with a proton coupled electron transfer as the initial step of the reaction. The oxidation of the relatively strong C-H bonds of the ligands is not surprising given the high oxidation potential of the complexes (where the oxidation potential and the pKa are the two thermodynamic driving forces for proton coupled electron transfers) [42].

From the non-electrolyzed complexes (but using the same treatment with concentrated KOH), no hydroxylated products or peaks assigned to the formation of LCHOHCN were observed on the residues, demonstrating that the changes to the ligand are linked to the generation of the Cu^{II}Cu^{III} species.

3. Materials and Methods

3.1. General

Reagents were purchased from commercial sources and were used without purification. The solvents were purified via standard methods before use. ESI mass spectra were recorded on an Esquire 3000 plus Bruker Daltonis with nanospray inlet. UV-Vis analyses were performed using a Cary 50 spectrophotometer operating in the 200–1000 nm range with quartz cells. The temperature was maintained at 25 °C with a temperature control unit. NMR solution spectra (^{1}H and ^{13}C) at 298 K were recorded on a unity Plus 400 MHz Varian spectrometer with the deuterated solvent as a lock. X-band EPR spectra were recorded in a range of 15–100 K with a Bruker EMX Plus spectrometer equipped with a nitrogen flow (or He flow) cryostat and operating at 9.4 GHz (X band). All spectra presented were recorded under non-saturating conditions. Simulation of EPR spectra was carried out with EasySpin program [39]. All electrochemical measurements were carried out under an argon atmosphere at room temperature using a Biologic SP-300 instrument. Experiments were performed with solutions of the complexes containing 0.1 M of the supporting electrolyte (TBA·ClO$_4$). For cyclic voltammetry, a standard three-electrode configuration was used consisting of a glassy carbon (d = 3 mm) working electrode, a platinum counter electrode and an Ag wire placed in an AgNO$_3$ (0.01 M in CH$_3$CN)/NBu$_4$ClO$_4$ (0.1 M in CH$_3$CN)) solution as a pseudo reference electrode. The system was systematically calibrated against ferrocene after each experiment and all the potentials are therefore given versus the Fc$^+$/Fc redox potential. Low-temperature electrolysis was carried out with a home-designed 3-electrode cell (WE: Carbon felt, RE: Pt wire, CE: Pt grid) dipped in an acetone/dry ice bath at −40 °C for 45 min. Samples for EPR and UV-Vis analysis of the mixed-valent complex were taken every 15 min. They were instantly frozen in liquid nitrogen for EPR measurements.

3.2. Ligands' Syntheses

Synthesis of ligand Ox$_4$. According to the literature procedures [22,24], the intermediates 2,7-dichloro-1,8-naphthyridine and BOX were synthesized. A total of 1.25 g (5.56 × 10^{-3} mol, 2.2 eq) of BOX dissolved in 40 mL of freshly distilled THF under argon and cooled to –60 °C was slowly added to 2.5 mL (5.56 × 10^{-3} mol, 2.2 eq) of 2.4 M n-BuLi. The colorless solution was stirred for 30 min and 0.526 g of 2,7-dichloro-1,8-naphthyridine was added, leading to a beige precipitate in a red-orange solution. The solution was stirred overnight and reached room temperature, resulting in the solubilization of the precipitate. After addition of water (5 mL), THF was evaporated and a 40 mL of water was added. The solution was extracted with dichloromethane (DCM) and dried over Na_2SO_4. The residue was purified via column chromatography (gradient of acetone/pentane) over silica to give products Ox_2Cl and Ox_4 (79% and 20%, respectively). ^{1}H-NMR (400 MHz, CD_3CN), δ_H (ppm): 8.27 (2H, d, J = 8.8 Hz, C_4H), 7.70 (2H, d, J = 8.8 Hz, C_3H), 3.97 (8H, s, $C_{10}H$), 1.95 (6H, s, C_7H), 1.23 (24H, s, $C_{11}H$); ^{1}H-NMR (400 MHz, $CDCl_3$), δ_H (ppm): 8.07 (2H, d, J = 8.8 Hz, C_4H), 7.66 (2H, d, J = 8.4 Hz, C_3H), 4.00 (8H, s, $C_{10}H$), 2.09 (6H, s, C_7H), 1.32 (24H, s, $C_{11}H$); ^{13}C-NMR (101 MHz, $CDCl_3$), δ_C (ppm): 165.44 (C_6), 162.71 (C_2), 154.16 (C_8), 136.10 (C_4), 122.14 (C_3), 120.66 (C_5), 79.47 (C_{10}), 67.36 (C_9), 50.58 (C_1), 27.93 (C_{11}), 23.15 (C_7); ESI-MS: m/z 575.4 [M + H]$^+$; elemental analysis: $C_{36}H_{42}N_6O_4 \cdot 1.75(H_2O)$. Theoretical (%): C: 63.40, H: 7.57 and N: 13.86. Obtained: C: 63.37, H: 7.24 and N: 13.77.

Synthesis of ligand Ox_2Py_2. In total, 0.70 g (0.38 × 10^{-3} mol, 1.1 eq) of 1,1-di-(2-pyridyl)ethane [23] was dissolved in 15 mL of freshly distilled THF under argon and cooled to −50 °C, followed by the slow addition of 0.26 mL (0.38 × 10^{-3} mol, 1.1 eq) of 1.4 M n-BuLi and the red solution was stirred for 30 min. Then, 0.133 g of 2,7-dichloro-1,8-naphthyridine was added using a powder finger. The solution was stirred overnight and allowed to reach room temperature, resulting in the formation of a precipitate. Then, 1 mL of water (1 mL) was added, dissolving the precipitate; the THF evaporated and water (20 mL) was added, and the solution was extracted with DCM (4 × 15 mL) and dried over $MgSO_4$. The residue was purified via column chromatography (gradient of acetone/pentane) over silica to give the ligand Ox_2Py_2 (140 mg, 76%). ^{1}H-NMR (400 MHz, CD_3CN), δ_H (ppm): 8.48 (2H, d, J = 4.4 Hz, $C_{16}H$), 8.22 (1H, d, J = 8.4 Hz, C_7H), 8.15 (1H, d, J = 8.4 Hz, C_4H), 7.67 (1H, d, J = 8.4 Hz, C_8H), 7.66 (2H, dt, J_1 = 8.0 Hz, J_2 = 2.0 Hz, $C_{14}H$), 7.40 (1H, d, J = 8.8 Hz, C_3H), 7.23 (2H, d, J = 8.0 Hz, $C_{13}H$), 7.20 (2H, dd, J_1 = 7.6 Hz, J_2 = 4.9 Hz, $C_{15}H$), 3.95 (4H, s, $C_{19}H$), 2.35 (3H, s, $C_{11}H$), 1.93 (3H, s, $C_{21}H$), 1.21 (12H, s, $C_{20}H$); ^{13}C-NMR (101 MHz, CD_3CN), δ_C (ppm): 170.72 (C_6), 166.57 (C_2), 166.21 (C_9), 163.50 (C_{17}), 154.85 (C_{12}), 149.47 (C_{16}), 137.65 (C_4), 137.24 (C_{14}), 136.99 (C_7), 124.79 (C_3), 124.56 (C_{13}), 123.14 (C_8), 122.60 (C_{15}), 121.00 (C_5), 80.09 (C_{19}), 68.20 (C_{18}), 62.02 (C_{10}), 51.41 (C_1), 30.69 (C_{11}), 28.16 (C_{20}), 27.80 (C), 26.27 (C), 23.06 (C_{21}); ESI-MS: m/z 557 [M + Na]$^+$, 535 [M + H]$^+$; elemental analysis: $C_{32}H_{34}N_6O_2 \cdot 0.6(CH_2Cl_2)$. Theoretical: C: 66.86, H: 6.06 and N: 14.35. Obtained: C: 66.91, H: 6.27 and N: 14.08.

3.3. Complexes Syntheses

Synthesis of complex $Cu^I{}_2Ox_2Py_2$. In a glove box, the ligand Ox_2Py_2 (0.073 g, 1.4 × 10^{-4} mol, 1 eq) was dissolved in 4 mL of distilled THF and added to a solution of [$Cu^I(CH_3CN)_4$]OTf (0.11 g, 2.9 × 10^{-4} mol, 2.1 eq) in 10 mL of THF. The resulting suspension was stirred for 12 h and then filtered. The solid was recovered and dried under vacuum giving a bright orange solid (0.127 g, 92%). ^{1}H-NMR (400 MHz, CD_3CN), δ_H (ppm): 8.57 (2H, br, $C_{16}H$), 8.35 (2H, m, $C_{4\&7}H$), 7.90 (2H, m, $C_{3\&8}H$), 7.77 (2H, br, $C_{14}H$), 7.51 (2H, br, $C_{13}H$), 7.37 (2H, br, $C_{15}H$), 4.07 (2H, d, J = 8.2 Hz, $C_{19}H$), 3.90 (2H, d, J = 8.2 Hz, $C_{19}H$), 2.43 (3H, s, $C_{11}H$), 2.13 (3H, s, $C_{21}H$), 1.39 (6H, s, $C_{20}H$), 1.32 39 (6H, s, $C_{20}H$); elemental analysis: $C_{36}H_{37}Cu_2F_6N_7O_8S_2 \cdot 2(H_2O) \cdot 1.5(CH_3CN)$. Theoretical: C: 42.64, H: 4.17 and N: 10.84. Obtained: C: 42.32, H: 4.03 and N: 10.82.

Synthesis of complex $Cu^I{}_2Ox_4$. In a glove box, the ligand Ox_4 (0.041 g, 7.5 × 10^{-5} mol, 1 eq) was dissolved in 4 mL of distilled THF and added to a solution of [$Cu^I(CH_3CN)_4$]OTf (0.057 g, 1.5 × 10^{-4} mol, 2.1 eq) in 10 mL of THF. The resulting suspension was stirred for 12 h, followed by filtering, and the solid was dried under vacuum, giving a bright orange solid (0.050 g, 70%). ^{1}H-NMR (400 MHz, CD_3CN), δ_H (ppm): 8.45 (2H, d, J = 8.4 Hz, C_4H), 7.61 (2H, J = 8.4 Hz, C_3H)), 4.10 (4H, J = 8.4 Hz, $C_{10}H$), 3.96 (4H, d, J = 8.4 Hz, $C_{10}H$), 2.08 (6H, s, C_7H), 1.40 (12H, s, $C_{11}H$), 1.36 (12H, s, $C_{11}H$); elemental analysis: $C_{36}H_{45}Cu_2F_6N_7O_{10}S_2 \cdot 3(H_2O) \cdot (CH_3CN) \cdot (CH_2Cl_2)$. Theoretical: C: 39.24, H: 4.70 and N: 9.51. Obtained: C: 39.16, H: 4.35 and N: 9.41.

Synthesis of the complex [(Cu$_2$(Ox$_4$)(μ-OH)$_2$](CF$_3$SO$_3$)$_2$ (**1**). Briefly, 130 mg of the ligand Ox4 (2.2 × 10^{-4} mol, 1 eq) was dissolved in 10 mL of THF. Triethylamine (66 μL, 4.7 × 10^{-4} mol, 2.1 eq) was added, followed by a solution of Cu(OTf)$_2$ (170 mg, 4.6 × 10^{-4} mol, 2.1 eq) in 5 mL THF. A blue precipitate gradually formed, and after 3 h, the solution was decanted and the solid washed in THF. The blue solid was recrystallized via slow diffusion of DIPE into a solution of CH$_3$CN to give blue crystals (160 mg, 71%) suitable for X-ray diffraction. ESI (acetonitrile): m/z 883 (M-OTf)$^+$, 359 (M-(OTf)$_2$)$^{2+}$. UV-Vis (acetonitrile) λ/nm (ε/M^{-1} cm^{-1}): 253 (9700), 305 (8300), 310 (8600), 317 (10000) 580 (94). Elemental analysis: C$_{34}$H$_{46}$Cu$_2$F$_6$N$_6$O$_{13}$S$_2$·H$_2$O. Theoretical: C: 38.82, H: 4.41 and N: 7.99. Obtained: C: 38.85, H: 4.51 and N: 8.34.

Synthesis of the complex [(Cu$_2$(Ox$_2$Py$_2$)(μ-OH)$_2$](CF$_3$SO$_3$)$_2$ (**2**). First, 50 mg of the ligand Ox$_2$Py$_2$ (9.4 × 10^{-5} mol, 1 eq) was dissolved in 5 mL of THF. H$_2$O (17 μL, 9.4 × 10^{-4} mol, 10 eq) and triethylamine (27 μL, 1.9 × 10^{-4} mol, 2.1 eq) were added, followed by a solution of Cu(OTf)$_2$ (71 mg, 2.0 × 10^{-4} mol, 2.1 eq) in 5 mL THF. A blue precipitate gradually formed, and after 3 h, the solution was decanted and the solid washed in THF. The blue solid was dissolved in 1 mL of CH$_3$CN, 40 mL of THF was added and the mixture placed in the freezer (−20 °C) for one week, after which crystals had formed. These were collected, and dried under vacuum (73 mg, 75%). X-ray diffraction quality crystals were obtained via slow diffusion of DIPE into a solution of CH$_3$CN. ESI (acetonitrile): m/z 843 (M-OTf)$^+$, 338 (M-(OTf)$_2$-H$_2$O)$^{2+}$. UV-Vis (acetonitrile) λ/nm (ε/M^{-1} cm^{-1}): 268 (8200), 303 (7400), 306 (8600), 317 (7300) 570 (100). Elemental analysis: C$_{34}$H$_{36}$Cu$_2$F$_6$N$_6$O$_{10}$S$_2$·0.5(H$_2$O). Theoretical: C: 40.72, H: 3.72 and N: 8.38. Obtained: C: 40.69, H: 3.60 and N: 8.58.

3.4. Crystallographic Studies

Crystals were mounted on a Kappa APEXII Bruker-Nonius diffractometer equipped with an Incoatec μsource with multilayer mirror mono-chromated Mo-Kα radiation (λ = 0.71073 Å) and a cryosystem Oxford cryostream cooler. Intensities were corrected for Lorentz and polarization (EVAL14) and for absorption (SADABS). Structural resolutions were carried out via direct method (SIR97) or the charge flipping method (Superflip) and refinement via full-matrix least squares on F2 (SHELX2013) [43] completed using the OLEX 2 analysis package [44]. The refinement of all non-hydrogen atoms was carried out with anisotropic thermal parameters. Hydrogen atoms were generated in idealized positions (excluding the hydroxido bridges, which were located on the difference Fourier map), riding on the carrier atoms, with isotropic thermal parameters. CCDC 2266887 and 2266888 contain the full data collection parameters and structural data for **2** and **1**, respectively.

Table 1. Crystallographic data for [(Cu$_2$(Ox$_4$))](CF$_3$SO$_3$)$_2$ (**1**) and [(Cu$_2$(Ox$_2$Py$_2$)(μ-OH)$_2$](CF$_3$SO$_3$)$_2$ (**2**).

Compound	1	2
Chemical Formula	[C$_{32}$H$_{44}$Cu$_2$N$_6$O$_6$](CF$_3$O$_3$S)$_2$	2[C$_{32}$H$_{36}$Cu$_2$N$_6$O$_4$)(CFO$_3$S)$_2$]·CH$_3$CN
Formula mass	1033.95	2028.83
Morphology	plate	plate
Color	blue	blue
Crystal size (mm)	0.48 × 0.3 × 0.1	0.45 × 0.2 × 0.1
Crystal system	monoclinic	triclinic
Space group	P1 2$_1$/n 1	P-1
a [Å]	10.332 (2)	12.377 (3)
b [Å]	30.427 (6)	14.361 (3)
c [Å]	13.415 (3)	23.657 (5)
$α$ [°]	90	84.30 (3)
$β$ [°]	92.80 (3)	82.98 (3)
$γ$ [°]	90	83.00 (3)
Unit-cell volume [Å^3]	4212.2 (15)	4127.1 (15)
D_x (g·cm^{-3})	1.63	1.633

Table 1. Cont.

Compound	1	2
T [K]	200	200
Z	4	2
μ [mm^{-1}]	1.202	1.222
Total reflections	68,415	77,009
Unique reflections	12,173	18,827
Obsd. reflections	9889 ($F > 2\sigma$)	11,658 ($F > 2\sigma$)
Rint.	0.0501	0.0961
R[a]	0.0400	0.0868
R(w)[a]	0.0873	0.2176
Goodness of fit S	1.085	1.061
Δρ$_{min}$/Δρ$_{max}$ (e·Å^{-3})	−0.641/0.574	−1.310/1.754
CCDC Number	2,266,888	2,266,887

[a] Refinement based on F where w = 1/[σ^2(Fo)^2 + (0.0288p)^2 + 5.5580 p] with p = (Fo^2 + 2Fc^2)/3 for **1**, w = 1/[σ^2(Fo)^2 + (0.1196p)^2 + 17.8205 p] with p = (Fo^2 + 2Fc^2)/3 for **2**.

3.5. Spectroelectrochemistry

Thin layer room-temperature UV-Vis-NIR spectroelectrochemistry was carried out with a specific home-designed cell in a reflectance mode (WE: platinum, RE: Pt wire, CE: Pt wire). The UV-Vis and Vis-NIR optic fiber probes were purchased from Ocean Optics. Time-resolved UV-Vis-NIR detection was performed with QEPro and NIRQuest spectrometers (Ocean Insight, Orlando, FL, USA). Spectroscopic data were acquired using the Oceanview software. A DH-2000-BAL light source (Ocean Optics) was used for these experiments. The potential of the spectroelectrochemical cell was monitored using an AUTOLAB PGSTAT 100 (Metrohm, The Netherlands) potentiostat controlled by the NOVA 1.11 software.

4. Conclusions

Two new bridged ligands bearing a naphthyridine spacer and symmetrical or unsymmetrical coordination environment including a bis-oxazoline arm were successfully synthetized. The related Cu$^I_2$ complex with four pyridine arms (**Py4**) [26] displayed no dioxygen activation, whereas the corresponding Cu$_2^I$ complexes from **Ox2Py2** and **Ox4** ligands were shown to bind dioxygen at −40 °C yielding μ-η^2:η^2-peroxido-Cu$^{II}_2$ species as clearly observed by using UV-Vis spectroscopy. For both, preliminary reactivity studies with sodium 2,4-di-*tert*-butylphenolate were performed. The resulting product (3,3′,5,5′-tetra-*tert*-butyl-2,2′-biphenol) indicated C-C coupling, whereas no *ortho*-hydroxylation was observed. The corresponding Cu$^{II}_2$ complexes **1** (**Cu$^{II}_2$Ox4**) and **2** (**Cu$^{II}_2$Ox2Py2**) have been prepared and characterized via single-crystal X-ray diffraction. Electrochemical mono-oxidation provided access to mixed-valent Cu$_2^{II,III}$ μ-hydroxido species with charge localization on one of the two copper ions. ESI-MS of the solution after electrolysis and demetallation show unequivocal evidence of intramolecular oxidation of the ligand through the bis-oxazoline moieties, contrary to complex **3** (**Cu$^{II}_2$Py4**), which is active in electrocatalysis at room temperature of exogenous substrate [18]. The present study therefore emphasizes that the electrochemically produced Cu^{II}Cu^{III} species are competent for aliphatic oxidation of C-H bonds (intramolecular or external substrate). In order to further advances towards generating a catalytic system (or oxidation of an external substrate), efforts on the synthesis of more robust ligands are currently being pursued.

Supplementary Materials: The following supporting information can be downloaded at: https://www.mdpi.com/article/10.3390/inorganics11080332/s1, Figure S1: ^{1}H NMR spectra of ligand Ox$_4$ and Ox$_2$Py$_2$; Figure S2: ^{13}C NMR spectra of ligand Ox$_4$ and Ox$_2$Py$_2$; Figure S3: ESI-MS spectra of ligand Ox$_4$ and Ox$_2$Py$_2$; Figure S4: ^{1}H-NMR spectra of Cu$^I_2$Ox$_4$ and Cu$^I_2$Ox$_2$Py$_2$; Figure S5: UV-vis spectrum after addition of O$_2$ to the complex Cu$^I_2$Ox$_2$Py$_2$; Figure S6: GCMS of the resulting solution after reaction of the μ-η^2:η^2-peroxo-Cu$^{II}_2$ species from complex Cu$^I_2$Ox$_4$ with 2,4-di-*tert*-butylphenol, Figure S7: Molecular structure of **1**, **2A** and **2B**; Figure S8: ESI-MS spectra of complexes Cu$^{II}_2$Ox$_4$ (**1**) and Cu$^{II}_2$Ox$_2$Py$_2$ (**2**); Figure S9: Plots of the normalized peak current $Iv^{-1/2}$ against the scan rate v for complexes **1**, **2** and [(Cu$_2$(Py$_4$))(μ-OH)$_2$](CF$_3$SO$_3$)$_2$; Figure S10: Plots of the I against $v^{1/2}$ for complexes **1**, **2** and [(Cu$_2$(Py$_4$))(μ-OH)$_2$](CF$_3$SO$_3$)$_2$; Figure S11: EPR spectrum of the mono-oxidized complex **1** after bulk electrolysis at −40 °C, and after heating to room temperature; Figure S12: UV-Vis-NIR spectroelectrochemistry data of **1** (a) Time-monitoring of the oxidation process; (b) Selected UV-Vis and NIR spectra at different time intervals; (c) CV of the complex during the spectroelectrochemical measurement; Figure S13: UV-Vis-NIR spectroelectrochemistry data of compound **2**; (a) Selected UV-Vis and NIR spectra from the spectroelectrochemical at different time intervals; (b) CV of the complex during the spectroelectrochemical measurement; Figure S14: Experimental and simulated of the NIR band obtained by spectroelectrochemistry of compound **2**; Figure S15: ESI-MS spectrum of the complex Cu$^{II}_2$Ox$_4$ after electrolysis, demetallation and purification; Figure S16: ESI-MS spectrum of the complex Cu$^{II}_2$Ox$_4$ after electrolysis, demetallation and purification; Figure S17: Proposed mechanisms; Table S1: Selected bond distances [Å] in **1**, **2A** and **2B** from X-Ray data; Table S2: Selected angles [°] in **1**, **2A** and **2B** from X-ray data.

Author Contributions: Formal analysis, investigation, J.A.I., N.L.P. and A.T.-P.; resources, J.A.I., G.G., F.M. and C.P.; data curation, N.L.P. and C.B.; writing—original draft preparation, C.B. and A.T.-P.; writing—review and editing, N.L.P., C.B. and A.T.-P.; project administration, N.L.P., C.B. and A.T.-P.; funding acquisition, N.L.P., C.B. and A.T.-P. All authors have read and agreed to the published version of the manuscript.

Funding: This research was funded by ANR (The French research agency), grant number ANR-22-CE07-0032 (COSACH) and Labex ARCANE (ANR-11-LABX-0003-01).

Data Availability Statement: Data is contained within the article or Supplementary Materials.

Acknowledgments: The authors gratefully acknowledge The French research agency (ANR) for support (ANR-22-CE07-0032). The authors are grateful to ICMG UAR 2607 for the analytical facilities (NMR, ESI-MS, EPR and X-ray). This work has been partially supported by the CBH-EUR-GS (ANR-17-EURE-0003) program, in the framework of which this work was carried out.

Conflicts of Interest: The authors declare no conflict of interest.

References

1. Elwell, C.E.; Gagnon, N.L.; Neisen, B.D.; Dhar, D.; Spaeth, A.D.; Yee, G.M.; Tolman, W.B. Copper−Oxygen Complexes Revisited: Structures, Spectroscopy, and Reactivity. *Chem. Rev.* **2017**, *117*, 2059–2107. [CrossRef] [PubMed]
2. Quist, D.A.; Diaz, D.E.; Liu, J.J.; Karlin, K.D. Activation of Dioxygen by Copper Metalloproteins and Insights from Model Complexes. *J. Biol. Inorg. Chem.* **2017**, *22*, 253–288. [CrossRef] [PubMed]
3. Keown, W.; Gary, J.B.; Stack, T.D.P. High-Valent Copper in Biomimetic and Biological Oxidations. *J. Biol. Inorg. Chem.* **2017**, *22*, 289–305. [CrossRef] [PubMed]
4. Trammell, R.; Rajabimoghadam, K.; Garcia-Bosch, I. Copper-Promoted Functionalization of Organic Molecules: From Biologically Relevant Cu/O$_2$ Model Systems to Organometallic Transformations. *Chem. Rev.* **2019**, *119*, 2954–3031. [CrossRef]
5. Garcia-Bosch, I.; Cowley, R.E.; Díaz, D.E.; Peterson, R.L.; Solomon, E.I.; Karlin, K.D. Substrate and Lewis Acid Coordination Promote O−O Bond Cleavage of an Unreactive L$_2$CuII$_2$(O$_2^{2-}$) Species to Form L$_2$CuIII$_2$(O)$_2$ Cores with Enhanced Oxidative Reactivity. *J. Am. Chem. Soc.* **2017**, *139*, 3186–3195. [CrossRef]
6. Magallón, C.; Serrano-Plana, J.; Roldán-Gómez, S.; Ribas, X.; Costas, M.; Company, A. Preparation of a Coordinatively Saturated μ-H$_2$:H$_2$-Peroxodicopper(II) Compound. *Inorg.Chim. Acta* **2018**, *481*, 166–170. [CrossRef]
7. Paul, M.; Teubner, M.; Grimm-Lebsanft, B.; Buchenau, S.; Hoffmann, A.; Rübhausen, M.; Herres-Pawlis, S. Influence of the Amine Donor on Hybrid Guanidine-Stabilized Bis(μ-Oxido) Dicopper(III) Complexes and Their Tyrosinase-like Oxygenation Activity towards Polycyclic Aromatic Alcohols. *J. Inorg. Biochem.* **2021**, *224*, 111541. [CrossRef]
8. Tahsini, L.; Kotani, H.; Lee, Y.-M.; Cho, J.; Nam, W.; Karlin, K.D.; Fukuzumi, S. Electron-Transfer Reduction of Dinuclear Copper Peroxo and Bis-μ-Oxo Complexes Leading to the Catalytic Four-Electron Reduction of Dioxygen to Water. *Chem.-Eur. J.* **2012**, *18*, 1084–1093. [CrossRef]

9. Li, S.T.; Braun-Cula, B.; Hoof, S.; Limberg, C. Copper(I) Complexes Based on Ligand Systems with Two Different Binding Sites: Synthesis, Structures and Reaction with O_2. *Dalton Trans.* **2018**, *47*, 544–560. [CrossRef]
10. Kodera, M.; Kano, K. Reversible O_2-Binding and Activation with Dicopper and Diiron Complexes Stabilized by Various Hexapyridine Ligands. Stability, Modulation, and Flexibility of the Dinuclear Structure as Key Aspects for the Dimetal/O_2 Chemistry. *Bull. Chem. Soc. Jpn.* **2007**, *80*, 662–676. [CrossRef]
11. Dalle, K.E.; Gruene, T.; Dechert, S.; Demeshko, S.; Meyer, F. Weakly Coupled Biologically Relevant Cu^{II}_2 (μ-η^1:η^1-O_2) *Cis*-Peroxo Adduct That Binds Side-On to Additional Metal Ions. *J. Am. Chem. Soc.* **2014**, *136*, 7428–7434. [CrossRef] [PubMed]
12. Karlin, K.D.; Lee, D.-H.; Kaderli, S.; Zuberbühler, A.D. Copper Dioxygen Complexes Stable at Ambient Temperature: Optimization of Ligand Design and Solvent. *Chem. Commun.* **1997**, *5*, 475–476. [CrossRef]
13. Lohmiller, T.; Spyra, C.-J.; Dechert, S.; Demeshko, S.; Bill, E.; Schnegg, A.; Meyer, F. Antisymmetric Spin Exchange in a μ-1,2-Peroxodicopper(II) Complex with an Orthogonal Cu-O-O-Cu Arrangement and $S = 1$ Spin Ground State Characterized by THz-EPR. *JACS Au* **2022**, *2*, 1134–1143. [CrossRef] [PubMed]
14. Börzel, H.; Comba, P.; Hagen, K.S.; Kerscher, M.; Pritzkow, H.; Schatz, M.; Schindler, S.; Walter, O. Copper−Bispidine Coordination Chemistry: Syntheses, Structures, Solution Properties, and Oxygenation Reactivity. *Inorg. Chem.* **2002**, *41*, 5440–5452. [CrossRef] [PubMed]
15. Brückmann, T.; Becker, J.; Würtele, C.; Seuffert, M.T.; Heuler, D.; Müller-Buschbaum, K.; Weiß, M.; Schindler, S. Characterization of Copper Complexes with Derivatives of the Ligand (2-Aminoethyl)Bis(2-Pyridylmethyl)Amine (Uns-Penp) and Their Reactivity towards Oxygen. *J. Inorg. Biochem.* **2021**, *223*, 111544. [CrossRef]
16. Jacobson, R.R.; Tyeklar, Z.; Farooq, A.; Karlin, K.D.; Liu, S.; Zubieta, J. A Copper-Oxygen (Cu_2-O_2) Complex. Crystal Structure and Characterization of a Reversible Dioxygen Binding System. *J. Am. Chem. Soc.* **1988**, *110*, 3690–3692. [CrossRef]
17. Isaac, J.A.; Gennarini, F.; Lopez, I.; Thibon-Pourret, A.; David, R.; Gellon, G.; Gennaro, B.; Philouze, C.; Meyer, F.; Demeshko, S.; et al. Room-Temperature Characterization of a Mixed-Valent μ-Hydroxodicopper(II,III) Complex. *Inorg. Chem.* **2016**, *55*, 8263–8266. [CrossRef]
18. Isaac, J.A.; Thibon-Pourret, A.; Durand, A.; Philouze, C.; Le Poul, N.; Belle, C. High-Valence $Cu^{II}Cu^{III}$ Species in Action: Demonstration of Aliphatic C–H Bond Activation at Room Temperature. *Chem. Commun.* **2019**, *55*, 12711–12714. [CrossRef]
19. Isaac, J.A. Conception et Synthèse de Catalyseurs de Cuivre Bio-Inspirés Pour l'activation de Liaisons C–H. Ph.D. Thesis, Université Grenoble-Alpes, Grenoble, France, 2018.
20. Desimoni, G.; Faita, G.; Jørgensen, K.A. C2-Symmetric Chiral Bis(Oxazoline) Ligands in Asymmetric Catalysis. *Chem. Rev.* **2006**, *106*, 3561–3651. [CrossRef]
21. Walli, A.; Dechert, S.; Bauer, M.; Demeshko, S.; Meyer, F. BOX Ligands in Biomimetic Copper-Mediated Dioxygen Activation: A Hemocyanin Model: BOX Ligands in Copper-Mediated Dioxygen Activation. *Eur. J. Inorg. Chem.* **2014**, *2014*, 4660–4676. [CrossRef]
22. Dagorne, S.; Bellemin-Laponnaz, S.; Welter, R. Synthesis and Structure of Neutral and Cationic Aluminum Complexes Incorporating Bis(Oxazolinato) Ligands. *Organometallics* **2004**, *23*, 3053–3061. [CrossRef]
23. Bechlars, B.; D'Alessandro, D.M.; Jenkins, D.M.; Iavarone, A.T.; Glover, S.D.; Kubiak, C.P.; Long, J.R. High-Spin Ground States via Electron Delocalization in Mixed-Valence Imidazolate-Bridged Divanadium Complexes. *Nat. Chem.* **2010**, *2*, 362–368. [CrossRef] [PubMed]
24. Newkome, G.R.; Garbis, S.J.; Majestic, V.K.; Fronczek, F.R.; Chiari, G. Chemistry of Heterocyclic Compounds. 61. Synthesis and Conformational Studies of Macrocycles Possessing 1,8- or 1,5-Naphthyridino Subunits Connected by Carbon-Oxygen Bridges. *J. Org. Chem.* **1981**, *46*, 833–839. [CrossRef]
25. Boelrijk, A.E.M.; Neenan, T.X.; Reedijk, J. Ruthenium Complexes with Naphthyridine Ligands. Synthesis, Characterization and Catalytic Activity in Oxidation Reactions. *J. Chem. Soc. Dalton Trans.* **1997**, *23*, 4561–4570. [CrossRef]
26. Davenport, T.C.; Tilley, T.D. Dinucleating Naphthyridine-Based Ligand for Assembly of Bridged Dicopper(I) Centers: Three-Center Two-Electron Bonding Involving an Acetonitrile Donor. *Angew. Chem. Int. Ed.* **2011**, *50*, 12205–12208. [CrossRef] [PubMed]
27. Davenport, T.C.; Tilley, T.D. Dinuclear First-Row Transition Metal Complexes with a Naphthyridine-Based Dinucleating Ligand. *Dalton Trans.* **2015**, *44*, 12244–12255. [CrossRef]
28. Mirica, L.M.; Ottenwaelder, X.; Stack, T.D. Structure and Spectroscopy of Copper Dioxygen Complexes. *Chem. Rev.* **2004**, *104*, 1013–1045. [CrossRef]
29. Hatcher, L.Q.; Karlin, K.D. Oxidant Types in Copper–Dioxygen Chemistry: The Ligand Coordination Defines the Cu_n-O_2 Structure and Subsequent Reactivity. *J. Biol. Inorg. Chem.* **2004**, *9*, 669–683. [CrossRef]
30. Lucas, H.R.; Li, L.; Sarjeant, A.A.N.; Vance, M.A.; Solomon, E.I.; Karlin, K.D. Toluene and Ethylbenzene Aliphatic C−H Bond Oxidations Initiated by a Dicopper(II)-μ-1,2-Peroxo Complex. *J. Am. Chem. Soc.* **2009**, *131*, 3230–3245. [CrossRef]
31. Solomon, E.I.; Heppner, D.E.; Johnston, E.M.; Ginsbach, J.W.; Cirera, J.; Quyyum, M.; Kieber-Emmons, M.T.; Kjaergaard, C.H.; Hadt, R.G.; Tian, L. Copper Active Sites in Biology. *Chem. Rev.* **2014**, *114*, 3659–3853. [CrossRef]
32. Addison, A.W.; Rao, T.N.; Reedijk, J.; van Rijn, J.; Verschoor, G.C. Synthesis, Structure, and Spectroscopic Properties of Copper(II) Compounds Containing Nitrogen–Sulphur Donor Ligands; the Crystal and Molecular Structure of Aqua [1,7-Bis(N-Methylbenzimidazol-2′-Yl)-2,6-Dithiaheptane]Copper(II) Perchlorate. *J. Chem. Soc. Dalton Trans.* **1984**, *7*, 1349–1356. [CrossRef]

33. Halvagar, M.R.; Solntsev, P.V.; Lim, H.; Hedman, B.; Hodgson, K.O.; Solomon, E.I.; Cramer, C.J.; Tolman, W.B. Hydroxo-Bridged Dicopper(II,III) and -(III,III) Complexes: Models for Putative Intermediates in Oxidation Catalysis. *J. Am. Chem. Soc.* **2014**, *136*, 7269–7272. [CrossRef]
34. Kochem, A.; Gennarini, F.; Yemloul, M.; Orio, M.; Le Poul, N.; Rivière, E.; Giorgi, M.; Faure, B.; Le Mest, Y.; Réglier, M.; et al. Characterization of a Dinuclear Copper(II) Complex and Its Fleeting Mixed-Valent Copper(II)/Copper(III) Counterpart. *ChemPlusChem* **2017**, *82*, 615–624. [CrossRef]
35. Thibon-Pourret, A.; Gennarini, F.; David, R.; Isaac, J.A.; Lopez, I.; Gellon, G.; Molton, F.; Wojcik, L.; Philouze, C.; Flot, D.; et al. Effect of Monoelectronic Oxidation of an Unsymmetrical Phenoxido-Hydroxido Bridged Dicopper(II) Complex. *Inorg. Chem.* **2018**, *57*, 12364–12375. [CrossRef] [PubMed]
36. Mabbott, G.A. An Introduction to Cyclic Voltammetry. *J. Chem. Educ.* **1983**, *60*, 697. [CrossRef]
37. Elgrishi, N.; Rountree, K.J.; McCarthy, B.D.; Rountree, E.S.; Eisenhart, T.T.; Dempsey, J.L. A Practical Beginner's Guide to Cyclic Voltammetry. *J. Chem. Educ.* **2018**, *95*, 197–206. [CrossRef]
38. Robin, M.B.; Day, P. Mixed Valence Chemistry-A Survey and Classification. In *Adv Inorg Chem Radiochem*; Elsevier: Amsterdam, The Netherlands, 1968; Volume 10, pp. 247–422. ISBN 978-0-12-023610-7.
39. Stoll, S.; Schweiger, A. EasySpin, a Comprehensive Software Package for Spectral Simulation and Analysis in EPR. *J. Magn. Reson.* **2006**, *178*, 42–55. [CrossRef]
40. Brunschwig, B.S.; Creutz, C.; Sutin, N. Optical Transitions of Symmetrical Mixed-Valence Systems in the Class II–III Transition Regime. *Chem. Soc. Rev.* **2002**, *31*, 168–184. [CrossRef]
41. Winter, R.F. Half-Wave Potential Splittings $\Delta E_{1/2}$ as a Measure of Electronic Coupling in Mixed-Valent Systems: Triumphs and Defeats. *Organometallics* **2014**, *33*, 4517–4536. [CrossRef]
42. Warren, J.J.; Tronic, T.A.; Mayer, J.M. Thermochemistry of Proton-Coupled Electron Transfer Reagents and Its Implications. *Chem. Rev.* **2010**, *110*, 6961–7001. [CrossRef]
43. Sheldrick, G.M. A Short History of SHELX. *Acta Cryst. A* **2008**, *64*, 112–122. [CrossRef]
44. Dolomanov, O.V.; Bourhis, L.J.; Gildea, R.J.; Howard, J.A.K.; Puschmann, H. OLEX2: A Complete Structure Solution, Refinement and Analysis Program. *J. Appl. Crystallogr.* **2009**, *42*, 339–341. [CrossRef]

Disclaimer/Publisher's Note: The statements, opinions and data contained in all publications are solely those of the individual author(s) and contributor(s) and not of MDPI and/or the editor(s). MDPI and/or the editor(s) disclaim responsibility for any injury to people or property resulting from any ideas, methods, instructions or products referred to in the content.

www.ingramcontent.com/pod-product-compliance
Lightning Source LLC
LaVergne TN
LVHW070429100526
838202LV00014B/1561

MDPI AG
Grosspeteranlage 5
4052 Basel
Switzerland
Tel.: +41 61 683 77 34

Inorganics Editorial Office
E-mail: inorganics@mdpi.com
www.mdpi.com/journal/inorganics

Disclaimer/Publisher's Note: The title and front matter of this reprint are at the discretion of the Guest Editors. The publisher is not responsible for their content or any associated concerns. The statements, opinions and data contained in all individual articles are solely those of the individual Editors and contributors and not of MDPI. MDPI disclaims responsibility for any injury to people or property resulting from any ideas, methods, instructions or products referred to in the content.